ISBN 978-0-260-44317-5
PIBN 10948454

THE PROCEEDINGS OF THE LINNEAN SOCIETY OF NEW SOUTH WALES.

FOR THE YEAR

1900.

Vol. XXV.

WITH FIFTY PLATES.

(Plates i.-xxxviii.-xxxviii.*bis*-xlix.)

Sydney:

PRINTED AND PUBLISHED FOR THE SOCIETY

BY

F. CUNNINGHAME & CO., 146 PITT STREET

AND

SOLD BY THE SOCIETY

1901.

SYDNEY:

F. CUNNINGHAME AND CO., PRINTERS

PITT STREET.

CONTENTS OF PROCEEDINGS, 1900.

PART I. (No. 97.)

(Issued August 8th, 1900.)

* To be issued separately with a later Volume.

PART II. (No. 98.)

(Issued September 25th, 1900.)

PART III. (No. 99.)

(Issued November 22nd, 1900.)

PART III. *(continued.)*

PART IV. (No. 100).

(Issued May 30th, 1901.)

PART IV. *(continued.)*

LIST OF PLATES.

PROCEEDINGS, 1900.

LIST OF PLATES *(continued)*.

CORRIGENDA.

of so useful a Society should be so comparatively small, a resolution was passed on 19th April, at his instance, that, during the residue of the year, the payment of an entrance fee should be suspended. The result was an accession of 38 new members, besides one associate, and the roll at the end of 1899 was increased to 116.

As it was thought that another innovation, in the shape of light refreshments at the close of each monthly meeting, would mitigate the objection sometimes made, that the Society's house was rather inaccessible to many of its members, the new arrangement was brought into operation during the Session, and appears to have worked satisfactorily.

As the Hon. Treasurer's accounts deal fully with the financial affairs of the Society, it is not necessary for me to enter into particulars, but I think it right to point out that the expenditure, in connection with Bacteriology, has, during the years 1898 and 1899, amounted to £1262 8s. 9d.

Members will no doubt think that, for this large amount, something more than two papers, read at the monthly meetings, might have been expected; but the facts must not be overlooked that Bacteriology is a new science, and that the difficulties of starting an institution, for its study and all necessary investigations, in this colony are much greater than in England or any part of the

During the year 40 papers on various subjects of interest were read at the monthly meetings, and various interesting exhibits were produced, as will fully appear in Volume xxiv. of the Society's Proceedings, three Parts of which have been already issued, and the fourth or final Part is now in the printer's hands.

As it is usual, on occasions like the present, for the President to address the members on some subject of interest, I have not thought it right to depart from the usual custom, although the preparation of such an address has been no small tax on my leisure hours.

The subject, which I have chosen and which has long been of interest to me, is the question of the age of Australia; a subject which, however, I admit, requires more knowledge and more careful handling than I have been able to bring to bear upon it.

Our great island continent having now been frequently crossed and recrossed in various directions, and pretty fully explored, at the expense of many valuable lives, and of an enormous amount of suffering by the less unfortunate explorers, and the crude notion of the existence of a great inland sea having been entirely dissipated, a not inconsiderable mass of evidence has become available for the consideration of the question of the age of Australia.

Far be it from me to pretend to deal authoritatively with this question, for the more I have thought over it, the more my incompetence to deal with it has been brought home to me. I trust, therefore, that what I am about to say may be looked upon merely as hints, thrown out with the view of inducing competent scientists to deal with the question. It will of course be so readily understood that I have been indebted to many authors for most of my facts, that I have only in a few instances quoted authorities

It is not necessary that we should go back to the very origin of the universe, but a few words, by way of preface, may, perhaps, not be thought inappropriate.

I therefore commence with the statement that, in the beginning of the present order of things, the whole universe was filled with matter of extreme tenuity, and generally supposed to have been intensely hot, though the existence of heat in the first instance does not seem to be an absolutely necessary postulate, it being quite conceivable that this immense mass of matter may have been able to generate sufficient heat, by means of the internal motion with which it was endowed.

The matter in question was a homogeneous mass of atoms, each being an exact counterpart of every other, all being in motion; but the homogeneity continued only until these atoms, in obedience to a law impressed upon them, began to arrange themselves in various manners, and thus to form molecules which, according to their arrangement among themselves and their distances from each other, formed the different elements with which we are acquainted, and no doubt others which we have not yet discovered.

The whole nebulous mass began to part with its heat, but, as there was at first no space not already filled with matter equally hot, it is difficult to determine whether the lost heat became latent, or was converted into motion, electricity, or light, or (if the theory that heat is not a condition of matter, but an actual fluid substance, is tenable) into some other form of matter, or what became of it.

However this may be, the diffused matter gradually cooling and shrinking, began to curdle into vast divisions, each of which

followed by the higher forms of vegetation, then marine and terrestrial animals, and ultimately intelligence, in the shape of man, appeared on the scene.

Cooling and consequential shrinkage divided the surface of the earth into land and water, and, by lateral thrust and folding, produced mountains, which gave rise to rivers; and these, with a constant heavy rainfall, wore down and denuded the mountains, and spread mud and débris over the beds of oceans, lakes, and depressed places, thus forming strata, which gradually built up and consolidated the whole of the earth's surface, including not only those parts which appeared above the water, but all the land covered by oceans, for these, being of insignificant depth compared with the whole mass of the earth, need not be taken into account.

The operations going on at this stage being so exactly the same everywhere, it may reasonably be asked why the whole surface of the earth was not converted entirely into dry land perfectly level, or else covered everywhere by water of an even depth; but the answer to this question must be that the earth revolved round the sun, not in a true circle, but in an ellipse, that its axis was not at right angles with the plane in which it revolved, and that the variations thereby effected would necessarily produce consequential variations in the operations and their effect, then moulding the surface.

During the turmoil of these troublous times, and apparently in the Cambrian epoch, some part of Australia arose above the ocean, and other parts appeared later in geological sequence.

Australia was originally very different in shape, size and features from what it is at the present day, and it is evident that, at no very distant geological date, not only Tasmania, but also New Guinea, formed part of the continent.

In the general instability of everything, Australia, after its birth, or the greater portion of it, must have been several times submerged beneath the waves before it began to settle down, in the Tertiary age, to its present form.

It is certain that, even then, it extended considerably more to the eastward than at present, and that in the west there existed a range of high mountains, from the wear and tear of which the strata of the lower lands were formed.

A great basin, over the centre of which Sydney now stands, and which extended from beyond Newcastle on the north to Shoalhaven on the south, and from beyond Hartley on the west to a distance of 70 miles or more to the eastward of the present coast line, then existed, and in this basin was formed the immense mass of coal which is of such incalculable value to New South Wales.

This basin gradually subsided, during the formation of the coal; and, in the Triassic period, sank beneath the Pacific Ocean. This subsidence may have been caused wholly or partially, firstly, by the pressure on the semifluid underlying matter, of the accumulations of vegetable matter, which were ultimately converted into coal, and secondly, by the constant sinkage of cool water, which would have the effect of contracting the heated matter lying below.

In the vicissitude of things, the eastern half of the submerged basin was cut off and permanently lost, though the residue, consolidated by the deposition of the Hawkesbury sandstone, afterwards rose again, but again sank so as to allow the deposition of

and worn down by the continuous roll of the Pacific Ocean, assisted by the revolution of the earth, which, co-operating with it, unceasingly hurls the whole of the eastern face, at the rate of about 500 miles an hour, against the waters of the ocean (which are not able to move as quickly as the solid land), and so increases their power.

By these means the inhospitable looking cliffs, which form a great part of our eastern boundary, are being continually thrown down and crushed into sand, available for the formation of the sandstone rocks of the future.

The loss of land, on the eastern side, was wholly or partially compensated, by the upheaval (?) on the south side, of a large tract of land, in the great Australian Bight, of over 200 miles in length with an extreme depth of 150 miles. It is, however, by no means clear that this was a case of upheaval, for it is possible that it, and many other so-called upheavals, are really due to the sinkage, into the body of the earth, of portion of the water forming the oceans lying on its surface, which are being thereby continually lowered.

Although it is evident that, after Australia first appeared as dry land, enormous disturbances took place on its eastern side, the far distant interior with its immense plains, inexpressibly dreary in their dull arid uniformity, does not appear to have undergone similar disturbances; and, being situated on a parallel unaffected by the comparative regularity of the weather of the Torrid and Arctic zones (the rains of which, as a general rule, exhaust themselves before reaching the desolate expanses which scarcely afford subsistence to the few miserable savages who roam over them) has only of late years been opened up to investigation by scientific men, and occupation by graziers.

A great portion of the Blue Mountain Range can hardly be considered as mountains, being really only an elevated plateau, through which great gorges have been cut by the rains of long ages; but, in the interior, many of the mountains, such for instance as the McDonnell Ranges, are quite entitled to the designation. These last are supposed, by some of the leading scientific men of

Australia, to be older than any other part of the continent, comprising as they do the so-called Larapintine limestone of Silurian formation. The oldest known rocks on our own mountains consist of slates and limestones, also of Silurian age.

In Mesozoic times, the Australian flora consisted mainly of *Filices*, *Lycopodiaceæ*, *Equisetaceæ*, *Coniferæ*, and *Cycadaceæ*, but, in Tertiary times, these were almost superseded by a very different flora, including among other trees the Oak (*Quercus*), Beech and Cinnamon. The *Equisetaceæ* have long since disappeared ; the *Lycopodiaceæ*, which formerly attained the height of 50 feet, have dwindled down to 12 inches; the *Coniferæ* are represented by *Araucaria*, *Podocarpus*, *Dammara*, and a few other genera, most of which are only found in the north ; various *Cryptogams* still flourish in such places as our now arid climate will permit, but the *Cycads* which flourished to an enormous extent in Oolitic times in Europe, and have now quite disappeared from that region, have managed, in the shape of *Zamias*, to survive all misfortunes, and still grow profusely in Australia, and are looked upon by botanists as examples of living fossils.

The old flora, except in some instances, the survival of which is difficult to account for, perished, but, when Australia, in the Tertiary period, finally arose from her watery bed, it was succeeded, in the later part of that period, by a strange flora, differ-

that would inevitably have destroyed all animal and vegetable life. Botanists would have been saved from a great deal of labour and uncertainty, in classifying the "gums," if parts of the region with which we are dealing had been submerged, so as to break the chain of variation, and by that means convert what now seem to be mere varieties into what would then be naturally considered distinct species.

It seems to me that the *Eucalypti* and kindred species of *Myrtaceæ* alone, without taking the *Leguminosæ*, the *Proteaceæ*, and other orders into account, conclusively prove Australia to be immensely old.

The origin of the Australian aborigines is as yet, and probably will be for ever, entirely unknown. Whether the ancestors of the present fast disappearing race arrived originally from Asia, Africa, or America, and whether immediately or mediately through some of the South Sea islands, it is not surprising that the barren inhospitable nature of their new home, unameliorated by the slightest attempt at cultivation, should have completed their degradation to almost the very lowest plane of humanity. Notwithstanding the vast space to be covered, they gradually spread themselves sparsely over the whole of the island, the faculties necessary for the prolongation of their miserable existence having been sufficiently sharpened by the necessities of the case.

The spread of these people over such a tract of three millions of square miles, a great part of which was an inhospitable desert, the development of scores of languages, comprising the most extraordinary inflections and complications, and the institution of so many barbarous rites and revolting customs, must have occupied an immense period, the length of which, in the entire absence of records and traditions, cannot even be guessed at.

Until the arrival of the white man, they were still in the stone age, and this fact alone carries them back many thousands of years (which, however, is a very short time when compared with geological epochs) and evidences the entire stagnation of civilization among them, if it is permissible to use such a word in connection with the state in which they existed.

It seems probable that the ancestors of the present aboriginal race were not the first human inhabitants of Australia, and that their predecessors were the ancestors of the lately extinct Tasmanian race. The Tasmanians, although resembling the Australians in colour, in their not constructing permanent dwellings, in their absolute nakedness, in the use of stone implements, wooden spears and waddies, and in other particulars, were evidently of another and an inferior race, for they had woolly hair, and were ignorant of the boomerang, the wommera and the tomahawk; and, there being no Dingoes in their island, they did not possess any pets until the introduction of dogs by the white man; while the shields and canoes which they used, if any, were of very inferior character.

Although there is no evidence of the fact, it seems likely that, when the Australians arrived, they found the Tasmanians in possession, and being a stronger and more domineering race they, in course of time, drove the latter gradually further south, as the Maori in New Zealand is supposed to have driven the Moriori to Stewart Island; but the Australians did not care to follow the Tasmanians across the Straits and extirpate them after the manner of the Maori.

her existing *Mammalia* but with no placentals, had been cut off from the rest of the world, the aboriginals had arrived with their dogs and driven off the old race, before they had been able to obtain any of these dogs from the last comers. If some such thing as this had not happened, and the Dingo had been indigenous, it is difficult to understand why it had not crossed to Tasmania.

If there is anything in these suppositions, man must have been in Australia while her now extinct volcanoes were active, and while her great fossil beasts were still in the flesh, but it is right to say that there is no proof of this, although carefully searched for.

I have taken little account of the bats and small rodents, as the former could easily have flown across from New Guinea or elsewhere, and the latter might as easily have been introduced by the new comers in their canoes, or have floated over on drift wood and rubbish.

It is well known that all Australian mammals, with the exceptions mentioned, but including many which being now extinct are only known by their fossil remains, fall within the order *Marsupialia*.

A few examples of this order have been discovered in America, the Malay Archipelago, Celebes, Amboyna, Banda, and Timor, but Australia, including Tasmania and New Guinea, is their stronghold, and has been characterised as their metropolis. A great number of Australian marsupial families are found nowhere outside this region. The brains of these creatures prove them to be of a very low order in the scale of animal life, and it is supposed that they everywhere preceded the placental mammals, none of which are known to have existed before the Tertiary era, although the Cretaceous rocks of North America and the underlying Jurassic or Oolitic rocks, both of that continent and of Europe, shew that marsupials, allied to the primitive carnivorous types now inhabiting Australia, were once widely spread over Europe: and the remains of like animals have now been discovered in Patagonia. As these marsupials were quite unknown in Europe after the Jurassic period, Lydekker assumes that during it, or sometime afterwards, they reached Australia, where they have ever since been completely cut off from all the rest of the old world. Being free from the competition of the higher types, they have flourished and developed to an extent which would otherwise have been impossible. One or two of the kangaroos, now found only in a fossil state, attained gigantic dimensions, and the *Diprotodon* must have been fully as large as the largest *Rhinoceros*.

Although the Emu is not unrepresented elsewhere, being related to the ostrich of Africa, and more closely to the lately extinct *Dinornis* of New Zealand, and also to the *Rhea* of America, it and the rarer Cassowary may be considered survivors of a perishing race.

Gould remarks that Australia comprises peculiarities unexampled in any other portion of the globe; that she possesses almost exclusively the *Marsupialia* and *Monotremata*, and many singular forms of birds especially adapted to find their existence among her very remarkable flora and equally remarkable insects.

Many years ago some strange fossils were discovered in the European rocks of the Secondary period, where they seemed to be extremely abundant. These consisted of the jaws of a fish,

scutes on the back as well as one on the belly, has recently been found living in some of our rivers.

Equally strange discoveries have been made with respect to shells, the first discovered being the beautiful *Trigonia*, known for a long time only as a fossil of the Secondary and Tertiary rocks, but this shell was afterwards found alive in Port Jackson and Tasmania. Buckland first pointed out the remarkable fact that these shells, being in their fossil state associated with *Cestracion*, the same association prevailed in their modern state.

I will only mention one other instance of survival in Australian waters, namely, that of the curious *Brachiopoda*, commonly known as "Lamp Shells," which, in Secondary and Tertiary times, were so abundant as to have formed in some instances thick strata of rock.

So many discoveries of "living fossils" having been made in Australia, it was not unreasonably expected, by the projectors of the Horn Expedition, that older forms of life than those prevailing elsewhere would be found in the more inaccessible parts of Australia, such as the McDonnell Ranges, which, being of Silurian age, were supposed to have existed as islands before the rest of the continent had accomplished its final emergence from the ocean; but in this expectation they were disappointed.

On careful consideration of all the facts which I have stated, I have little doubt that, after Australia had been cut off from the old world, Tasmania and New Guinea were cut off from Australia, but, although various changes afterwards took place, there always remained above the ocean, since their arrival, a home for the plants and animals mentioned, which have therefore been able to survive to the present day, while their less fortunate congeners, located in other lands, being cut off by the submergence of their homes, have perished and left no other record of their former existence than their fossil remains.

Australia is unique as being the home of so large an array of plants and animals, which have very appropriately been called "Living Fossils," and, for this reason, strengthened by facts which I have mentioned, I cannot doubt that, counting as geologists do from her last upheaval, which took place in the early part of the Tertiary era, she comprises perhaps the oldest land on the globe.

If this absence of submersion since Triassic times be conceded, and if any reliance can be placed on Goodchild's calculation, or I should perhaps rather say surmise, as to the length of the Tertiary period, the age of Australia may be stated at upwards of 93,420,000 years, while he calculates that the time which has elapsed since life first appeared on the earth amounts to no less than 704,235,000 years.

On the motion of Mr. Henry Deane an appreciative vote of thanks was accorded to the President for his interesting Address.

The Hon. Treasurer presented his balance sheet, and moved its adoption, which was carried. The Society's total income for the financial year ending December 31st, 1899, was £1,549 16s. 1d.; the total expenditure, £2,018 14s 9d., which, with a credit balance of £878 5s. 3d. at the beginning of the year, and with £350 placed at fixed deposit in the Commercial Bank, left the Society's ordinary account with a credit balance of £21 14s. 3d., and the Bacteriology account with a credit balance of £37 12s. 4d.

No other nominations having been received, the President

ENDOWMENT (CAPITAL).

	£	s.	d.		£	s.	d.
Amount received from Sir William Macleay in his life-time	14,000	0	0	Loan A	3,000	0	0
Further Sum bequeathed by Will £6,000, less Probate Duty	5,700	0	0	Loan B	5,000	0	0
				Loan (portion of Loan C, secured with other money by mortgage for £24,000)	11,700	0	0
	£19,700	0	0		£19,700	0	0

BACTERIOLOGY (CAPITAL).

	£	s.	d.		£	s.	d.
Legacy of £12,000, bequeathed by Sir William Macleay to the University of Sydney (less £600 probate duty), paid by the University into Court and ordered to be paid out to the Society	11,400	0	0	Loan (portion of Loan C)	11,400	0	0
Amount (out of interest received) ordered by the Council to be added to Principal	900	0	0	Also £900, from interest, included in Loan C	900	0	0
Further Amount (out of interest received) ordered by the Council to be added to Principal	700	0	0	Cash in Bank awaiting investment ...	700	0	0
Further Amount on fixed deposit	350	0	0	On fixed deposit, in Commercial Bank, due June, 1900	350	0	0
	£13,350	0	0		£13,350	0	0

March 15th, 1900.

Audited and found correct

E. G. W. Palmer.

P. N. Trebeck Hon. Treasurer.

NCE SHEET.

ty of New South Wales.

Cr.

d.	£	s.	d.		£	s.	d.
				Dec. 31, 1899.			
	878	5	3	By Ground Rent, Rates and Taxes ...	88	6	10
				,, Bank Exchange, 7s. 6d.; Bank Fee,			
	2	2	0	10s.; Cheque Books, 16s. 8d. ..	1	14	2
0				,, Printing, &c.	293	16	8
0				,, Plates and Illustrations	142	19	6
6				,, Stationery, &c.	2	18	6
	119	15	6	,, Purchase of Books	16	3	2
	1,292	5	0	,, Salaries and Wages	478	0	0
				,, Advertisements	6	15	0
	100	0	0	,, Crockery and Sundries	2	4	10
	11	10	7	,, Refreshments and Attendance ...	4	19	9
				,, Postcards	1	17	6
	3	3	0	,, Postage	15	11	8½
				,, Gas	1	19	3
				,, Petty Cash	8	9	5½
	6	5	6	,, Telephone	9	0	0
				,, Hon. Treasurer's Petty Cash ...	2	0	0
				,, Bacteriology Account ... £795 16 5			
				,, Interest 504 5 0			
					1,300	1	5
				,, Repairs, &c., to Hall	13	4	1
				,, Shipping Charges	1	10	9
				,, Balance in Bank	21	14	3
	£2,413	6	10		£2,413	6	10

rect.

G. W. Palmer.

P. N. Trebeck, Hon. Treasurer.

BACTERIOLOGY (INCOME).

	£ s. d.
To Balance from last year	703 10 5
" Interest on £19,300	[illegible] 0 0
" " " £700	91 0 0
" Fee for Bacteriologist	9 9 0
" Over credited to Bacteriology Account	10 3 0
	£1,391 1 5

	£ s. d.	£ s. d.
By Fittings, Apparatus and Chemicals	[illegible] 10 0	
" Books and Journals	94 17 0	
		400 8 8
" Salary and Wages		410 0 0
" Insurance		1 0 0
" Petty Cash		10 9 10
" Gas Account		8 18 0
" Bank Fee		0 10 0
" Commercial Bank Fixed Deposit		350 0 0
" Fee to Bacteriologist		9 9 0
" Dugald & Co. Freight, Shipping Charges		4 0 0
" Balance in Bank		87 19 4
		£1,391 1 5

March 15th, 1900.

I have audited the books of account, and seen the securities.

E. G. W. Palmer.

P. N. Trenerry, Hon. Treasurer.

WEDNESDAY, MARCH 28TH, 1900.

The Ordinary Monthly Meeting of the Society was held at the Linnean Hall, Ithaca Road, Elizabeth Bay, on Wednesday evening, March 28th, 1899.

The Hon. James Norton, LL.D., M.L.C., President, in the Chair.

Mr. P. G. BLACK, North Sydney, Dr. W. H. CRAGO, College Street, the Hon. HENRY C. DANGAR, M.L.C., Potts' Point, Mr. GEORGE ELLIOTT, O'Connell Street, and Mr. JOHN A. FERGUSON, Glebe Point, were elected Members of the Society.

DONATIONS.

(Received since the Meeting in November, 1899.)

Department of Agriculture, Sydney—Agricultural Gazette of New South Wales. Vol. x. Part 12 (Dec., 1899); Vol. xi. Parts 1-3 (Jan.-March, 1900). *From the Hon. the Minister for Mines and Agriculture.*

New South Wales Chamber of Mines, Sydney—Journal. Vol. i. Nos. 3-6 (Dec., 1899-March, 1900). *From the Chamber.*

Pamphlet—"On a Fern *(Blechnoxylon talbragarense)*, with Secondary Wood, &c." (Records of Aust. Museum. Vol. iii.). By R. Etheridge, Junr. *From the Author.*

Royal Society of New South Wales—Abstract of Proceedings, December 6th, 1899. *From the Society.*

The Surveyor, Sydney. Vol. xii. No. 12 (Dec., 1899); Vol. xiii. Nos. 1-3 (Jan.-March, 1900). *From the Editor.*

Three Entomological Pamphlets (from Agricultural Gazette of N.S.W., being Miscellaneous Publications Nos. 327, 337 and 348). By W. W. Froggatt, F.L.S. *From the Author.*

Two Botanical Pamphlets (from Agricultural Gazette of N.S.W., being Miscellaneous Publications No. 282 and 331, 1899). By J. H. Maiden, F.L.S. *From the Author.*

Two Pamphlets—"La Vésicule Germinative et les Globules Polaires chez les Batraciens" (1899). Par J. B. Carnoy et H. Lebrun; and "*Peripatus leuckarti*" (1898). By Frank Paulden *From T. Steel, Esq., F.C.S., F.L.S.*

Australasian Association for the Advancement of Science, Melbourne—Handbook, 1890 and 1900. *From Professor Baldwin Spencer, M.A.*

Australasian Journal of Pharmacy, Melbourne. Vol. xiv. No. 168 (Dec., 1899); Vol. xv. Nos. 169-171 (Jan.-March, 1900). *From the Editor.*

Department of Mines, Victoria: Geological Survey—Monthly Progress Report. Nos. 3-7 (June-Oct., 1899): Reports on the Victorian Coal-Fields. No. 7 (1900). By James Stirling, F.L.S. *From the Secretary for Mines.*

Field Naturalists' Club of Victoria—Victorian Naturalist. Vols. xvi. Nos. 8-11 (Dec., 1899-March, 1900). *From the Club.*

Institute of Mining Engineers, Melbourne—Proceedings. Annual Meeting, January, 1900. *From the Institute.*

University of Melbourne—Annual Examination Papers, October and December, 1899: Matriculation Papers, November, 1899. *From the University.*

Public Library, Museum, and Art Gallery, Adelaide, S.A.—Report of the Board of Governors for the Year 1898-99. *From the Director.*

Royal Society of South Australia—Transactions. Vol. xxiii Part ii. (Dec., 1899). *From the Society.*

South Australian School of Mines and Industries: Technological Museum—Remarks on the Tertiaries of Australia; together with Catalogue of Fossils (1892). By G. B. Pritchard. *From C. Hedley, Esq., F.L.S.*

Woods and Forests Department, Adelaide, S.A.—Annual Progress Report for the Year 1898-99. *From W. Gill, Esq., F.L.S., Conservator of Forests.*

Department of Agriculture, Perth, W.A.—Journal. December,

Geological Survey of Canada—Annual Report for the Year 1897, with Maps: Contributions to Canadian Palæontology. Vol. iv. Part i. (1899). *From the Director.*

Hamilton Association—Journal and Proceedings. No. xv. (1899). *From the Association.*

Montreal Society of Natural History—Canadian Record of Science. Vol. vi. Nos. 1-2 (1894); Vol. viii. No. 2 (1899). *From the Society.*

Royal Society of Canada—Proceedings and Transactions. Second Series. Vol. iv. (May, 1898). *From the Society.*

Bernice Pauahi Bishop Museum, Honolulu—Memoirs. Vol. i. No. 1 (1899): Fauna Hawaiiensis. Vol. i. Parts i.-ii (1899). *From the Museum.*

Academy of Natural Sciences, Philadelphia — Proceedings, 1899. Parts i.-ii. *From the Academy.*

Academy of Science, St. Louis—Transactions. Vol. viii. Nos. 8-12 (July, 1898-Jan., 1899); Vol. ix. Nos. 1-5 and 7 (Feb.-June, 1899). *From the Academy.*

American Academy of Arts and Sciences, Boston—Proceedings. Vol. xxxiv. Nos. 21-23 (May-June, 1899); Vol. xxxv. Nos. 1-3 (July-Aug., 1899). *From the Academy.*

American Geographical Society, New York — Bulletin. Vol. xxxi. Nos. 4-5 (1899). *From the Society.*

American Museum of Natural History, New York—Annual Report for the Year 1898: Bulletin. Vol. xi. Part ii. (Oct., 1899); Vol. xii. Articles xi.-xx. (pp 161-264; Oct.-Dec., 1899): Memoirs. Vol. i. Parts iv. and v. (Oct., 1899); Vol. ii. Anthropology i. (May, 1899). *From the Museum.*

American Naturalist (Cambridge). Vol. xxxiii. Nos. 395-396 (Nov.-Dec., 1899); Vol. xxxiv. No. 397 (Jan., 1900). *From the Editor.*

American Philosophical Society, Philadelphia. Proceedings. Vol. xxxviii. No. 159 (Jan., 1899). *From the Society.*

California Academy of Sciences, San Francisco—Proceedings. Third Series. *Botany.* Vol. i. Nos. 6-9 (April-July, 1899): *Geology.* Vol. i. Nos. 5-6 (March, 1899): *Zoology.* Vol. i. Nos. 11-12 (Dec.,1898, and May, 1899): Occasional Papers. No. vi. (1899). *From the Academy.*

Essex Institute, Salem, Mass.—Bulletin. Vol. xxviii. Nos 7-12; Vol. xxix. Nos. 7-12; Vol. xxx. Nos. 1-12 (1896-98). *From the Institute.*

Field Columbian Museum, Chicago—Botanical Series. Vol. i. No. 5 (Aug., 1899): Geological Series. Vol. i. Nos. 3-6 (March-April, 1899): Zoological Series. Vol. i. Nos. 11-15 (Feb.-May, 1899): The Birds of North America, Water Birds. Part i. (1899). *From the Museum.*

Journal of Applied Microscopy, New York. Vol. ii. No. 12 (Dec., 1899). *From the Editor.*

Missouri Botanical Garden, St. Louis—Tenth Annual Report, 1899. *From the Director.*

Museum of Comparative Zoology, Harvard College, Cambridge—Bulletin. Vol. xxxiii. (May, 1899); Vol. xxxiv. (Sept, 1899); Vol. xxxv. Nos. 3-7 (Oct.-Dec., 1899). *From the Curator.*

Smithsonian Institution, Washington: U.S. National Museum

Wagner Free Institute of Science, Philadelphia—Transactions. Vol. vi. (May, 1899). *From the Institute.*

Washington Academy of Sciences — Proceedings. Vol. i. pp. 111-159, 161-187, 189-251, T.p., &c. (Dec., 1899-Feb., 1900). *From the Academy.*

Wisconsin Academy of Sciences, Arts, and Letters, Madison — Transactions. Vol. xii. Part 1 (1898). *From the Academy.*

Société Scientifique du Chili, Santiago—Actes. Tome viii. 5ᵐᵉ Liv. (1898). *From the Society.*

Museo Nacional de Buenos Aires—Anales. Tomo vi. (1899): Comunicaciones. Tomo i. Nos. 3-5 (May-Dec., 1899). *From the Museum.*

Museo de La Plata—Revista. Tomo ix. (1899). *From the Museum.*

Museo Nacional de Montevideo—Anales. Tomo ii. Fasc. xii. (1899). *From the Museum.*

College of Science, Imperial University of Tokyo—Journal. Vol. xi. Part iv. (1899). *From the University.*

Dictionary of the Lepcha Language, compiled by the late General G. B. Mainwaring, revised and completed by Albert Grünwedel (8vo. Berlin, 1898). *From the Bengal Secretariat Book Depôt, Calcutta.*

Geological Survey of India, Calcutta—Palæontologia Indica. Series xv. Himálayan Fossils. Vol. i. Part 2 (1899); Vol. ii. Title Page, &c. (1897); New Series. Vol. i. Parts 1-2 (1899). *From the Director.*

Indian Museum, Calcutta—Annual Report for the Year 1898-99: Descriptive Catalogue of the Indian Deep-sea Fishes in the Indian Museum (4to. 1899): An Account of the Deep-sea Brachyura collected by the R.I.M.S.S. "Investigator" (4to. 1899): Materials for a Carcinological Fauna of India. No. 5 (1899): Natural History Notes from the R.I.M.S.S. "Investigator." Series iii. No. 2 (in two Parts) [1899]. *From the Museum.*

Perak Government Gazette. Vol. xii. Nos. 31-40 (Oct.-Dec., 1899); Vol. xiii. Nos. 1-6 (Jan.-Feb., 1900). *From the Government Secretary.*

British Museum (Natural History), London—Catalogue of Welwitsch's African Plants. Vol. ii. Part 1. By A. B. Rendle, M.A., F.L.S, &c. (8vo. 1899): Hand-List of the Genera and Species of Birds. Vol. i. By R. B. Sharpe, LL.D. (8vo. 1899); List of the Genera and Species of Blastoidea. By F. A. Bather, M.A., F.G.S. (8vo. 1899). *From the Trustees.*

Cambridge Philosophical Society — Proceedings. Vol. x. Parts 3-4 (1899-1900). *From the Society.*

Conchological Society of Great Britain and Ireland, Manchester —Journal of Conchology. Vol. ix. No. 9 (Jan., 1900). *From the Society*

Entomological Society of London—Transactions, 1899. Part 4. *From the Society.*

Entomologists' Record and Journal of Variation, London. Vol. xi. No. 12 (Dec., 1899). *From the Editor.*

Geological Society of London—Quarterly Journal. Vol. lv. Part 4 (No. 220; Nov. 1899); Vol. lvi. Part 1 (No. 221; Feb., 1900); List of Members, &c. (Nov., 1899). *From the Society.*

Ten Conchological Pamphlets. By Edgar A. Smith, F.Z.S. (8vo. 1898-99). *From the Author.*

Zoological Society of London—Abstract, Nov. 14th and 28th, Dec. 19th, 1899; Jan. 23rd and 6th Feb., 1900: Transactions. Vol. xv. Part 4 (Dec., 1899). *From the Society.*

Royal Irish Academy, Dublin—Proceedings. Third Series. Vol. v. No. 3 (Nov., 1899). *From the Academy.*

Archiv fur Naturgeschichte, Berlin. lx. Jahrgang. ii. Band. 3 Heft (1894): lxiii. Jahrg. ii. Band. 2 Heft (1897): lxvi. Jahrg. i. Band. 1-3 Heft (1899-1900). *From the Editor.*

Gesellschaft für Erdkunde zu Berlin—Verhandlungen. Band xxvi. Nos. 2-6 (1899): Zeitschrift. Band. xxxiii. No. 6 (1898): Band xxxiv. Nos. 1-2 (1899). *From the Society.*

Kaiserliche Leopoldino-Carolinische deutsche Akademie der Naturforscher, Halle—Leopoldina. Heft. xxxiv. (1898): Nova Acta. Band xxxvii. Nos. 1-3: xxxviii. Nos. 1 and 4; xxxix. Nos. 2, 5-7: xl., Nos. 4, 6 and 8: xli. (1) Nos. 1-3 and 5: xli. (2) Nos. 1 and 7: xlii. Nos. 1, 4 and 6: xliii. Nos. 4 and 5: xliv. Nos. 1, 3 and 4: xlv: xlvi. Nos. 1-3; xlvii. Nos. 1 and 4-6: xlviii. Nos. 2-3: xlix Nos. 1-3: l. Nos. 3-5: li. Nos. 3-4: lii. Nos. 1-3 and 6-7: liii. Nos. 4-5: liv. No. 2: lv. Nos. 2, 5 and 7: lvi. No. 1; lvii. Nos. 1, 2 and 6: lviii. Nos. 2-3 and 5-7: lx. Nos. 1 and 3; lxi. Nos. 1-4: lxii.: lxiii.: lxiv. Nos. 1 and 3-5: lxv. No. 1: lxvi. Nos. 1 and 4: lxvii. Nos. 1-3: lxix. Nos. 2-3: lxx.: lxxi.: lxxii.: lxxiv. Nos. 1-4 Titel zu den Nova Acta. Bd. xxxvii-xliv., xlvi-lxi., lxiv-lxix. *From the Academy.*

Naturforschende Gesellschaft zu Freiburg i. Br.—Berichte. ix. Band. 1 Heft (1899). *From the Society.*

Naturhistorische Gesellschaft zu Nurnberg—Abhandlungen. xii. Band (1899). *From the Society.*

Naturhistorischer Verein in Bonn — Verhandlungen. lvi. Jahrgang. Erste Halfte (1899): Sitzungsberichte der Niederrheinischen Gesellschaft fur Natur- und Heilkunde zu Bonn, 1899. Erste Halfte *From the Society.*

Verein für vaterländische Naturkunde in Württemberg—Jahreshefte. 55 Jahrgang (1899). *From the Society.*

Zoologischer Anzeiger, Leipzig. xxii. Band. Nos. 601-604 (Nov.-Dec., 1899); Band xxiii. Nos. 605-607 (Jan.-Feb, 1900). *From the Editor.*

Königliche Böhmische Gesellschaft der Wissenschaften, Prag—Jahresbericht für das Jahr 1898: Sitzungsberichte, 1898. *From the Society.*

K.K. Naturhistorisches Hofmuseum, Wien—Annalen. Band xiii. Nos. 1-3 (1898). *From the Museum.*

K.K. Zoologisch-botanische Gesellschaft in Wien—Verhandlungen. xlviii. Band (1898); xlix. Band (1899). *From the Society.*

Muséum National Hongrois, Budapest — Természetrajzi. Anhangsheft zum xxi. Bande (1898); Vol. xxii. Partes i.-ii. (1899). *From the Museum.*

L'Académie Royale des Sciences, &c., de Belgique, Bruxelles—Annuaire. Années 1898 et 1899: Bulletin. 3me Série. Tomes xxiv.-xxvi. (1897-98): Tables Générales du Recueil des Bulletin. 3me Série. Tomes i. à xxx. (1881-1895). *From the Academy.*

Société Belge de Microscopie, Bruxelles—Annales. Tome x[illegible]i.

Journal de Conchyliologie, Paris. Vol. xlvii. No. 4 (1899). *From the Director.*

Société Entomologique de France—Bulletin. Année 1895. *From the Society.*

Nederlandsche Dierkundige Vereeniging, Helder—Tijdschrift. 2de Serie. Deel vi. Afl. 2 (Aug., 1899): Aanwinsten van de Bibliotheek, 1st. Aug., 1897-31st. Dec., 1898. *From the Society.*

Nederlandsche Entomologische Vereeniging, Hague—Tijdschrift voor Entomologie. xlii. Deel. 3 Afl. (1898). *From the Society.*

Société Hollandaise des Sciences à Harlem—Archives Néerlandaises. Série ii. Tome iii. 2e Liv. (1899). *From the Society.*

Académie Impériale des Sciences de St. Pétersbourg—Annuaire du Musée Zoologique, 1899. Nos. 1-3: Bulletin. v^{e} Série. Tome viii. No. 5 (1898); T. ix. Nos. 1-5 (1898); T. x. Nos. 1-4 (1899). *From the Academy.*

Russisch-Kaiserliche Mineralogische Gesellschaft zu St. Petersburg—Materialien zur Geologie Russlands. Band xix. (1899): Verhandlungen. Zweite Serie. xxxvi. Band. ii. Lief. (1899); xxxvii. Band. i. Lief. (1899). *From the Society.*

Société Impériale des Naturalistes de Moscou—Bulletin. Année 1899. No. 1. *From the Society.*

Société Entomologique à Stockholm—Entomologisk Tidskrift. Arg. 20. Häft. 1-4 (1899). *From the Society.*

Royal University of Upsala—Bidrag till en Lefnadsteckning ofver Carl von Linné, No. viii. Af Th. M. Fries (1898); Caroli Linnæi Hortus Uplandicus med Inledning och Förklaringar. Af Th. M. Fries (8vo. Upsala, 1899): Ofversikt af Faunistiskt och Biologiskt Vigtigare Litteratur Rörande Nordens Fåglar. Af J. M. Hulth (4to. Stockholm, 1899): One Dissertation (8vo. Stockholm, 1898). *From the University.*

Académie Royale des Sciences et des Lettres de Danemark, Copenhague—Oversigt, 1899. Nos. 4-5. *From the Academy.*

Naturhistoriske Forening i Kjöbenhavn — Videnskabelige Meddelelser for Aaret 1899. *From the Society.*

Museo Civico di Storia Naturale di Genova—Annali. Serie 2ª. Vols. xii.-xix. (1892-99). *From the Museum.*

Museo di Zoologia, &c., della R. Università di Torino—Bolletino. Vol. xiv. Nos. 354-366, T.p., &c. (July-Dec., 1899). *From the Museum.*

La Nuova Notarisia, Padova. Serie xi. (January, 1900): Sylloge Algarum. Vol. iv. Sectio ii. Familiæ i.-iv. (1900). *From the Editor, M. le Doct. G. B. de Toni.*

R. Università degli Studi di Siena—Bulletino del Laboratorio ed Orto Botanico. Vol. ii. Fasc. iii.-iv. (1899). *From the University.*

South African Museum, Cape Town—Annals. Vol. i. Part 3 (1899). *From the Trustees.*

Forty-two South African Government Reports, chiefly Geological. *From the South African Philosophical Society, Cape Town.*

DESCRIPTIONS OF NEW AUSTRALIAN LEPIDOPTERA.

By Oswald B. Lower, F.E.S., Lond.

SYNTOMIDIDÆ.

Syntomis cremnotherma, n.sp.

♂. 22 mm. Head, thorax, abdomen and palpi orange, abdomen with two or three black segments beneath, anal tuft black. Antennæ fuscous, pectinations 1. Legs orange, tibiæ and tarsi blackish. Forewings elongate-triangular, considerably dilated posteriorly, costa straight, hindmargin rounded; orange, with black markings; a narrow line along costa to about end of cell, thence becoming confluent with a thick costal streak, continued around apex and along hind margin to anal angle, broader towards apex and narrowed towards anal angle, and with a moderate, somewhat cuneiform, indentation above anal angle; a thick oblique bar at end of cell; a thick line along inner margin throughout, emitting at ¼ a moderately erect band, reaching more than half across wing, much flattened on inner margin, and somewhat curved inwards on upper half; all veins outlined with blackish: cilia blackish. Hindwings with hindmargin rounded; orange; an irregular blackish spot at end of cell; a moderately broad black hindmarginal band, with a well defined obtuse projection near anal angle; veins more or less outlined with blackish; cilia as in forewings.

Irrapatana, South (Central) Australia; in November.

Very distinct from any of its Australian congeners.

NOCTUINA.

CARADRINIDÆ.

Leucania stenographa, n.sp.

♂♀. 38-40 mm. Head, palpi, thorax, legs and abdomen ochreous-fuscous, face with two narrow interrupted fuscous bars, thorax with two transverse anterior bands, anterior one very fine, posterior very broad, coxæ densely hairy, posterior pair mixed with fuscous. Antennæ fuscous-ochreous. Abdomen beneath more ochreous. Forewings elongate, moderately dilated, costa nearly straight, hindmargin gently rounded; ochreous, strongly infuscated throughout with fine fuscous and dark fuscous lines, becoming edged by an equal width of groundcolour on hindmarginal area, which gives the appearance of alternating ochreous and fuscous lines; lower edge of cell becoming very strongly infuscated, sometimes more or less continued as a thick streak to apex; a fine whitish-ochreous spot at end of cell; an outwardly curved row of fine black dots from $\frac{5}{6}$ of costa to inner margin before anal angle: cilia ochreous-fuscous, with a hindmarginal row of black intervenal spots, more pronounced on underside. Hindwings with hindmargin very slightly waved; iridescent-whitish; a fine fuscous hindmarginal line; cilia whitish.

towards hindmargin outlined with black, and with small white dots at hindmarginal extremity, sometimes absent: cilia dark fuscous. Hindwings grey-whitish, fuscous-tinged along hindmargin and more strongly along apical area: cilia whitish, with a fuscous subbasal line, becoming blackish round apex.

Semaphore, South Australia; two specimens, in September.

METAPTILA (?) PTILOMELA, n.sp.

♀. 35 mm. Head, thorax and antennæ ashy-grey-whitish. Palpi white, terminal joint infuscated. Thorax with a dense posterior crest, and with a narrow transverse curved anterior black line, and two similar lines, one in middle, and one anterior to crest, obscure. Abdomen silvery-white. Legs whitish, tarsi fuscous, obscurely ringed with whitish. Forewings elongate, moderate, costa arched, hindmargin oblique, hardly rounded: ashy-grey-whitish, with blackish markings: a short narrow line above middle, from base to before first line: first line hardly waved, from $\frac{1}{4}$ costa to $\frac{1}{4}$ inner margin, obliquely outwards on upper half, thence angulated inwards: second line irregularly dentate, from $\frac{3}{4}$ costa to just beyond middle of inner margin, curved inwards on upper $\frac{2}{3}$, and with a broad obtuse angulation above middle, thenced curved inwards, acutely angulated above inner margin: orbicular whitish, well defined, and with some raised black scales in centre and encircled with a fine black line: a similar, yet with a more distinct spot of raised scales below and beyond: another indistinct mark above and beyond, preceded by an obscure black spot on costa: reniform whitish, encircled with a fine black line, indented posteriorly, followed beneath by a similar rounded coloured spot: subterminal obscure, blackish, deeply indented: veins towards hindmargin outlined with blackish: cilia ashy-grey, darker at base. Hindwings white, fuscous-tinged around apex and hindmargin to middle: cilia white, fuscous-tinged around apex and upper half of hindmargin.

Melbourne, Victoria: two specimens: I have seen others from Stawell and Sale.

Recalls in some respects a large specimen of *Uraba lugens*, Walk., (*Bombycina*). A new genus will no doubt be required to receive it.

CARADRINA PELTALOMA, n.sp.

♀. 25 mm. Head, thorax and palpi dull whitish, finely irrorated with blackish, terminal joint of palpi fuscous. Abdomen ochreous-yellow. Legs whitish-fuscous, tarsi with whitish rings. Antennæ fuscous, lighter at base. Forewings elongate-triangular, costa nearly straight, hindmargin gently bowed; dull whitish, finely irrorated with blackish, except on hindmarginal area, which is brownish; four obscure fuscous elongate marks on posterior $\frac{2}{3}$ of costal edge; a fine blackish line, starting from second costal mark, continued obliquely outwards towards hindmargin, thence curved inwards beneath and finely traceable to near anal angle; a large fuscous cuneiform spot on inner margin at $\frac{2}{3}$, edged on either side with clear whitish, upper portion mixed with black, its apex imperfectly defined, inclining towards hindmargin; a hindmarginal row of blackish dots: cilia fuscous, somewhat chequered, darker towards base, and with a fine waved whitish basal line. Hindwings with hindmargin faintly crenulate; pale ochreous-yellow; hindmarginal and apical area infuscated, especially towards base; cilia whitish, at base yellow, and with a fuscous parting line, and with two or three fuscous teeth around

tions orange, about 10. Forewings elongate-triangular, costa straight, hindmargin waved, bowed, dull whitish; a dentate black transverse line, edged posteriorly by its own width of orange throughout, from costa at about $\frac{2}{5}$ to inner margin at about $\frac{1}{3}$; a second, similar, yet more dentate line parallel to first, edged anteriorly by its own width of orange throughout, from $\frac{3}{4}$ of costa to $\frac{2}{3}$ of inner margin: cilia dull whitish. Hindwings with the hindmargin rounded; fuscous, becoming darker on basal half; cilia as in forewings.

Broken Hill, N.S.W.; two specimens, in September.

SORAMA DRYINA, n.sp.

♀. 40-44 mm. Head, thorax, palpi, antennæ, legs and abdomen grey-whitish. Forewings elongate, moderate, costa moderately arched, hindmargin oblique; whitish, strongly mixed with ochreous and minutely irrorated throughout with blackish scales; an ill-defined ochreous basal patch, outer edge from before $\frac{1}{4}$ of costa to $\frac{1}{8}$ inner margin, limited by an obscure waved line; a moderate V-shaped dark ochreous mark at end of cell, followed by a broad ochreous-fuscous fascia narrowed on inner margin, more or less scalloped on posterior edge, and followed by a clear white fascia, obscure on inner margin; a strongly waved fuscous line in middle of ochreous fascia, hardly traceable on upper third; a thick ochreous submarginal line; sometimes the whole of the wing is more or less obscured with dark fuscous: cilia ochreous, with greyish teeth at extremities of veins. Hindwings greyish-ochreous, with a fine fuscous hindmarginal line; cilia greyish.

Mackay and Brisbane, Queensland; two specimens, in March.

NOTODONTINA.

SELIDOSEMIDÆ.

ECTROPIS HIEROGLYPHICA, n.sp.

♂. 25 mm. Head, thorax, palpi, antennæ and abdomen dull greyish-ochreous, very minutely dusted with blackish-fuscous, face with an obscnre dark fuscous median bar, thorax with narrow transverse blackish anterior and posterior bands, abdomen with blackish segmental bands becoming whitish internally, second more or less wholly fuscous. Legs fuscous, middle and posterior pair strongly mixed with whitish; tibiæ and tarsi with whitish rings. Forewings elongate, moderate, costa straight, hindmargin not waved, oblique, faintly sinuate inwards on lower $\frac{1}{3}$; dull greyish-ochreous, very finely dusted and strigulated with fuscous and dark fuscous; costa finely strigulated throughout with blackish; a fine blackish line, slightly curved inwards, from costa at $\frac{1}{5}$ to inner margin at $\frac{1}{6}$; a black transverse discal mark at end of cell; a well defined fine blackish line, starting on costa just beyond $\frac{3}{4}$, slightly curved inwards and reaching half across wing, thence twice strongly scalloped and reaching nearly to beneath middle of cell, thence returning and ending on inner margin before anal angle, and with a short angulation just above termination; subterminal whitish, obscure, edged anteriorly on upper half by a blackish shade; hindmarginal area more blackish than rest of wing; veins between subterminal and hindmargin outlined with blackish; a fine black hindmarginal line, more or less interrupted on veins: cilia greyish-ochreous, chequered with blackish. Hindwings with hindmargin as in forewings; colour as in forewings, but inner margin more strongly marked; first line and

segments. Anterior legs blackish, middle and posterior pair greyish. Forewings elongate-triangular, hindmargin hardly waved, bowed: ashy grey-whitish, mixed with blackish: costal edge finely spotted with black: first line slender, from $\frac{1}{6}$ inner margin, reaching half across wing, immediately followed by a second similar line: median shade indistinct: a moderately well defined oblique black line from $\frac{3}{4}$ of costa to just beyond middle of inner margin, with two or three sharp on upper fourth: a moderate thick fuscous parallel shade beyond: subterminal dentate, whitish, preceded by a similar dark fuscous line: veins preceding subterminal more or less outlined with black: a waved black hindmarginal line: cilia dark fuscous, with a black subbasal line. Hindwings with the hindmargin waved: colour as in forewings, excepting basal half of wing, which is much lighter; median shade moderate, distinct, blackish, upper third hardly traceable; hindmarginal line and cilia as in forewings.

Broken Hill, N.S.W.: three specimens, in April and August.

SELIDOSEMA PETROZONA, n.sp.

♂. 30 mm. Head, thorax, palpi, antennæ and abdomen yellowish-ochreous, face (fuscous) rubbed, antennal pectinations 6 at greatest length, abdomen with blackish segmental rings. Legs fuscous, posterior pair lighter. Forewings elongate-triangular, hindmargin gently waved, rounded: light ochreous, fuscous-tinged: costal edge narrowly fuscous throughout: first line hardly traceable, lower half very obscure: median shade fuscous, traceable on lower half only: a broad oblique shade of fuscous, from middle of inner margin to hindmargin below apex, darker posteriorly: subterminal whitish, posteriorly edged by its own width with fuscous: a fine fuscous waved hindmarginal line: cilia ochreous-fuscous. Hindwings with hindmargin waved, rounded: colour as in forewings, excepting basal area, which is lighter: first line absent: median shade obscurely fuscous, nearly straight: an elongate, blackish discal dot between shade and second line; second line blackish, twice sinuate, followed by a moderately broad fuscous parallel shade; subterminal and hindmarginal lines as in forewings; cilia as in forewings.

Melbourne, Victoria; two specimens, in January.

In the *excursaria* group.

SELIDOSEMA DIAGRAMMA, n.sp.

♂. 23 mm. Head, palpi, thorax and abdomen dull ochreous-fuscous, finely and densely irrorated with whitish, abdomen with blackish segmental bands, finely edged posteriorly with whitish, face fuscous. Antennæ fuscous, ciliation 4, apical fourth simple. Legs fuscous, tibiæ and tarsi with whitish rings. Forewings elongate-triangular, hindmargin faintly waved; fuscous or dull ochreous-fuscous, finely and densely irrorated with whitish; lines fine, black, distinct; first waved, from costa at $\frac{1}{4}$ to base of inner margin, angulated outwards beneath costa, and anteriorly throughout by a fine fuscous shade; median shade obsolete; a blackish discal dot at end of cell; second line gently waved, from costa at $\frac{3}{4}$, curved inwards to beneath and beyond discal spot and ending on inner margin in middle, posteriorly edged by a fine fuscous shade throughout; a small mark in middle of disc touching second line; subterminal waved, obscurely whitish; a fine black hindmarginal line, becoming somewhat dot-like on veins : cilia ochreous-fuscous. Hindwings with hindmargin waved; colour as

blackish scales; markings black, obscure; a moderate line, twice dentate, from costa at $\frac{1}{4}$ to inner margin at $\frac{1}{3}$; a narrow median shade, strongly curved outwards above middle, from costa in middle to middle of inner margin; second line shaped as median shade, somewhat interrupted on and above curve, from costa at about $\frac{3}{4}$ to inner margin at $\frac{2}{3}$ and gently curved inwards above inner margin; subterminal dot-like, obscure; a fine waved hind-marginal line, sometimes interrupted on veins: cilia blackish-fuscous. Hindwings with hindmargin sinuate above and below middle, causing anal angle to become somewhat prominent; whitish, somewhat fuscous-tinged around hindmargin and apex; a blackish discal dot; a fine blackish hindmarginal line; second line blackish, obscurely indicated by a row of fuscous dots, more pronounced on underside; cilia whitish.

Broken Hill, N.S.W.; seven specimens during April and May.

PYRALIDINA.

SEDENIA XEROSCOPA, n.sp.

♂♀. 19-22 mm. Head, thorax, palpi and antennæ light ochreous-fuscous, posterior half of thorax and abdomen grey-whitish, palpi beneath whitish. Legs ochreous-fuscous, posterior pair greyish. Forewings elongate-triangular, costa somewhat sinuate from base to middle, hindmargin obliquely rounded beneath; light ochreous-fuscous, darkest along costa and hind-margin; an obscure, moderately straight dark fuscous line, from $\frac{3}{4}$ of costa, gently curved to beneath cell, thence proceeding to inner margin at $\frac{2}{3}$, more obscure on lower half; a dull whitish spot on costa, preceding line; an obscure fuscous discal spot at end of cell: cilia greyish-ochreous, with fuscous median, terminal, and basal lines. Hindwings greyish; a fuscous discal spot· a gently curved fuscous line from costa at $\frac{3}{4}$, towards inner margin in middle but not reaching it; cilia greyish, fuscous-tinged at base. Markings of wings strongly produced on underside.

Broken Hill, N.S.W.; four specimens, in September.

Sedenia polydesma, n.sp.

♂. 16 mm. Head, thorax and palpi whitish, mixed with ochreous-fuscous, patagia ochreous-fuscous. Antennæ fuscous, obscurely annulated with white. Legs fuscous, posterior pair whitish, tibiæ with one or two obscure fuscous bands, anterior coxæ greyish-ochreous. Abdomen light ochreous. Forewings elongate-triangular, costa nearly straight, hindmargin rounded; grey-whitish, sprinkled with ochreous-fuscous scales, which tend to form 3 moderate transverse fasciæ; 1st from $\frac{1}{3}$ costa to $\frac{1}{3}$ inner margin, mixed with blackish and constricted on inner margin; 2nd more obscure, from about $\frac{2}{3}$ of costa to inner margin just before anal angle; 3rd irregularly waved, starting from a black spot on costa at $\frac{5}{6}$, finely attenuated at $\frac{3}{4}$, and appearing to join second line above anal angle: cilia greyish-ochreous, mixed with blackish, more pronounced round apex and middle of hindmargin. Hindwings light fuscous, with two darker fuscous transverse fasciæ, obscure, one from $\frac{3}{4}$ of costa, the other from middle of costa, both reaching half across wing, thence lost in general ground colour; a fine line of fuscous scales along hindmargin; cilia greyish, with a faint fuscous median line.

Broken Hill, N.S.W.; three specimens, in October.

whole of the markings are lost in general ground colour, excepting the two orange spots, which then become more pronounced: cilia grey-whitish, chequered with blackish scales at base, and with a fuscous median line. Hindwings fuscous, darker round hind-margin; cilia whitish.

Broken Hill, N.S.W.; five specimens, in May and June.

Easily known by the orange-red spots on forewings; it is allied somewhat to *lichenopa*, Lower.

PSYCHINA.

PSYCHIDÆ.

Oiketecis gymnophasa, n.sp.

♂. 20 mm. Head, thorax and palpi dull ochreous-fuscous. Antennæ dark fuscous, pectinations blackish, about 6 at greatest length. Legs ochreous-fuscous. Abdomen ochreous. Forewings elongate, posteriorly slightly dilated, costa nearly straight, hind-margin rounded; whitish, faintly ochreous-tinged: a fine fuscous costal streak, attenuated at extremities, from base to near apex: cilia whitish. Hindwings whitish; cilia as in forewings

Broken Hill, N.S.W.; two specimens, in March.

ZEUZERIDÆ.

Zeuzera neuroxantha, n.sp.

♂. 36-40 mm. Head and face pale ochreous. Palpi black Thorax whitish, strongly mixed with black on posterior half, patagia whitish. Abdomen whitish, with moderately broad suffused segmental rings, anal tuft black. Antennæ blackish, whitish on basal half, pectinations blackish. Legs blackish, middle and posterior pair with obscure whitish tarsal rings. Forewings very elongate, costa nearly straight, hindmargin gently rounded, oblique; white, with black markings; all veins more or less outlined with yellow; a moderate cuneiform spot on costa at ½, its apex reaching ¼ across wing, between this and base are 3 moderate costal spots, nearly equidistant; a second, somewhat outwardly oblique, more or less cuneiform spot, apex obtuse, from costa at $\frac{2}{3}$, reaching nearly $\frac{1}{2}$ across wing, preceded on costa by 2 equidistant spots; 2 or 3 irregular spots on costa between apex and second cuneiform spot; lower half of wing more or less completely irrorated with large spots, not reaching apex or base, leaving lower $\frac{1}{2}$ of hindmargin clear; a moderate elongate clear whitish space above inner margin at $\frac{1}{3}$; 3 or 4 narrow blackish rings on lower $\frac{2}{3}$ of hindmargin: cilia whitish, chequered with black at extremities of veins. Hindwings with hindmargin rounded, somewhat sinuate in middle; whitish, strongly irrorated throughout with fuscous, except on base of costa and basal third of inner margin; markings towards hindmargin more blackish; cilia as in forewings, except inner marginal $\frac{2}{5}$ which is white.

Broken Hill, N.S.W.; two specimens, in October.

TINEINA.

ŒCOPHORIDÆ.

Eulechria photinopis, n.sp.

♂♀. 24-25 mm. Head, thorax, palpi and antennæ fuscous, second joint of palpi internally grey-whitish, apical third of terminal joint ochreous-yellowish. Legs fuscous, tibiæ and tarsi

In the *melesella* group. Recalls species of *Phlœopola* in general appearance; the strong antennal pecten, however, removes it from that genus.

EULECHRIA ERIOPA, n.sp.

♂. 16 mm. Head, thorax, palpi and antennæ white, antennæ faintly annulated with fuscous. Abdomen greyish. Legs fuscous, strongly mixed with whitish, posterior pair greyish. Forewings elongate, moderate, costa gently arched, hindmargin obliquely rounded; white; costa narrowly fuscous on anterior third; a fine short line from base to ⅓, very obscure; some fuscous scales beneath middle of cell; hindmarginal and apical areas mixed with fuscous, especially immediately before apex on costa: cilia whitish, mixed with a few fuscous scales, ochreous at base. Hindwings grey, fuscous-tinged; cilia greyish ochreous.

Broken Hill, N.S.W.; two specimens, in October.

EULECHRIA PENTASPILA, n.sp.

♂♀. 19-22 mm. Head, thorax, palpi and antennæ ashy-grey-whitish, palpi beneath mixed with whitish. Legs grey-whitish, anterior and middle pair infuscated. Abdomen greyish-ochreous, with greyish segmental rings. Forewings elongate, moderate, rather narrow, costa gently arched, hindmargin oblique: ashy-grey-whitish, with dark fuscous markings: five moderate spots, first on fold at ⅓, second just below, third at end of cell, fourth just below third, fifth before hindmargin in a direct line with first and third; sometimes a fine blackish line along fold and lower margin of cell, but this is generally obsolete: cilia ashy-grey-whitish, obscurely chequered with fuscous at extremities of veins. Hindmargin greyish-fuscous; cilia greyish, somewhat ochreous at base.

Broken Hill, N.S.W.; two specimens, in May.

LINOSTICHA PUDICA, n.sp.

♂. 20 mm. Head, thorax and palpi whitish, faintly ochreous-tinged. Antennæ fuscous, mixed with whitish. Legs grey-whitish, anterior and middle pair somewhat infuscated. Abdomen ochreous, with silvery-grey segmental bands. Forewings elongate, moderate, costa gently arched, hindmargin obliquely and gently rounded; ochreous-whitish ; costal edge obscurely and narrowly white; extreme costal edge towards base fuscous; a fuscous dot in disc at $\frac{1}{3}$ from base; a second, very obscure, below and before first, and a third, larger, at end of cell : cilia whitish-ochreous, tips greyish. Hindwings whitish; cilia whitish, slightly ochreous at base.

Broken Hill, N.S.W.; two specimens, in August.

Judging from the description of *L. cycnoptera*, Meyr., this species is in the immediate neighbourhood.

TRACHYNTIS EREBOCOSMA, n.sp.

♂. 26 mm. Head and palpi pale flesh-colour, second joint of palpi externally fuscous at base, terminal joint fuscous-tinged. Thorax pale reddish. Abdomen ochreous. Legs fuscous, tibiæ somewhat reddish-tinged, posterior pair ochreous Forewings elongate, slightly dilated posteriorly, costa slightly arched at base, thence straight, apex. rounded, hindmargin obliquely rounded ; ochreous, more or less suffused with reddish-pink, except beneath costa on basal $\frac{2}{3}$, which is ochreous; a moderate elongate black dot on fold before $\frac{1}{3}$; a second, larger and roundish above and beyond first; a third obliquely placed at end of cell; a

abdomen fuscous, basal joint of antennæ reddish. Legs ochreous, somewhat infuscated. Forewings elongate, moderate, costa gently arched, hindmargin obliquely rounded; pale flesh, more or less suffused with patches of reddish ferruginous, especially towards apex and hindmargin, and very finely irrorated throughout with blackish; a suffused blackish mark at base; a moderate whitish spot in middle, at $\frac{1}{3}$ from base, more or less encircled with reddish-fuscous; a similar, more distinct, spot at end of cell, beneath which is an irregular blackish mark going towards anal angle; a suffused blackish mark along hindmargin near base; an oblique suffused blackish mark from costa at $\frac{5}{8}$ reaching half across wing; a suffused row of blackish dots along hindmargin and apical fifth of costa: cilia greyish (imperfect). Hindwings dark fuscous; cilia dark fuscous.

Warrego, N.S.W.; one specimen *(Coll. Lyell).*

Obscurely marked but quite distinct.

Up to the present the genus *Trachyntis* has not been observed outside of Western Australia.

NEPHOGENES BASATRA, n.sp.

♂♀. 28-30 mm. Head, palpi and thorax fleshy-pink. Abdomen greyish-ochreous. Anterior and middle legs dark fuscous, posterior pair greyish-ochreous; all with more or less whitish tarsal rings. Forewings elongate, moderate, costa gently arched, hindmargin obliquely rounded: fleshy-red, minutely irrorated throughout with fuscous; a very sharply defined short black elongate mark, on inner margin close to base, appearing somewhat raised; a moderate fuscous dot in disc at $\frac{1}{3}$ from base, slightly above middle; a second, similar, immediately below and before it, and a third at end of cell; in some specimens with a fleshy spot immediately below; sometimes the whole of the forewings are infuscated, obliterating the discal dots but leaving the fleshy spot well defined and surrounded with fuscous: cilia fleshy-red, more or less mixed with fuscous, sometimes ochreous-fuscous. Hindwings pale fuscous; cilia ochreous-grey, with a blackish-fuscous basal line.

Broken Hill, N.S.W.; not uncommon in August, September and October. It is a sluggish insect and usually falls to the ground when disturbed.

Although showing some variation in the ground colour of the forewings, it is an easily recognised species by the sharply defined black basal mark, which is always present.

NEPHOGENES CRASSINERVIS, n.sp.

♂♀. 18-22 mm. Head, palpi, antennæ and thorax ashy-grey-fuscous, palpi beneath darker fuscous, basal joint of antennæ somewhat whitish, ciliations 1. Anterior and middle legs dark fuscous, posterior pair whitish-ochreous. Abdomen greyish, segmental margins whitish. Forewings elongate, moderate, costa gently arched, hindmargin oblique, gently rounded; ashy-grey-whitish, becoming more whitish in cell and towards hindmargin and apical area; a blackish line from base, continued along lower margin of cell, thence right around to upper margin, sometimes the whole of the cell is outlined with black, but the upper margin is generally obscure; a moderate, generally distinct, elongate spot in cell at $\frac{1}{5}$ from base, sometimes very obscurely continued right through cell; two or three more or less confluent dark fuscous elongate marks on upper margin of posterior end of cell; veins towards hindmargin more or less finely outlined with fuscous:

orange-yellow. Legs dark fuscous, posterior pair yellow. Forewings elongate, moderate, costa gently arched, hindmargin oblique, gently rounded; dark ashy-fuscous; sometimes a darker fuscous patch on costa at 3/4, obscurely continued across wing to anal angle, more pronounced at end of cell: cilia ashy-grey-whitish, tips more whitish; a few yellowish scales at base near anal angle. Underside of wing strongly mixed with yellow; median area blackish. Hindwings bright orange-yellow; apex blackish, shortly continued along hindmargin and costa; a small patch of black scales at base; cilia yellow, more or less mixed with fuscous.

♀. 22-24 mm. Forewings as in ♂, but more whitish and with a fine distinct whitish costal streak throughout, indications of a fine white line from base in middle to 1/3: cilia as in ♂. Hindwings dull ochreous, more or less infuscated on apical half; cilia as in ♂.

Broken Hill, N.S.W.; common in August; beaten from *Bassia biflora*, the leaves of which are an admirable imitation of the forewings of the species; the ♀ is much more retired in habit. The only known species with yellow hindwings.

I have dedicated this pretty and interesting little species to my mother, not so much for her entomological knowledge, but in grateful recognition of her unvarying kindness in attending to my large collection during my sojournings in the pursuit of entomology.

NEPHOGENES ATRISIGNIS, n.sp.

♂♀. 28-30 mm. Head, palpi, thorax, antennæ and abdomen ashy-fuscous, base of second joint of palpi whitish and with a white ring at apex, terminal joint mixed with whitish. Legs dark fuscous, tibiæ with whitish rings, posterior legs ochreous-whitish. Abdomen fuscous, becoming blackish on anal segments, anal tuft pale yellowish-ochreous. Forewings elongate, moderate, costa gently arched, apex rounded, hindmargin obliquely rounded; dark ashy-grey-fuscous, sprinkled with white, and with ferruginous on costa and hindmargin; a row of black disconnected subcostal spots along upper median vein; a similar row along lower median; a black lunate mark at end of cell; three elongate black longitudinal marks, from near base above inner margin; a large blotch of dull white immediately beyond lunate mark, bounded posteriorly by a band of dull ferruginous along hindmargin; veins between blotch and hindmargin neatly outlined with black : cilia fuscous, mixed with ferruginous, and with a greyish median line. Hindwings elongate-ovate, slightly sinuate on inner margin near base; pale greyish; cilia greyish, with fuscous basal and subterminal lines.

Broken Hill, N.S.W.; two specimens at light, in May.

Allied to the following species.

NEPHOGENES EREBOMORPHA, n.sp.

♂♀. 38-42 mm. Head, palpi, thorax and antennæ dark fuscous, palpi beneath greyish. Legs dark fuscous, obscurely ringed with whitish except coxæ, posterior pair greyish-ochreous. Abdomen ochreous-grey, anal tuft yellowish. Forewings very elongate, moderate, costa gently arched, hindmargin obliquely rounded; blackish-fuscous, sparsely mixed with whitish; a fine whitish line along upper margin of cell, dotted throughout with fine black spots; a somewhat interrupted black line along lower margin of cell, and continued at end of cell to meet extremity of first-mentioned line; veins faintly indicated in black towards hindmargin; a hindmarginal row of elongate black dots : cilia blackish-fuscous, somewhat mixed with grey. Hindwings och-

gently arched, apex round-pointed, hindmargin obliquely rounded; ashy-grey, densely suffused with white, especially in disc and towards hindmargin; a blackish streak along lower median vein, but not nearly reaching base; an indistinct blackish crescentic mark at end of cell; a dull suffused fuscous patch immediately below, absent in some specimens; all veins more or less outlined with fuscous: cilia greyish, mixed with fuscous, and with a darker irregular basal line. Hindwings pale greyish; cilia pale greyish-ochreous.

Broken Hill, N.S.W.; five specimens, in May and June.

MACROBATHRA PHILOPSAMMA, n.sp

♂. 20, ♀. 22 mm. Head dull ochreous; palpi ochreous, apex of second and median third of terminal joint fuscous. Abdomen ochreous. Antennæ and thorax fuscous, patagia fuscous. Legs ochreous-white, strongly banded with fuscous. Forewings elongate-lanceolate; dark bronzy-fuscous, with whitish-ochreous markings; a small spot at base, sometimes absent; three somewhat cuneiform spots on costa, at $\frac{1}{4}$, $\frac{1}{2}$, and about $\frac{3}{4}$, from each of which proceeds an oblique irregularly suffused dentate line towards inner margin, sometimes reaching it, but generally lost in general ground colour at $\frac{3}{4}$ of length: cilia dark fuscous, tips greyish. Hindwings ochreous-fuscous, somewhat darker towards apex; cilia ochreous-fuscous, becoming darker round apex.

Semaphore, South Australia; five specimens, in November.

GELECHIADÆ.

ATASTHALISTIS EUCHROA, n.sp.

♂. 15 mm. Head and thorax pale fleshy-ochreous. Abdomen fuscous, anal tuft yellowish. (Antennæ broken). Anterior legs dark fuscous, posterior and middle pair ochreous-whitish. Palpi fuscous. Forewings elongate, dilated posteriorly, costa slightly arched at base, thence straight, apex round-pointed, hindmargin nearly straight, sinuate in middle; pale fleshy-ochreous, with a few scattered blackish strigulæ; a well defined, somewhat elongate, cuneiform, blackish patch, edged with whitish, on costa before middle, hardly reaching $\frac{1}{4}$ across wing, attenuated anteriorly; costa between base and patch spotted with fuscous; an interrupted series of about 10 fine elongate blackish spots, from costa at about $\frac{3}{4}$, around hindmargin, and ending on anal angle, those on hindmargin being placed on a much paler ground colour: cilia pale fleshy-pink, with 3 darker lines throughout, basal thickest. Hindwings with apex prominent; bright orange, posterior half of wing dark fuscous, darkest at apex; cilia fuscous, paler on inner margin.

Brisbane, Queensland; one specimen, in December.

GELECHIA (?) MESOLEUCA, n.sp.

♀. 20 mm. Head and palpi white, terminal joint of palpi blackish, second joint nearly smooth, not grooved. Antennæ $\frac{3}{4}$. black faintly annulated with white, basal $\frac{1}{2}$ blackish laterally. Thorax blackish, with two broad whitish longitudinal stripes throughout. Abdomen silvery-grey. Anterior and middle legs blackish, strongly suffused with whitish, posterior pair whitish. Forewings narrow, elongate; costa hardly arched, apex pointed; blackish; a broad white longitudinal median streak from base to apex; an elongate, somewhat indistinct, blackish spot before middle, resting on lower half of ground colour; a well defined

fuscous apical and basal rings on each joint. Antennæ somewhat annulated with fuscous, all legs with fuscous bands somewhat indistinct. Forewings elongate, moderate, costa gently arched, hindmargin obliquely rounded; fleshy-ochreous, with fuscous purple markings, more or less edged with black; six costal spots between base and apex, 3 anterior largest, equidistant, posterior 3 much smaller and closer together; a large quadrate spot at base, reaching to ⅛ along inner margin, much constricted on costa; a quadrate spot in middle of disc at ⅓ from base; a similar but smaller spot at end of cell; an obscure fuscous suffusion towards apex; an obscure fuscous hindmarginal line, somewhat interrupted, edged internally with a fine line of ground colour: cilia fleshy-ochreous, fuscous-tinged. Hindwings with the apex hardly pointed; termen hardly sinuate; greyish-fuscous; cilia greyish, with a faint fuscous subbasal line.

Parkside, South Australia; five specimens, in August.

Allied to *G. hæmaspila*, Lower.

Gelechia dictyomorpha, n.sp.

♂. 20 mm. Head, palpi and thorax light ochreous, terminal joint of palpi with fuscous subapical ring, thorax with a whitish posterior spot. Antennæ nearly ¾, fuscous. Abdomen ochreous, posterior third fuscous-blackish. Legs ochreous-fuscous, posterior pair ochreous, slightly infuscated. Forewings elongate, moderate, costa gently arched, apex hardly pointed, hindmargin oblique; light ochreous, finely reticulated with fuscous and dark fuscous, which coalesces in disc at ⅓ so as to form a fuscous blotch, indicating median portion of a transverse fascia; a similar, yet more obscure, fascia at end of cell, margins indistinct; costal edge from base to termination of second fascia more fuscous; a narrow obscure fuscous hindmarginal band: cilia greyish. Hindwings with termen slightly sinuate; greyish; cilia greyish, ochreous-tinged at base.

Broken Hill, N.S.W.; two specimens, in September.

Gelechia orthanotos, n.sp.

♀. 12 mm. Head, antennæ, palpi, thorax and legs whitish-ochreous, terminal joint of palpi, and posterior $\frac{2}{3}$ of second fuscous beneath, patagia fuscous. Abdomen blackish, anal tuft whitish-ochreous. Forewings elongate, moderate, apex hardly pointed, hindmargin oblique; whitish-ochreous; a thick straight blackish line from base in middle, continued direct through to $\frac{3}{4}$, with very faint indications of its continuance to hindmargin below apex, where it becomes less obscure, and edged beneath by a pale yellowish patch; a few obscure fuscous scales along costa and hindmargin: cilia ochreous-whitish, mixed with a few fuscous scales. Hindwings with apex pointed, termen not sinuate beneath; dull bronzy-fuscous; cilia about $\frac{1}{2}$, whitish-ochreous.

Stawell, Victoria; one specimen, in November.

Xenolechia peltosema, n.sp.

♀. 9 mm. Head, palpi, antennæ and thorax ochreous, antennæ annulated, fuscous, second joint of palpi strongly infuscated externally, terminal joint with apical and median fuscous rings. Legs dark fuscous, tibiæ and tarsi banded with white. Abdomen greyish. Forewings elongate, moderate, costa hardly arched, apex pointed, hindmargin very oblique; dull ochreous; a

Thorax yellowish, mixed with fuscous-purplish, patagia fuscous-purplish, with an orange spot posteriorly. Forewings moderate, slightly dilated posteriorly, costa hardly arched, apex somewhat pointed, hindmargin nearly straight, oblique; fuscous-purplish; markings dull orange; a narrow somewhat dentate, inwardly oblique fascia, from costa beyond $\frac{1}{3}$, to inner margin at $\frac{1}{3}$, emitting an obscure cuneiform spot from anterior edge below middle towards base; a large roundish spot in disc at $\frac{3}{4}$, almost reaching margins; a small elongate spot on costa near apex; a few blackish scales at apex: cilia yellow, with a row of more or less interrupted black marks at base. Hindwings with apex produced; termen sinuate beneath apex; dull bronzy-fuscous; cilia fuscous.

Rockhampton, Queensland; two specimens, in January.

PLUTELLIDÆ.

HOMADAULA COSCINOPA, n.sp.

♀. 14 mm. Head, thorax, palpi and antennæ blackish. Abdomen fuscous. Legs fuscous, anterior and middle pair mixed with whitish, tarsi with whitish rings, posterior pair greyish. Forewings elongate, moderate, costa gently arched, hindmargin oblique; whitish, irrorated with small black spots, more or less arranged in transverse series, becoming more dense on margins, leaving a basal patch of ground colour much whiter; an obscure elongate, somewhat cuneiform spot of black, from costa at $\frac{1}{3}$ to apex, reaching more than half across wing: cilia dark fuscous. Hindwings fuscous; somewhat bronzy; cilia light fuscous.

Broken Hill, N.S.W.; three specimens, in March.

DESCRIPTIONS OF NEW SPECIES OF AUSTRALIAN *RHOPALOCERA*.

By G. A. Waterhouse, B.Sc., B.E.

(Plate i.)

Ogyris ianthis, sp.nov.

(Plate i., figs. 1-4.)

♂. 36 mm. Head fuscous; palpi fuscous-grey; antennæ fuscous, becoming fuscous-red at their distal ends, slightly lighter on the underside and faintly annulated with white; thorax black, with long grey hairs at the sides.

Abdomen.—Upperside purplish-fuscous, underside fuscous-grey.

Upperside.—Light silvery electric blue margined with blackish-fuscous.

Anterior wing.—A band bounded by the costa and subcostal vein blackish-fuscous, beginning from the base of the wing and

Abdominal groove grey, becoming darker near the anal angle, speckled with light blue. The remainder of wing light electric blue.

Underside.—Anterior wing with the costal margin from base to apex, and the termen ¾ from apex lilac-fuscous, widest at apex. Cell fuscous, the dorsum fuscous-grey. Markings as follows—in the cell close to the base an oval fuscous spot faintly margined with white ; in middle of cell a large oblong black spot extending across the cell, margined with pale electric blue; just at the termination of cell another oblong black spot margined with electric blue. Below each of the foregoing spots is a blackish-fuscous blotch without a border and outside the cell. At ¾ of the distance from base is a series of five small fuscous-black spots arranged parallel to the termen, all but the lowest being margined with white; a faint irregular fuscous band situated beyond and parallel to these last.

Posterior wing.—General colour lilac-fuscous, darker fuscous along the termen, a dark fuscous blotch at anal angle. The whole wing is marked with faint fuscous marks arranged concentrically with the margins in one or more rows; these last are ill defined and variable in different specimens. Cilia fuscous.

♀ 37 mm. Head, palpi, antennæ, thorax and abdomen as in ♂.

Upperside.—Pale orange-yellow, very broadly margined with dark fuscous-black.

Anterior wing.—The dark bands as in ♂, but much broader. extending along the costa halfway into the cell, much broader along the termen and tornus or lower angle, and extending along the greater part of the dorsum, the base fuscous-black. The remainder of the wing orange-yellow, except the distal end of cell which is fuscous-black.

Posterior wing.—Fuscous-black, with a very faint orange colouring just below the cell towards the anal angle, the abdominal groove fuscous-grey. Dorsum prolonged outwards along the veins.

Underside.—Anterior wing with the base, costal margin, and termen to ¾ from apex lilac-fuscous. Tornus and ½ the dorsum fuscous-grey, the rest of the wing orange-yellow, with the markings as follows—the oval fuscous spot margined with white near base of cell, a blackish spot margined with electric blue in the middle of cell, and a similar one at end of cell as in ♂. The blotches below the cell are not shown. The first series of spots is represented by three blue-margined black spots, and the outer series is faintly visible.

Posterior wing.—Base of wing and ¼ from base lilac-fuscous, faintly bordered with white from middle of cell to dorsum, a whitish suffusion on dorsum near anal angle. The rest of wing fuscous, mottled with fuscous-black, termen within the cilia marked with a black line, fuscous-black at termination of the veins, slightly lighter between them.

Loc.—Como, near Sydney, N.S.W.; seven specimens (♂ 6; ♀1, which is probably undersized): in February and March. Types in author's collection. The insects were all caught flying round the tops of small Eucalypts, about 16 feet from the ground.

The sexes of this insect correspond to one another in the same way as do those of *O. abrota.* The purple of the upper side of *O. abrota* (♂) is replaced by the Morpho-blue of *O. ianthis*, (♂), and the dark margin is narrower in *O. abrota* (♂). The pale round spot of *O. abrota* (♀) is represented by an oval orange-yellow

white more conspicuous below. Thorax black; abdomen black above, with alternate bands of fuscous and ochreous on each body segment below.

Upper side.—Anterior wing with costa nearly straight, apex rather pointed; termen oblique. Colour fuscous, with a faint ochreous suffusion over the wing, more marked close to the base. Five light ochreous-yellow hyaline spots arranged as follows—three small, adjacent, arranged in a transverse series beneath the costa towards the apex; one large, trapezoidal, occupying distal half of cell; the last small, situated obliquely below the end of cell; the black oblique transverse bar characteristic of the male commencing between these last spots, and reaching the middle of dorsum. Cilia on termen fuscous at terminations of veins, ochreous-fuscous between them.

Posterior wing.—Anal angle rather sharply rounded. Colour dark blackish-fuscous with one large opaque ochreous spot, broadest towards the termen, occupying the centre quarter of the wing, a few long ochreous hairs towards the base. Cilia long, fuscous at terminations of the veins, otherwise deep ochreous.

Underside.—Anterior wing with hyaline spots as above, a pale opaque yellow spot below the lowest spot visible on upperside. Costa cream at base, then fuscous, with four cream marks towards the apex; basal third of wing blackish-fuscous, apical third cinnamon-fuscous, with a large subquadrate cream spot divided by a fuscous vein, and two small cream spots, one above and one below this, all towards the termen. Dorsum pale yellowish, widest near the middle of the wing Termen marked by a pale cream border, interrupted by the veins, which are fuscous; cilia somewhat lighter than on the upperside.

Posterior wing.—General pattern similar to the apical third of the anterior wing. Base and basal half of costa cream; from apex extending across the wing to the middle of dorsum a wide cream band interrupted in the centre of the wing, and on the abdominal fold by reddish-fuscous; in this band close to apex a single circular black spot. Below this another cream band, narrower and more interrupted towards anal angle. The rest of wing reddish-fuscous, with the abdominal fold fuscous except at the middle of dorsum, which is cream. Termen marked by an interrupted cream line as in anterior wing. Cilia long, pale fuscous-ochreous, with four fuscous patches at terminations of veins. Legs reddish-fuscous.

♀ 46 mm. Head, palpi, antennæ and legs as in ♂; thorax dark fuscous, with long pale yellowish-fuscous hairs at sides. Abdomen fuscous, with the segments marked by pale yellowish-ochreous above and cream in median line below, with bands of fuscous and cream at sides.

Upperside as in ♂, with the spots somewhat larger.

Anterior wing fuscous, with the five hyaline spots well marked and somewhat more ochreous; a sixth ochreous-yellow hyaline spot, subquadrate, below the fifth, and below this again an elongated opaque ochreous spot reaching nearly to the dorsum; at $\frac{1}{3}$ near the dorsum a small ochreous oval blotch, with short ochreous hairs extending from it to the base. Costa only faintly suffused with ochreous. The shape of the wing more rounded than in ♂. Cilia as in ♂.

Posterior wing.—Much more rounded than in ♂, especially at anal angle. The central ochreous spot much larger. The rest

This species belongs to the *H. picta* and *H. ornata* types of *Hesperilla*. It is closest to *H. picta*, Leach, but differs from it in having lighter-coloured forewings, fewer spots on the forewing of the male, and the central ochreous spot of the hindwing much brighter and larger. On the undersurface they are similar in having the apical third of the forewing and the whole hindwing of the same general pattern, but these general patterns are totally distinct, and constitute the most marked difference between the two species. The pattern of *H. picta* is fairly sharply defined, while that of *H. Mastersi* is mottled, and not at all well defined.

I have to record the presence in New South Wales of *Ogyris genoveva*, Hew., and *O. olane*, Hew. Of the former I caught several males flying round the tops of Eucalypts, about 25 feet from the ground, at Como during March. The latter was caught by Mr. N. W. Hansard at Lawson, Blue Mts., in January.

EXPLANATION OF PLATE.

Ogyris ianthis, ♂.

Fig. 1.—Upperside.
Fig. 2.—Underside.

O. ianthis, ♀.

Fig. 3.—Upperside
Fig. 4.—Underside.

Hesperilla Mastersi, ♂.

Fig. 5.—Upperside.
Fig. 6.—Underside.

H. Mastersi, ♀.

Fig. 7.—Upperside.
Fig. 8.—Underside.

ON THE SKELETON OF THE SNOUT AND OS CARUNCULÆ OF THE MAMMARY FŒTUS OF MONOTREMES.

By Professor J. T. Wilson, M.B., Ch.M.

For the research three specimens were utilised. One was the fœtal *Ornithorhynchus*, whose external characters were described by the writer in a previous paper before the Society. Another was a more advanced specimen of *Ornithorhynchus*, whilst the third was an *Echidna* of about the same stage as the earlier of Professor W. N. Parker's specimens. All the stages were more advanced than those of *Echidna* lately investigated by Seydel. Wax-plate reconstructions of the anterior snout region were exhibited together with serial photographs of the younger *Ornithorhynchus*.

The following features are revealed and illustrated by the models :—(1) The complete continuity of the nasal floor cartilage and the extensive marginal cartilage of the upper lip, which in the adult are separated by the premaxillæ. (2) As a result of

proved to be a perfectly distinct element—a true anterior vomer. (5) Anteriorly, the ventral premaxillary splints turn up dorsally in front of the anterior extremity of the snout in both *Ornithorhynchus* specimens, in the form of rather attenuated trabeculæ, lodged in the notch between the alar expansions of the rostral cartilage. Above this plane they fuse and are continued dorsally into a remarkable osseous mass which forms a definite skeletal foundation for the caruncle, and may therefore be named the os carunculæ. This is at its maximum development in the younger stage of *Ornithorhynchus*, and is undergoing resorption in the older; whilst in the *Echidna* model it is only represented by a small nodule of bone which has lost all connection with the premaxillæ. From Seydel's figures of earlier stages it is evident that the *Echidna* condition is originally identical with that of *Ornithorhynchus*, though it would appear to exist in a less exaggerated form. (6) The cartilaginous septum of both Monotremes exhibits an oval "internasal fenestra" immediately behind its anterior termination at the prerostral notch. A similar fenestra, according to W. K. Parker, is "a common feature in low Eutheria."

CATALOGUE OF THE DESCRIBED MOSSES OF NEW SOUTH WALES.

By Rev Walter W. Watts and Thomas Whitlegge, F R M S.

To be issued separately with one of the later Parts of this Volume.

NOTES AND EXHIBITS.

Mr. J. H. Maiden exhibited the olive-green gum-resin and fruits of *Gardenia Aubryi*, Vieillard, from New Caledonia. The fruit is remarkable for its large calyx-limbs; and the resin, which profusely exudes, has formed the subject of an exhaustive research by Heckel and Schlagdenhauffen in the Répertoire des Pharmacie for 1893. Also specimens of a Fig from the National Park, near Sydney, which answers well to the description of the Queensland *Ficus Henneana*, Miq., subject to examination of male flowers which could not be detected in any of the over-ripe specimens available this season.

Mr. Waterhouse exhibited a collection of the species of the genera *Ogyris* and *Hesperilla* in illustration of his paper.

Mr. Cheel exhibited an interesting form of the fern *Blechnum cartilagineum*, Sw., collected at Cundletown, Manning River, showing many of the segments to be pinnatisect, giving the frond the appearance of being bipinnatifid.

Mr. Palmer contributed a note describing his experiences after being bitten by a black snake in February last. He also exhibited a snake which had attacked a member of his family; and a number of insects from Lawson.

Mr. Fletcher exhibited several specimens of a small freshwater crab which Mr. Whitelegge had kindly examined, and identified as *Hymenosoma lacustris*, Chilton. The species was originally described from New Zealand, but was subsequently obtained in Lord Howe Island by Mr. Whitelegge, who was a member of the Australian Museum party which visited the Island in 1887. Of this interesting addition to the Tasmanian fauna, the specimens exhibited were forwarded by Mr. E. Stuart Dove, who collected them in the north of Tasmania, in a creek at Flowerdale, near Table Cape, and also in Barnard's Creek, a tributary of the Tamar. In the first-mentioned locality the crabs live in the cracks and crevices of submerged decaying wood, and in colour so much resemble their surroundings that they are hardly noticeable until they move. A Gammarus-like crustacean was abundant in the vicinity, and the crabs appeared to be lying in wait for these. In the second locality the crabs were found among tangled masses of waterweed in company with some molluscs.

WEDNESDAY, APRIL 25TH, 1900.

The Ordinary Monthly Meeting of the Society was held at the Linnean Hall, Ithaca Road, Elizabeth Bay, on Wednesday evening, April 25th, 1900.

The Hon. James Norton, LL.D., M.L.C., President, in the Chair.

The President announced that under the provisions of Rule xxv., the Council had elected Dr. J. C. Cox, F.L.S., Prof. T. W. E. David, B.A., F.G.S., Mr. Henry Deane, M.A., F.L S., and Prof. J. T. Wilson, M.B., Ch.M., to be VICE-PRESIDENTS; and Mr. Prosper N. Trebeck, J.P., to be HON. TREASURER for the current year.

Mr. JAMES J. WALKER, F.L.S., F.E.S., R.N., H.M.S. Ringa-

Sydney Observatory – Records. No. 154 (Pamphlet—Current Papers. No. 4, 1899). By H. C. Russell, B.A., C.M.G., F.R.S. *From the Director.*

The Surveyor, Sydney. Vol. xiii. No 3 (March, 1900). *From the Editor.*

Field Naturalists' Club of Victoria — Victorian Naturalist. Vol. xvi. No. 12 (April, 1900). *From the Club.*

Zoological and Acclimatisation Society of Victoria—Thirty-sixth Annual Report (1899). *From the Society.*

Department of Agriculture, Perth, W.A.—Journal. March, 1900. *From the Secretary of Agriculture.*

New Zealand Institute, Wellington—Mangareva Dictionary, Gambier Islands (1899). By Edward Tregear. *From the Institute.*

Entomological Society of London—Transactions, 1899. Part 5 *From the Society.*

Manchester Literary and Philosophical Society—Memoirs and Proceedings. Vol. xliii. Part 5 (1898-99); Vol. xliv. Part 1 (1899-1900). *From the Society.*

Royal Microscopical Society, London—Journal, 1900. Part 1 (February). *From the Society.*

Royal Society, London—Proceedings. Vol. lxvi. No. 425 (March, 1900). *From the Society.*

Zoological Society of London—Abstracts, February 20th and March 6th, 1900. *From the Society.*

Royal Physical Society, Edinburgh—Proceedings. Vol. xiv. Part 2. (Session 1898-99.) *From the Society.*

Scottish Microscopical Society, Edinburgh—Proceedings. Vol. ii. No. iv. (1898-99). *From the Society.*

American Museum of Natural History, New York—Bulletin. Vol. xii. Art. xxii. (pp. 265-327 ; Feb., 1900), T.p., &c.. *From the Museum.*

American Naturalist (Cambridge). Vol. xxxiv. No. 398 (Feb., 1900). *From the Editor.*

Washington Academy of Sciences—Proceedings. Vol. ii. pp. 1-30 (March, 1900). *From the Academy.*

Wisconsin Natural History Society, Milwaukee — Bulletin New Series. Vol. i. No. 1 (Jan., 1900). *From the Society.*

U.S. Department of Agriculture—Division of Entomology: Bulletin. New Series. No. 22 (1900). *From the Secretary of Agriculture.*

Instituto Geológico de México—Boletin. Num. 12-13 (1899). *From the Institute.*

Perak Government Gazette. Vol. xii. T.p., &c. (1899); Vol. xiii. Nos. 7-8 (March, 1900). *From the Government Secretary.*

Asiatic Society of Bengal, Calcutta—Journal. Vol. lxviii. Part ii. Nos. 2-3, 1899: Proceedings, 1899. Nos. viii.-x and Extra No. xi.; 1900. No. i. *From the Society.*

Instituto Botanico dell' Universitá di Pavia—Atti. II. Serie. Vol. ii. (1892). *From the Institute.*

Zoologischer Anzeiger, Leipzig. xxiii. Band. Nos. 608-610 (February-March, 1900). *From the Editor.*

Société des Sciences Naturelles de l'Ouest de la France—

THE FLOCCULATION OF BACTERIA.

By R. Greig Smith, M.Sc., Macleay Bacteriologist.

Small particles of clay or finely divided chemical precipitates may remain suspended in water for a long time. When, however, certain salts are added to the water, the microscopic particles are seen to settle to the bottom of the liquid in which they were formerly suspended, with greater or less rapidity according to the kind of salt, the amount of salt per volume of liquid, and the temperature. In a previous paper* I have shown that the suspension of the particles is caused by the molecular pressure of the individual water molecules upon the surfaces of the suspended solid, with the result that there is, as it were, a hydrate formed. By reason of its superior attraction for water, the saline flocculating agent causes the withdrawal of the water molecules which had crowded upon the surfaces of the particles. What was formerly a surface pressure, now becomes a surface tension, which, being exerted upon all the particles, causes them to run together into little clumps which quickly gravitate to the bottom of the liquid.

The flocculating action is to be traced chiefly to the metallic portion of the salt, and in a small degree only to the acid radicle. The metals vary in the intensity of their action, some being strong, others weak; as an instance, calcium is about 160 times more powerful than potassium. These agents not only cause the coagulation of particles which are visible with the microscope, but they also induce substances which are in solution to precipitate. The latter are in what is called "pseudo-solution"—that is, they consist of molecular aggregates which are just retained in solution and no more. The flocculating agent induces a further coalescence of the molecular aggregates, and these being no longer able to remain in solution separate out as a precipitate.

If we look upon bacteria growing in a culture fluid as particles in suspension, it seems reasonable to hope that they might be sensitive to the action of flocculating agents precisely like particles

* Journ. Soc. Chem. Industry, xvi. 872; xvii. 117.

of clay. If they behave like suspended particles, flocculation may be utilised as a means of obtaining them more readily from the solutions in which they have been grown. But better than this, flocculation as in pseudo-solution might be the means of separating those bacteria which are supposed to be ultra-microscopical—as, for example, the organism of pleuro-pneumonia (Nocard and Roux), which is hardly visible under the highest powers of the microscope. It might also lead to the elucidation of some questions connected with the agglutination of bacteria by active sera.

With regard to the choice of flocculating agents, it must be borne in mind that salts of the heavy metals would coagulate the constituents of the culture media and of the organism. We are, therefore, deprived of the strongest agents. The salts of the zinc and iron metals are too prone to form basic salts, and accordingly a choice of the metals of the alkalies and alkaline earths remain. Of these metals, calcium has the highest flocculating power, and, therefore, calcium chloride was employed in my experiments. The density of bacteria as shown by Almquist is about 1·4, and that of most inorganic particles much more—say from 1·8 to 2·6; it was, therefore, to be expected that a strong flocculating agent would be required.

Preliminary experiments showed that a flocculation was obtained with calcium chloride, and also indicated what should be the approximate strength of the flocculating solution. At first a

mention that testing by means of the hanging drop is not advisable when using lime salts, unless the platinum loop is cleaned in hydrochloric acid after each ignition.

Precipitation of bacteria by calcium chloride.—The results will probably be best seen by looking at the following table, in which the relative precipitation is indicated by numbers running from 1 to 6; 1 represents uniform turbidity with a very slight precipitate, whilst 6 indicates a complete precipitation with a clear supernatant liquid. The intermediate numbers represent intermediate stages of precipitation. A zero means that no change occurred, while a plus shows that flocculation was visible when the liquid was examined with a hand lens:—

c.c. Calcium chloride added to 2 c.c. suspension.		Calcium chloride expressed as milligrams of calcium per 100 c.c. of total suspension.	*B. prodigiosum.*		*B. coli commune.*		*B. typhi.*	
			½ hour.	1 hour.	½ hour.	1 hour.	½ hour.	1 hour.
1/10 gram-molecule.	0·2	36	0	1	2	3	4	6
	0·4	67	2	4*	2	3	4	6
	0·6	92	3	5†	2	4*	4	6
	0·8	114	3	5	3	5†	5	6
	1·0	133	3	5	3	5	5	6
1/25 gram-molecule.	0·2	14			1	2	—	—
	0·4	27			1	2	5	6
	0·6	37			1	2	5	6
	0·8	46			2	3	—	—
	1·0	53			2	3	—	—
1/50 gram-molecule.	0·2	7					0	0
	0·4	13					0	+
	0·6	18					+	4*
	0·8	23					3	5†
	1·0	27					5	6

It will be seen that the bacteria exhibit different susceptibilities to the precipitating action of calcium chloride. *Bact. typhi* is more sensitive than *Bact. prodigiosum*, and the latter than *Bact.*

coli commune. The bacteria were thoroughly precipitated, and presented similar appearances with the following amounts of calcium in milligrams per 100 c.c. of total suspension (marked with a † in the table).

Bact. typhi	23
„ *prodigiosum*	92
„ *coli commune*	114

A proportion roughly as 1 : 4 : 5.

Bearing in mind the utility of being able to separate the organisms quickly by filtration through paper, I filtered the cultures through paper of fine texture like that used for fine precipitates such as barium sulphate. To avoid the possible passage of the bacteria over the inside and down the outside of the filter, the margin of the paper was painted with vaseline. The smallest quantities of calcium necessary to give a clear filtrate were as follows (marked with a * in table):—

Bact. typhi	18
„ *prodigiosum*	67
„ *coli commune*	92

A proportion again roughly as 1 : 4 : 5.

With the object of ascertaining whether or not the bacteria were retained on the filter paper, the *Bact. prodigiosum* precipi-

The action of sodium salts.—I have previously found common salt to have rather a weak action, but as it occurs in most media, and is frequently quoted as a flocculating agent, it was used upon these experimental bacteria. There was no flocculation, however, with small quantities. Larger quantities gave a similar result even when the salt was added in proportions varying from 550 to 7,700 milligrams of sodium per 100 c.c. and the tests were allowed to stand for 24 hours. Since this is equal to nearly 20 per cent. of common salt, it is evident that sodium salts do not flocculate bacteria.

The action of potassium and ammonium salts.—Although the action of these is much superior to sodium salts, no flocculation was obtained, and in the case of potassium chloride even when 24 per cent. was present. It can, therefore, be said that these salts do not flocculate bacteria.

The action of peptone.—This is not recognised as a flocculating agent, but since bouillon cultures contain 1 per cent. and bacteria are generally found more or less precipitated in bouillon cultures, a trial with it seemed advisable. As no coagulation appeared even when the culture contained 10 per cent., its use as a flocculating agent may be discounted.

The effect of temperature.—It is a well-known fact that heating causes the rapid precipitation of many chemical precipitates. Extreme temperatures cannot be employed in working with living bacteria, and the range between room temperature and blood heat was found to be too narrow to show any difference in the precipitation of the cultures with calcium chloride.

The action of lime water.—Whilst the hydrates of potassium and sodium prevent the flocculation of inorganic particles, hydrate of calcium greatly assists flocculation. This fact was remembered when calcium chloride was chosen in these experiments. When a solution of calcium hydrate, containing 2 milligrams of calcium per c.c., was gradually added to a culture of *Bact. prodigiosum*, a precipitation occurred. This was complete in half-an-hour when a volume equal to the volume of culture had been added.

Several circumstances were noted in the lime-water experiments. The bacterial culture when neutralised to phenolphthalein with sodium hydrate, was still alkaline when an equal and corresponding quantity of calcium hydrate had been added. Taking 10 c.c. portions, neutrality was reached with 2·35 c.c. tenth-normal soda and 6·0 c.c. tenth-normal lime. Sterile bouillon made neutral to soda was found to be still alkaline to lime, and as the latter was gradually added a flocculent precipitate continued to appear until neutrality was reached. This curious behaviour of lime and soda suggested the presence of phosphates of the alkalies, a suspicion that was confirmed by testing the precipitate, which proved to be tricalcium phosphate. The alkaline phosphates are derived from the meat which forms the basis of the culture media, and these are not completely precipitated when the media are neutralised with soda.

It is evident that this raises the whole question of precipitation by calcium salts, since it is probable that the coagulum obtained in the experiments consisted of bacteria entangled in a matrix of calcium phosphate. This is all the more probable, since the precipitates were certainly more voluminous than could have been expected from a simple flocculation of bacteria. On the other hand, however, microscopical examination showed the

To obtain a medium free from the disturbing influence of phosphoric acid, bouillon was shaken up with slaked lime for several hours and filtered; washed carbon dioxide was passed through the filtrate for some time and the precipitated carbonate filtered off. The filtrate was then boiled to decompose the dissolved bicarbonate and the fluid again filtered to separate the precipitated carbonate. The resulting neutral medium contained no phosphoric acid and no lime, which was shown by testing with ammonium molybdate and ammonium oxalate. Neither did it give a precipitate with calcium salts, even with the addition of small quantities of sodium hydrate.

The three experimental bacteria grew slowly in the phosphate-free bouillon. When they had made some headway portions were tested, and the bacteria were found to be entirely unaffected by the addition of calcium chloride, potassium, ammonium or sodium salts. When the phosphoric acid was restored to the medium by the addition of traces of potassium phosphate, calcium chloride resumed its flocculating power. On treating the phosphate-free cultures with calcium chloride and alkali (0·5 c.c. tenth-normal sodium hydrate to 2 c.c. culture) a fine precipitate was obtained which very slowly gravitated. The deposit when examined microscopically was found to contain no bacterial floccules, and the organisms were free in the supernatant liquid. Accordingly it seems probable that the precipitate was the calcium salt of an organic acid elaborated by the bacteria. The precipitate is more marked when the bacteria are killed and partly disintegrated by boiling.

These experiments have shown that a pure flocculation of bacteria by means of the usual flocculating agents cannot be obtained. The reason for this is undoubtedly because the salt diffuses quickly through the bacterial cell and no surface pressure is occasioned. In working with dilute solutions of calcium chloride or better calcium bicarbonate and cultures containing phosphates, a flocculation of the bacteria is noticed, and this is quite apart from the entangling action of the calcium phosphate. It is worthy of emphasis that the amorphous particles of recently precipitated tricalcium phosphate are so like large clumps of bacteria that they might readily be mistaken for such. In a mixture of phosphate particles and bacterial clumps, differences are to be seen in the smaller clumps where the individual bacteria can be recognised. That nascent calcium phosphate should flocculate bacteria is to be expected from the fact that, as well as being non-diffusible, it has an affinity for loose water molecules, and forms with them hydrated calcium phosphate. Soon after formation it becomes less and less hydrated, and when added to cultures at this stage no flocculation of bacteria occurs.

We can now refer to the differing susceptibility of the three experimental organisms to the action of calcium chloride. Microscopical examination of the precipitates did not indicate anything unusual, because the precipitated calcium phosphate looked precisely like large clumps of bacteria. There are two causes that might be brought forward—(1) The bacteria which have the most flagella become sooner entangled in the tricalcium phosphate coagulum; and (2) the organisms may elaborate substances which are precipitated by the calcium salt. With regard to the first cause, it is to be noted that the amount of calcium necessary to produce a certain effect in the cultures is inversely proportional to the number of flagella on the organisms. The greater the number of flagella, the more firmly will the organisms be retained

media, the differences in the acid-content of the cultures were very small. Since acid in the culture would dissolve a certain amount of calcium phosphate, the culture that contained most acid would require the addition of most calcium chloride to produce a certain effect. The relative acidities of the cultures, however, were not sufficiently distinctive to account for the difference in the amounts of calcium chloride necessary for complete precipitation The nature of the acid radicles in the culture will probably explain the chief reason of the differing susceptibility. That the bodies of the bacteria have only a small function in the phenomenon is to be seen from the behaviour of filtered cultures. Three cultures grown in ordinary neutralised bouillon were filtered through porcelain filters and 2 c.c. portions were treated with calcium chloride ($\frac{1}{10}$ gram-molecule per litre) clear supernatant fluids were obtained in one hour with the following amounts of solution in c.c.:—

Bact. typhi	0·4
" *prodigiosum*	0·8
" *coli commune*	1·0

The differences are sufficient to indicate that it is to the product of the bacteria that the phenomenon is due. I am of the opinion that the cause may be traced to *Bact. typhi* withdrawing less phosphoric acid from the medium than the other two organisms which take up more and replace what they have taken with other acids These acid products of metabolism form with calcium, insoluble salts which have less tendency to coagulate into floccules than tricalcium phosphate.

The differing susceptibility of *Bact. typhi* and *Bact. coli commune* to calcium salts can be utilised to distinguish between them. The method consists in pipetting two c.c. of a two or three days' bouillon culture into a narrow test tube and adding one c.c. of calcium chloride solution containing one gram crystallised calcium chloride per 100 c.c The mixture is shaken and allowed to stand for an hour. At the end of this time *Bact. typhi* shows a well-defined precipitate, and in an almost clear supernatant fluid several large floccules adhering to the walls of the tube. *Bact.*

coli commune, on the other hand, has an ill-defined precipitate and a very turbid supernatant liquid.

From these numerous experiments I have shown :—

1. That bacteria are not flocculated by salts of potassium, sodium or ammonium like particles of suspended inorganic matter, and consequently that a pure flocculation or coagulation cannot be employed as a means of separating bacteria from cultures or of causing ultra-microscopical bacteria to cohere into visible cell-aggregates.
2. That salts of lime form a precipitate of calcium phosphate with the phosphoric acid of the medium.
3. That, since all ordinary media contain phosphates, and the organisms grown therein always retain traces of phosphoric acid, any substance capable of forming an insoluble phosphate will, when added to bacterial suspensions, cause a precipitate to form, and this, by entrapping the bacteria, will produce an apparent flocculation of the organisms. Microscopical examination may not indicate the presence of a precipitate because some insoluble phosphates, as for instance tricalcium phosphate, appear like large bacterial clumps.
4. That bacteria when grown in ordinary media exhibit

THE MECHANISM OF AGGLUTINATION.

By R. Greig Smith, M.Sc., Macleay Bacteriologist.

Pfeiffer and his pupils about 1894 discovered that when an animal is repeatedly inoculated with certain organisms, its serum has the power of causing the organisms in a bouillon culture to become altered and to cohere or agglutinate into microscopical masses or clumps. The serum only reacts in this manner or is active with the bacteria with which the animal has been inoculated, and this fact caused the reaction to be used as a diagnostic for that particular organism. Widal inverted the reaction and used a culture of typhoid bacteria to discover whether or not a serum was active, and especially in human practice to determine if a patient had typhoid fever. To apply the test a drop or loop of the blood serum which has separated from the clot is added to about 30 drops or loops of bouillon containing typhoid bacteria. Should the bacteria collect into clumps in half-an-hour the reaction is positive, and is by some considered as a proof of typhoid fever, by others (1) as a symptom of that disease. The test has been extended to other diseases.

The phenomenon of agglutination forcibly recalls that of flocculation or coagulation of inorganic particles, where instead of adding an indefinite active serum there is added a definite chemical substance, and a natural assumption would be that they are brought about by the same causes. One difference, however, between the two phenomena is that the bacteria are living and sensitive, while inorganic particles are insensitive. Before the act of agglutination they are actively motile; the agglutinine in the serum causes them to lose their motility in great part or entirely; they become immobilised. After a variable time, it may be hours or days, the bacteria regain their motility and the clumps break up. It is clear that the loss of motility or suspension of vitality is a necessary factor in the phenomenon.

The action of the serum upon the bacteria has been variously explained. Pfeiffer saw the bacteria swell just as they did in the peritoneal fluids of immune animals, and Gruber (2) considered this to be the cause of agglutination, the outer membranes of the bacteria becoming gelatinous and sticking to one another.

Kraus (3) obtained a precipitate on adding anti-cholera serum to a filtered culture of cholera vibrions. This is an agglutination of the soluble products of the metabolism or of the disintegration of the bacteria. Nicolle (4) showed that these agglutinable substances were excreted by the bacteria during life as well as being contained in the products of their disintegration. These bodies are not affected by a temperature of 150° C. When inert chemical substances such as talc, or as Nicolle showed, foreign bacteria are introduced into the filtured culture they are entrapped in the precipitate produced by the active serum and appear to agglutinate. Nicolle and also Paltauf considered that this gelatinous precipitate surrounded the bacteria and caused them to adhere together. Dineur considered that the precipitation took place on the flagella, which becoming adhesive caused an entanglement of the bacteria.

Bordet considers that the cause of agglutination is also the cause of the coagulation of casein, of the precipitation of chemical substances and of the agglutination of blood corpuscles. To disprove Gruber's hypothesis that the swelling of the bacterial capsule causes agglutination, he added a small quantity of an active serum to a suspension of cholera vibrions in normal saline. The clumps which formed were separated from the normal saline by centrifuging and subsequent treatment with water. The bacteria were shaken with the water until a homogeneous suspension was obtained. This was divided into two portions, to one of which common salt solution was added and to the other distilled water. Clumping occurred in the former case, but not in the latter. That common salt should cause the agglutination of the immobilised bacteria shows that a swelling of the membranes, if such occur, is not a necessary factor in the phenomenon. It is contended by Bordet that the agglutination of bacteria by active sera is identical in principle with the coagulation of casein by rennet. In the furtherance of this idea he found that the serum of animals inoculated with milk contained an enzyme that coagulated milk after the manner of rennet. Since both phenomena appear to be similar, he considered that the name agglutination should be changed to coagulation, and the agglutinines, of which there are many varieties—each capable of clumping its particular organisms—should be called coagulines. The agglutinines he believes to be enzymes, an opinion which is shared by Emmerich and Löw (6). The mechanism of the process, as explained by Bordet, consists primarily in the enzyme altering the relations between the bacteria and the solution, and secondly, as a result of the alteration the bacteria gather themselves into clumps.

It is claimed by those who have experimented with the mechanism of agglutination that clumping is caused by a precipitate forming on the organisms (Nicolle, Paltauf) and making them adhesive, by the organisms swelling (Gruber), or by the agglutinating enzyme causing them to flocculate (Bordet). Neither the formation of a precipitate on the bacteria, as Bordet has pointed out, nor the swelling of the organism explains the reason of their gathering together. With regard to the action of the enzyme in causing them to run together, Bordet has shown that once the bacteria have been acted upon by the active serum they are flocculated by common salt. It has been shown in my former paper that bacteria are not flocculated by common salt, from which it is to be concluded that the organisms have through the action of the active serum become altered into or have been endowed with some substance that is capable of being coagulated or flocculated. That it is not the action of the enzyme purely, is shown by Bordet's experiment, but as I shall show he has wrongly interpreted the phenomenon. He undoubtedly considers agglutination to be the work of the enzyme alone and confirms it by the action of rennet on milk, apparently forgetting that rennet does not coagulate casein in the absence of salts of lime. The casein is altered by the rennin into paracasein and an albumose, but the paracasein is only coagulated in the presence of lime.

The question then arises, what is the action of the active serum? It is apparently not a coagulation of the protoplasmic albuminoids, since bacteria, the albumen of which has been coagulated by heat, are, as I have found, not flocculated by salts. The immobilisation would seem to indicate an alteration of the protoplasm. But since bacteria killed by heat are not flocculated,

bouillon or serum. *It is the precipitate that is clumped; the bacteria are carried with it mechanically.* By adopting this view we have an agreement between the numerous observers. It is also easily understood why dead typhoid bacteria agglutinate like living ones. The products of metabolism being in contact with the dead cells are precipitated by the active serum upon their surfaces and in the medium. The precipitate is flocculated by the saline constituents of the solution, and both dead cells and precipitate gather together into floccules. The fact that dead typhoid bacteria may be employed in Widal's test bears out the theory of a coagulable surface precipitate, and agrees with Nicolle's experiments, which showed that foreign bacteria in a filtered culture of *Bact. typhi* were clumped by active typhoid sera.

Gruber (7), writing recently, considers that Kraus' precipitate is quantitatively too small to explain agglutination, and thinks it probable that in the act of agglutination certain substances in the bacterial membranes are made more insoluble. A shrinkage and separation follow whereby glutinous masses are formed on the bacterial surfaces. This appears to be very similar to his old hypothesis of the swelling of the membranes, to which Bordet pointed out that there was no reason given for the approach of the bacteria. Again, the shrinkage and formation of sticky masses on the bacteria is an idea, while Kraus' precipitate is a fact.

Radzievsky (8), in a preliminary paper, objects to the precipitate idea apparently because he obtained no precipitate in young cultures in which the bacteria clumped normally. Both writers apparently forget that the bacteria must be saturated with the precipitable substance before it is given off into the culture medium, and in a young culture while the organisms are saturated there may be but an infinitely small amount in the culture fluid.

I have tested the validity of Gruber's and of Radzievsky's objections and cannot agree with them. The precipitate may be and undoubtedly is very small in amount, but it is still appreciable. A twenty-four hours' bouillon culture of *Bact. typhi* was filtered through a Kitasato filter, and an agar culture of *Bact.*

coli commune was distributed in a small portion of the filtrate. This suspension was treated with active typhoid serum in the proportion of 20 parts of suspension to 1 of the serum. The bacteria, which before the addition of the active serum had been uniformly distributed in the *Bact. typhi* filtrate, had after an hour become collected into clumps. As it seemed possible that some objection might be made to the use of *Bact. coli commune* on the ground that the serum might have been obtained from a case of mixed *Bact. typhi* and *Bact. coli commune* infection, a second experiment was made with *Bact. Hartlebii.* Agglutination occurred precisely as when *Bact. coli commune* had been employed. These experiments show that Gruber's and Radzievsky's objections are groundless, and they are in agreement with Nicolle's, who showed that foreign bacteria suspended in the filtrate of a *Bact. typhi* culture were agglutinated by active sera.

Since agglutination is essentially the coagulation of a precipitate, it will be prevented by the presence of anti-coagulating agents such as the alkaline citrates and acetates. Winterberg (9) in a recent paper showed that the so-called agglutinines were destroyed by acetates, as evidenced by the absence of clumping. It is clear that the non-clumping was due to the acetates preventing the flocculating action of the serum and bouillon salts and not to the destruction of the enzyme.

to eliminate from the fluid traces of precipitable salts. Generally a slight precipitate was obtained, and above this a uniform suspension of the bacteria. The nitrates of potash, soda and ammonia failed to produce a flocculation when added to the suspension, even when the emulsions were centrifuged (2,500 revolutions per minute). The continued addition of dilute silver nitrate in place of the alkali nitrates produced very slight precipitates. Strong silver nitrate, however, produced complete precipitation.

The failure of the dilute silver nitrate to effect complete flocculation shows that either no silver salt had formed on the surfaces of the bacteria, or, if one had formed, the silver nitrate or the alkali nitrates were too weak to induce flocculation. It is probable that no surface film had been formed. The absence of flocculation by so strong a flocculating agent as dilute silver nitrate emphasises the fact that bacteria when freed from bouillon salts and their by-products, are not coagulated like inorganic particles.

The bacteria after being flocculated by the strong silver nitrate were seen to be in clumps. Careful examination of these clumps, and especially after they had been exposed to the light, showed that the bacteria were enclosed in a matrix which undoubtedly consisted of a silver compound of the intracellular salts which had diffused out from the cells under the influence of the strong silver nitrate. It is evidently impossible to obtain a pure flocculation or agglutination of bacteria, and when such an appearance is presented the failure to reveal the presence of a flocculated matrix is due entirely to our instruments or methods of demonstration.

Some experiments of Malvoz (10) are frequently quoted to show that agglutination of typhoid bacteria may be obtained by the addition of certain chemical reagents and stains. Since I have shown that true chemical agglutination does not occur, it seemed advisable to repeat his experiments. In one of these experiments clumping occurs when strong alcohol or strong formalin is added to an equal volume of a suspension of bacteria in water. This appears to be due to a dehydration rather than to a flocculation in which the loose water molecules are withdrawn. But such as it is, these strong reagents produce the nearest approach to true flocculation that can be obtained with bacteria. Another of his agglutinating agents is dilute mercuric chloride. This salt undoubtedly acts like calcium chloride in producing a precipitate of the culture salts that is flocculated together with the bacteria. The case of a dilute solution of saffranin promised to be different. This stain when in dilute solution (1-1000) and added to an equal volume of bacterial suspension produced an apparent agglutination of the bacteria. A test of the stain, however, with sterile bouillon showed the formation of an immediate precipitate which was found microscopically to resemble clumps of bacteria and cocci. This shows that the case of saffranin is no exception to the rule that in agglutination a precipitate is first formed in the fluid. On investigating the constituent of the bouillon that is precipitated by saffranin, it was found to be among those that are precipitated by lime, since no agglutination was obtained with bacteria that had been grown upon or in media that had been treated with lime to remove phosphoric acid. It does not appear to be a phosphate, because neither ammonium nor potassium phosphate forms a precipitate with the dilute stain. Malvoz

effect of dil

LITERATURE.

1.—FISCHER, Zeitschr. für Hygiene, xxxii. 407.
2.—GRUBER, Centralblatt für Bakteriologie, xix. 579.
3.—KRAUS, Abstract in Centralblatt für Bakteriologie, xxiii. 297.
4.—NICOLLE, Ann. Institut Pasteur, xii. 161.
5.—BORDET, Ann. Institut Pasteur, xiii. 225.
6.—EMMERICH and Löw, Zeitschr. für Hygiene, xxxii. 1.
7.—GRUBER, Abstract in Centralblatt für Bakt., xxvii. 285.
8.—RADZIEVSKY, Centralblatt für Bakt., xxvi. 753.
9.—WINTERBERG, Zeitschr. für Hygiene, xxxii. 375.
10.—MALVOZ, Ann. Institut Pasteur, xi. 582.

ON A NEW SPECIES OF *ANGOPHORA*.

By R. T. Baker, F.L.S., Curator, Technological Museum, Sydney.

Angophora melanoxylon, sp.nov.

"Coolabah."

(Plate ii.)

A medium-sized tree, from 40-50 feet high, with a diameter up to 3 feet; the bark somewhat similar to a "Box" bark, much less fibrous than that of *A. subvelutina*, F.v.M., or *A. intermedia*, DC. Branchlets glabrous or minutely pubescent, with or without bristles.

Leaves much more numerous than in the other species of *Angophora*, mostly under two inches long, rarely exceeding 2½in., and under ½in. broad; lanceolate or cordate at the base, with rounded auricles, sessile or almost so, nearly always opposite and decussate; blue-green on the upper side, pale yellow-green on the lower or under side, which is occasionally minutely pubescent;

Timber.—The timber of this species appears to be quite different from that of the coastal species, inasmuch as it is of a very dark brown (in fact, almost black) colour, and not pale coloured as is the case with other Angophoras. It has a pretty wavy figure, and in colour and hardness is almost identical with American Walnut, and is thus particularly suitable for cabinet work.

Kino.—The kino is in brownish-coloured masses, having a dull fracture. It is very friable, so much so that it crumbles to an ochrey-coloured powder between the fingers. It is but little soluble in cold water, forming a whitish turbid solution, the turbidity disappearing when boiled, the solution again becoming turbid on cooling. The substance causing the turbidity is removed by extracting with ether, and the reactions show it to be aromadendrin. No eudesmin is present. The presence of this substance in the kinos of the Angophoras shows a chemical connection between these trees and the Eucalypts. Eudesmin appears to be the more common in the Eucalypts, but in the kinos of some species both eudesmin and aromadendrin are present; while only in one species as yet has aromadendrin alone been found (*E. calophylla* of West Australia).

The tannin present in the kino of this Angophora gives a green coloration in a very dilute aqueous solution with one drop of ferric chloride, and in this respect differs from the kino of *E. calophylla*, which gives a blue coloration under like conditions (Henry G. Smith).

Fodder.—This is a tree that should be extensively cultivated on the arid land of the interior, as it is drought-resisting, and the leaves are much relished by cattle. A great point in its favour is that it is very foliaceous.

That this tree should in the past have missed recognition is rather strange, as it is well known throughout the area indicated by the above localities. The rare shape of some of the leaves (the lanceolate form) connects it with *A. intermedia*, whilst the rounded auricular base of the predominant-shaped leaf gives it some affinity to *A. subvelutina*.

The nature of its bark and timber differentiates the tree at once from any described species. The compact, terminal panicle, as well as the larger flowers, also differentiate it from *A. subvelutina.*

From *A. cordifolia* it differs also in its smaller flowers, fruits, and leaves.

It appears to have little affinity with *A. lanceolata.*

In botanical sequence it is placed between *A. subvelutina* and *A. intermedia,* as the leaves have the form of those of both these species. The inflorescence more nearly approaches *A. lanceolata*; but as stated above it is differentiated from this species by the nature of its bark and timber.

In passing it may be mentioned that the town and railway station on the main western line, 424 miles west of Sydney, are named after this tree, where a cluster of them stands at present near the railway platform; these are known for miles around as "Coolabah."

The name "Coolabah" is also given to two or three species of Eucalypts, and it comes rather as a surprise to a botanist travelling west to be shown these particular trees as "Coolabahs," and then to discover that they are Angophoras.

It is mostly a crooked tree; and it is from this feature that the aboriginal name "Coolabah" is derived (W.B. Under the

STUDIES ON AUSTRALIAN MOLLUSCA.

PART I.

By C. HEDLEY, F.L.S.

(Plates iii.-iv.)

As the result of collecting on holidays and examining material in hours not engaged in official duties, a considerable amount of information comes into my possession. In previous communications some of this was embodied: a further contribution is now tendered, and I trust that under this title I may offer more in the future.

I am constrained to apologise for the disconnected and fragmentary state of the various items: they are but leaves from the journal of a working naturalist, and their order is that in which chance may present facts.

The first requisite of my fellow students and myself is to assure ourselves of the identity of the species we handle; questions of structure and of higher classification, though of great importance, cannot be approached until specific identity is assured. Certain English writers who have dealt with our fauna have presented us with brief Latin descriptions of species unaccompanied by figures. The earlier Australian conchologists unfortunately selected this style for their model. Personally I have failed to identify species from writings of this class; I hear from correspondents the same confession, and observe that species thus unfigured and briefly diagnosed have suffered reduplication of names at the hands of the most distinguished European specialists.

It has been therefore my first aim to fix by illustration the identity of such unfigured species as I can procure. The occurrence of species on the coast of N.S. Wales not previously recorded thence will be given prominence.

Sirius, gen.nov.

(Plate iii., fig. 8.)

A genus of the *Trichotropidæ*, differing by a turbinate instead of conical shape, thin, without (as far as my information goes) the characteristic epidermis of the type. Especially is it distinguished by a concave, expanded pillar, broadening anteriorly to an abrupt termination, and failing to reach the siphonal notch. There is considerable similarity between *Crossea* and *Sirius* in general shape and in the features of the base.

Type *Raulinia badia*, Tenison-Woods.

The type of my new genus was specifically described by Tenison-Woods in these Proceedings (Vol. ii. 1876, p. 264), and discussed at some length. His reference of a living species to the genus *Raulinia*, Mayer, created for a European Miocene fossil has received the attention of subsequent textbooks. The arguments advanced by Woods fail to convince me. I can trace no sequence of family or genus between *R. alligata* and *Sirius badius*. The salient character of the fossil is a sharp transverse median fold on the columella; to this the broad, smooth columella of *Sirius badius* presents no counterpart. In support of this statement a figure of the Australian species (fig 8) is now submitted for contrast with that of *R. alligata*, Deshayes (Journ.

Fossarus sydneyensis, sp.nov.

(Plate iii., fig. 12.)

Shell broadly ovate, rather solid and narrowly perforate, body whorl large, spire short and turreted. Colour dead white (? bleached). Whorls four, flattened for a space below the suture, a little inflated at the periphery and gently rounded to the base. Sculpture: the first two whorls are smooth, the next has three raised spiral cords, while the last is encircled by eleven, sharply elevated, narrow, spiral cords, separated by interstices of twice or thrice their own breadth, two on the shoulder and two on the base are more prominent than the rest, the basal ones wind obliquely into the narrow umbilical fissure, minute striæ in the direction of growth lines decussate the troughs between the smooth-topped ridges. Suture impressed. Aperture oblique, ovate, exceeding half the length of the shell, angled above, rounded beneath, furrowed within by the print of the external sculpture; outer lip sharp, denticulated by the sculpture. Columella arched, broad, above plastered on the body whorl and over the axial perforation, below spreading and reflected; at its anterior termination is the faint rudiment of a channel. Length 4·5, breadth 3 mm.

Hab.—Balmoral Beach, near Sydney; several specimens among shell sand.

Type to be presented to the Australian Museum.

Couthouyia aculeata, sp.nov.

(Plate iii., fig. 10.)

Shell ovate, with a slender acuminate spire and inflated body whorl. Colour dead white (? bleached). Whorls six, rapidly increasing, divided by a narrowly but deeply grooved suture. The last whorl just previous to the aperture, is free from its predecessor. Sculpture: the last whorl is encircled by nine narrow, sharp, projecting, spiral ridges; passing from suture to base these ridges gradually and proportionately grow larger and farther apart. The lowest overhangs the umbilical fissure. Interstitial threads develop in the two lowest furrows. All the ridges and furrows are crossed by fine, sharp, obliquely ascending threads, of which there are on the body whorl about thirty-three. The major spiral, and minor transverse, lines enclose deep square pits; at their intersection arise sharp little prickles. A corresponding sculpture occurs on the upper whorls and fades gradually away towards the apex. Aperture oblique, almost D-shaped, but rounded off at the angles. Outer lip sharp, frilled by the sculpture. Inner lip distant from the whorl, nearly straight, edge a trifle curled, broadened at the anterior corner, where the circum-umbilical ridge arches in to meet it. Here is the shallow impression of a rudimentary siphonal notch. Umbilicus a long, deep, narrow cavity whose inner wall is smooth. Length 4, breadth 2·5 mm.

Hab.—Off Bet Island, Torres Straits; two specimens dredged by Mr. J. Brazier in 11 fathoms.

Type to be presented to the Australian Museum.

The genus *Couthouyia* (Ann. Mag. N.H. ser. 3, v., May, 1860, p. 410) has not been previously seen in Australian waters.

MENON, gen.nov.

A genus of the *Eulimidæ*. Shell perforate, solid, dull, com-

grooves occur on the two topmost whorls and there cease, the rest of the shell is faintly and irregularly ribbed by arcuate growth lines; from the aperture to the apex and upon the opposite side of the shell a series of strong cord-like varices project and mount the shell perpendicularly. The end of the penultimate varix is thrust into the aperture. The suture is impressed and ragged from the irregular longitudinal sculpture. Aperture ovate-oblong, above solute, subangled and channelled; beneath rounded. Outer lip sharp, curving forward anteriorly; columella long and very straight, above narrow and appressed to the body whorl, beneath broadened and reflected. Umbilicus a narrow triangular deep pit, walled in by the reflected columella, the penultimate and incipient varices. Length 5, breadth 2 mm.

Hab.—Little Coogee Bay, near Sydney; in shell sand. I am indebted to Mr. J. Brazier for a specimen which he collected in July, 1895.

Type to be presented to the Australian Museum.

My illustrations show the shell seen in front, the apex from behind, and the base, with varix in profile, from beneath, all enlarged.

This genus appears to link the bizarre form *Hoplopteron* (Fischer, Journ. de Conch. xxiv., 1876, p. 232) to the more normal *Eulimidæ*. The difference between the varices of *Menon* and of *Hoplopteron* is rather one of degree than one of kind. Indistinct varices occur in *Eulima* proper. On a specimen of *E. tessellata*, Sowerby, I find a series of inconspicuous varices on the upper whorls.

The apex described above is likely to prove a plug formed in life before the loss of the apical whorls.

Seila attenuata, n.sp.

(Plate iii., figs. 9, 9a.)

Shell dextral, very tall and slender, gently tapering, varies a little in proportion. Whorls thirteen, gradually increasing, rounded, contracted at the sutures. Colour varying from deep chocolate to pale ochre, the primary whorls always darker than the remainder. Sculpture: four, evenly spaced, spiral, sharp cords ascend the whorls, midway in their interspaces on the upper whorls threads appear, which increasing in more rapid proportion rival on the last whorl the primary cords; one specimen before me thus exhibits on the last whorl nine equally sized and spaced cords. Both cords and interspaces are crossed by coarse, irregular, arcuate growth-striæ. The apical whorls are obliquely, longitudinally ribbed, thus recalling normal *Cerithiopsis* sculpture. A keel appears, the ribbing diminishes, and by gradual transition the adult sculpture is attained The base is concave, above smooth, below faintly spirally ribbed. Columella arched, canal short and straight. The dark specimen figured measures—length 9 mm, breadth 2 mm. A pale, more strongly ribbed shell measures—length 10 mm., breadth 2·5 mm.

Hab.—Balmoral Beach, Middle Harbour, near Sydney. I have collected a dozen dead specimens.

No members of *Seila* (A. Ad.; Ann. Mag. N. H. [3], vii. 1861, p 130) have been yet noticed in Australia. The present species differs from the type *S. dextroversa*, Ad. & Rv., (Voy. Samarang, Zool. pl. xi. p. 31) by more lyræ, more rounded whorls, and a straighter canal.

Type to be presented to the Australian

ZEIDORA TASMANICA, Beddome.

Beddome, Proc. Roy. Soc. Tas. 1882 [1883], p. 169.

In the note quoted under the preceding species, Mr. Henn published the first occurrence of this species in our waters. I can again confirm his discovery, having taken a young specimen in sand in a cleft of the cliffs a mile south of the South Head

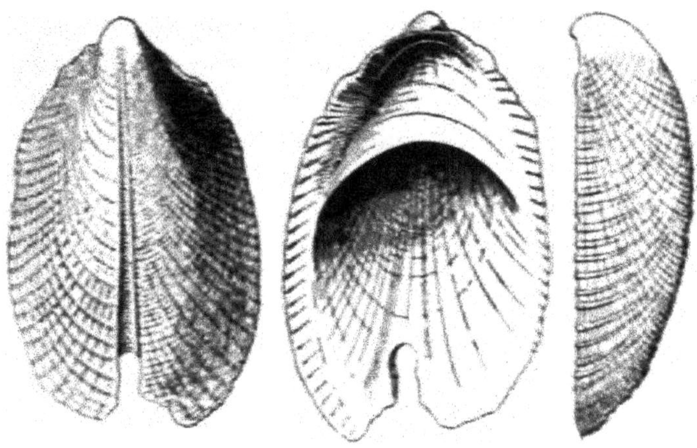

Lighthouse. Mr H. L. Kesteven has shown me a specimen which he collected at Botany Head As the species has never been figured, and as my example is too young to use, I have derived an illustration from an authentic Tasmanian example, kindly lent me by Miss Lodder.

Professor Tate recognised the occurrence of this genus in South Australia in a species probably conspecific with the Tasmanian (Trans. Roy. Soc S A. xviii. 1894, p. 118). A strict comparison between specimens yet remains to be instituted.

HELIACUS FOVEOLATUS, Tate.

Tate. Trans. Roy. Soc. S.A. xvii. 1893, p. 191, pl. i. figs. 13, 13a.

I have lately taken several examples of this species upon Balmoral Beach. The identification is based upon comparison with specimens kindly forwarded by the author. It had not been previously noticed beyond the borders of South Australia.

AMAUROPSIS MOERCHI, Adams & Angas.

(Plate iii., fig. 4.)

Adams & Angas, Proc. Zool. Soc. 1863, p. 423.

Only two examples of this rare species were taken by Angas; he found them "adhering to the under surface of a large stone, at Watson's Bay, just inside Port Jackson Heads, during an unprecedently low tide" (P.Z.S. 1867, p. 198). Another was collected by Brazier under a stone at Point Piper (Proc. Roy. Soc. N.S.W. xxiii. 1889, p. 259). The fourth known specimen occurred to me at Twemlow's Reef, Middle Harbour, under a stone on muddy ground in the mangrove (*Avicennia*) zone, in company with such mud-loving species as *Phenacolepas cinnamomea*, Gould, *Plecotrema bicolor*, Pfr., and *Columbella regulus*, Souverbie.

Mr. Brazier is acquainted with the species by sight, having derived his information direct from Angas. He has kindly confirmed my determination, indeed from literature alone no identification could be made. To assist future observers I now tender a drawing of my specimen, which is 5 mm. in diameter. Angas describes his as 5 lines in length. The youth of my exam le is

with Victorian specimens from Western Port, kindly supplied by Mr. J. H. Gatliff.

The species is omitted from Tryon's Manual, and the original description is insufficient for the recognition of the species. To figures of one of my specimens, therefore, I add the following description :—

Shell minute, above almost flat, below rounded and widely umbilicate. Whorls two and a half, very rapidly increasing, so that the outline of the shell approaches the figure of the Arabic numeral 6. Apical whorls smooth, one and a half. Sculpture: throughout closely, finely, spirally grooved; above is one and on the base are two prominent spiral lyræ. The whole shell is crossed by transverse sculpture, which is nearly suppressed for most of its course, but appears in a series of fine denticules below the suture, in beading on the major lyræ, and in basal ribs projecting teeth into the umbilicus. A small varix appears at a whorl behind the aperture. Aperture very oblique, a round, trumpet mouth, with a double, widely expanded lip, fortified behind with a heavy varix. Major diameter 1·16, minor ·84; height ·64 mm.

Fissuridea lineata, Sowerby.

(Plate iii., fig. 11.

This species was included by the earlier writers in the genus *Fissurella*. Pilsbry in the Manual of Conchology placed it in *Glyphis*, but he afterwards abandoned that name in favour of *Fissuridea* of Swainson (The Nautilus, v., Jan. 1892, p. 104).

The animal has not been yet described. It is not uncommon in Sydney Harbour, living on rocks and among piles of loose stones in clear sea water. Its movements are very sluggish; in crawling the animal raises its shell to a considerable height as if holding an umbrella over itself. When fully extended no part of the animal envelops the shell. In striking contrast to the asymmetrical forward position of the shell in *Lucapinella nigrita* (vide Proc. Roy. Soc. Vic. 1894, pl. [illegible]) is the central position of the shell in *F. lineata*. The animal can quite conceal itself within the shell; when entirely retracted the shell touches the ground at the anterior end, but a slight gape remains behind. When fully exserted, as in my figure, a series of filaments project from beneath the margin of the shell, each answering to a radial of the shell sculpture; beneath and beyond these, the mantle skirt depends; the latter is often puckered into waves and can be extended to about twice the length of the filaments. Below the mantle is the usual row of epipodial filaments. Muzzle slightly bilobed. Tentacles moderately stout, slightly tapering, sharply pointed, frequently engaged in searching the ground with a slow sweeping motion. Tail short and blunt. Anal tube very little exserted, surrounded by a dozen small papillæ. Mantle and filaments pale yellow, remainder of body creamy white, anal tube dark brown.

Cerithium tenue, Sowerby.

Sowerby, Thesaurus Conch. ii. 1855, p. 876, pl. clxxxiv. fig. 212 (202).

This species was originally described from Port Lincoln, South Australia. An examination of South Australian specimens induces me to unite with it *Bittium variegatum*, Brazier, described in these Proceedings (Second Series, ix. 1894, p. 172, pl. xiv. fig. 9).

behind the aperture to a few crowded riblets. Base flattened. Umbilcus narrow, semicircular, deep, oblique, bounded by a heavy, outstanding, obliquely-entering funicle. Aperture very oblique, ovate. Lip thickened within. Columella united to the umbilical funicle and spreading above a callous pad on the preceding whorl. Major diameter 6, minor 5, height 2 mm.

Hab.—Port Darwin; one specimen, collected from the beach by Mr. Spalding, was communicated to me by Mr. J. Brazier. Two young shells dredged in 11 fathoms off Bet Island, Torres Straits, by Mr. Brazier appear to belong to the same species.

Type to be presented to the Australian Museum.

In the last volume of these Proceedings (p. 433) I described as *Teinostoma starkeyæ* the first Australian member of the subgenus *Solariorbis*. Professor Tate immediately followed with a second one named *Cyclostrema caperatum* (Trans. Roy. Soc. S.A. 1899, p. 216, pl. vii. figs. 1a-b). The present is the third. Its affinities are with *T. starkeyæ*, from which it differs by spiral sculpture and greater size, opacity and solidity.

Notarchus glaucus, Cheeseman.

(Plate iv.)

About five years ago I collected from time to time, at the edge of a *Zostera* flat on the west side of Rose Bay, near Sydney, several examples of the species now depicted. Most of them were rolled up dead or dying from the injurious effects of the volume of fresh water which heavy rains had poured into the Bay. One, however, was found in a healthy and expanded condition; from it my drawing was immediately made.

I have referred, but not with certainty, the species to *Aclesia glauca*, Cheeseman, judging it from the account given in the Proc. Zool. Soc. (1878, p. 277, pl. xv. fig. 4). Further information is required upon several points not noticed there, to supply which I have endeavoured without success to procure specimens from New Zealand, where I am told the species is rare.

I have not had an opportunity of late years of searching the locality, and reclamation works now proceeding there are likely to ruin the Bay for a collecting ground. Since there is no immediate prospect of learning more about the animal and as it is an addition both generically and specifically to the Australian fauna, it seems best to no longer reserve the little information now given.

MEGALATRACTUS ARUANUS, Linn.

Hanley has shown (Ipsa Linnæi Conchylia, 1855, p. 301, and Journ. Linn. Soc. Zool. iv. 1860, p. 78) that the Linnean species "*Murex aruanus*" referred by nomenclature, description and bibliography to two species. One, an American shell, was separated in 1788 from *aruanus* by Gmelin as "*Murex carica.*" The other, an Australian shell, was renamed "*Fusus proboscidiferus*" by Lamarck in 1822.

Unless we consent to altogether erase the Linnean name, it is obvious that the names of both Gmelin and Lamarck cannot be maintained.

Gmelin's classification has been accepted by Dillwyn (Decr. Cat. Recent Shells, ii. 1817, p. 723), by Binney (Journ. Nat. Hist. Boston, i. 1833, p. 67), and by Swainson (Exotic Conchology, 1841, p. 6).

There are several other Linnean specific names which concern the Australian student, such as *ianthina*, *delphinula* and *lima*, which Lamarckian usage has banished from their rightful place.

The nepionic shell of *M. aruanus* has a literature of its own. Tryon described it (Manual Conch. ix. p. 142, pl. xxvi. fig. 16) as *Cerithium brazieri*. Pilsbry followed (Nautilus, viii. June, 1894, p. 17) by erecting for its reception a new genus *Perostylus* and adding a supposed second species *P. fordianus*. Tryon's error was recognised by Tate, who showed in these Proceedings (Second Series, viii. 1893 [1894], p. 244) that "it is nothing more than the embryo of *Fusus proboscidiferus*." Pilsbry then in an article entitled "*Perostylus*, the embryo of *Megalatractus*" (Nautilus, viii. p. 67) at once withdrew his genus and species.

Megalatractus was proposed as a subgenus for this species by Fischer (Manual, 1884, p. 623); Pilsbry remarks that its use as a full genus is justified by the remarkable apex.

Melvill and Standen mention (Journ. Linn. Soc. Zool. xxvii. 1899, p. 158) "a mass of nidamental capsules" as being produced by this mollusc. Further information on this subject would be acceptable.

As little has appeared in literature about the size of this gigantic shell, it may not be amiss to say that a specimen now in the collection of my friend, Mr. P. G. Black, though imperfect at both extremities, still measures 22 inches in length and 9½ in breadth. Brazier has in these Proceedings (Vol. ii., 1878, p. 368) mentioned a maximum length of 24 inches.

The Australian range of the species is from the Dampier Archipelago in the north-west continuously to the Great Barrier Reef in the north-east.

EXPLANATION OF PLATES.

Plate [illegible]

Figs. 1-3.—*[illegible] [illegible]*, Tenison-Woods: in different aspects

Fig. 4.—*[illegible] [illegible]*, Adams & Angas.

Fig. 5.—*[illegible] [illegible]*, Hedley

Fig. 6.—*Menon anceps*, Hedley, viewed in profile from the base to show the varix.

Fig. 7.—*Menon anceps*, Hedley; apex.

Fig. 8.—*Sirius badius*, Tenison-Woods.

Fig. 9.—*Seila attenuata*, Hedley.

Fig. 9a.—*Seila attenuata*, Hedley; apex.

Fig. 10.—*Couthouyia aculeata*, Hedley.

Fig. 11.—*Fissuridea lineata*, Sowerby: animal.

Fig. 12.—*Fossarus sydneyensis*, Hedley.

Figs. 13-15.—*Teinostoma orbitum*, Hedley; in different aspects.

Fig. 11 slightly reduced, the rest enlarged in various proportions.

Plate iv.

Fig. 1.—*Notarchus glaucus*, Cheeseman; the animal from life; reduced by one-third.

NOTES FROM THE BOTANIC GARDENS, SYDNEY.

No. 6.

By J. H. Maiden and E. Betche.

SAPINDACEÆ.

Dodonæa filifolia, Hook.

Penrith (J. L. Boorman, January, 1900).

The second New South Wales locality recorded. Previously collected in this colony only in the low land between Double Bay and Bondi. (See Part 1 of this series, Proceedings, 1897, p. 147.)

MYRTACEÆ.

Leptospermum scoparium, Forst., var. rotundifolium, var. nov.

Tallwong, South of Shoalhaven River (W. Forsyth, January 1900).

A very distinct-looking variety with shortly petiolate, nearly orbicular leaves, about two lines in diameter, and with large fruits and flowers. We have been in doubt whether this form should be united with *L. flavescens*, to which the obtuse leaves and the long style point, or with *L. scoparium*; but we failed to find any connecting links in the series of forms of *L. flavescens*, while we have evidence in the Herbarium of broad-leaved forms of *L. scoparium* passing into the variety under notice. The characteristic pungent points of the leaves of *L. scoparium* disappear gradually with the increase in breadth of the leaves.

COMPOSITÆ.

Helichrysum rosmarinifolium, Less., var stenophyllum. var. nov.

Jenolan Caves (W. F. Blakely, February, 1900).

Differs from the type chiefly in the very narrow and acute leaves, about 1 inch in length, with closely revolute margins concealing the white under surface. Florets often only 4 or 5 in the flower-heads.

EPACRIDEÆ.

EPACRIS HAMILTONI, sp. nov.

A flattened shrub up to 3 feet high (see below), covered all over with rather long, soft, white hairs, especially on the young stems, more sparingly on the leaves. Leaves very shortly petiolate, broad, lanceolate, 3- to rarely 5-veined underneath, pungent-pointed, about 5 lines long. Flowers white, nearly sessile in the uppermost axils, often apparently terminal, forming short sparingly leafy heads. Bracts and sepals acute, glabrous but generally minutely ciliate. Corolla-tube attaining 3 lines in length, shortly exceeding the sepals, the lobes rather large but shorter than the tube. Hypogynous scales short and broad. Ovarium glabrous, style glabrous, much exserted, 5 to above 6 lines long, with a dilated stigma.

Blackheath, Blue Mountains (A. A. Hamilton, January, 1900).

Allied to *E. paludosa* and *E. Calvertiana*, but chiefly differing from the former in the long style, and from the latter in the shape of the corolla and in the inflorescence, and from both by the remarkable degree of hairiness, so rare in the genus. Mr. Hamilton's specimens are not in flower, but from the withered

New for New South Wales. Previously recorded from Western, South and North Australia and Queensland.

CYPERACEÆ.

LEPIDOSPERMA TORTUOSUM, F.v.M.

Blue Mountains (Blackheath [?]; E. Betche, December, 1885).

New for New South Wales. Previously recorded from Victoria (Mount Wellington, Gippsland). *L. falcatum*, Rodway, (Trans. R. Soc. Tas. 1895, p. 103), seems to us merely a smaller form of *L. tortuosum*, so that Tasmania should also be recorded within the range of this species.

CYPERUS ERAGROSTIS, Vahl.

Yarrangobilly Caves (E. Betche, February, 1897).

New for New South Wales. A very dwarf form, not exceeding two inches in height, identical with Victorian specimens from the Upper Hume River, but not previously collected in this colony, so far as we know.

OBSERVATIONS ON THE EUCALYPTS OF NEW SOUTH WALES.

Part VII.

By Henry Deane, M.A., and J. H. Maiden.

(Plates v.-vii.)

Eucalyptus affinis, sp.nov.

(Plate v.)

A tree of moderate size, attaining a height of 80 feet, and a diameter of 2 feet 6 inches.

Vernacular names.—"Tallow Wood" at Murrumbidgerie, owing to the greasy nature of its wood, and "Black Box" at Stuart Town; "White Ironbark" and "Ironbark Box" at Grenfell and above Mt. Macdonald, at the junction of the Abercrombie and Lachlan Rivers, according to Mr. Cambage; "Bastard Ironbark" at Minore (J. L. Boorman).

Our first complete series of specimens were received from Mr. Andrew Murphy in October, 1899. Mr. J. L. Boorman collected it in February, 1899.

Sucker leaves.—Alternate, ovate, obtuse, slightly emarginate and mucronate (in our specimens); about 3 inches long by 1¾ broad; intramarginal vein at a considerable distance from the edge.

Mature leaves.—Lanceolate, slightly falcate; pale-coloured, dull on both sides, rather coriaceous, usully 2 to 3 inches long; veins at an angle of about 30° with the midrib, but inconspicuous except the midrib and the thickened margin; intramarginal vein inconspicuous and at some distance from the edge.

Peduncles axillary, flattened at first but nearly terete when the fruit is ripe; with 3 to 7 flowers.

Buds.—Shaped like a tip-cat, to use a homely expression, *i.e.*, tapered equally towards base and operculum; somewhat angular, the operculum attenuate. Calyx-tube likewise attenuate, tapering into a short pedicel. Anthers in the bud all folded; stamens white, the outer ones seemingly all fertile; anthers opening in terminal pores. Style and stigma as figured at fig. 6, *E. hemiphloia*, in the Eucalyptographia.

Fruits.—Ovate-truncate, tapered at the base, somewhat contracted at the orifice, about 3 lines in diameter, the rim narrow, slightly convex and dark-coloured; the capsule depressed.

Range.—Between Wellington and Dubbo, towards Molong and Parkes, Grenfell, and in other parts of the Western districts.

The species appears to possess resemblances to the imperfectly known Ironbark, *E. drepanophylla*, F.v.M. The fruits of the latter are, however, sub-cylindrical, the orifice not constricted, the rim different and the valves slightly exserted when the capsule is perfectly ripe; the leaves are narrower and the veins finer and more parallel. Further observations on this head may be deferred until *E. drepanophylla* is more perfectly known.

The true affinities of our species are, in our opinion, with *E. sideroxylon*, A. Cunn., and *E. hemiphloia*, F.v.M. Roughly speaking, it resembles the inflorescence of *E. hemiphloia*, the fruits of *E. sideroxylon*, while its timber and bark partake of the characters of both.

Mr. Cambage is of opinion that the tree is a hybrid between *E. hemiphloia*, v. *albens*, and *E. sideroxylon*, a view which had already occurred to us. It certainly seems only to be found when the other two trees are present. There are difficulties in the way of recognising hybridism in Eucalypts, and as we propose to treat this subject later on we refrain from being dogmatic on the present occasion.

Below we give an account of some trees which partake of the characters of both an Ironbark, probably in this case *E. siderphloia*, and a Box, *E. hemiphloia*, and here hybridism again suggests itself. It is curious that in *E. affinis* we have a tree also partaking of the characters of an Ironbark and a Box, but in this case *E. sideroxylon* and *E. hemiphloia*, var. *albens*, apparently combine.

In view of the imperfect evidence of hybridisation before us we think it safer to give to *E. affinis* specific rank.

EUCALYPTUS CAMBAGEI, sp.nov.

(Plates vi.-vii.)

A small or medium-sized tree.

Vernacular names.—"Bundy" of Burraga, Rockley and some other places; called also "Bastard Apple, "Bastard Box," and "Grey Box" in different localities throughout the Bathurst and Orange districts. The glaucous form is "Rough-barked Mountain

and even emarginate, also cordate and passing in later growth through all the stages of broad-lanceolate to lanceolate, very glaucous. The sucker leaves have been confused with those of *E. dives;* the latter are, however, more acuminate and possess a stronger odour of peppermint. The sucker stems are markedly round.

Mature leaves.—Lanceolate, more or less falcate, generally very long, up to 8 or 10 inches, usually pale-coloured, with the pellucid dots rather conspicuous (as in *E. goniocalyx*), usually more or less besprinkled with small blackish dots (as in the case of *E. goniocalyx*, *E. Stuartiana* and other species); veins oblique and numerous, but not close, at an angle of about 30 to 40° with the midrib; the intramarginal one at a distance from the edge.

Buds.—Sessile, the peduncles short, thick and flat, each with three to seven or more flowers. For a figure of some buds see fig. 2, plate xli., Proceedings, 1899.

Calyx-tube 3 to 4 lines long, about 2 to 2½ lines in diameter, with 2 to 4 prominent angles; operculum shortly pointed or hemispherical, much shorter than the calyx-tube. Stamens about 3 lines long, folded in the bud.

Fruits.—Ovoid or broadly ovoid, truncate valves generally well exserted (as at fig. 1 of plate xli., Proceedings, 1899), more so than *E. goniocalyx*. In the var. *pallens* the valves are apparently not so much exserted.

Range.—It is found in the southern districts (see localities under var. *pallens*, Part v. p. 463, 1899). It has not been found on the Blue Mountains, though very close to them (*e.g.*, at Hassan's Walls, J.H.M.). It occurs as far west as Mt. Bulaway (3,450 feet, Coonabarabran district; W. Forsyth). It is common in the Bathurst and Orange districts, and possibly it may be found on any of the ranges lying between the southern and western railway lines of the colony.

Affinities.—Its affinities are undoubtedly with *E. goniocalyx* in the first place, and *E. Stuartiana* in the second. There is not the same tendency to confusion of this species in the field with *E. goniocalyx* as with *E. Stuartiana*, which in bark and general growth it resembles. The peduncles of *E. Cambagei* are shorter, thicker and flatter than those of *E. goniocalyx* usually are, not to mention other differences. It may be compared with an interesting form of *E. goniocalyx* from Mt. Wilson and other places in the Blue Mountains with peduncles often 1 inch in length, with smaller and often spreading calyx, valves much exserted, and with *pointed* buds.

The leaves of *E. Cambagei* are usually much longer than those of *E. Stuartiana* and the tree is easily distinguished. The buds and fruit are quite distinct, as may be seen from the figure.

Mr. Cambage writes :—" A typical Bundy grows on Silurian slate ridges, though it will also grow on hills of igneous rock, but in such cases the wood seems to me to be slightly softer, and in more than one instance I have noticed that the inflorescence seems better, the flowers more numerous, when the trees grow on slate ridges. The Apple *(E. Stuartiana)* grows on the flats and near creeks and thins out towards the top of a hill, giving place to the Bundy, and bushmen have argued with me that the tree simply goes by easy gradations from soft Apple on the low lands to hard Bundy on the high. I am satisfied that this is not so, as I have followed the Apple till it ceases up the side of a hill and it does not change, while the Bundy begins at once as a hard wood. By special search I have found more than once the top

is known locally as "White Ash;" its fruits are nearly sessile and usually in threes. The only other Eucalypt growing with it on the very summit is *E. Sieberiana*, F.v.M. We have also received,

(*a*) From Mr. A. Murphy in November, 1899, collected from the top of the Penang Range, near Gosford, specimens of a tree very like a "Peppermint" in appearance, but having more flaky bark on the trunk and white smooth limbs. The fruit is like that of *E. hæmastoma* in shape, but the rim is sunk as in *E. stricta*. The trees, which are not abundant, show a transition towards *E. hæmastoma*, but we consider they should be grouped under *E. stricta*.

(*b*) From Mr. R. H. Cambage we have also received specimens which we take to be a somewhat aberrant form of *E. stricta*. The tree has the appearance of a "Peppermint." The base of the capsule is remarkably constricted, the whole being pear-shaped; the rim is broad and somewhat sunk. The specimens were collected in January, 1900, at Burrill, Milton, and on Pigeon-house Mountain, Milton, about 100 feet from the summit.

Eucalyptus eugenioides, Sieb.

(For a previous reference, see Proceedings, 1896, 803.)

"Bastard Stringybark" (Penrith; J. L. Boorman, January, 1900). We desire to invite attention to an interesting form of this species. The opercula are more pointed; the fruits are smaller than is usual and nearly globular. They are on nearly filiform pedicels of about 2 lines; the common peduncle is twice that length and more. The plant is indubitably *E. eugenioides*, though from examination of the fruits alone it might reasonably be supposed to be *E. hæmastoma*, var. *micrantha*.

Eucalyptus Stuartiana, F.v.M., var. parviflora, var.nov.

(For a previous reference, see Proceedings, 1899, 628.)

It seems desirable to indicate, by some name, a small-fruited form of *E. Stuartiana* which has been found near Hassan's Walls, at Young, and in several other parts of the colony. The sucker or seedling-foliage is small, very glaucous, cordate and stem clasping; the fruits resemble those of a small-fruited form of *E. tereticornis* as much as that of *E. Stuartiana*, but the buds and timber sharply separate it from the former, and there is no doubt that it is referable to the latter species.

Eucalyptus squamosa, D. & M.

(For a previous reference, see Proceedings, 1897, 561; 1899, 629.)

Upper Bankstown and Cabramatta (J. L. Boorman; February, 1900).

Eucalyptus quadrangulata, D. & M.

(For a previous reference, see Proceedings, 1899, 451.)

Tillowrie, Milton (R. H. Cambage; January, 1900); the second locality for this rare species.

Eucalyptus pulverulenta, Sims.

(For a previous reference, see Proceedings, 1899, 465.)

This species is widely diffused in the Goulburn district, but is so rare in the Western district that it is usually looked upon as exclusively Southern. Yet Allan Cunningham collected it at Cox's River on 8th October, 1822, and noted it in his journal as "a large shrub about 8 feet high, and a species of Eucalyptus

mention that a young tree, bearing a profusion of lanceolate leaves, very glaucous, being almost white and even silvery, and gracefully pendulous, is one of the most beautiful plants we have, and a fit emblem of purity.

On Apparent Hybridisation between *E siderophloia*, Benth., and *E. hemiphloia*, F.v.M.

For many years certain trees partaking of the characters of both the above have been known to local residents in the County of Cumberland, and have received distinctive names. The Rev. Dr. Woolls (Proceedings, v. 504) has drawn attention to them. We have specimens selected from six different trees, which we may call *a*, *b*, *c*, *d*, *e* and *f*, and will describe their characteristics.

The anthers of *a* and *b* doubtless are like var. *E. siderophloia;* these open in parallel slits, not in pores as do those of *c*, *d*, *e*, and *f*.

(a) is the normal *E. siderophloia*, Benth. The tree grows pretty plentifully throughout the Bankstown and Cabramatta district, and is the only form recognised as Ironbark by the residents. The fruit is a little shorter and broader than that figured in the *Eucalyptographia*.

b) is the Black Box of the Bankstown and Cabramatta district. Mr. J. L. Boorman describes the bark as rough (somewhat like Box), but sometimes it is scaly; the colour seems to vary considerably from black to pale brown. The timber is yellow and paler than that of *E siderophloia*, interlocked in grain, exceptionally heavy and is very superior for all purposes; in the neighbourhood of Penrith it is sought after for making mauls. The buds are blunter than those of *E. siderophloia*, the anthers are similar, but the valves of the capsules are less exserted than those of *E. siderophloia* growing in the same district. The tree grows to a fair size, with long, pendulous, acuminate, glossy leaves.

In the specimens *c*, *d*, *e* and *f* the resemblance of flowers and fruit to *E. hemiphloia* is much greater than to *E. siderophloia*. The anthers open in pores, the fruits are broader and less cylindrical, and the valves are scarcely exserted. The bark is, however, more of an "Ironbark" in character except in the case of *f*.

(*c*) is called "Bastard Box" or "Ironbark." It is found on the Waterloo Estate near Upper Bankstown and St. John's Park.

(*d*) is "Bastard Box" or "Box." The trees are of large size, the bark rough to the tips of the branches.

(*e*) has a bark rougher than the preceding, but is otherwise identical.

(*f*) grows at Cabramatta and is called "Black Box" by Mr. T. Shepherd. The bark is fibrous and persistent, "the wood hard and bad for burning," according to the late Rev. Dr. Woolls. Dr. Woolls, who had an intimate and extensive knowledge of the Eucalypts of the colony, expressed himself to be much puzzled with these trees, and marked specimens of this particular form at one time as *E. siderophloia* and at another as *E. paniculata*. The original specimens are in the Herbarium of the Sydney Botanic Gardens and illustrate the difficulty that these supposed hybrids have given rise to in times gone by.

It will be seen how the above forms show different steps in the gradation between *E. hemiphloia* and *E. siderophloia*. A suspicion of a resemblance to *E. sideroxylon* is visible in at least one of the specimens. The case is very different in *E. affinis*, which,

E. PUNCTATA, DC.

This tree is called "Black Box" at Capertee owing to the darkness of its bark. Specimens have been received from Mr. W. Heron, collected at Conjola. This is apparently the most southerly locality hitherto recorded. In the Flora Australiensis the species is stated to range as far north as the Macleay River. Most of the northern specimens are, however, our *E. propinqua*, allied to but now separated from *E. punctata*, DC. We have now received indubitable specimens of the latter from Mr. H. L. White collected on the main southern spur of the Wollooma Mountain in the Scone district, while Mr. W. Bäuerlen has collected it as far north as Lismore.

EXPLANATION OF PLATES.

Plate v.—*E. affinis.*

Fig. 1.—Sucker leaf.
Fig. 2.—Flowering twig.
Fig. 3.—Fruiting twig.
Figs. 4-7.—Fruits.
Fig. 8.—Stamens.

Plate vi.—*E. Cambagei.*

Fig. 1.—Flowering twig with leaves and fruits.
Fig. 2.—Twig with buds in various stages.
Figs. 3-5.—Fruits.
Figs. 6-7.—Stamens.

Plate vii.—*E. Cambagei.*

Figs. 1-2.—Sucker leaves; reduced in size.

NOTES AND EXHIBITS.

Mr. Baker exhibited herbarium and timber specimens of *Angophora* in illustration of his paper; a "fire-stick" used by the Aborigines in the neighbourhood of Grafton, N.S.W.; and two undescribed fungi (*Colus* sp., and *Calostemma* sp.) from Katoomba (Mr. T. Steel, F.C.S.).

Mr. Froggatt exhibited a fine mounted series of specimens of twelve species of Australian Ticks, determined by Professor Neumann, of Toulouse, as follows:—

1. *Ixodes holocyclus*, Neum., the common Bush Tick. *Hab.*—near Sydney (upon man); Godwin Island, Clarence River (upon a phalanger); Queanbeyan (upon a calf).

2. *I. ornithorhynchi*, Lucas, the Platypus Tick. *Hab.*—Guntawang (from the Platypus; Mr. Alex. G. Hamilton).

3. *Amblyomma moreliæ*, L. Koch, the Snake Tick. *Hab.*—Coonamble (upon a kangaroo); Narrabri and Coonabarabran (upon horses).

4. *A. triguttatum*, C. L. Koch, the common large Cattle Tick. *Hab.*—Narrabri and Dubbo (upon cattle); Narrabri (upon dogs);

11. ***Rhipicephalus annulatus***, Say, var. ***australis***, Neum., a Queensland Cattle Tick. *Hab*.—The Island of Neui, off New Britain (upon cattle said to have been shipped originally from Cooktown to New Caledonia, and thence to Neui, according to information supplied by the sender of the specimens.)

12. ***Argas americanus***, Packard, the Fowl Tick. *Hab.*—Various localities (upon fowls).

Of the above, Nos. 8, 10, and 11 were previously undescribed; and with the exception of No. 11 all the specimens were collected in New South Wales.

Mr. C. W. Darley exhibited a portion of the timbers of a punt, showing the depredations of an Isopod which had been determined by the authorities of the Australian Museum to be the destructive "Gribble," *Limnoria lignorum*, Rathke (= *L. terebrans*, Leach), not previously recorded from Australian waters, and therefore presumably introduced from Europe or America.

Mr. Stead exhibited an ant (*Iridomyrmex purpureus*), and a piece of quartz, relatively much bulkier than the animal, which it been observed to lift bodily.

Mr. Palmer exhibited a very perfect cast skin of a snake from the Blue Mountains.

Mr. Trebeck showed a good specimen of the rattle of *Crotalus* sp., from British Columbia.

Mr. Fred. Turner sent for exhibition a specimen of the grass *Eragrostis nigra*, Nees, var. *trachycarpa*, from near Armidale, N.S.W., with the inflorescence infested with a parasitic fungus—the first instance of this valuable pasture grass in this condition known to him. (For records of twenty other species similarly affected, see Proceedings, 1897, p. 686, and 1899, p. 194.) Also a variegated form of *Kennedya rubicunda*, Vent., collected at Pennant Hills, near Sydney, in January last, which had not previously come under his notice, though the species is common and widely distributed.

The President exhibited a piece of sandstone, found lying in the open at Springwood, partially covered with an organic structure of an undetermined nature.

Mr. Fletcher exhibited five specimens (♂ 2; ♀ 3) of a *Peripatus* with fourteen pairs of walking legs, the males with white papillæ on the legs of the posterior nine pairs, from the North Island of New Zealand. The specimens were obtained by Mr. C. T. Musson near Te Aroha in the early part of last January. They will probably prove to be referable to the species for which Professor Dendy [Nature, March 8th, 1900, p. 444] has recently proposed the name *P. viridimaculatus*, founded on specimens collected at the head of Lake Te Anau in the South Island. The (spirit) specimens exhibited, however, do not in their present condition seem to show the "fifteen pairs of green spots arranged segmentally" which Dr. Dendy describes as characteristically present in the specimens from the South Island.

WEDNESDAY, MAY 30TH, 1900.

The Ordinary Monthly Meeting of the Society was held at the Linnean Hall, Ithaca Road, Elizabeth Bay, on Wednesday evening, May 30th, 1899.

The Hon. James Norton, LL.D., M.L.C., President, in the Chair.

DONATIONS.

Department of Agriculture, Brisbane—Queensland Agricultural Journal. Vol. vi. Part 5 (May, 1900). *From the Secretary for Agriculture.*

Royal Society of Queensland, Brisbane—Proceedings. Vol. xi. Pt. ii. (1895-96); Index to Vols. xii.-xiv.; Vol. xv. (1899). *From the Society.*

Australian Museum, Sydney—Memoir iii. The Atoll of Funafuti. Part [illegible] (May, 1900); Memoir iv. Scientific Results of the Trawling Expedition of H.M.C.S. "Thetis." Part 2 (May, 1900). *From the Trustees.*

Department of Agriculture, Sydney—Agricultural Gazette of New South Wales. Vol. xi. Part 5 (May, 1900). *From the Hon. the Minister for Mines and Agriculture.*

Five separates—"New Species of Australian Mosses." Parts i.-v. By V. F. Brotherus. [illegible] Helsingfors, 1890-98. *From the Author, per W. Forsyth, Esq.*

One Separate—"On Plague and its Dissemination." By Frank Tidswell, M.B., Ch.M., D.P.H. (4to. Sydney, 1900). *From the Author.*

Royal Society of New South Wales, Sydncy—Abstract, May 2nd, 1900. *From the Society.*

The Surveyor, Sydney. Vol. xiii. Nos. 4-5 (April-May, 1900). *From the Editor.*

The Wealth and Progress of New South Wales, 1898-99. Twelfth Issue (1900). *From the Government Statistician.*

Australasian Journal of Pharmacy, Melbourne. Vol. xv. No. 173 (May, 1900). *From the Editor.*

Field Naturalists' Club of Victoria, Melbourne—Victorian Naturalist. Vol. xvii. No. 1 (May, 1900). *From the Club.*

Department of Mines, Victoria: Geological Survey—Monthly Progress Report. New Series. Nos. 8-10 (November-December, 1899, and January, 1900). *From the Secretary for Mines.*

Royal Society of Victoria—Proceedings. Vol. xii. Parts i.-ii. (August, 1899-April, 1900). *From the Society.*

Zoological Society, London—Abstracts, March 20th and April 3rd, 1900. *From the Society.*

American Academy of Arts and Sciences, Boston—Proceedings. Vol. xxxv. Nos. 4-9 (Oct.-Dec., 1899). *From the Academy.*

American Geographical Society, New York — Bulletin. Vol. xxxii. No. 1 (1900). *From the Society.*

American Museum of Natural History, New York—Bulletin. Vol. xiii. Articles ii.-vii. (pp. 19-86; April, 1900). *From the Museum.*

American Naturalist (Cambridge). Vol. xxxiv. Nos. 399-400 (March-April, 1900). *From the Editor.*

Boston Society of Natural History, Boston—Proceedings. Vol. xxix. Nos. 1-8 (June-Oct., 1899). *From the Society.*

Buffalo Society of Natural Sciences, Buffalo, N.Y.—Bulletin. Vol. vi. Nos. 2-4 (1899). *From the Society.*

Chicago Entomological Society—Occasional Memoirs. Vol. i. No. 1 (March, 1900). *From the Society.*

Cincinnati Society of Natural History, Cincinnati—Journal. Index to Vol. xviii.: Vol. xix. No. 5 (Jan., 1900). *From the Society.*

Field Columbian Museum, Chicago—Zoological Series. Vol. i. Nos. 16-17 (Oct.-Nov., 1899): The Birds of Eastern North America. Part ii. Land Birds (1899). *From the Museum.*

New York Academy of Sciences—List of Members, &c. (1899). *From the Academy.*

Smithsonian Institution, Washington: U.S. National Museum—Annual Report for the Year ending June, 1897. *From the Secretary.*

U.S. Department of Agriculture, Washington:—*Division of Agrostology* Bulletin No. 21 (March, 1900). *From the Secretary for Agriculture.*

Washington Academy of Sciences—Proceedings. Vol. ii. pp. 31-40 (March, 1900). *From the Academy.*

Perak Government Gazette, Taiping. Vol. xiii. Nos. 9-12 (March-April, 1900). *From the Government Secretary.*

Gesellschaft für Erdkunde zu Berlin—Verhandlungen. Band xxvi. No. 7 (1899): Zeitschrift. Band xxxiv. No. 3 (1899). *From the Society.*

Senckenbergische Naturforschende Gesellschaft, Frankfurt a. M. —Abhandlungen. xx. Band. ii. Heft (1899); xxvi. Bd. i. Heft (1899): Bericht, 1899. *From the Society.*

Verein für Erdkunde zu Leipzig—Wissenschaftliche Veröffentlichungen. Zweiter Band (1895). *From the Society.*

Zoologischer Anzeiger, Leipzig. xxii. Band. No. 600 (Oct., 1899); xxiii. Bd. Nos. 611-613 (March-April, 1900). *From the Editor.*

K.K. Naturhistorisches Hofmuseum, Wien—Annalen. Band xiii. No. 4 (1898). *From the Museum.*

Musée National Hongrois à Budapest — Természetrajzi. Füzetek. Vol. xxii. Partes iii.-iv. (1899). *From the Museum.*

Société Belge de Microscopie, Bruxelles—Annales. Tome xxiv.

Nederlandsche Dierkundige Vereeniging, Helder—Aanwinsten van de Biblioteek, Jan.-Dec., 1899; Tijdschrift. 2de Serie. Deel vi. Afl. 3 (Dec., 1899). *From the Society.*

L'Académie Royale des Sciences et des Lettres de Danemark, Copenhague—Oversigt, 1899. No. 6; 1900. No. 1. *From the Academy.*

La Nuova Notarisia, Padova. Serie xi. (April, 1900). *From the Editor, Dr. G. B. De Toni.*

Nature. Vol. liii., Nos. 1364-1383; liv.; lv. (except No. 1428); lvi.; lvii.; lviii.; lix., Nos. 1514-1522 (Dec. 19th, 1895-Dec. 29th, 1898, except March 11, 1897). *From T. Steel, Esq., F.C.S.. F.L.S.*

A NEW BACILLUS PATHOGENIC TO FISH.

By R. Greig Smith, M.Sc., Macleay Bacteriologist.

(Plates viii.-ix.)

A few fishes which were supposed to have died of some obscure disease were received from the Fisheries Department towards the end of March. They came from Lake Illawarra, an inlet from the sea about 33 miles south from Sydney. The flesh of the fishes was of a greenish hue and somewhat congested in the neighbourhood of the main blood vessels. The vessels of the stomach and intestine were also much congested, but beyond these appearances there was nothing abnormal to be seen. When, however, portions of the various muscles and organs were examined microscopically a large rod-shaped bacterium with rounded ends was invariably found in all the fishes. In some cases the organisms had completely filled the smaller blood vessels. Portions of fluid, juice or pulp from the parts examined gave rise, when inoculated into culture media, to many bacterial colonies, among which those containing the large rod-shaped organism predominated. The organism readily produced spores, and accordingly by the nomenclature of Lehmann and Neumann and others it is a

It died towards the end of the second day. The two mullets developed reddish discoloured patches at first on an area around the point of inoculation and afterwards further forward on that side and also on the other. The scales fell off these places and the body became swollen. The smaller mullet died in 45, the larger in about 100 hours.

In the *post mortem* examination of the trevally the muscle beneath the epidermal scar was seen to be of a cream-white colour and brittle, both in appearance and texture like boiled cod roe. Portions of this when pressed between cover-glasses were found to be swarming with motile bacilli. The muscle on the side reverse to that inoculated was apparently healthy and free from bacteria. The blood vessels were congested. The peritoneal cavity contained a quantity of yellow serous fluid in which floated large white cheese-like masses. The liver was pale and mottled, the gall bladder distended, the stomach slightly congested. The kidneys, spleen and heart were apparently normal. The intestine, especially in the posterior portion, was much congested. A number of inoculations were made on sloped agar and these showed the organisms in pure culture in the heart blood, the gangrenous muscle near the site of inoculation and the apparently healthy muscle on the reverse side. The liver and peritoneal fluids gave negative results.

The smaller mullet showed very few lesions. The intestine was congested and filled with gas, the liver was dark and mottled. The other organs were apparently normal, and the muscle, although discoloured near the point of inoculation, did not show the localised gangrenous appearances of the other fishes. Pure cultures of the organism were obtained from the liver, the heart-blood and the muscle from the inoculated as well as the reverse side.

The larger mullet showed several haemorrhagic areas over the surface of the epidermis. The scales had fallen from [illegible] [illegible] patches of a purplish or grey [illegible] with [illegible] streaks. The muscles had three gangrenous spots, one at the site of inoculation, another further forward on the reverse side and near the epidermis, while a third was found deep down in the tissue near the vertebral column on the reverse side and about midway between the first two lesions. The peritoneal cavity was distended with a quantity of a whitish fluid. The intestines were swollen with gas and had the blood-vessels congested. The liver was mottled, the gall-bladder distended. The blood vessels of the stomach were slightly congested; those inside the swim bladder showed much congestion. The heart, kidneys and spleen appeared normal. The colour of the muscles generally was bluish with a pink tinge. The gangrenous portions of the muscles were cream-white and brittle as in the case of the trevally, and they swarmed with bacilli. When small portions of the heart-blood and of the three gangrenous localities were inoculated into liquefied culture media, there were obtained colonies of the bacillus and no other organism. The spleen and peritoneal fluids contained a mixture of organisms, including the bacillus. The liver was sterile.

Characteristics of the Organism.

Microscopical appearance.—The bacilli appear as large rods, with rounded ends, and measure 0·8 : 2-3·6 μ in bouillon culture. They grow singly, in pairs, and short chains in liquid media. On solid media they grow as long chains or filaments,

Relation to media, temperature and oxygen.—The organism grows quickly in the usual culture media at 20° as well as at 37°. In the absence of air, the growth is very scanty indeed.

Gelatine plate.—The colonies appear at first as white circular points, which develop into circular colonies. The medium slowly liquefies, forming a shallow, saucer-like depression, in which the colony assumes a more or less zonal or annular appearance. There may be a white centre, around which are arranged circles or zones of varying intensity of white, or the centre may be clear, the margin clear, and midway between centre and margin a broad ring, which may appear homogeneous or striped with radial bands. As liquefaction proceeds the growth gathers towards the centre and the margin, when an appearance is presented like that figured in Lehmann and Neumann's Diagnostik, i. 37, iv. This stage is reached in about 4 days, at 22° C.

When examined with a 60-fold magnification, the colonies near the surface and deep in the medium are circular, granular and brownish-black in colour, with a rough margin as if beset with short hairs. In thickly sown plates the deep colonies are irregular and generally studded with opaque root-like fibres. The colonies absolutely on the surface spread at first irregularly like a fragment of twisted or crumpled paper. The rosette-like colony then sends out straight and sinuous granular processes, which are visible as long as liquefaction has not proceeded very far. At the later stages, when the colony has assumed the annular appearance, the structure is seen to consist of tufts of short interlacing threads.

Gelatine stab.—At first there is an uncharacteristic white filiform growth along the track made by the inoculating needle. The growth widens a little more at the surface than in the depth; an air-bubble then appears at the top of the canal. Liquefaction proceeds in the upper layers of the gelatine, producing a funicular or napiform liquefied space, and at the same time a white film covers the surface of the solid or liquefied gelatine. With a napiform liquefaction the surface film remains indented round the point of inoculation. As the liquefaction proceeds downwards the original white cord like growth may persist more or less, or it may break up into floccules which slowly gravitate, leaving the upper liquefied gelatine clear, the lower layers turbid. The appearances are never like 36, i. or 36, ii., but approach 38, iv. (Lehmann and Neumann, Diagnostik, i.). These appearances are similar with 15 and 20% nutrient gelatine.

Glucose-gelatine shake.—There is no gas developed.

Agar plate.—The colonies are white and may assume various shapes. The surface colonies may be rounded with an irregularly lobed margin and a white central point where the surface growth has originated from a subsurface colony, or when the growth has always been upon the surface the centre may be dark. In cases where the agar surface is moist, the colony may send out amœboid processes which gradually extend over the whole surface after the manner of *Bact. vulgare*. The colonies just below the surface may send out processes into the agar medium. The organisms may also grow as a film between the solidified agar and the glass of the Petri dish. When examined with a 60-fold magnification the colonies are seen to be highly granular. The margins of the surface colonies are sinuous or waved, and are seen to consist of filaments The subsurface colonies are irregular in outline and beset with fibrous projections. The processes sent out by the subsurface colonies into the agar medium are seen to be composed of interlacing filaments.

Bouillon.—At first there appears a slight turbidity, which, while increasing, slowly gravitates, leaving the upper portions clear, the middle layers turbid, and a voluminous flocculent white precipitate at the bottom of the medium. A film grows upon the surface, and after a time becomes indented in places. There is no indol produced.

Potato.—A dry white layer is formed, which slowly spreads over the surface. It remains flat, and is never raised, puckered, nor wrinkled when grown at 22°. At 30° a slight wrinkling may appear. The centre becomes of a slightly brown tinge, and in the old cultures (12 days) this brown colour has spread over the expansion.

Blood serum.—A spreading grey-white layer is formed on the solidified serum. The margin of the growth is irregular and bristly when viewed with a lens. The serum is not liquefied.

Milk.—The casein is first coagulated and then dissolved.

Germination.—For observing this process an old agar culture

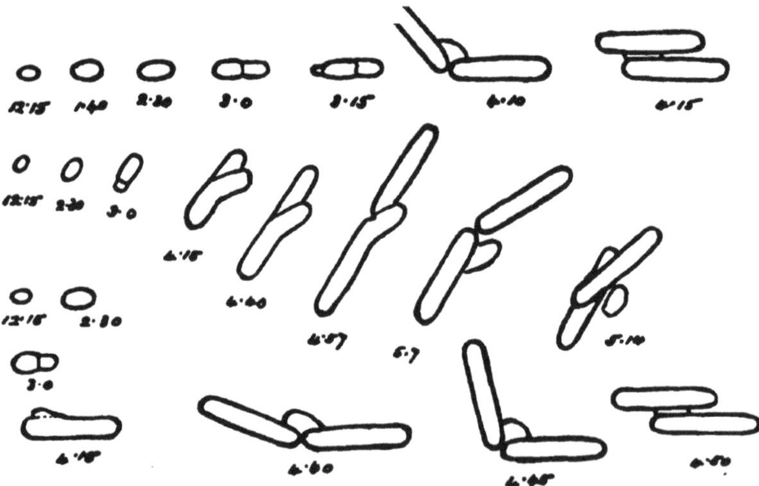

Germination of three spores in nutrient gelatine; the numbers indicate the times of observation.

was sown in gelatine and a loopful smeared over a sterile cover-glass, which was then inverted over a hollow glass slide. The highly refractile spore slowly enlarged, and at the same time became less and less refractile. The rod emerged from one of the poles and grew outwards until the protruded end was as long as the spore case. By this time the distinction between rod and spore had practically disappeared. A protrusion then appeared at the other end of the spore case, and growth proceeded in two directions. The spore case was then thrown off, and this was followed by a division of the lengthened rod into two parts. The daughter bacillus, formed from the end last to emerge, slowly bent round until it was at an angle of 45° with the other rod, when it suddenly slipped along the side of the latter. Germination occurred in 3½ hours at laboratory temperature (20°); an hour later there were two cells and in another, four bacilli. The dimensions of the mature rod after division and whilst thus imbedded in gelatine were 1·5 : 7·5 μ.

The early gelatine cultures.—The characteristics that have been described appear to be permanent, for the characters of the twelfth transfer or crop were identical with the third. The first and second crops in gelatine differed markedly from the later. The gelatine stab infected from the first agar plate produced an arborescent growth along the needle track—an appearance similar to that of anthrax. Cultures from this, however, failed to reproduce a similar appearance. There developed instead the uncharac-

are to be seen in the polar germination of the spores and the pathogenicity. The method of germination distinguishes it from the members of the *Bac. subtilis* and the *Bac. vulgatus* groups, and from the thermophile bacteria. The rosette-like folded appearance of the gelatine surface colonies closely resembles that figured by Winkler as *Tyrothrix distortus* (Centralblatt für Bakt. 2 Abt. i. 609), but differences occur in the gelatine stab, the agar stroke, and potato culture, as well as in the method of germination of the spores. The uniform breadth of the sporulated rods shows that it is not a clostridium, the members of which group are chiefly anaerobes. The aerobic character marks it off from the *Bac. œdematis* group. There remains the anthrax group, which consists of spore-forming bacteria. They are frequently united to form threads, and are stained by Gram's method of staining. The distinguishing feature of this group is the polar germination of the spores as contrasted with the equatorial germination of the closely allied *subtilis-vulgatus* group. The organism differs markedly from *Bac. anthracis* in shape, motility and growth; from *Bac. anthracoides* in motility; and from *Bac. pseudanthracis* in agar culture. The size, the colour of the liquefied gelatine, and the reaction towards blood serum show that it is not *Bac. apicum*, to which is closely allied the eel-disease bacillus of Canestrini and the poisonous fish bacillus of Fischel and Enoch. The separation of the capsule from the germinated rod differentiates it from *Bac. gracilis* and *Bac. leptosporus*. Burchard* has published diagrams of the germination of 20 species of bacilli, only one of which—viz., *Bac. bipolaris*—bears any likeness to the organism. The cultural characteristics generally are different from those which I have described. The organism, therefore, appears to be a new species, and I propose for it the name *Bacillus piscicidus bipolaris*, a title which indicates the pathogenicity and the bipolar germination of the spores.

The bacteria which have been recorded as being pathogenic to fish are few in number. Emmerich and Weibel have described a

* Arbeiten Bact. Instit. Hochschule, Karlsruhe, i., 1.

non-sporulating organism, *Bac. salmonicida*,* which is pathogenic to salmon. Wyss† found *Bac. vulgare* to be the cause of a disease in freshwater fish. Sieber‡ describes an organism under the name of *Bac. piscicidus agilis*, which was responsible for the death of fish in an aquarium. This organism differs from *Bac. piscicidus bipolaris* in its production of gas in gelatine and agar culture, its growth on potato, and its pathogenicity to certain rodents.

EXPLANATION OF PLATES.

Plate viii.

Fig. 1.—Section through stomach wall of original fish (Bream), stained with methylene-blue (× 500).

Fig. 2.—Bacilli from a 48 hours' at 20° C. agar culture, stained with methylene-blue (× 1000).

Plate ix.

Fig. 3.—Young surface colony on gelatine plate (× 50).

Fig. 4.—Various forms of the surface and sub-surface colonies in agar plate culture, 48 hours at 20° (× $\frac{7}{10}$).

AUSTRALIAN *PSELAPHIDÆ*.

By A. Raffray.

(Communicated by Arthur M. Lea.)

(Plate x.)

A good many Australian species of *Pselaphidæ* have already been published by different authors, principally the following:—

Westwood, J. O., Trans. Ent. Soc. Lond. 1856 and 1870.

King, Rev R. L., Trans. Ent. Soc. N.S.W. 1862, 1863, 1865, 1873.

Schaufuss, Dr. D., Ann. Soc. Ent. Belg., reprinted in Nunq. Otios.; Tijds. voor Ent. 1886 and 1887.

Macleay, William, Trans. Ent. Soc. N.S.W. 1871, 1873.

Sharp, Dr. D., Trans. Ent. Soc. Lond. 1874; Ent Mo. Mag. 1892.

Raffray, A., Revue d'Ent. 1883, 1887, 1890, 1898.

Blackburn, Rev. T., Trans. R. Soc. South Aust. 1889, 1891.

The species mentioned in these different works amount to about 200 species, which is already a fair number, but certainly is only a small part of the *Pselaphidæ* existing on the Australian continent.

Mr. Arthur M. Lea kindly sent me a large collection of *Pselaphidæ*, which, together with those hitherto undescribed which my collection contains, enables me to add 73 new species.

The production of a complete Catalogue of known Australian *Pselaphidæ* would prove a useful and interesting work, but as I have not seen the types of the Rev. R. L. King, such a work is practically impossible, and I am thus compelled to narrow the scope of this paper and to restrict it to descriptions of new species, with remarks on such little known and doubtful species, the authenticated types of which are in my possession.

Tribe EUPLECTINI.

Genus Paraplectus, Raffray.

Rev. d'Ent. 1898, p. 269.

This genus is closely allied to *Euplectops*, Reitter (Verh. Nat. Ver. Brünn, xx. p. 197), the type of which is *Euplectus odewahni*, King.

In addition to *P. punctulatus*, Raffr., (*loc. cit.*), from Victoria, and *P. setulosus*, Raffr., (*loc. cit.*), from Tasmania, Mr. Lea has sent me the following new species :—

Paraplectus biplagiatus, n.sp.

(Plate x., fig. 33.)

Elongatus, subparallelus et subdeplanatus, castaneus, inter strias suturalem et dorsalem in elytris basi macula oblonga nigra, totus creberrime punctatus, subopacus, minutissime subvelutino-pubescens, antennis pedibusque rufis. Caput trapezoidale, antice attenuatum, lateribus obliquis, postice retusum, medio brevissime sulcatum, inter oculos foveis duabus, sulcis obsoletis, fronte utrinque supra antennas nodoso. Oculi magni. Antennæ breves, articulis duobus primis majoribus, subovatis, 3 subconico, 4-10 monilibus, longitudine decrescentibus, 8 leviter, 9 et 10 valde

truncato. Pedes breves, femoribus anticis incrassatis et infra compressis. ♀. Long. 1·90-2·00 mm.

Swan River, W.A. (Mr. A. M. Lea).

The colour of this species is very peculiar for a Pselaphid; it bears on each elytron a large and oblong black spot close to the suture. It differs from *P. punctulatus*, Raffr., by the head being longer, with the sides more oblique, the sulci of the prothorax more shallow, and the punctures which are not so strong but are confluent.

Paraplectus infuscatus, n.sp.

Præcedenti simillimus. Elytris basi et juxta suturam tantummodo infuscatis, vix perspicue punctatus, subsericeo-pubescens. Caput magis elongatum et antice magis attenuatum. Antennarum articulis 9-10 magnis, transversis. Prothorax valde cordatus, latitudine maxima multo ante medium, lateribus antice valde rotundatis, postice sulco transverso incisis, sulcis longitudinalibus validioribus. Metasternum obsolete sulcatum; segmentis ventralibus 2, 3, 4 subæqualibus, 5 valde angusto, 6 majore, arcuatim profunde emarginato, 7 magno, rhomboidali, operculo ornato, parum notato. Femoribus anticis et intermediis incrassatis, tibiis anticis et intermediis apice intus minute calcaratis. ♂. Long. 1·70 mm.

Pinjarrah, W.A. (Mr. A. M. Lea).

This species is very closely allied to *P.* [illegible]; the head is longer, the prothorax more regularly cordate, with the sides in front more rounded and the black spot at the base of the elytra very diffuse.

Tribe TRICHONYCHINI.

Genus M[illegible], Raffray.

Rev. d'Ent. 18[illegible], p. [illegible].

The type of this genus is *[illegible]* King, which I formerly included in the genus *Trichonyx* Chaud.

Mesoplatus tuberculatus, n.sp.

(Plate x., fig. 32.)

Elongatus, parum convexus, totus castaneus, subnitidus, pedibus rufis, palpis tarsisque pallide testaceis, pube brevi, grisea, sparsa. Caput rugoso-punctatum, transversum, deplanatum, supra fere concavum, juxta oculos foveis duabus parum profundis. Antennæ validæ, articulis 1 parum elongato, 2 subquadrato, 3-8 minoribus, monilibus, 9-10 transversis, crescentibus, 11 majori, ovato. Prothorax capite plus duplo longior et paulo latior, ovatus, antice posticeque subæqualiter attenuatus, parce punctatus, sulco longitudinali integro, transverso, sinuato, foveis lateralibus magnis. Elytra latitudine sua vix longiora, subquadrata, disperse sed valde punctata, humeris obliquis, basi quadrifoveata, sulco dorsali obsoleto ad medium evanescenti. Abdomen elytris longius sat convexum, segmento 1° sequenti haud duplo longiori, basi impresso, carinulis duabus divergentibus, parum elongatis et plus quam tertiam partem disci includentibus.

♂. Trochanteribus intermediis inflatis, segmentis ventralibus 3 basi media tuberculo acuto et valido, 4-5 transversim depressis. Long. 2·10 mm.

Clarence River, N.S.W. (Mr. A. M. Lea).

The differences between this species and *M. barbatus*, King, are:

thorace latiora et paulo breviora, subquadrata, leviter convexa, humeris rotundata, basi quadrifoveata, sulco dorsali obsoleto et brevi, segmento 1° dorsali sequenti fere duplo longiori. ♀. Long. 1·60 mm.

Clarence River, N.S.W. (Mr. A. M. Lea.)

This species differs from *M. barbatus*, King, by the shape of the body, which is more elongate and more attenuate in front, and by the prothorax, which is opaque and has a dense and rugose punctuation.

MESOPLATUS MASTERSI, n.sp.

Elongatus, sat crassus, ferrugineus, nitidus, elytris rubris, antennis pedibusque rufis, tarsis testaceis, pube brunnea, brevi. Caput transversum supra deplanatum, rugoso-punctatum. Antennæ validæ, articulis 1° parum elongato, 2 ovato, 3-6 latitudine sua paulo longioribus, 7 quadrato, 8 leviter transverso, 9-10 multo majoribus, 9 subquadrato, 10 leviter transverso, 11 breviter ovato, acuminato. Prothorax ovato-cordatus, antice plus, postice minus attenuatus, lateribus pone medium leviter sinuatus, irregulariter sat dense punctatus, sulco longitudinali in disco vix perspicuo, transverso, valido et medio angulato, foveis lateralibus profundis. Elytra prothorace longiora, præsertim latiora, latitudine sua paulo longiora, humeris subobliqua, disperse punctata, basi quadrifoveata, sulco dorsali obsoleto et brevi. Abdomen leviter convexum. Segnento 1° dorsali sequenti dimidio longiori, basi transversim impresso.

♂. Femoribus anticis infra medio tuberculo minuto præditis; trochanteribus intermediis inflatis et subangulatis. Segmento 3° ventrali tuberculo brevi compresso prædito. Long. 2·00 mm.

New South Wales.

Closely allied to *M. barbatus*, King, but larger; the antennæ are thicker and the club is stronger.

I have much pleasure in naming this insect after Mr. Masters, to whom I am indebted for it.

MESOPLATUS NITIDUS, n.sp.

Elongatus, leviter convexus, læte ferrugineus, elytris disco pallidioribus, totus nitidus, antennis pedibusque rufis, palpis tarsisque testaceis, pube brevi, grisea. Caput postice sat convexum, læve, antice punctatum et transversim depressum, cum foveis quatuor minutis juxta oculos, postice foveis duabus majoribus sed parum profundis, medio juxta collum, sulco longitudinali valido. Antennæ crassæ, articulis 1° parum elongato, 2 subgloboso, 3-8 monilibus, clava parum conspicua, 9-10 transversis, parum crescentibus, 11 magno, ovato, acuminato. Prothorax elongato-ovatus, sat convexus, disperse et irregulariter punctatus, sulco longitudinali valido, integro, transverso profundo et valde sinuato, foveis lateralibus magnis et profundis. Elytra prothorace latiora et paulo longiora, latitudine sua longiora, humeris subnodosa, disco leviter convexa, basi quadrifoveata, sulco dorsali obsoleto et brevi. Abdomen elytris paulo longius, segmento 1° sequenti vix longiori, basi obsolete transversim impresso.

♂. Metasternum impressum, segmentis ventralibus 3 medio tuberculo valido, compresso, 4 depresso, 6 longitudinaliter impresso, trochanteribus intermediis leviter nodosis. Long. 1·70 mm.

Victoria.

This species will be distinguished at once by the sculpture of

about the middle of the head; prothorax always more or less longitudinally sulcate; the shoulders on the elytra are always more or less dentate or at least notched; the first dorsal segment of the abdomen is only a little larger than the following ones which are very conspicuous and not abruptly declivous, the first one bears always on the side two carinules at least at the base, the body is elongate.

Even when so restricted, the genus *Batrisus* includes a large number of species very irregularly distributed in the different parts of the world. Although very numerous in the Indo-Malayan region, it seems rather scarce in Australia.

Batrisus cyclops, King.

Trans. Ent. Soc. N.S.W. 1866, p. 306 : *giraffa*, Schfs , Soc. Ent. Bel. 1880, p. 31; Nunq. Otios. iii. p. 507.

(Plate x., fig. 28.)

Rubro-piceus, elytris antennis pedibusque rubro-ferrugineis, disperse ochraceo-setosus. Caput leviter transversum, irregulariter, grosse punctatum. Antennæ validæ, articulis cylindricis et latitudine sua fere duplo longioribus, 5 paulo longiori, 8 paulo breviori, 9 ovato-truncato, majori, 10 subtrapezoidali, majori, latitudine et longitudine subæquali, 11 breviter ovato, basi truncato, apice abrupte acuminato. Prothorax cordatus, disperse et minute tuberculosus, lateribus medio dentatus, sulco medio longitudinali subintegro et bicarinato, utrinque carinis duabus alteris sinuatis postice in spinam validam convergentibus, lateribus pone medium foveatis et basi ipsa quadrifoveolata. Elytra disperse, forte punctata, latitudine sua paulo longiora, humeris obliquis, carinatis et dentatis, sulco dorsali obsoleto et brevi. Segmento 1° dorsali cæteris haud multo longiori, lateribus bicarinato, basi transversim triimpresso, impressione media minori et carinis duabus brevibus limitata. Pedes validi.

♂. In vertice fovea magna transversa, fronte medio longitudinaliter ovato et convexo, postice supra foveam transversam breviter producto, vertice posterius cristato, ista crista medio sinuata et quadratim tuberculata, utrinque, lateribus area ovata crebre rugoso-punctata. Metasternum late impressum, segmento ultimo ventrali late impresso et aspero-punctato; trochanteribus posticis apice acute dentatis.

♀. Caput minus transversum, post oculos foveis duabus et sulco obsoleto transverso, sulcis duobus validis antice conjunctis, lateribus depressis et bicarinatis, caput longitudinaliter medio fere totum carinatum. Metasternum sulcatum. Long. 2·10-2·30 mm.

Pine Mountain, Brigham [? Brisbane.]

I have not seen any type of *B. cyclops*, King, but according to the description I have no doubt that it is the same species as *giraffa*, Schfs.

The above description is drawn up from a type of *giraffa*, Schfs.

BATRISUS URSINUS, Schaufuss.

Soc. Ent. Belg. 1880, p. 31; Nunq. Otios. iii. p. 507.

(Plate x., fig. 27).

Elongatus, gracilis, brunneus, subopacus, totus confertim punctatus et setis brevibus, depressis, ochraceis coopertus. Caput latitudine æquilongum, temporibus postice obliquis et leviter rotundatis, subrugoso-punctatum. Antennæ crassæ, validæ,

tata. Metasternum leviter depressum et fundo sulcatum. Pedes validi elongati.

♂. Capitis vertice juxta collum longitudinaliter cristato, ista crista antrorsum recurva et impressione transversa utrinque in foveam post oculos sitam desinenti, capitis disco inter oculos leviter gibboso et summo medio minute impresso, utrinque sulcis duobus obsoletis fronte connexis. Trochanteribus posticis apice breviter, recurve dentatis.

♀. Capitis crista longitudinali minori, impressione transversa obsoletissime interrupta. Long. 2·40-2·50 mm.

Wide Bay, Q.

This species, which is the very type of Dr. Schaufuss, will be recognised at once by its strong punctures and the short depressed setæ which cover the body.

Batrisus asperulus, n.sp.

(Plate x., fig. 26.)

Præcedenti valde affinis et similis, differt attamen colore dilutiori, antennis gracilioribus, et magis elongatis, capite magis quadrato, et temporibus magis rotundatis, crista longitudinali integra usque ad frontem extensa et antrorsum bifurcata; segmento 1° dorsali inter impressiones laterales, basi, medio, subgibboso: trochanteribus posticis longe et valde apice productis et recurvis. ♂. Long. 2·20 mm.

This species has been considered by Dr. Schaufuss as being the ♀ of *ursinus*. This is an undoubted mistake, as I have both sexes of each species.

Very likely found with *ursinus*.

Batrisus hirtus, Macleay.

Bryaxis hirta (?) Macleay.

Elongatus, subcylindricus, rufus, minutissime punctatus, breviter dense rufo-pubescens. Caput quadratum, vertice convexo lateribus et antice sulco profundo limitato, fronte, inter antennas, haud impresso. Antennæ paulo breviores, articulis latitudine sua vix duplo longioribus, 8 fere subquadrato, 9 brevissime ovato, 10 breviter obconico, 11 ovato, acuminato. Prothorax subovatus, lateribus simplicibus, sulco longitudinali medio integro, transverso medio interrupto, postice tuberculis duobus magnis, acutis, carinulis duabus lateralibus delicatulis et medium superantibus. Elytra sat elongata, basi attenuata, humeris obtuse carinatis, sulco dorsali obsoleto sed plicatura longitudinali medium fere attingenti limitato. Abdomen elytris longius, sat convexum, basi leviter angustatum, segmento 1° dorsali subæqualiter transversim impresso.

♂. Metasternum late impressum, trochanteribus posticis apice longe productis, isto processu compresso, apice abrupte et valde recurvo, acuto. Long. 2·70 mm.

Gayndah, Q.

I received the insect above described from Mr. Masters under the name of *Bryaxis hirta*, Macleay. I cannot say if it is the true *Bryaxis hirta*, Macleay, but it does not at all resemble a *Bryaxis*, and it is difficult to admit that Mr Macleay made such a mistake; his description, however, may apply to many insects.

BATRISUS HAMATUS, King.

Trans. Ent. Soc. N.S.W. 1863, p. 45, tab. 16, fig. 6c.

Oblongus, ferrugineus, elytris pedibusque dilutioribus, ochraceo-

elongata, ad basin attenuata, humeris obliquis et dentatis, sulco dorsali obsoleto. Abdomen elytris longius, haud angustius, segmento dorsali 1° sequentibus vix longiori, basi impressionibus tribus transversis inter se æqualibus. Metasternum sulcatum et postice inter coxas tuberculatum.

♂. Trochanteribus intermediis medio dente conspicuo, brevi, acuto armatis, posticis apice longe productis, isto processu recurvo et apice truncato; segmento ventrali ultimo deplanato, lævi. In capite medio elevatione longitudinali, cariniformi, transversa antice limitata et utrinque fovea minuta. Long. 2·60 mm.

Sydney, Parramatta, N.S.W.

I have found in the Schaufuss collection under the name *hamatus* the insect from which the above description has been made; I think that it is the true *hamatus*, King.

Batrisus speciosus, King.

Trans. Ent. Soc. N.S.W. 1863, p. 45.

Oblongus, totus rufus. Caput longitudine sua paulo latius, fere deplanatum, sulcis parum profundis, carinula media valde obsoleta. Antennarum articulis 3-7 latitudine sua plus duplo longioribus, 8 latitudine sua paulo longiori, 9 subovato, 10 trapezoidali, 11 ovato, acuminato. Prothorax sicut in *hamato*, sed paulo breviori, sulco medio et carinulis obsoletis. Elytra antrorsum et apice æqualiter attenuata, lateribus leviter rotundatis. Segmento dorsali 1° basi transversim triimpresso, impressione media lateralibus angustiori.

♂. Trochanteribus intermediis medio spina brevi armatis, posticis paulo ante apicem dente valido sed brevi compresso, recurvo et truncato præditis. Long. 2·60 mm.

Forest Reefs, N.S.W. (Mr. A. M. Lea).

B. speciosus was considered by the Rev. R. L. King as a mere variety of *hamatus*. The insect above described is at any rate different from *hamatus*; it corresponds as well as possible to the few words of description given by the Rev. R. L. King, and if it proves—as I suppose it will—to be identical with *speciosus*, King, the latter is a very good species.

Batrisus Leai, n.sp.

Totus rufus, caput latitudine sua longius carinula media longitudinali, valida, simplici, antrorsum, inter oculos, abbreviata, sulcis latis et parum profundis, fronte inter antennas late, transversim depresso. Antennæ sicut in *specioso*. Prothorax breviter cordatus, antice abrupte attenuatus, lateribus muticus, sulco medio longitudinali lato, cum carinulis lateralibus medio abrupto. Elytra latitudine sua longiora, humeris valde obliquis et notatis, sulco dorsali fere inconspicuo. Segmento dorsali 1° basi valde transversim triimpresso, impressione media paulo majori, et utrinque carinula intus recurva, valida, limitata. ♀. Long. 2·70 mm.

Tamworth, N.S.W. (Mr. A. M. Lea).

This is a species very closely allied to both *B. hamatus* and *B. speciosus;* I received it from Mr. Lea under the name of *hamatus*, and although I know only the ♀ I do not hesitate in considering it as different. The prothorax has no lateral tooth; the median sulcus and the lateral carinules are abbreviated towards the middle; the antennæ are much more elongate, the 9th joint being ovate and the 10th longer than broad; the abdomen is longer and somewhat contracted at the base, the median impression at the base of the first dorsal segment is wider than the lateral ones, (whilst it is equal in *B. hamatus* and smaller in *speciosus*) and

lateribus rotundatis, sulco dorsali obsoletissimo. Abdomen elytris paulo angustius et basi constrictum, segmento dorsali 1° basi triimpresso, impressione media paulo minori et carinulis duabus rectis.

♂. Trochanteribus intermediis apice leviter et acute productis, posticis apice longissime productis, isto processu recurvo et apice obtuso. Metasternum late impressum et fundo obsolete sulcatum. Long. 2·80 mm.

Clarence River, N.S.W. (Mr. A. M. Lea); Brigham (? Brisbane).

This species resembles very much *B. hamatus*, but the prothorax has no tooth on the sides, the elytra are somewhat shorter, the metasternum is broadly and deeply impressed with a furrow at the bottom of the impression; the intermediate trochanters instead of having a median tooth as in *B. hamatus*, are somewhat concave and produced at the apex externally.

Batrisus rugicollis, n.sp.

Oblongus, plus minusve dilute ferrugineus, elytris dilutioribus, rufo-pubescens. Caput longitudine sua paulo latius, sat deplanatum, sulcis duobus latis antrorsum in depressione frontali angulatim conjunctis, carinula media longitudinali apice bifurcata et tuberculis duobus minutis. Antennæ validæ, rugoso-punctatæ, articulis cylindricis, 3-7 latitudine sua duplo longioribus, 8 tantummodo longiori, 9 breviter ovato, 10 breviter obconico, 11 ovato, acuminato. Prothorax disperse et minute tuberculatus, cordatus, antice abrupte constrictus, lateribus parum rotundatis, sulco medio valido, fere integro, carinulis lateralibus brevissimis, tuberculis posticis obtusis, sulco transverso obsoleto. Elytra latitudine sua paulo longiora, humeris obliquis, valde notatis, sulco dorsali obsoleto. Abdomen elytris longius, basi breviter angustatum, segmento dorsali basi obsolete triimpresso, impressione media lateralibus fere dimidio angustiori. Metasternum totum deplanatum, et fundo profunde sulcatum, segmentis ventralibus 2, 3, 4 medio deplanatis, trochanteribus posticis apice longissime productis, isto processu cylindrico, gracili, apice abrupte recurvo et acuto. ♂. Long. 3·50 mm.

New South Wales (Mr. A. M. Lea).

This species is larger than the others; the longitudinal carinules of the prothorax are very short and obsolete; the antennæ are very strong and densely rugose

BATRISUS FALSUS, n.sp.

Oblongus, totus rufus, sat dense fulvo-pubescens. Caput latitudine et longitudine æquale, vertice subconvexo, delicatule carinato, foveis duabus et sulcis parum profundis antice angulatim junctis. Antennæ mediocres, articulis 3-7 latitudine sua haud duplo longioribus, leviter ovatis, 8 brevissime ovato, 9-10 multo majoribus, brevissime obconicis, 11 ovato, acuminato. Prothorax disperse et minute tuberculosus, cordatus, antice abrupte attenuatus, lateribus vix rotundatis, minutissime dentatis, sulco medio longitudinali obsoleto, subintegro, carinulis ante medium abbreviatis, spinis posticis brevibus, acutis. Elytra disperse punctata, latitudine sua longiora, basi attenuata, humeris valde obliquis, elevatis, lateribus rotundatis, sulco dorsali obsoleto. Abdomen elytris angustius, basi haud angustatum, segmento dorsali 1° basi transversim triimpresso, impressione media lateralibus fere dimidio minori.

♂. Metasternum sulcatum; trochanteribus intermediis minute medio angulatis, posticis apice productis, isto processu parum

hardly carinate at base on the sides, is much larger than all the others together, which are generally very little conspicuous when seen from above and are abruptly declivous; the prothorax bears generally three, or at least two, longitudinal grooves.

This genus is very largely represented in the Indo-Malayan region and extends westwards as far as Africa and eastwards to Japan: it seems to be very scarce in Australia.

BATRISODES TIBIALIS, King.

Batrisus tibialis, King, Trans. Ent. Soc. N.S.W. 1863, p. 171.

(Plate x., fig. 29.)

Oblongus, ferrugineus, elytris rubris, antennis pedibusque rufis, ochraceo-pubescens. Caput grosse sed disperse punctatum, elongato-quadratum, deplanatum, lateribus leviter sinuatum, inter oculos foveis duabus mediocribus et sulcis duobus vix arcuatis et antice liberis. Antennae elongatae, articulis 3, 4, 6 latitudine sua fere triplo, 5 et 7 plus quam triplo, 8 duplo longioribus, 9-10 paulo crassioribus, leviter ovatis, 11 fere fusiformi. Prothorax cum punctis aliquot dispersis, subovatus, sulco medio longitudinali deficiente, utrinque fovea magna laterali et ante basin tuberculis duobus obtusis. Elytra latitudine sua longiora, basi valde attenuata, humeris obliquis notatis. Abdomen elytris longius et paulo angustius, basi aequaliter transversim triimpressum. Metasternum sulcatum, femoribus omnibus incrassatis, trochanteribus intermediis et posticis medio obtuse angulatis, tibiis anticis irregularibus, intus lobatis, isto lobo antrorsum acuto, supra sulcatis et fasciculatis, intermediis apice intus calcaratis. ♂ Long. 2·30 mm.

Clarence River, N.S.W. (Mr. A. M. Lea).

I received the insect above described from Mr. Lea under the name of *B. tibialis*, King, and it answers exactly to the description. The dilatation and emargination of the fore tibiae are very remarkable.

I do not know the ♀

Batrisodes Mastersi, n.sp.

(Plate x., fig. 30.)

Oblongus, capite abdomineque plus minusve piceis, elytris et prothorace obscure rubris, antennis pedibusque dilute ferrugineis, sat dense ochraceo-pubescens, totus præsertim capite et elytris grosse punctatus. Caput quadratum, inter oculos foveis duabus magnis et sulcis duobus subrectis antice cum sulco transverso subquadratim uno junctis. Antennæ irregulares, articulis 1° valido, 2° subcylindrico, latitudine sua paulo longiori, 3° breviter obconico, 4 transverso, 5 maximo, irregulariter trapezoidali et intus obtuse producto, 6 magno, irregulariter ovato, intus medio leviter angulato, 7 obconico et latitudine sua paulo longiori, 8 simili, attamen paulo breviori, 9 oblongo-ovato, præcedenti paulo crassiori et plus duplo longiori, 10 subconico, nono paululum breviori, 11 breviter fusiformi. Prothorax capite (cum oculis) fere latior, cordatus, lateribus rotundato-lobatis, antice sat abrupte angustatus, postice sinuatus, disco leviter gibboso, sulcis lateralibus intus valde incurvis, sulco medio valido, juxta basin sulco transverso. Elytra subquadrata, humeris valde obliquis et acute dentatis, sulco dorsali ad medium evanescenti. Abdomen elytris

. haud angustius segmento do

Tribe BRYAXINI.

Genus Batraxys, Reitter.

Verh. Zool. Bot. Ges. Wien, 1881, p. 464; *Batrisomorpha*, Raffray, Rev. d'Ent. 1882, p. 38.

This genus, which is specially abundant in the Indo-Malayan region, has but few representatives in Australia.

It belongs really to the tribe *Bryaxini*, having one single tarsal claw, but it has more the general appearance of the *Batrisini*.

The type of the Australian species is *Bryaxis Armitagei*, King, which Dr. Schaufuss took to be a *Batrisus*.

In the Rev. d'Ent., 1882, I established the genus *Batrisomorpha* for some new species of the Indo-Malayan region and of New Guinea, and I included in it *Bryaxis Armitagei*, King, but Mr. Reitter had created previously the genus *Batraxis* for a European insect from Greece which proved to be generically identical, and the name *Batrisomorpha* is therefore synonymous with *Batraxis*.

The maxillary palpi are rather elongate, with the last joint fusiform; the intermediate coxæ are approximate and the posterior ones much less distant from each other than is usually the case amongst *Bryaxini*; the antennæ are strong, with the club evidently two-jointed; the prothorax is devoid of sulci, the elytra have no dorsal stria; the first dorsal segment of the abdomen is very large, more or less, but always briefly carinate on the sides at the base and without lateral margin.

Batraxys Armitagei, King.

Trans. Ent. Soc. N.S.W. i. 1864, p. 104, tab. 7, fig. 15.

Elongata, castanea, minute subrugosa, pube brevissima, depressa. Caput leviter transversum, temporibus rotundatis, fronte transversim valde sulcato, utrinque fere ante oculos fovea mediocri a sulco transverso parum remota. Antennæ validæ, crassæ, articulis 3-9 cylindricis, inter se subæqualibus, 10 majori, transverso, 11 magno, subtriangulari et extus oblique sulcato. Prothorax capite vix latior sed longior, convexus, latitudine maxima anteriori, lateribus leviter obliquis, juxta basin medio fovea punctiformi. Elytra subquadrata, convexa, lateribus leviter rotundata, humeris rotundatis sed leviter callosis, basi bifoveata, stria suturali integra, dorsali nulla. Abdomen elytris longius, postice declive, segmento 1° dorsali maximo, cæteris supra visis fere inconspicuis, basi angustato et utrinque lateribus breviter carinato, carinulis duabus alteris validis parallelis, quartam partem disci vix attingentibus, plus quam tertiam includentibus, inter eas disco vix depresso. Metasternum angustatum, longitudinaliter depressum et utrinque delicatule carinatum. Pedes validi, elongati, femoribus omnibus medio incrassatis.

♂. A fœmina differt metasterno angustiori, magis depresso et fortius carinato. Long. 1·80-1·90 mm.

Sydney, Parramatta, N.S.W.

This species will be at once distinguished by its very short and depressed pubescence, whilst the other species are generally entirely glabrous and shining.

BATRAXYS LÆVIGATA, n.sp.

Præcedenti valde affinis sed brevior, colore dilutiori, pube minutissima et vix conspicua, antennis crassioribus et brevioribus, elytris latitudine sua paulo longioribus; abdomine breviori, basi minus angustato, lateribus usque ad medium carinato, carinis mediis brevissimis tertiam partem disci, inter eas de ressi

latitudine sua paulo longiora, humeris obliquis et subcallosis. Abdomen elytris paulo angustius et leviter brevius, basi haud angustatum, lateribus longius carinatis, segmento 1° breviori, fere transverso, carinulis mediis leviter intus arcuatis et fere convergentibus et mediam partem disci includentibus, disco inter eas magis depresso.

♂. Metasternum longitudinaliter impressum, utrinque delicatule carinatum et juxta apicem minute foveolatum. Long. 1·80 mm.

Forest Reefs, N.S.W. (Mr. A. M. Lea).

Genus Rybaxis, Saulcy.

Spec. ii. p. 96; Raffray, Rev. d'Ent. 1890, pp. 118 and 123.

The name of *Rybaxis* was given by de Saulcy as a subgenus of *Bryaxis* to some insects of the palæarctic fauna which must be considered as generically distinct.

The prothorax bears always a transverse furrow more or less angulate in the middle and joining together the lateral and basal foveæ when the latter one exists; the elytra have always a well marked dorsal stria and a furrow on the deflexed side of the elytron, close to the lateral margin.

This genus proves to be extensively distributed, being met with in every part of the world. It is particularly numerous in Australia, while the genera *Bryaxis* and *Reichenbachia* have not been found there, as far as I know.

The Australian species of *Rybaxis* may be easily divided into two groups: in the first one the transverse furrow of the prothorax is well defined and the median fovea very small and even sometimes wanting, at the base of the elytra there are only two foveæ and the general form of the body is short, broad and convex; in the second group the transverse furrow of the prothorax is very faint and seems even interrupted, the median fovea is very strong and generally somewhat transverse, at the base of the elytra are generally four foveæ, the body is more elongate, parallel and depressed, the prothorax more decidedly cordate.

I should have been inclined to consider this second group as a distinct genus were it not for the existence of a species which is really intermediate and forms a transition from the first to the second group.

First Group.

RYBAXIS HYALINA, Schaufuss.

Nunq. Otios. iii. p. 502; *hyalinipennis*, Schaufuss, *ibid.* iii. p. 501.

(Plate x., fig. 41.)

Sat crassa, castanea vel furruginea, minute pubescens. Caput latitudine sua multo longius antrorsum attenuatum, inter oculos foveis duabus validis, fronte medio depresso et minute unifoveato. Antennæ validæ, articulis duobus primis validioribus, 3, 4 latitudine sua dimidio longioribus, 5 paulo longiori, 6 tertio simili, 7 paulo breviori, 8 quadrato, cæteris in utroque sexu variabilibus. Prothorax haud punctatus, longitudine sua paulo latior, breviter cordatus, foveis lateralibus magnis a margine distantibus, sulco transverso medio angulato, fovea media minutissima. Elytra obsoletissime et disperse subrugosula, parum elongata, et basi leviter attenuata, humeris subnodosis, stria dorsali ante apicem terminata, introrsum nonnihil arcuata et apice extrorsum perparum recurva. Abdomen breve, apice obtusum, segmento 1°

♀. Antennarum articulis 9-10 obconicis, crescentibus, simplicibus, 11 ovato, basi truncato. Metasternum medio longitudinaliter late sed parum profunde sulcatum et utrinque convexum, segmento ventrali ultimo obsoletissime biimpresso. Femoribus posticis supra, basi, perparum emarginatis. Long. 1·80-2·00 mm.

This species will be easily distinguished in the ♂ by the shape of the last three joints of the antennæ and the emargination of the posterior femora which make them appear abruptly constricted between the base and the middle; such a conformation is still noticeable in the ♀, but in a lesser degree. The development of the last joints of the antennæ and of the emargination of the posterior femora varies.

From a comparison of the types, I cannot see the slightest difference between *R. hyalina*, Schfs., and *R. hyalinipennis*, Schfs.; those two species are certainly synonymous.

I think the species referred by the Rev. T. Blackburn to *R. hyalina*, Schfs., (Trans. R. Soc. S. Austr. 1891, p. 79) is a very different one on account of the coarse punctuation on the elytra.

R. hyalina, Schfs., seems to have a wide range in Australia; I have it from Eastern Creek and Clyde River, N.S.W.; and Tasmania; I received it from Mr. Masters from Gayndah, Q ; from Mr. Lea from Upper Ord River, E. Kimberley, W A.; the specimens from the last locality are more developed.

RYBAXIS ISIDORÆ, Schaufuss.

Nunq. Otios. iii. p. 500; *Harti*, Blackburn, Trans. R. Soc. S. Austr. 1991, p. 78.

Dr. Schaufuss' type is a unique specimen and a female; but the insect does not seem very rare. I have both sexes from Sydney, Melbourne, and Victoria. The accurate description of the Rev. T. Blackburn (*loc. cit.*) and the peculiarity of the intermediate tibiæ of the ♂ leave no doubt as to the identity of *Isidoræ*, Schfs., and *Harti*, Blackb. The Rev. T. Blackburn has omitted, however, to mention the following points: the carinules at the base of the first dorsal segment of the abdomen are very short, a little divergent and very approximate to each other; the dorsal stria on the elytra is nearly straight, very little curved outside at the end and terminates somewhat far from the posterior margin. In the ♂ the intermediate trochanters have a strong and little curved but short tooth exactly at the base.

RYBAXIS FLAVIPES, Schaufuss.

Nunq. Otios. iii. p. 502.

Sat crassa, ferruginea, elytra, sutura et margine apicali exceptis dilutioribus, pedibus et antennis rufo-testaceis, istarum articulis ultimis plus minusve infuscatis, sublente, minute pubescens. Caput latitudine sua longius, antice leviter sed abrupte coarctatum, supra antennas subnodosum et fronte media fovea suboblonga, valida, inter oculos foveis duabus paulo majoribus. Antennæ validæ, articulis 1° subcylindrico, 2 ovato, 3 obconico, 4-7 latitudine sua dimidio longioribus, 7° et præsertim 5° cæteris perparum longioribus, 8 quadrato, cæteris in utroque sexu diversis. Prothorax longitudine sua et capite latior, antice plus, postice minus attenuatus, lateribus rotundatus et latitudine maxima anteriori, foveis lateralibus magnis, a margine distantibus, sulco transverso vix angulato, fovea media minuta. Elytra obsoletissime et disperse subrugoso-punctato, latitudine sua paulo longiora, lateribus vix rotundatis, basi parum attenuata, humeris subno-

♀. **Antennarum articulis 9 octo vix dimidio longiori et paulo crassiori, 10 multo majori, obconico-truncato, latitudine sua paulo longiori, 11 ovato, basi truncato. Elytrorum margine postica recte truncata. Metasternum late sed parum profunde sulcatum. Femoribus omnibus leviter incrassatis. Long. 2·00-2·10 mm.**

I have the types of Dr. Schaufuss (♂♀) from Sydney and Clyde River, N.S.W.; and I have received it from Mr. Lea from Tweed River, N.S.W., under the name of *hortensis*, King. I do not think, however, that it is *hortensis*, as the Rev. R. L. King would have certainly mentioned the swollen intermediate femora of the ♂ if he had had a specimen. This sex will be at once recognised, but I confess that I can hardly find any difference between the females of *flavipes*, Schfs., and what I consider as being *hortensis*, King. It is not uncommon to find that in *Bryaxis*, *Rybaxis* and *Reichenbachia* it is practically impossible to discriminate the ♀ of two closely allied species of which the ♂'s are otherwise very different.

RYBAXIS HORTENSIS, King.

Trans. Ent. Soc. N.S.W. i. 1863, p. 47.

I do not possess any authenticated specimen of this species, and the description of the Rev. R. L. King may apply to the ♀ of several species; but I have received from Mr. Lea a good number of specimens which seem to me to be the true *R. hortensis*, King.

The ♀ is practically similar to the ♀ of *R. flavipes*, Schfs.; but the antennæ are more slender, more infuscated at the apex, and the 9th joint is more elongate.

The ♂ has the anterior and intermediate tibiæ thicker than in the ♀, but not swollen as in *flavipes*; the last three joints of the antennæ are not so large: the posterior margin of the elytra is straight: the metasternum has a broad and longitudinal furrow, but is not prominent between the posterior coxæ as is the case in *R. flavipes*; the ventral segments are not flattened; the intermediate tibiæ are not incised on the upper face, and the spur is very different: it consists in a long spine obliquely inserted inside, at a certain distance from the apex; this spur may vary to a certain extent. In New South Wales specimens from Tamworth, Tweed River, Clarence River and Windsor it is thick at the top and bears some short setæ. In other specimens from Bunbury and Swan River, W.A., the spur is thick but sharpened at the top, and the antennæ are more largely infuscated. I have a ♀ from Cape York, N.Q., which hardly differs except that the antennæ are but very little infuscated and the median fovea of the prothorax is hardly visible. It is quite possible that *R. flavipes*, Schfs., may be a variety (♂) of *hortensis*, King, larger and more developed; such is the case for the European species *R. sanguinea*, Linn., and its variety *R. laminata*, Mots.; I have not at my disposal a sufficient number of *flavipes* and *hortensis* from different localities to settle the question.

RYBAXIS RECTA, Sharp.

♀. *Bryaxis recta*, Sharp, Trans. Ent. Soc. Lond. 1874, p. 496; ♂. *B. bison*, Schfs., Nunq. Otios. iii. 1880, p. 499.

(Plate x., fig. 42.)

I have the types of both Dr. D. Sharp and Dr. W. Schaufuss, and there cannot be any doubt as to these two species being synonymous. The striking characters common to both sexes are: the colouration of the antennæ which is ferrugineous from the

truncato, leviter infuscato, 2 leviter infuscato, transverso, 3-4 latitudine sua paulo longioribus, 5 paulo longiori et præsertim crassiori, 6 quarto simili, 7 eadem longitudine, intus paulo producto, 8 subtransverso, obtuse producto, 9 et 10 majoribus, trapezoidalibus, 9 leviter, 10 magis transversis piceis, 11 breviter ovato, testaceo, basi truncato, apice acuminato. Femoribus anticis paulo et intermediis magis incrassatis, posticis supra basi leviter constrictis et medio paululum inflatis; trochanteribus anticis basi dente valido acuto, intermediis medio dente valido, compresso, apice obtuso præditis. Metasternum longitudinaliter late impressum. Long. 1·80-2·40 mm.

N.W. Australia; King George's Sound, Champion Bay, Bunbury and Swan River, W.A.

Rybaxis antilope, n.sp.

(Plate x., fig. 43.)

I have one ♂ only of this species which is similar to *R. recta*, Sharp, as far as colouration of the body and the antennæ, size, general form, dorsal stria of the elytra and carinules of the abdomen are concerned, but the sexual characters of the ♂ are very different.

♂. Caput quadratum, depressum, nitidum, inter oculos foveis duabus magnis, fronte medio depresso et minute bifoveolato; epistomate magno, rhomboidaliter bicarinato. Antennæ crassæ, articulis duobus primis maximis, 1 basi constricto, latere interno obtuse dilatato, 2 quadrato, 3 obconico, longitudine sua parum longiori, 4 eadem longitudine, ovato, 5 paulo longiori et præsertim crassiori, 6 quarto simili, 7 quadrato, 8 leviter transverso, 9, 10 piceis, inter se subæqualibus, trapezoidalibus et subtransversis, 11 testaceo, ovato, basi truncato. Femoribus omnibus incrassatis, intermediis tumefactis, posticis supra, basi constrictis; trochanteribus anticis basi obtuse dentatis, intermediis compressis; tibiis anticis et intermediis intus pone medium dente valido, brevi, acuto præditis. Metasternum breve, late subtriangulare, valde impressum. Long. 2·10 mm.

I am indebted for this insect to Mr. G. Lewis; it is labelled Australia without any other locality.

RYBAXIS GRANDIS, n.sp.

Sat crassa, rufo-castanea, lævis, parcissime pubescens, capite et antennarum articulis penultimis infuscatis. Caput latitudine sua multo longius et antrorsum vix attenuatum, sed utrinque lateribus leviter incisum, fronte medio impressione magna et inter oculos foveis duabus validis. Antennæ mediocres, articulis 1° cylindrico, 2 ovato, 3-7 oblongis, 8 quadrato, 9 obconico, majori, latitudine sua paulo longiori, 10 majori, trapezoidali, latitudine et longitudine subæquali, 11 ovato, basi truncato et apice acuminato, testaceo. Prothorax capite multo et longitudine sua paulo latior, antice plus et postice minus attenuatus, sulco transverso medio obtuse angulato, fovea media inconspicua, foveis lateralibus magnis. Elytra magna, basi leviter attenuata, humeris notatis, stria dorsali recta, quartam posticam partem disci attingenti, margine postica leviter rotundata. Abdomen breve, basi leviter angustatum. Abdominis segmento 1 dorsali basi minute impresso et fasciculato, carinulis deficientibus. Metasternum convexum obsolete sulcatum; segmento ultimo ventrali biimpresso. ♀. Long. 2·20 mm.

Australia (locality unknown).

oblongis, 7 paulo breviori, 8 subquadrato, 9 paulo majori, breviter obconico, 10 majori, trapezoidali, leviter transverso, infuscato, 11 ovato, testaceo. Prothorax cordatus, sulco transverso medio vix angulato, fovea media nulla, lateralibus magnis. Elytra subelongata, ad basin attenuata, humeris obliquis, parum notatis, stria dorsali recta, ante apicem abbreviata. Abdominis segmento 1° dorsali basi, medio impresso et setoso, absque carinulis. Metasternum late impressum. Pedes elongati, tibiis subrectis, ad apicem incrassatis.

♂. Metasternum multo magis impressum, antennarum articulis paulo majoribus; femoribus anticis et intermediis magis incrassatis. Long. 1·60-1·90 mm.

Sydney, Clarence River, N.S.W.

RYBAXIS ATRICEPS, W. Macleay.

Trans. Ent. Soc. N.S.W. 1873, ii. p. 152.

Oblonga, sat convexa, obscure ferruginea, elytris dilutioribus, capite interdum piceo, sublente brevissime pubescens. Caput latitudine sua multo longius, antrorsum leviter attenuatum, trifoveatum, fovea anteriori sulciformi. Antennæ validæ, elongatæ, articulis duobus primis majoribus, 2 ovato sequenti breviori, 3-6 elongatis, (5 paulo longiori), 7 sexto paulo breviori, 8 quinto dimidio breviori, 9-10 suboblongo-ovatis, crescentibus, 11 subfusiformi, intus leviter emarginato. Prothorax subovatus et latitudine sua paulo longior, capite vix latior, sulco transverso medio rotundatim angulato, fovea media nulla, foveis lateralibus magnis. Elytra subelongata, basi parum attenuata, humeris notatis, stria dorsali subrecta, ante apicem desinenti. Abdomen elytris brevius, segmento dorsali 1° basi medio carinulis duabus brevibus, approximatis et valde divergentibus, inter eas impresso. Pedes elongati, femoribus intermediis et anticis leviter incrassatis; tibiis omnibus rectis et ad apicem incrassatis. Metasternum convexum. Segmentis ventralibus basalibus abbreviatis et valde convexis, ultimo transversim valde impresso. ♀. Long. 2·10-2·50 mm.

Gayndah, Q.; Clarence River, N.S.W.

This species bears some analogy to *R. insignis*, King, but it is larger, the colouration is very different, and the antennæ are much longer.

I received this insect from Mr. Lea under the name of *atriceps*, Macleay, and it agrees with the description.

RYBAXIS ADUMBRATA, n.sp.

Oblonga, sat convexa, nigro-picea, pedibus rufo-castaneis, elytris rubro-castaneis, basi et sutura plus minusve infuscatis, antennis ferrugineis, articulo ultimo testaceo. Caput latitudine sua multo longius, antrorsum leviter attenuatum, valde trifoveatum, fovea anteriori sulciformi. Antennæ validæ, articulis 1° cylindrico, 2 cylindrico, minori, 3-7 oblongis, (5 paulo longiori, 7 breviori), 8 quadrato, 9 majori, obconico, latitudine sua paulo longiori, 10 magno, trapezoidali, leviter transverso, 11 ovato. Prothorax cordatus, capite multo latior, latitudine et longitudine subæqualis, sulco transverso, medio obtuse angulato, foveis lateralibus magnis, media minuta sed conspicua. Elytra subelongata, ad basin leviter attenuata, humeris notatis, margine apicali utrinque medio rotundato et membranaceo-dilatato, stria dorsali subrecta et ante apicem desinenti. Abdomen magnum, convexum, segmento 1° dorsali magno, carinulis duabus brevibus, divergentibus, tertiam

♀. Metasternum leviter depressum, utrinque longitudinaliter convexum. Segmentis ventralibus basalibus abbreviatis et convexis, ultimo valde transversim biimpresso. Tibiis anticis intus ante apicem leviter emarginatis. Long. 2·20-2·50 mm.

Clarence River, Tweed River, N.S.W. (Mr. A. M. Lea).

This fine species resembles somewhat *R. insignis* and *R. atriceps*, but is larger, and its peculiar colouration leads to its identification.

It is worthy of note that in some Australian species of *Rybaxis* the female bears on the last ventral segments impressions which are generally a characteristic of the male, but at the same time the ventral segments are very convex, a character which never occurs in the male. *R. adumbrata*, both sexes of which are known, proves undoubtedly that such impressions belong to the female.

Dr. Sharp (Trans. Ent. Soc. Lond. 1874, p. 496) has been certainly misled by those impressions when he considers that his type of *optata* was a male; it is certainly a female.

Second Group.

This group includes certainly *R. lunatica*, King, and *R. electrica*, King; unfortunately I have not seen any authentic specimen of these species. However, I will give the descriptions of what I consider as being very likely *R. lunatica* and *R. electrica*, which I found in Dr. Schaufuss' collection under those respective names. I will add a new species which forms a transition between the first and second groups.

RYBAXIS QUINQUEFOVEOLATA, n.sp.

Oblonga, subconvexa, minus parallela, plus minusve obscure ferruginea, elytris paulo dilutioribus, sat dense et longe brunneo-pubescens. Caput subquadratum, quinquefoveolatum, foveis tribus anticis minoribus et fronte medio declivi. Antennæ validæ, articulis duobus primis majoribus, 3-5 subelongatis et inter se æqualibus, 6 paululum breviori, 7 breviori, 8 latitudine sua tantummodo longiori, 9 paulo majori, obconico, latitudine sua paulo longiori, 10 multo majori, trapezoidali, vix transverso, 11 magno, ovato, acuminato. Prothorax capite multo et longitudine sua paulo latior, valde cordatus, latitudine maxima ante medium, postice sinuato-angustatus, foveis tribus subæqualibus, sulco transverso obsoleto sed conspicuo. Elytra parum elongata, ad basin attenuata, humeris subrotundatis sed valde notatis, basi quadrifoveata, stria dorsali minus arcuata, ad angulum suturalem procul a margine desinenti. Segmenti primi dorsalis carinulis duabus validis, leviter divergentibus, mediam partem disci attingentibus et tertiam includentibus. Pedes validi.

♂. Antennæ paulo longiores. Metasternum late impressum et fundo sulcatum. Segmento ultimo ventrali minute impresso. Femoribus præsertim anticis et intermediis inflatis; tibiis anticis apice obtuse calcaratis; trochanteribus intermediis basi obtuse et minute dentatis.

♀. Metasternum parum profunde sulcatum. Long. 2·20-2·30mm.

Forest Reefs, N.S.W. (Mr. A. M. Lea).

This species is much larger than the following ones; the body is more convex and much less parallel.

RYBAXIS LUNATICA, (?) King.

Trans. Ent. Soc. N.S.W. i. 1863, p. 48, tab. 16, fig. 8b.

Oblonga, depressa, obscure castanea, elytris dilutioribus,

dorsali arcuata ad angulum suturalem, paulo ante apicem, desinenti. Segmento primo dorsali magno, basi triimpresso et fasciculato, impressione media transversa, carinulis duabus leviter divergentibus quartam partem disci includentibus et attingentibus. Metasternum totum profunde sulcatum; segmento ultimo ventrali toto late transversim impresso. Pedes mediocres, femoribus anticis et intermediis leviter inflatis. ♂. Long. 1·80 mm.

New South Wales.

RYBAXIS ELECTRICA, (?) King.

Trans. Ent. Soc. N.S.W. i. 1863, p. 48, tab. 16, fig. 9b.

Oblonga, subdepressa, parallela, castanea vel rufo-castanea, vix perspicue pubescens. Caput latitudine sua longius, antrorsum haud attenuatum, quinquefoveatum. Antennæ mediocres, articulis duobus primis majoribus, 3-6 subelongatis, 5 paulo longiori, 7 latitudine sua tantummodo longiori, 8 subtransverso, 9 paulo majori, subtransverso, 10 multo majori, trapezoidali, transverso, 11 ovato acuminato, præcedenti haud latiori. Prothorax capite multo et longitudine sua vix latior, valde cordatus, lateribus medio rotundatis, dein ad apicem angustatus, foveis tribus subæqualibus, sulco transverso delicatulo sed conspicuo. Elytra elongata, ad basin leviter attenuata, humeris obliquis et notatis, basi foveis quatuor, stria dorsali arcuata, ad angulum suturalem, paulo ante apicem, desinenti, margine postico leviter sinuato, interdum minutissime punctulata. Segmento 1° dorsali magno, carinulis duabus longis et valde divergentibus, inter se basi quintam partem disci includentibus. Metasternum totum profunde sulcatum.

♂. Metasternum latius et profundius sulcatum; trochanteribus intermediis basi dente minuto, recurvo et obtuso armatis. Long. 1·30-1·40 mm.

Australia (without locality); Swan River, W.A.; Tasmania.

This species is smaller than the preceding one, the antennæ are a little shorter, the pubescence is hardly visible, the prothorax is much more rounded in the middle and more abruptly narrowed before the base; the carinules of the first dorsal segment are longer and more approximate.

I have two specimens from Swan River which are much darker, with a fine but subrugose punctuation on the elytra. I do not think that they can be considered as a distinct species.

Genus BRIARA, Reitter.

Verh. Naturf. Ver. Brünn, xx. p. 90; *Gonatocerus*, Schfs., Nunq. Otios. p 506.

This genus is closely allied to both *Rhybaxis* and *Eupines*, and differs from them in the following points: the body is more elongate and more parallel, the prothorax bears only two lateral foveæ. the median one is wanting, the elytra have a decided dorsal stria, the first dorsal segment is large and bears two approximate and parallel carinules, the mesosternum is strongly carinate between the intermediate coxæ, which are not quite approximate; the first antennal joint is always longer than the second one, more especially in the male in which it is more or less irregular and toothed, in the male sex the forehead and epistoma are more or less produced and armed.

This genus is peculiar to Australia.

The name of *Gonatocerus*, Schfs., being preoccupied, has been changed by Reitter to *Briara*.

(Plate x., fig. 34.)

Oblongus, sat crassus, totus ferrugineus, pedibus antennisque rufis. Caput magnum, transversum, antrorsum leviter attenuatum, inter oculos foveis duabus magnis et in fronte duabus multo minoribus et subapproximatis, fronte valde retuso. Antennarum articulis 2 subovato, 3 et 5 latitudine sua duplo, 4, 6, 7 tantummodo paulo longioribus, 8 subgloboso, 9-10 multo majoribus et crescentibus, globosis, 11 ovato, acuminato. Prothorax cordatus, capite paulo angustior, foveis lateralibus validis. Elytra latitudine sua longiora, lateribus leviter rotundata, humeris subrotundatis sed notatis, stria dorsali subrecta fere integra. Segmenti primi abdominis dorsalis carinulis duabus parallelis, parum distantibus et brevibus. Metasternum convexum; segmento ventrali 2° (primo conspicuo) subtiliter et longitudinaliter fere toto carinato. Pedes validi elongati, tibiis leviter ad apicem incrassatis

♂. Fronte transversim producto, epistomate medio minute tuberculato et trifasciculato. Antennarum articulis 1° valde elongato, quatuor sequentibus simul sumptis longitudine fere æquali, leviter sinuato et angulo apicali interno leviter producto, clava majori. Metasternum apice profunde impressum; trochanteribus intermediis medio minute et obtuse tuberculatis.

♀. Fronte mutico, simpliciter retuso; antennarum articulo 1° parum elongato, duobus sequentibus simul sumptis haud longiori, subcylindrico, æquali; metasternum vix apice impressum. Long. 1·30-1·60 mm.

Clyde River and Clarence River, N.S.W.

BRIARA FRONTALIS, n.sp.

(Plate x., fig. 31.)

Subelongata, et subparallela, ferruginea, antennis pedibusque rufis. Caput leviter transversum et antrorsum attenuatum, retusum, inter oculos foveis duabus magnis, fronte utrinque juxta latera foveola punctiformi. Antennæ validæ sicut in specie præcedenti, articulis nono attamen minori, isto et decimo leviter transversis. Prothorax latitudine sua longior, capite paulo latior, antice posticeque constrictus, lateribus medium versus rotundatus. Elytra latitudine sua multo longiora, lateribus vix rotundata, humeris obliquis, parum notatis, stria dorsali subintegra, leviter sinuata, carinulis abdominalibus parallelis, parum elongatis et quartam partem disci includentibus. Segmenti primi ventralis carinula paulo breviori. Pedes validi, elongati, tibiis apicem versus incrassatis.

♂. Fronte quadratim valde producto, isto processu medio impresso et summo sericeo, epistomate utrinque obtuse et medio acute producto; antennarum articulis 1° tribus sequentibus simul sumptis longitudine fere æquali, crasso, subcylindrico, insuper leviter sinuato et intus apicem versus sulcato, clava majori; metasternum fere totum valde impressum; trochanteribus intermediis medio obtuse angulatis.

♀. Fronte simplici, retuso, antennarum articulo 1° secundo vix duplo longiori; metasternum subconvexum apice leviter transversim impressum. Long. 1·4-1·60 mm.

Windsor and Clarence River, N.S.W. (Mr. A. M. Lea).

This species differs from *B. basalis* by the more elongate and more parallel body; the processus of the head is different, the first joint of the antennæ is shorter and thicker.

the metasternum is convex, and bears at the apex a moderate ovate impression, the margins of which are somewhat carinate. Long. 1·60 mm.

Rockhampton, Q.

BRIARA CAPITATA.

B. basalis, Schfs.,‡ nec King; Tijds. voor Ent. xxix. p. 281.

(Plate x., fig. 35.)

Subovata, crassa, ferruginea, antennis pedibusque rubro-rufis. Caput in utroque sexu variabile. Antennæ breves, crassæ, articulis 2° ovato, 3 et 5 latitudine sua paulo longioribus, 4, 6 et 7 fere quadratis, 8 subtransverso, 9-10 valde crescentibus, transversis, 11 breviter ovato. Prothorax capite latior et latitudine sua vix longior, lateribus rotundatus, antice posticeque valde angustatus. Elytra latitudine sua paulo longiora, lateribus leviter rotundata, humeris subrotundatis sed notatis, stria dorsali subintegra et subrecta. Carinulis abdominis parallelis et brevibus, parum distantibus. Segmento 2° ventrali (1° conspicuo) fere toto carinato. Pedes validi, elongati, tibiis apicem versus leviter incrassatis.

♂. Caput vix transversum, antrorsum attenuatum, longitudinaliter gibbosulum, fronte subrotundato, retuso, infra subquadratim producto, utrinque ante oculos foveis duabus maximis, oblongis; antennarum articulis 1° valido, intus leviter bisinuato, infra impresso; metasternum convexum, apice impressum; tibiis anticis ad medium magis incrassatis.

♀. Caput magis transversum et antice minus attenuatum, fronte retuso, inter oculos foveis duabus magnis sed parum profundis et antice foveolis duabus minutis inter se parum distantibus; antennarum articulo 1° simplici, secundo vix duplo longiori; metasternum apice vix impressum. Long. 1·40 mm.

Clyde River (type, Schaufuss); Clarence River, N.S.W. (Mr. A. M. Lea).

This species is smaller, and especially shorter and broader than the preceding ones.

BRIARA DOMINORUM, King.

Trans. Ent. Soc. N.S.W. i. 1865, p. 173; *breviuscula*, Schfs., Nunq. Otios. iii. p. 503.

Sat crassa, prothorace et elytris obscure castaneis, capite abdomineque piceis, pedibus et antennis (articulis ultimis obscure castaneis exceptis) pallide rufis, sat dense breviter griseo-pubescens. Caput subtransversum, antrorsum attenuatum, lateribus obliquis, fronte truncato, insuper minute transversim biimpresso et inter oculos foveis duabus magnis. Antennæ parum elongatæ, articulis 1° valido, subcylindrico, 2° paulo breviori, ovato, ambobus majoribus, 3 leviter obconico, 4, 5, 6 ovatis, 5 paulo majori et præsertim crassiori, 7 quadrato, 8 leviter transverso, 9 paulo majori, leviter transverso, 11 ovato, basi truncato. Prothorax subconvexus, capite latior, latitudine et longitudine subæqualis, antice plus et postice paulo minus attenuatus, basi recta, truncata, lateribus medio valde rotundatis, foveis lateralibus mediocribus. Elytra latitudine sua paulo longiora, subconvexa, lateribus vix rotundata, humeris rotundatis, basi foveis duabus, striis suturali integra, dorsali ante apicem attenuata et leviter bisinuata. Segmenti primi dorsalis carinulis duabus brevissimis et approximatis. Metasternum parum convexum, simplex; segmento 2 ventrali (1° conspicuo) disperse punctulato et fere toto longitudinaliter

segmento 1° dorsali brevi, carinulis vix conspicuis, valde distantibus. Mesosternum haud carinatum, coxis intermediis approximatis.

This new genus is intermediate between *Briara* and *Eupines*, but is more closely allied to the latter, from which it differs by the strong sutural and dorsal striæ of the elytra: the first joint of the antennæ in both sexes is normal and shorter than the second one. The sexual characters of the abdomen resemble much those of *Eupines*. It includes only one species.

ANARMOXYS SIMPLICIFRONS, n.sp.

Convexa, nitidissima et lævis, rufo-ferruginea, capite, abdomine et elytrorum sutura leviter infuscatis, pedibus dilute rufescentibus, setis aliquot dispersis. Caput quadratum, deplanatum, fronte retusa, insuper transversim sulcato, isto sulco medio profundiori, inter oculos foveis duabus magnis, magis inter se quam ab oculis distantibus. Antennæ elongatæ, graciles, articulis 1° brevi, 2 paulo majori, ovato, 3 subelongato, obconico, 4 paulo breviori, subcylindrico, 5 cylindrico et multo longiori, 6 quarto simili, 7 præcedenti paulo longiori sed quarto breviori, 8 quadrato, 9 paulo majori obconico, 10 majori, trapezoidali, longitudine et latitudine æquali, 11 oblongo-ovato. Prothorax subcordatus, simplex, foveis lateralibus minutis et in latere ipso sitis. Elytra magna, lateribus leviter rotundata, humeris obliquis et callosis, basi bifoveata, stria dorsali fere integra et subrecta. Segmento primo dorsali transverso, carinulis duabus minutissimis, magis inter se quam a margine distantibus. Metasternum late sulcatum. Pedes valde elongati.

♂. Antennæ magis elongatæ, metasternum latius et profundius sulcatum, utrinque obtuse carinatum; segmentis ventralibus 2° (1° conspicuo) ad apicem medio lamina transversa, subquadrata, longissime fasciculata et recta prædito, ultimo minute impresso, femoribus præsertim anticis magis incrassatis.

♀. Antennæ breviores, articulis intermediis brevioribus, 9 et 10 leviter transversis, 11 paulo breviori. Long. 1·50 mm.

Tamworth and Clarence River, N.S.W. (Mr. A. M. Lea).

Genus EUPINODA, nov.gen.

Oblonga, plus minusve antrorsum attenuata. Elytra abdomine tantummodo paulo longiora, foveis et stria dorsali deficientibus. Prothorax absque foveis et sulcis. Abdomen lateribus late marginatum. Mesosternum haud carinatum; coxis intermediis subapproximatis.

This new genus resembles much *Eupines*, the prothorax and elytra being destitute of foveæ and dorsal stria, but the body is much more elongate, the elytra compared with the abdomen much shorter, and the lateral margin of the abdomen is much thicker; the intermediate coxæ are nearly contiguous.

EUPINODA LEANA, n.sp.

(Plate x., fig. 15.)

Oblonga, antrorsum attenuata, plus minusve obscure castanea, elytris interdum rubro-castaneis, antennis pedibusque rufis, parce setosa. Caput latitudine sua longius, lateribus subrectis, angulis posticis obtusis, leviter convexum et simplex, utrinque super antennarum insertionem leviter impressum. Antennæ validæ, elongatæ, articulis 1° multo majori, subcylindrico, 2 sequentibus paulo majori, subcylindrico, cæterum in utroque sexu variabiles.

externo producto et acuto. Metasternum late concavum et fundo longitudinaliter carinatum.

♀. Antennarum 3 obconico, 4 subcylindrico, fere dimidio breviori, 5 cylindrico, duobus praecedentibus simul sumptis paulo longiori, 6 subcylindrico, latitudine sua paululum longiori, 7, 8 subquadratis, 9 paulo majori, obconico, latitudine sua vix longiori, 10 multo majori, obconico, praecedenti duplo longiori, 11 majori, ovato. Long. 1·60 mm.

Clarence River, N.S.W.

The facies of this insect is singular owing to the shape of the body, being much narrowed in front, and the elytra having oblique sides.

I received this insect from Mr. Lea under the name of *Tychus nigricollis*, King, which is a totally different insect, and belongs to the genus *Eupines*. The figure of the antennæ of *nigricollis* given by the Rev. R. L. King is quite sufficient to make certain the identity of this species.

I have much pleasure in naming this fine insect after Mr. Arthur M. Lea, who discovered it.

Eupinoda amplipes, n.sp.

E. diversicolor (♀)‡, Schfs., Nunq. Otios. iii. p. 499.

(Plate x., figs. 12-13.)

Suboblonga, nitida, lævis, rubro-castanea, capite, abdomine et antennis (articulis duobus ultimis testaceis exceptis) plus minusve infuscatis, setis aliquot longis dispersis. Caput latitudine sua longius, haud antice attenuatum nec impressum. Antennæ elongatæ, articulis duobus primis majoribus, subcylindricis, 3 obconico, 4 paulo breviori, 5 duobus præcedentibus simul sumptis paulo longiori, subcylindrico, 6 obconico, tertio longiori, 7 præcedenti longiori, 8 quadrato, 9 majori, obovato, 10 majori, trapezoidali, latitudine æquilongo, 11 ovato. Prothorax cordatus, convexus, capite multo latior et lateribus valde rotundatus. Elytra elongata, basi parum attenuata, lateribus vix rotundata. Pedes validi, femoribus crassis.

♂. Antennæ magis elongatæ. Metasternum late impressum, utrinque supra coxas posticas longitudinaliter carinatum et fasciculatum; pedium anticorum trochanteribus apice valde dentatis, femoribus supra inflatis, tibiis intus medio incrassatis et maxime dentatis, tibiis posticis ad apicem intus gradatim ampliatis et compressis, ante apicem emarginatis et valde ciliatis.

♀ Antennæ similes sed articulis omnibus brevioribus, 8, 9, 10 leviter transversis. Metasternum longitudinaliter sulcatum. Long. 1·60-1·80 mm.

King George's Sound, W.A.

This species is more regularly oblong and less attenuate in front than the preceding one.

I found in Dr. Schaufuss' collection four insects recorded under the name of *diversicolor*, one of them being labelled ♀ var. Two of those four insects answer very well to the description of *Bryaxis diversicolor*, Schfs. (♂). I consider them as being the true *diversicolor*, Schfs. (♂), and they come in the genus *Eupines*. The two other ones have been considered as the ♀ of *B. diversicolor* and described as such by Dr. Schaufuss, but they are really very distinct; one of them, considered the typical ♀, is a ♂ and the one labelled ♀ var., is really the ♀ of the preceding ♂.

They are very distinct from *diversicolor*, so much so that they do not come in the same genus, and the specimens considered by Dr.

et latitudine sua duplo longiori, 6, 7 quadratis, 8 transverso, 9 magno, subquadrato, angulo interno apicali valde producto et appendiculato, 10 minori transverso, intus infra obliquo et appendiculato, 11 ovato, basi truncato. Prothorax breviter cordatus, longitudine sua fere latior. Elytra mediocriter elongata, basi leviter attenuata, lateribus parum rotundata. Metasternum postice longitudinaliter valde impressum et utrinque tuberculo fasciculato. Segmentis ventralibus 2° (primo conspicuo) fere toto longitudinaliter delicatule carinato et postice tuberculis duobus transversis et fasciculatis, ultimo transversim impresso. Femoribus crassis, anticis inflatis; trochanteribus anticis basi minute dentatis; tibiis posticis ad apicem gradatim ampliatis, compressis, intus dense breviter ciliatis, ante apicem calcaratis et dense ciliatis. ♂. Long. 1·70 mm.

Clarence River, N.S.W. (Mr. A. M. Lea).

This species is stouter than *E. amplipes*.

Genus Eupines, King.

Trans. Ent. Soc. N.S.W. 1866, p. 310; *Patranus*, Raffray, Rev. d'Ent. 1890, p. 118; *Brabaxys*, Raffray, *ibid.* p. 119.

In the Ann. Soc. Ent. France, 1896, pp. 256 and 258, I have pointed out that the genera *Patranus*, Raffr., *Brabaxys*, Raffr., and *Abryxis*, Raffr., are synonymous with *Eupines*, King.

Since that time, I have accurately examined each species of those genera in my possession, and I have come to the conclusion that *Patranus* and *Brabaxys* are certainly synonyms of *Eupines*, but that *Abryxis* is really distinct. This opinion is based on characters of the mesosternum which I had not observed previously, and which I will explain further on.

As to the characters of the under part of the head which I had used to establish those different genera, they have no value, being exceedingly variable. Sometimes there are three carinæ, a median and two lateral ones, more or less sinuate, so that the under part of the head appears to be concave, but while the median carina is constant, the edges of the head, which are very sharp in the extreme types, become more or less rounded, and the under part of the head is convex. Between those two extreme forms every degree of transition is to be found.

The mesosternum is composed of two parts, an anterior one more or less declivous, for the resting of the anterior coxæ, and the posterior one which is flat and more or less produced between the intermediate coxæ.

In *Eupines*, *Patranus* and *Brabaxys*, the anterior part of the mesosternum is simply declivous, the edges are more or less sharpened but never carinate; in *Abryxis* as well as in some other American genera, closely allied, this anterior part of the mesosternum is not only declivous but more or less strongly concave and carinate all round. This character, together with the relative length of the elytra, induces me to look on *Abryxis* as a really distinct genus.

In *Eupines* and its synonymous forms, the posterior part of the mesosternum varies to a certain extent. In the extreme type, it is absolutely flat and broadly produced between the intermediate coxæ, which are distant from each other.

In other types the posterior part is still flat, but not so broad between the intermediate coxæ, which are but a little distant from each other; in others this posterior part of the mesosternum bears a more or less pronounced carina which is always broad

tantummodo deplanata, nec concava, nec circa carinata, parte posteriori variabili, simplici vel obtuse carinata, semper inter coxas intermedias distantes plus minusve late producta. Caput infra uni- vel tricarinatum.

It may be, however, very convenient to use the characters of the under part of the head and of the posterior part of the mesosternum to divide the genus *Eupines* into different groups.

The species which are known to me and described or mentioned further on may be distributed as follows :—

Head with three carinæ; mesosternum broad and flat.

E. picta, fuscicornis, aurora, polita, tuberosa, bicolor, nodicornis, transversa, pallipes.

Head with three carinæ; mesosternum narrow and flat.

E. biclarata.

Head with three carinæ; mesosternum carinate.

E. nigriceps.

Head with one carina; mesosternum broad and flat.

E. melanocephala, soror, nigricollis.

Head with one carina; mesosternum narrow and flat.

E. megacephala, longicornis, capitata, diversicolor.

Head with one carina; mesosternum carinate.

E. pumilio, spreta, dubia, pectoralis, liliputana, concolor, triangulata, nodosa, sternalis, compressinoda, lævifrons, Elizabethæ, globulifer.

EUPINES PICTA, Schaufuss.

Nunq. Otios. iii. p. 497; var. *frontalis*, Schfs., *loc. cit.*, p. 497; var. *verticalis*, Schfs., *l.c.*, p. 498; var. *ebenifer*, Schfs., *l.c*, p. 498; var. *Æthiops*, Schfs., *l.c.*, p. 498; *læviceps*, Schfs., *l.c.*, p. 503.

This species is rather elongate, generally unicolorous, but sometimes the head or the last joints of the antennæ are darker. On the head are four foveæ exceedingly variable, two very small ones on the anterior part, on each side of the forehead, and two larger ones between the eyes. They are mere varieties to which Dr Schaufuss thought fit to give names, but which have no specific value.

The antennæ are the same in both sexes, rather stout with the intermediate joints moniliform, 9 a little stronger, 10 large, trapezoid and transverse, 11 ovoid. The sexual characters are very important: in the ♂, all the femora, more especially the anterior ones, are thicker; the metasternum is broadly concave and nearly carinate all round; the 2nd ventral segment bears, in the middle, two small brushes of hairs, short, very compressed, broad, and bending forwards; between those brushes the surface of the segment is a little depressed; in the ♀ the metasternum is convex with a more or less prominent sulcus.

E. læviceps, Schfs., the unique type of which is fortunately a ♂, does not differ except by the cephalic foveæ, which are not so deep.

This species seems abundant in Australia and Tasmania. The type of *læviceps*, Schfs., is labelled East Creek.

EUPINES FUSCICORNIS, n.sp.

Ovata, testacea vel rufo-testacea, pedibus pallidioribus, capite antennisque apice plus minusve infuscatis, nitida, lævis, glabra, interdum in elytris juxta suturam punctis duobus piliferis. Caput quadratum, in fronte et inter oculos foveis quatuor obsoletis. Antennæ elongatæ, articulis duobus primis majoribus, 3-7 sub-oblongis, 5 paulo majori, 8 quadrato, 9 paulo majori, quadrato, 10 multo majori, subgloboso, leviter transverso, 11 magno ovato,

EUPINES AURORA, Schaufuss.

Nunq. Otios. iii. p. 496; *affinis*, Schfs., *loc. cit.*, p. 501.

This species is very closely allied to *E. picta*. The antennæ, however, are much more slender, with the 10th joint much smaller, this makes the 11th appear proportionally larger.

The sexual characters are quite different; the metasternum is longitudinally depressed; the 2nd ventral segment is provided in the middle, on the posterior margin, with a compressed tubercle bending forwards, the last ventral segment shows a large fovea with two long setæ, and at the base a small, more or less elongate tubercle; the anterior tibiæ are thickened and toothed in the middle, and the femora much thickened.

E. affinis, Schfs., is entirely identical. I received it from Mr. Lea under the name of *clavatula*, King, but I do not think the identification to be correct. According to King's description *E. clavatula* has the 9th joint of the antennæ elongate; in *E. aurora* this 9th joint is hardly longer than broad

Australia; Tasmania.

Mr. Lea has sent me recently a specimen which he considers, after comparison with the type of King, as the true *E. polita*, King. This makes the question rather complicated. Mr. Lea's insect is entirely rufous, while, according to King's description, *E. polita* is piceous. Mr. Lea's supposed co-type of *E. polita*, King, is absolutely identical with *aurora*, Schfs., and bears the same sexual characters, which is most important for the identification of these minute insects. If the type of King should bear the same sexual characters we should have to admit that the description of King, as far as the colour is concerned, is erroneous. *E. aurora*, Schfs., = *affinis*, Schfs., would be synonymous with *polita*, King, and the following species, which I consider as being the true *E. polita*, King, would be a new species.

I suppose that Mr. Lea has been misled by the anterior legs, which are similar in both species—*aurora*, Schfs., and *polita*, King,—the tibiæ being slightly toothed in the middle and the femora much thickened.

EUPINES POLITA, King.

Trans. Ent. Soc. N.S.W. i. 1863, p. 49.

The description of the Rev. R. L. King makes the identification of this species nearly a certainty.

It resembles very much *E. picta* and *aurora*, but the body is of a darker colour, with the legs pale testaceous; the antennæ are similar to those of *E. picta*, but a little longer and more slender, and the club is not so strong.

♂. The anterior tibiæ are thickened and obtusely toothed inside at the middle, the anterior femora are thicker; the 2nd ventral segment bears in the middle of the posterior margin a transverse carinula bent forwards and shortly ciliate, the last segment has a deep and slightly semicircular exavation.

This species having the anterior tibiæ similar to those of *E. aurora*, differs from this last one by the colour, much darker (piceous in *polita*, King, rufous in *aurora*, Schfs.), and the sexual characters of the abdomen.

Windsor, N.S.W. (Mr. A. M. Lea).

EUPINES TUBEROSA, n.sp.

Ovata, ferruginea, capite antennisque apice infuscatis, pedibus antennisque basi testaceis, nitida, glabra, lævis, in elytris punctis duobus vel tribus piliferis. Caput quadratum, angulis anticis

This species resembles the preceding one, but the antennæ are somewhat more slender and the sexual characters are different.

EUPINES BICOLOR, n.sp.

Ovata, picea, elytris rubrescentibus, pedibus et antennis testaceis, istarum articulo ultimo infuscato, nitida, glabra, lævis. Caput subquadratum et subconvexum, in fronte, ad angulum externum, puncto minuto, inter oculos punctis duobus inter se magis quam ab oculis distantibus. Antennæ mediocres, articulis duobus primis majoribus, 3-7 latitudine sua paulo longioribus, 5 paulo longiori, 8 quadrato, 9 vix majori, leviter transverso, 10 multo majori, transverso, 11 magno, ovato. Prothorax leviter ovatus. Elytra subovata, basi attenuata, humeris fere nullis. Metasternum rufescens, late profundeque sulcatum; segmentis ventralibus 2° prope marginem posticam lamella minuta, apice rotundata et antrorsum declinata prædito, ultimo transversim et arcuatim profunde impresso. ♂. Long. 1·10 mm.

Clarence River and Tamworth, N.S.W. (Mr. A. M. Lea).

I have received this species from Mr. Lea under the name of *Bryaxis dominorum*, King, but this last species belongs to the genus *Briara*.

E. bicolor resembles *E. polita*, but the colour is darker and the lamella of the second ventral segment is smaller, not wider than long and rounded at the tip.

EUPINES NODICORNIS, n.sp.

Ovata, sat crassa, rubro-castanea, capite nigro-picco, antennis pedibusque testaceis, disperse et longe pilosa. Caput subquadratum, latitudine sua paulo longius, inter oculos foveolis duabus. Antennæ crassæ, articulis 3-4 breviter ovatis, longitudine et latitudine subæqualibus, 5 multo majori, ovato, 6-8 moniliformibus, leviter transversis, 9-10 valde transversis, isto intus leviter producto, ultimo breviter ovato et crassiori. Prothorax subcordatus. Elytra breviter ovata, lateribus leviter rotundata, humeris notatis. Metasternum late sulcatum, utrinque pone medium tuberculo minuto, acuto; segmentis ventralibus 2° medio tuberculo minuto, rotundato et penicillato prædito, ultimo testaceo et medio impresso. Femoribus anticis et intermediis crassioribus, trochanteribus anticis basi spina tenui subrecta. ♂. Long. 1·20 mm.

New South Wales (without locality).

This species will be at once distinguished amongst the allied species of the same group by the shape of the antennæ in the ♂, which is the only sex known.

EUPINES TRANSVERSA, King.

Trans. Ent. Soc. N.S.W. 1866, p. 311.

It is with some doubt that I identify with *transversa*, King, a ♀ specimen which I received from Mr. Lea, under this name: the 10th joint of the antennæ is really transverse, being a little dilated forward, the general form is less convex, the elytra less attenuate at the base with the shoulders a little marked; the metasternum has a trace of a longitudinal groove and the second ventral segment is entirely simple. The colour is piceous-black with the elytra reddish, the antennæ ferrugineous, short, and the legs rufous.

Windsor, N.S.W. (Mr. A. M. Lea).

EUPINES PALLIPES, n.sp.

Of this species the ♀ only is known; it is much differentiated by the more elongate and less convex body and by the 10th joint of the antennæ, which is broader than the 11th.

Eupines biclavata, n.sp.

(Plate x., fig. 16.)

Obovata, rufo-testacea, brevissime pubescens. Caput subquadratum et subconvexum, angulis anticis leviter rotundatis, inter oculos punctis duobus et in fronte duobus alteris vix perspicuis. Antennæ breves, crassæ, articulis duobus primis magnis, 3-4 quadratis, 5 magno, globoso, infra tuberculo minutissimo prædito, 6-8 moniliformibus, leviter transversis, 9 valde transverso, 10 magno, leviter transverso, intus latiori, angulo interno apicali appendice lamellato minuto prædito, 11 præcedenti fere angustiori, breviter ovato, basi truncato. Prothorax cordatus. Elytra latitudine sua longiora, basi paululum attenuata, lateribus vix rotundata, humeris subnodosis. Metasternum postice late longitudinaliter impressum. Segmentis ventralibus 2° basi strigoso et medio carinula longitudinali vix perspicua, ultimo obsolete impresso. Pedes breves, femoribus crassis. ♂. Long. 1·20 mm.

Forest Reefs, N.S.W. (Mr A. M. Lea).

This species much resembles *nodicornis*, but the colour is different and the antennæ are still shorter and stouter.

Eupines nigriceps, n.sp.

(Plate x., fig. 11.)

Ovalis, rufo-castanea, capite et antennarum articulis tribus ultimis nigro-piceis, pedibus rufis, setis aliquot dispersis. Caput quadratum, angulis anticis truncatis et foveatis, fronte medio obsolete impresso, inter oculos foveolis duabus minutis. Antennæ crassæ, articulis duobus primis majoribus, 3 obconico, 4 quadrato, 5 multo majori, elongato-ovato, 6-7 quadratis, 8 fere transverso, 9 majori, transverso, intus producto, 10 maximo, leviter transverso, apice nonnihil obliquo, 11 breviter ovato. Prothorax cordatus, latitudine sua paulo longior. Elytra magna, basi attenuata, et lateribus rotundata, humeris parum notatis. Metasternum postice delicatule sulcatum; segmentis ventralibus 2° medio et ultimo fere toto transversim impressis. Pedes sat elongati et graciles. ♂. Long. 1·20 mm.

Clarence River, N.S.W. (Mr. A. M. Lea).

This species will be known at once by the colouration and the size of the last joints of the antennæ.

EUPINES MELANOCEPHALA, Schaufuss.

Nunq. Otios. iii. p. 494.

This species resembles much *E. picta*, Schfs., but belongs to another group. The sexual characters are different: in ♂ the metasternum is broadly concave, and at the top of this concavity is a small oblong tubercle; the 2nd ventral segment bears—not in the middle of the segment as in *picta*, but on the very edge of the posterior margin—two small brushes compressed and bent forward; the last segment has a small transverse depression.

Tasmania.

I know only the type example of Dr. Schaufuss.

EUPINES SOROR, n.sp.

improbable. According to King's description *æquata* is "picea," while *soror* is "rufa." *E. soror* has short and thick antennæ.

EUPINES NIGRICOLLIS, King.

Tyrhus nigricollis, King, Trans. Ent. Soc. N.S.W. i. 1864, p. 103.

Ovata, sat crassa, rubro-castanea, elytris dilutioribus, capite nigro-piceo, prothorace infuscato, brevissime et parce pubescens. Caput subquadratum, leviter convexum, antice leviter rotundatum, inter oculos punctis duobus. Antennæ breves, crassæ, articulis duobus primis majoribus, 1° brevi, 2° ovato, 3 brevissime obconico, 4 moniliformi, subtransverso, 5 maximo, ovato, lateraliter dente truncato prædito, 6-8 moniliformibus, subtransversis, 9 latiori, transverso, 10 maximo, irregulariter trapezoidali, angulo apicali externo appendiculato, margine [illegible] interna emarginata et appendiculata, 11 fere globoso, apice attamen leviter acuminato, angulo basali interno infra dentato, latere interno supra late foveato. Prothorax cordatus, latitudine sua longior. Elytra latitudine sua paulo longiora, basi leviter attenuata, lateribus rotundata, [illegible]. Metasternum [illegible] [illegible] depressum et medio sulcatum. Segmentis ventralibus 2° basi [illegible] [illegible] [illegible] [illegible] [illegible] transversim impresso. Pedes [illegible] [illegible] [illegible] [illegible] apice [illegible] [illegible]. ♂. Long. [illegible] mm.

Melbourne, Vic.

By the shape of the antennæ this species bears much analogy to [illegible] and [illegible].

EUPINES [illegible]

[illegible] convexa, ferruginea, antennis apice [illegible] [illegible], pedibusque pallide testaceis, [illegible] [illegible]. Caput magnum, quadrato-transversum, angulis [illegible] rotundatis, [illegible] in fronte [illegible] [illegible] et [illegible], inter [illegible] foveis duabus [illegible] et minutis. Antennæ breves crassæ, articulis 1 et 2 paulo majoribus, ovatis, 3 [illegible] et [illegible], 4-8 moniliformibus.

9 paulo majori, transverso, 10 majori, transverso, 11 magno, ovato. Prothorax breviter cordatus, lateribus rotundatis. Elytra convexa, lateribus rotundata. Metasternum simplex, subconvexum, segmento 2° ventrali punctato. ♀. Long. 1·10 mm.

Clarence River, N.S.W. (Mr. A. M. Lea).

I have not seen the ♂ of this species, which will be very easily distinguished by the size of the head and the last joint of the antennæ, which is abruptly larger, the club appearing as being formed by one joint only.

EUPINES CAPITATA, King.

Trans. Ent. N.S.W. i. 1866, p. 311.

I have two specimens answering pretty well to King's description, more especially on account of the marked transverse line of the frontal part, above the insertion of the antennæ which are themselves short and thick, the joints 4-9 being transversely moniliform, 9 not larger than the preceding, whilst 10 is much larger, transverse, 11 truncate at base and briefly ovate; the elytra are rather elongate, attenuate at base, with the sides rounded and the shoulders prominent and carinate; the scattered punctuation is hardly visible, but still it exists; the metasternum and base of the second ventral segment (first visible) are finely and rugosely

anticis obtusis et emarginatis, inter oculos punctis duobus minutis et fronte medio leviter deplanato. Antennæ elongatæ, graciles, articulis 1° elongato, cylindrico, 2° paulo minori, 3 oblongo, obconico, 4 paulo breviori, obconico, 5 præcedenti plus dimidio longiori, 6 quarto æquali, 7 paulo longiori, tertio æquali, 8 subquadrato, 9 paulo majori, latitudine sua paulo longiori, 10 multo majori, trapezoidali, longitudine latitudineque subæquali, 11 mediocri, ovato, basi truncato. Prothorax antice plus et postice minus attenuatus, latitudine maxima media. Elytra sat elongata, lateribus vix rotundata et basi parum attenata, humeris notatis. Metasternum leviter sulcatum. Pedes mediocres, tibiis subrectis et ad apicem leviter incrassatis. ♀. Long. 1·40 mm.

Swan River, W.A. (Mr. A. M. Lea).

This species is very different from the two preceding ones, being more elongate and with slender antennæ. The unique specimen is unfortunately a ♀. I think that, according to the size of the 5th joint in the ♀, the same joint in the ♂ must be much larger.

EUPINES DIVERSICOLOR, Schaufuss.

Nunq. Otios. iii. p. 498.

(Plate x., fig. 9.)

Subelongata, nitida, lævis, prothorace et elytris rubro-castaneis, capite et abdomine infuscatis, antennis castaneis, pedibus testaceis, setis aliquot longis dispersis. Caput quadratum et deplanatum, inter oculos foveis duabus, in fronte transversim impressum et medio punctis duobus approximatis. Antennæ elongatæ, articulis duobus primis majoribus, 1° subcylindrico, 2 ovato, 3-4 subobconicis, 4° paulo breviori, 5 multo majori, oblongo, 6 quarto subæquali, 7 præcedenti nonnihil longiori, 8 quadrato, 9 magno, intus longe antrorsum producto et apice appendiculato, 10 magno, valde compresso, supra viso oblongo, lateraliter viso transverso, 11 basi truncato, ovato. Prothorax cordatus, convexus, lateribus valde rotundatis. Elytra cum punctis aliquot dispersis setam longam ferentibus, elongata, basi attenuata, lateribus rotundata.

humeris subnodosis. Pedes sat elongati, tibiis subrectis, ad apicem leviter incrassatis. Metasternum valde totum impressum et fundo longitudinaliter sulcatum. Segmento 2 ventrali (1° conspicuo) basi tuberculis duobus penicillatis et approximatis prædito, ultimo basi transversim impresso. ♂. Long. 1·10-1·20 mm.

King George's Sound, W.A.

I found in Dr. Schaufuss' collection four very different insects under the same name of *diversicolor*. I consider as being the true *diversicolor* the specimen which answers best to the description; the other one belongs to the genus *Eupinoda*, and has been described under the name of *Eupinoda amplipes*, Raffr.

E. diversicolor will be at once recognised by the peculiar shape of the last joints of the antennæ.

EUPINES PUMILIO, Schaufuss.

Nunq. Otios. iii. p. 504.

This species resembles much *E. picta*, *aurora* and *polita*, but the head is much more quadrate and bears a transverse impression on the forehead; the antennæ are thicker, with the 5th joint cylindrical and twice as long as the preceding one, the 9th square, 10th a little transverse; all along the body are long scattered

on the second ventral segment the punctuation is strong but remote.

In the ♂ the metasternum is longitudinally concave with a groove in the bottom, the second ventral segment has on the base a small tubercle and on the median part of the posterior margin a small brush of hairs compressed and bent forward; the last segment is transversely depressed.

Victoria (type, Sharp); Parramatta, Tamworth and Clarence River, N.S.W. (Mr. A. M. Lea).

EUPINES DUBIA, n.sp.

Brunnea, elytris obscure ferrugineis, antennis pedibusque rufis. Caput subquadratum, latitudine sua paulo longius, in fronte vix impressum, inter oculos foveis duabus. Antennæ breves, crassæ, articulis duobus primis majoribus, ovatis, 3 obconico, 4 quadrato, 5 multo longiori, ovato, 6-8 moniliformibus, quadratis, 9 paulo latiori, transverso, 10 fere duplo majori, transverso, 11 ovato. Prothorax breviter cordatus. Elytra lateribus minus convexa, humeris subquadratis. Metasternum apice profunde sulcatum; segmento 2° ventrali basi plicatulo et medio carinula delicatula segmentum longitudine fere æquant. ♀. Long. 1·20 mm.

Australia (without locality).

In the Schaufuss collection I found a specimen of this species labelled "*polita*, King, ex Janson." This identification does not seem to be correct. This so-called type of *polita* is a ♀, but differs from the ♀ of *polita*, the ♂ of which has characteristic dilatated anterior tibiæ.

In *dubia* the antennæ are much shorter and thicker, the 5th joint is more developed, and the 10th is decidedly transverse, while it is nearly globose in *polita*.

EUPINES PECTORALIS, n.sp.

Subovata, rufa, antennis pedibusque testaceis, sparsim albido-pubescens, elytris disperse punctatis. Caput minus transversum, antice attenuatum, fronte transversim sulcato, inter oculos foveis duabus magnis. Antennæ mediocres, articulis duobus primis ovatis, majoribus, 3 breviter obconico, 4-9 moniliformibus, 9 leviter transverso, vix majori, 10 paulo majori, transverso, 11 magno, ovato. Prothorax breviter cordatus. Elytra magna ovata, humeris notatis. Metasternum late profundeque sulcatum; segmentis ventralibus 1° basi punctato, ultimo testaceo. Tibiis posticis elongatis, gracilibus, et ante apicem leviter curvatis.

♂. Metasternum profundius sulcatum. Long. 1·10 mm.

Tamworth, N.S.W. (Mr. A. M. Lea).

The head is small, the antennal club is formed almost exclusively of the last joint: the elytra are large with a strong but scattered punctuation.

There is very little difference between the male and the female.

EUPINES LILIPUTANA, n.sp.

Castanea, antennis pedibusque testaceis, breviter sat dense pallide pubescens. Caput mediocre, antrorsum leviter attenuatum. inter oculos punctis duobus minutis, fronte supra antennarum insertionem foveis duabus magnis et transversis, medio fere contiguis Antennæ crassiores, articulis duobus primis majoribus, 3 leviter conico, 4-8 monilibus, 9 vix latiori, sed transverso, 10

EUPINES CONCOLOR, Sharp.

Trans. Ent. Soc. Lond. 1874, p. 502.

(Plate x . fig. 20.)

I have received a typical specimen from Dr. D. Sharp.

The appearance of the antennæ may be different, according to the way of viewing them. The description of Dr. D. Sharp and the figure I give are taken when the antenna is viewed from above: when viewed from the side, the 10th joint is very oblique on the under part, whilst the upper part is straight: the legs are rather long and slender : the general form of the body much resembles *pectoralis*, the elytra being very large in comparison with the head and prothorax.

Dr. D. Sharp's type comes from Victoria; I have another specimen from Gawler in which the 10th joint of the antennæ is slightly shorter.

The following four species are very similar to *concolor*, and differ nearly exclusively by the shape and size of the last two joints of the antennæ; it is quite possible that further discoveries may prove that they are mere local or individual variations of *concolor*.

EUPINES TRIANGULATA, n.sp.

(Plate x., fig. 17.)

E. concolori simillima, differt articulis duobus ultimis antennarum majoribus, 10 supra viso curvilineatim triangulari, latitudine et longitudine subæquali cum margine superna medio angulata et, lateraliter viso, margine superna medio valde producta, ultimo brevissime ovato, basi truncato, apice subturbinato; metasternum et abdomen sicut in *concolore*. Long. 1·10 mm.

Tamworth, N.S.W. (Mr. A. M. Lea).

This species differs simply from *concolor* by the last joints of the antennæ; the shape of the 10th joint is very different if it is viewed from above or from the side: from above it resembles a shield; from the side an irregular trapeze, the one angle of which would be much produced, the expansion covering the base of the last joint.

EUPINES NODOSA, n.sp.

(Plate x., fig. 18.)

Præcedenti simillima; foveolà media frontali deficienti; antennarum articulo 10 irregulariter trapezoidali, angulo interno apicali acuto et producto, externo apicali obtuso, apice oblique truncato et latere interno magis obliquo. Punctis in elytris deficientibus. Metasternum sulcatum, segmentis ventralibus 2° medio minutissime tuberculato, ultimo profunde transversim excavato; tibiis posticiis apice leviter incurvis. ♂.

♀. Antennæ paulo graciliores, articulo 10 præcedenti majori, transverso. Metasternum vix perspicue sulcatum. Long. 1·00 mm.

Tamworth, N.S.W. (Mr. A. M. Lea).

EUPINES COMPRESSINODA, n.sp.

(Plate x., fig. 19.)

Præcedenti simillima, sed colore dilutiori et forma nonnihil magis elongata. Antennarum articulis 4-9 transversis, 10 transverso et ovali, sequenti multo latiori, infra angulo interno com-

angustiori, ovato, acuminato, basi truncato. Prothorax breviter cordatus. Elytra subtiliter et disperse punctulata, subovata, basi leviter attenuata, humeris notatis. Metasternum obsolete depressum, dense rugoso-punctatum et minute squamosum. Segmentis ventralibus 2 (primo conspicuo) basi dense rugoso-punctato et minute squamoso, delicatule longitudinaliter carinato, ultimo transversim fere toto impresso. Pedes sat elongati et graciles. Long. 1·20 mm.

Moreton Bay, Q.; Eastern Creek, N.S.W.

This species is closely allied to *concolor*, but the antennæ are shorter and thicker, with the 10th joint decidedly transverse.

Eupines lævifrons, n.sp.

(Plate x., fig. 22.)

Ovata, sat crassa, obscure castanea, antennis pedibusque testaceis, tota lævis et glabra. Caput latitudine sua longius, leviter convexum, antrorsum leviter rotundatum, inter oculos foveis duabus. Antennæ breves, crassæ, articulis duobus primis majoribus, subcylindricis, 3 breviter subobconico, 4-8 moniliformibus, leviter transversis. Prothorax capite et longitudine sua paulo latior, antice posticeque æqualiter attenuatus, lateribus rotundatus. Elytra parum elongata, basi vix attenuata, lateribus parum rotundata et humeris obtuse notatis. Pedes breves.

♂. Antennarum articulis 8 magis transversis, 9 sublenticulari, 10 magno, transverso, subhemispherico, 11 maximo, globoso. Metasternum obsolete sulcatum; segmentis ventralibus 2 ad apicem tuberculo minuto predito, ultimo transversim sulcato.

♀. Antennarum articulis 9 transversis, 10 paulo majori, valde transverso, 11 magno, [illegible] quam in mare minori. Metasternum convexum, simplex. Long. [illegible] mm.

Clarence River, N.S.W. (Mr. A. M. Lea).

The difference in colour between the body, which is rather dark, and the legs and antennæ, which are pale testaceous, and the size of the last two joints of the antennæ in the ♂ will help in discriminating this species.

EUPINES GLOBULIFER, Schaufuss.

Bryaxis globulifer, (♂) Schfs., Nunq. Otios. 1880, p. 504; *Tychus politus*, (♂) Schfs., Tijds. v. Ent. xxix. 1886, p. 206; *Patranus* (Raffray) *globulifer*, (♂) Schfs., Rev. d'Ent. 1890, pp. 118 and 123, pl. ii. fig. 33; *Bryaxis ampliventris*, (♀) Schfs., Nunq. Otios. p. 505.

(Plate x., fig. 14.)

Ovalis, rufo-testacea, antennis pedibusque dilutioribus, disperse sed longe setosa. Caput latitudine sua paulo longius et antrorsum vix attenuatum, inter oculos foveis duabus, fronte medio impresso. Antennæ in utroque sexu variabiles. Prothorax cordatus, latitudine sua paulo longius. Elytra sat elongata, basi attenuata, lateribus leviter rotundata, humeris oblique notatis. Pedes sat graciles, elongati.

♂. Antennarum articulis duobus primis majoribus, 3 et 4 obconicis, latitudine sua tantummodo longioribus, 5 magno, ovato, intus ante medium appendiculato, isto appendici apice oblique truncato, 6-7 inter se subæqualibus et latitudine sua paulo longioribus, 8 fere quadrato, 9 paulo majori, leviter transverso, 10 multo majori et leviter transverso, 11 oblongo-ovato, basi truncato. Metasternum valde et profunde sulcatum, utrinque postice obtuse

politus, Schfs., comes from Champion Bay, W.A.; the type of *globulifer*, Schfs., comes from Melbourne, Vic.; the type of *amplicentria*, Schfs., from Sydney, N.S.W. Mr. Lea sent it to me from Tamworth, N.S.W.; he has found also a ♀ with a small ant, the name of which is unknown to me.

Genus Eupinopsis, Raffray.

Ann. Soc. Ent. Fr. [illegible], pp. [illegible] and [illegible].

Ovata, convexa. Abdomen brevissimum, anguste marginatum. Elytra magna, stria dorsali brevi. Prothorax basi media subfoveatus. Mesosternum simplex. Coxis intermediis distantibus. Palporum articulo ultimo fusiformi.

This new genus is very closely allied to *Eupines*, from which it differs by the abdomen being extremely short and transverse, while the elytra are much longer and the presence of a prothoracic fovea and dorsal stria on the elytra. In the two species which are included in this genus the elytra are strongly punctured.

Eupinopsis perforata, Schaufuss.

Eupines perforata, Schfs., Nunq. Otios. [illegible]

Ovata, convexa, castanea, breviter et dispersе pallide pubescens. Caput breve, antrorsum attenuatum, inter oculos minute bifoveatum, sulcis [illegible] subparallelis, fronte medio depressa. Antennae elongatae, graciles, articulis oblongis, [illegible] et [illegible] brevioribus, [illegible] paulo crassior, [illegible] magno [illegible] acuminato. Prothorax cordatus, basi [illegible] lineato-punctata et medio minute foveata. Elytra [illegible] punctata, [illegible] [illegible] [illegible] [illegible] integra [illegible]. Pedes [illegible] [illegible] profunde [illegible].

♂. Penultim[illegible] minute [illegible] [illegible] minute [illegible] [illegible] emarginata, ultimo valde impresso. Metasternum magis impressum. Long. 1·60-1·70 mm.

Clarence River and Tamworth, N.S.W., (Mr. A. M. Lea); Rope's Creek, N.S.W., (Mr. Masters); Clyde River, N.S.W.; Tasmania.

EUPINOPSIS PUNCTATA, n.sp

Breviter ovata, convexa, castanea, breviter et densius pallide pubescens. Caput antrorsum attenuatum, inter oculos minute bifoveatum, fronte subretuso et haud impresso. Antennæ breves, crassæ, articulis 3-9 subquadratis, 10 multo latiori, transverso, 11 magno, breviter ovato, acuminato Prothorax subglobosus, basi ipsa vix punctata, fovea media valida. Elytra disperse, parum profunde punctata, breviter ovata, basi minus attenuata, humeris notatis, stria suturali integra, sulco dorsali brevi. Abdomen breve. Metasternum simplex Pedes validi, tibiis posticis leviter incurvis.

♂. Abdominis segmento 1° dorsali apice leviter sed late producto et retuso, utrinque parum profunde emarginato. Long. 1·10-1·30 mm.

This species is shorter and smaller than *perforata;* the antennæ are short and thick, and the punctuation on the elytra is not so strong.

emarginato, 7 (♂) rhomboidali, magno, convexo, longitudinaliter sulcato. Pedes breves, tarsorum articulis 1° minutissimo, 2 magno, subconico, 3 minori cylindrico, ungue singulo.

This new genus, which belongs to the tribe of the *Bythinini*, owing to its large basal ventral segment, resembles *Bythinoderes*, Reitt., (from Sumatra and Borneo) but differs by the intermediate coxæ which are approximate, whilst the posterior ones are much less distant and the first dorsal segment which is larger than the others. It is more closely allied to *Sunorfa*, Raffr., (from Northern New Guinea) but differs by the first joint of the antennæ which is shorter, the shoulders which are dentate with a strong susepipleural sulcus, and the first dorsal segment which is larger.

The tribe of *Bythinini* seems to have very few representatives in Australia. The insect described by the Rev. R. L. King under the name of *Bythinus impressifrons* is not known to me, but it belongs certainly to a totally different genus. I do not think that *Bythinus niger*, which also I have not seen, is a true *Bythinus*.

GNESION RUFULUM, n.sp.

Totum rufum, sat dense pallide pubescens. Caput leviter transversum, inter oculos foveis duabus et sulcis duobus rectis, brevibus, in sulco frontali transverso desinentibus, lateribus ipsis isto sulco transverso angulatim emarginatis, fronte utrinque supra antennas nodoso, medio depresso: epistomate producto et apice recte truncato. Antennæ sat elongatæ, articulis duobus primis multo majoribus, ovatis et inter se subæqualibus, 3 leviter et breviter obconico, 4-9 moniliformibus, 10 vix latiori, sed leviter transverso, 11 magno, breviter ovato, acuminato. Prothorax capite latior, leviter transversus, cordatus, pone medium lateribus fovea laterali emarginatis, ista fovea magna, altera media maxima et valde transversa cum foveis lateralibus obsolete juncta, basi ipsa medio transversim subgibbosa. Elytra subquadrata, basi foveis duabus maximis, oblongis. Metasternum obsolete impressum. Pedes breves, femoribus omnibus, præsertim intermediis, r incrassatis, tibiis ad apicem leviter incrassatis. ♂. Long. 1·10 mm.

Australia (without locality).

Tribe PSELAPHINI.

Genus Pselaphus, Herbst.

Käf. iv. 1792, p. 106.

The insects belonging to this genus are to be found all over the world, and seem to be specially abundant in Australia and New Zealand.

The character of the true genus *Pselaphus* resides in the maxillary palpi, which are always very long and sometimes even longer than the antennæ; the 1st joint is elongate; between the 1st and 2nd joints there is only a suture and no articulation; the 2nd, elongate and slender, is pedunculate at base, and is clavate at apex; the 3rd is very small; the 4th is very long, slender at base and ends in a strong club; this club is generally sulcate at the apex, and in this groove is inserted the membranous appendage; in some species the club has no groove at tip, but is covered with small papillose tubercles. This last form has not yet been met with in Australia.

The following is a synopsis of the Australian species which I

F 2. Longitudinal furrow of the head extending as far back as the neck; ♂ 2nd ventral segment bearing in the middle of the posterior edge two small tubercles .... *lineatus*, King.

F 1. Longitudinal furrow of the head not extending back further than the middle of the vertex.

G 2. Head shorter and broader, a little dilated in the middle; the fovea on the vertex is large, rounded behind, broader than the longitudinal furrow; close to the eyes, on each side, is a strong blunt tubercle; maxillary palpi shorter than the antennæ; ♂ second ventral segment with an oval impression. Long. 2·10-2·30 mm.. *antipodum*, Westw.

G 1. Head longer and more slender, the fovea on the vertex smaller, angulate behind, not broader than the longitudinal furrow; the tubercle close to the eyes, on each side, is hardly conspicuous; maxillary palpi as long as the antennæ. Long. 1·80-1·90 mm......... *leanus*, n.sp.

E 1. Pubescence rather dense, squamose, more especially on the head, two punctures at the bottom of the fovea of the vertex; maxillary palpi shorter, club of the 4th joint as long as the peduncle......... .... *brevipalpis*, Schfs.

D 1. Base of the prothorax behind the transverse furrow smooth and shining, very narrow; frontal furrow not extending on the vertex and ending abruptly between the eyes and two sharp tubercles; vertex somewhat depressed with a very small tubercle in the middle; maxillary palpi very slender, as long as the antennæ, club small....... .... *tuberculifrons*, n.sp.

C 1. Transverse furrow of the prothorax interrupted in the middle by a median somewhat square fovea limited on each side by a very short carinule.

D 2. Lateral fovea of the prothorax very large, triangular, distant from the edge which is somewhat carinate; maxillary palpi not longer than the antennæ........................ ... *insignis*, Schfs.

D 1. Lateral fovea of the prothorax rounded, smaller, situated close to the edge which is rounded.

E 2. Maxillary palpi very long, slender and nearly straight; antennæ slender, joints longer than broad; vertex neither convex, nor ampliate... *elongatus*, **n.sp.**

E 1. Maxillary palpi shorter, 4th joint very arcuate; antennæ short and thick, joints short; vertex very convex and much broader than the anterior part of the head.. .... *crassus*, **n.sp.**

A 1. Pubescence hair-like, long and black.

B 2. Tubercles between the eyes large, rounded; lateral edge of the prothorax rounded; elytra hardly longer than broad at apex, dorsal stria geminate.. *pilosus*, **n.sp.**

B 1. Tubercles between the eyes very small, lateral edges of the prothorax carinate behind the middle; elytra much longer than broad with the dorsal stria simple............................... *longepilosus*, **Schfs.**

PSELAPHUS TRIPUNCTATUS, Schaufuss.

Tijds. v. Ent. xxix. 1886, p. 252.

The characteristics of this species are the three independent foveæ of the prothorax, and the absence of the transverse furrow; the colour is rather dark reddish, with the legs and palpi pale testaceous; the palpi are moderately long, little sinuate, the peduncle in the 4th joint is a little longer than the club.

Melbourne, Vic.; one specimen (♀) only, type of Dr. Schaufuss.

This species seems to be common. I have it from Parramatta, Sydney, Windsor, Tamworth and Clarence River, N.S.W.; Melbourne and Victoria.

There has been a certain amount of confusion between *lineatus*, King, and *antipodum*, Westwood. The Rev. R. L. King was not quite sure that his *lineatus* was different from *antipodum*, and he was doubtful about the correctness of Westwood's description and figure. I have seen no type of *antipodum*, Westw., but I take it for granted that the descriptions and figures of this celebrated entomologist are perfectly correct.

This confusion is illustrated by an insect which I found in Reiche's collection: it had two labels, one written by Reiche and bearing the name *antipodum*: the other in a writing unknown to me, but likely that of King himself, and bearing the name *lineatus*, King, Parramatta. This insect is the true *lineatus*, King.

I have specimens of the same species which I received from Dr. D. Sharp under the name of *lineatus*, King. Mr. S. Reitter confounded the two species. Dr. Schaufuss mistook it for a new species and described it again under the name of *frontalis*.

Pselaphus antipodum, Westwood.

Trans. Ent. Soc. Lond. iii. 1856, p. 274, Pl. 16, fig. 8.

This species is closely allied to *lineatus*, King, but very different; the colour is generally lighter, the antennæ and palpi are more elongate: the head is a little shorter and broader, the longitudinal furrow extends as far as the vertex and ends in a somewhat round large depression, but it does not reach the neck: the elytra are shorter. The size is a trifle larger (*lineatus*, 1·80-2·00 mm.; *antipodum*, 2·00-2·30 mm.). In the ♂ the metasternum is slightly depressed in the middle, the second ventral segment bears, in the middle, a large longitudinally ovate impression.

This species is much rarer than *lineatus*. Westwood records it from Melbourne: I have received it from Mr. Lea from the Tweed and Clarence Rivers, N.S.W.

PSELAPHUS LEANUS, n sp.

Sat convexus, totus rufus, antennis pedibusque dilutioribus. Caput elongatum et angustum, sulco longitudinali lato, inter oculos abrupto, lateribus vix ampliatis et vix gibbosis, in vertice impressione parum profunda, minuta et subangulata. Palpi maxillares elongati, articuli quarti parte pedunculata basali clava fere triplo longiori. Antennæ parum elongatæ, articulis sub-oblongis, clava triarticulata, parum conspicua. Prothorax subovatus antice plus attenuatus, sulco transverso, arcuato, lato, parum profundo, parte basali reticulata, opaca. Elytra elongata, basi attenuata; humeris vix obliquis, stria suturali alteraque dorsali. Abdomen elytris brevius, disperse breviter pubescens, postice obtuse acuminatum. Metasternum juxta coxas posticas leviter depressum. ♀. Long. 1·70-1·75 mm.

This species differs from *antipodum* by the head being more elongate, hardly dilated between the eyes; the impression on the vertex is much smaller, the maxillary palpi are longer, the peduncle of the 4th joint being nearly three times longer than the club.

PSELAPHUS BREVIPALPIS, Schaufuss.

Pselaphus tuberculifrons, n.sp.

Sat convexus, totus rufus, antennis pedibusque dilutioribus, setis aliquot brevibus, pallidis, dispersis. Caput parum elongatum, sulco longitudinali lato sed parum profundo, inter oculos abrupto, utrinque tuberculo acuto et in vertice medio tuberculo minuto. Palpi valde elongati, graciles, articuli quarti parte pedunculata basali clava minuta fere quadruplo longiori. Antennæ parum elongatæ, articulis 3-6 latitudine sua vix et 7-8 paulo longioribus, 9-10 oblongis, 11 subfusiformi. Elytra parum elongata, basi attenuata, humeris fere nullis, lateribus rotundatis, stria suturali alteraque dorsali. Abdomen elytris subæquale. Metasternum convexum inter coxas posticas declive. ♀. Long. 1·70 mm.

This species is shorter than the preceding ones; the cephalic furrow is very shallow, between the eyes are two sharp tubercles and on the vertex another small one; the maxillary palpi are very long and slender, with the club small; the antennæ are comparatively short.

Bridgetown, W.A. (Mr. A. M. Lea).

Pselaphus insignis, Schaufuss.

Tijds. v. Ent. xxix. 1886, p. 249; *bipunctatus*, Schfs., *loc. cit.* p. 252.

I have the types of the two above-named species and I cannot detect the slightest difference between them. The head is rather broad, with the longitudinal furrow shallow, ending between the eyes, where the tubercles are replaced by two foveæ: the palpi are long, the peduncle of the 4th joint is nearly twice as long as the club: the antennæ are short, thick and the prothorax is nearly exactly hexagonal, the posterior transverse furrow is deep, interrupted in the middle by a fovea which is carinate on each side, the lateral foveæ are strong; the elytra are less attenuate at base, with the shoulders oblique.

Eastern Creek, N.S.W.; Rockhampton, Q.

PSELAPHUS ELONGATUS, n.sp.

Elongatus et gracilis, rufus, pedibus antennisque dilutioribus. Caput parum elongatum, sulco longitudinali parum profundo, inter oculos abrupto et bifoveato, vertice convexo; palpi valde elongati, graciles, articuli quarti parte basali pedunculata clava ovata mediocri vix triplo longiori. Antennæ parum elongatæ sed graciles, articulis omnibus suboblongis, clava parum conspicua. Prothorax ovatus, sulco transverso arcuato, parum profundo medio interrupto, foveis lateralibus validis, parte basali minutissime reticulata. Elytra elongata, basi attenuata, lateribus subobliquis, humeris nullis. Abdomen elytris paulo brevius, leviter convexum. Metasternum convexum postice declive. ♀. Long. 1·60 mm.

This species has the head very much like that of *insignis*, but the maxillary palpi are much longer; the antennæ are not very long and slender, and each joint is longer than broad; the elytra are much longer, the sides are oblique, and there are no shoulders.

Tweed River, N.S.W. (Mr. A. M. Lea).

PSELAPHUS CRASSUS, n.sp.

Totus pallide rufus, setis aliquot pallidis, dispersis. Caput ante oculos constrictum, juxta antennas ampliatum, inter oculos bigibbosum, vertice multo latiori, convexo, antrorsum medio declivi

The posterior part of the head is much broader and more raised than the anterior part; the antennæ are thick and short.

Australia (without locality).

PSELAPHUS LONGEPILOSUS, Schaufuss.

Tijds. v. Ent. xxix. 1886, p. 248.

This species, together with the following one, is remarkable for the long black hairs covering the body, which are even longer on the posterior tibiæ; the head is rather long and narrow, with the furrow extending back as far as the neck, and bears no tubercles, but two obsolete oblique carinules between the eyes; the palpi are long, slender and little arcuate; the prothorax is broadest in front of the middle, the sides are somewhat flattened and carinate, the transverse furrow is very much arcuate and interrupted in the middle by a bicarinate fovea; the elytra have a strong but simple dorsal stria.

Rockhampton, Q.

I know only the type specimen of Dr. Schaufuss, which has no antennæ left.

PSELAPHUS PILOSUS, n.sp.

Læte rufo-castaneus, longe, disperse nigro-setosus. Caput crassum, inter oculos tuberculis duobus magnis, sulco longitudinali angusto, profundo, inter oculos abrupto, vertice parum convexo. Palpi elongati, articulo quarto arcuato, clava crassa, parte basali pedunculata clava haud triplo longiori. Antennæ crassæ, validæ, articulis latitudine sua paulo longioribus, 9 breviter obconico, 10 brevissime ovato, 11 magno, ovato, acuminato. Prothorax breviter ovatus, sulco transverso arcuato, parum profundo, medio vix interrupto. Elytra brevia subtrigona, lateribus leviter rotundatis, humeris nullis, stria dorsali minute geminata. Abdomen elytris subæquale. ♀. Long. 2·10 mm.

This species differs from the preceding one by the head being shorter and bearing strong tubercles; the palpi are more arcuate; the prothorax is much shorter and regularly ovate, with the sides rounded, not carinate; the elytra are shorter, with a finely geminate dorsal stria.

It must be very closely allied to *P. geminatus*, Westwood, which unfortunately I have not seen, but according to this author's description and figure the antennæ are much more elongate and slender, the prothorax bearing three well marked foveæ; for these reasons it seems to me evident that *pilosus* is a very different species.

Bridgetown, W.A. (Mr. A. M. Lea).

Genus PSELAPHOPHUS, Raffray.

Rev. d'Ent. 1890, pp. 137 and 139.

This genus is very closely allied to *Pselaphus* and differs only in the following points—head broader and shorter; maxillary palpi much shorter, last joint fusiform or flagelliform, with hardly any basal peduncle, the long, slender basal part so striking in *Pselaphus* wanting; the prothorax more cordate.

It has been confounded by Dr. Schaufuss with *Curculionellus* from which it is different.

It includes *Bryaxis atriventris*, Westwood, of which I have not seen an authentic specimen, but I have received several times from British entomologists, under the name of *Bryaxis atriventris*, Westw., insects which answer exactly to the description and figure of Westwood, and belong to the genus *Pselaphophus*.

This genus is peculiar to Australia.

arly trans-
o the base.

grolhoraceque
elongatum,
nodosum, sulco
orupto et pro-
nuo graciles,
et latitudine
paulo longiori-
us, II magno,
r, antice plus
medium;
arcuatis,
arcuatis et

ngitudinaliter
gilliam; tibi-
arcuatis

elytris vix latius, longitudine æquali, segmento primo dorsali valde transverso, mediocri.

♂. Trochanteribus anticis nodosis. Metasternum obsolete totum longitudinaliter depressum. Segmenti secundi ventralis margine apicali medio tuberculo minuto compresso prædito. Femoribus incrassatis. Long. 1·80-1·90 mm.

This species is closely allied to *bicolor*, and the difference of colouration would be of very little value; but the antennæ are longer and the 9-10 joints obconic, whilst they are nearly globose in *bicolor;* the elytra are longer, much less narrowed at the base; the abdomen is longer and not so broad; the last joint of the maxillary palpi has no trace whatever of a peduncle, but is more elongate and more cylindrical than in *bicolor*, but not flagelliform as in *atriventris*. I think it is really a distinct species.

Clarence River, N.S.W. (Mr. A. M. Lea).

I have a specimen from Tasmania which does not differ materially, yet the prothorax is a little more regularly ovate and the femora are stouter.

Genus CURCULIONELLUS, Westwood.

Trans. Ent. Soc. Lond. 1870, ii. p. 127.

This genus is very distinct, having the maxillary palpi shorter, with the last joint nearly cylindrical; the first joint of the

Curculionellus semipolitus, Schaufuss.

Tijds. v. Ent. xxix. 1886, p. 255.

The antennæ are short and thick, with the joints nearly transverse; the dorsal stria is simple, not geminate even at the base. The unique specimen known is a ♀.

Rockhampton, Q.

Curculionellus riparius, n.sp.

Ferrugineus, palpis testaceis, lævis, nitidus, capite prothoraceque minute reticulatis et opacis exceptis. Caput valde elongatum, antice attenuatum, juxta antennas transversim subnodosum, sulco longitudinali parum profundo, lato, inter oculos abrupto et profundiori, in vertice obsoletissime extenso. Antennæ graciles, elongatæ, articulis 1° longo, cylindrico, 2 cylindrico et latitudine sua longiori, 3-6 breviter obconicis, latitudine sua paulo longioribus, 7-8 longioribus, 9-10 sexto fere duplo longioribus, 11 magno, suboblongo-ovato. Prothorax latitudine sua longior, antice plus postice minus attenuatus, latitudine maxima ante medium, lateribus postice nonnihil sinuatus, sulco transverso arcuato, obsoleto. Elytra parum elongata, lateribus leviter arcuatis et humeris perparum obliquis, stria dorsali basi geminata.

♂. Metasternum valde sulcatum et utrinque longitudinaliter gibbosum; segmento 2° ventrali medio valde gibboso; tibiis anticis intus, pone medium, leviter emarginatis et arcuatis.

♀. Metasternum minus sulcatum et gibbosum, segmento ultimo dorsali apice leviter obtuse producto; trochanteribus intermediis apice minute spinosis. Long. 1·90 mm.

Tweed River, N.S.W. (Mr. A. M. Lea).

Genus Tyraphus, Sharp.

Trans. Ent. Soc. Lond. 1874, iv. p. 489.

This genus is very distinct owing to the maxillary palpi which are short, thick, with the last joint more or less triangular and rounded at apex.

It is more abundant in Australia than anywhere else, but is found also in Tonkin, Manila and Sumatra.

I have described one species as coming from Brazil, but very likely the indication of locality is erroneous and the insect comes from Australia. I will include it therefore in the following synopsis :—

A 2. Prothorax smooth and shining.

B 2. Transverse basal furrow of the prothorax much interrupted in the middle by a strong fovea; lateral foveæ strong; antennæ short, joints transverse, 9-10 larger, very transverse, last one not longer than the two preceding ones together and very obtuse at apex; last joint of the palpi triangular, rounded at apex—Rockhampton.......................... *proportionalis*, Schfs.

B 1. Transverse furrow of the prothorax hardly interrupted, the median fovea being hardly visible.

C 2. Rather convex, prothorax irregularly ovate, being broader in front, but the sides are well rounded; antennæ not very thick, joints 3-8 moniliform, 9-10 broader and transverse, 11 regularly ovate; dorsal stria simple, even at the base—Australia.......................... *Howitti*, King.

C 1. Rather flat, prothorax nearly cordate, a little

C 1. Rather convex and ovate; antennæ not so thick, joints 3-8 moniliform, little transverse; 9-10 larger, more transverse; 11 much broader, ovate-acuminate; dorsal stria nearly entirely geminate—Tamworth, Windsor... ... *rugicollis*, n.sp.

B 1. Antennæ much longer and more slender, with all the joints, or at least 9-10, quadrate.

C 2. Joints of the antennæ 3-8 a little transverse, 9-10 quadrate, 11 ovate, rather elongate; head flattened and enlarged, rounded between the eyes, longitudinal furrow narrow, ending between the eyes, where are two foveæ very approximate; prothorax cordate, with the transverse furrow very obsolete—Adelaide... *sobrinus*, Schfs.

C 1. Antennæ longer and slender, joints 3 obconic, 4-8 quadrate, 9-10 longer than broad, 11 oblong-ovate; head not so flattened and not so broad—Clyde River..... .... *umbilicaris*, Schfs.

TYRAPHUS RUGICOLLIS, n.sp.

Cinnamomeus, capite prothoraceque subtilissime reticulatis et subopacis, setis aliquot ochraceis, sparsis. Caput ante oculos elongatum et parallelum, sulcatum, inter oculos leviter ampliatum, late impressum et fundo foveis duabus. Oculi magni. Palpi maxillares crassi, articulo ultimo rotundatim securiformi. Antennæ crassæ, articulis 1° magno, 2 ovato, 3 brevissime obconico, 4-8 transversis, 9-10 paulo majoribus, transversis, 11 magno, ovato. Prothorax subcordatus, lateribus obtuse carinatus, foveis lateralibus sulcum obsoletissimum longitudinalem emittentibus, sulco dorsali transverso obsoleto, fovea media lata sed obsoleta. Elytra latitudine sua longiora, subconvexa, lateribus subrotundatis, humeris bene notatis, stria dorsali arcuata, minutissime geminata, margine postica nigro-setosa. Metasternum late sed parum profunde longitudinaliter impressum. Long. 1·50 mm.

This species is closely allied to *planus*, Sharp. The antennæ are not quite so thick, the head is a little longer, the lateral foveæ are smaller and not so deep; the elytra are much more convex and rounded on the sides, with the dorsal stria bearing outside a second stria exceedingly fine, but distinct.

Tamworth and Windsor, N.S.W. (Mr. A. M. Lea).

Tribe CTENISTINI.

Genus CTENISOPHUS, n.g.

This new genus is closely allied to *Ctenistes*, but differs in the following points—body more robust, broader and shorter; the last joint of the maxillary palpi is transverse or transversely globose and provided externally with a long appendage, its internal apical angle is more or less sharp and provided with a very small appendage; in the under face of the head, close to the neck, is a strong transverse carina ended on each side by a long, sharp, more or less curved spine, distant from the eyes and quite independent of them; the antennæ of the ♂ have a long club of four joints, but this club is always much shorter than in *Ctenistes*, and is hardly as long as the half of the antenna.

Ctenisophus resembles much more *Desimia* (from Europe and Asia) and *Ctenisis* (from America), but in those two genera the last joint of the maxillary palpi is not transverse, but oblong-ovate and strongly acuminate, and the infracephalic spines are not distant from the eyes but inserted on the inferior margin of the eyes.

basi curvato, apice curvatim incrassato, 3 latitudine sua maxima haud longiori, lateribus curvato et extus ante medium producto et sat longe appendiculato, 4 paululum minori, triangulari, lateribus curvatis, angulo interno apicali obtuso, minutissime appendiculato, angulo externo leviter producto et longe appendiculato. Prothorax transversus, lateribus rotundatus, subæqualiter antice et postice attenuatus, foveis lateralibus minutis, media brevi simplici, disco æquali.

♂. Antennæ elongatæ, graciles, articulis 3-7 latitudine sua paulo longioribus, 8-10 longitudine æqualibus sed latitudine crescentibus, 8 latitudine sua duplo longiori, 11 decimo crassiori, decimo et nono simul sumptis paulo breviori, irregulariter ovato, ad apicem incrassato, obtuse acuminato; segmento 3 ventrali late transversim impresso, istæ impressionis margine erecta.

♀. Antennæ breves, graciles, articulis 3-8 latitudine sua paulo longioribus, 9 præcedenti vix longiori, 10 nono fere duplo longiori, paulo crassiori, 11 magno, ovato, decimo triplo longiori. Long. 1·40-1·60 mm.

Tasmania; Australia; Port Augusta, S.A.; Swan River, W.A.; Melbourne, Vic.

The antennæ vary to a certain extent, the joints of the club being more or less elongate (more elongate in the specimens from Tasmania), but all the other characters, and more especially the

♀. Antenuarum articulis 3-7 latitudine sua longioribus, 7 cæteris longiori, 8-9 quadratis, 10 majori, fere quadrato, 11 magno oblongo-ovato, ad apicem leviter incrassato, duobus præcedentibus simul sumptis longiori. Long. 1·50 mm.

Swan River, W.A. (Mr. A. M. Lea).

Ctenisophus vernalis (?), King.

Trans. Ent. Soc. N.S.W. i. 1863, p. 40; ♀, *hesperi*, King, *l c.*, p. 40.

I have no type of *vernalis*. Mr. King's description is very short; it is possible that the true *vernalis* may be another insect than the one I consider as being so and whose description follows.

Head narrower than in the other species, with the sides nearly parallel between the eyes, which are very large; foveæ nearly transverse ; maxillary palpi rather elongate, resembling those of *Kreusleri*, but shorter, joints 2 angulate, 3 a little longer than broad, the external expansion which is very acute and long, situated a little before the middle, the upper part of the joint nearly cylindrical, the appendage very long ; the 4th large, triangularly transverse, the outside and inferior margins curved and convex, the upper one curved but concave, with the internal apical angle acute and sharp and having a very short appendage, the external one much produced, with a long appendage: pro-

The rather numerous species range from North America to Japan, and occur also in Indo-Malay region, Africa and Australia.

TMESIPHORUS KINGII, Macleay.

Trans. Ent. Soc. N S.W. 1873, ii. p. 151.

This is a fine insect; the head and prothorax are densely rugoso-punctate; the elytra are much less punctate, more especially in the depressions which are nearly smooth, the abdomen is more evenly punctate than the elytra; in the ♂ the 9th antennal joint is rather obconical, 10th more cylindrical, but emarginate beneath, 11th large, subglobose, and somewhat gibbose beneath in the middle; in the ♀ 9th and 10th are similar to each other and more regularly subovate, 11th is a little larger and ovate.

Gayndah, Q.; Brigham [? Brisbane]; N S. Wales.

TMESIPHORUS MACLEAYI, King.

Trans. Ent. Soc. N.S.W. i. 1863, p. 40; 1864, p. 102: *Sintectes carinatus* (?) Westw., Trans. Ent. Soc. Lond. 1870, p. 130; Thes. Ent Oxon. Pl 4, f. 10.

I do not know this species, which, according to Macleay, is lighter in colour and not so rough nor so strongly sculptured as *Kingii*.

This species is probably identical with *Sintectes carinatus*,

recorded again. In any case *Sintectes carinatus*, Westw., belongs to the genus *Tmesiphorus*, the figure and the description leaving no doubt about it.

TMESIPHORUS FORMICINUS, Macleay.

Trans. Ent. Soc. N.S.W. ii. 1863, p. 370.

I received from Mr Lea two specimens which he considers as co types of *T. formicinus*, Macl. The punctuation is dense, fine, and rugose on the head and prothorax which are opaque, on the elytra and the abdomen it is not rugose and not so dense; the pubescence is rather long and fine; the last joint of the maxillary palpi is shorter and more rounded externally; the joints of the antennæ 4-8 are more and more transverse, 9-10 much larger, nearly transverse, with the sides a little rounded and both of the same size, the 11th is ovate, acuminate, not truncate at the base; the prothorax has the disc evenly rounded and the sides not sinuate, the foveæ are much smaller; the carinæ on the elytra are more obsolete and shorter; the carinæ of the abdomen are much smaller and conspicuous only on the first segment and on the base of the second one; the metasternum and the second ventral segment are only a little depressed in the middle; in the ♂ the 10th joint of the antennæ is somewhat notched in the middle beneath; the metasternum and ventral segment more impressed.

TMESIPHORUS TERMITOPHILUS, n.sp.

Castaneus, nitidus, elytris dilutioribus, brevissime et rude ochraceo-pubescens. Caput grosse et confertim punctatum, antice attenuatum, fronte sulcatum, inter oculos foveis duabus magnis, temporibus pone oculos acute prominentibus. Palpi testacei, articulis 2 apice incrassato, 3 suboblongo, medio extus leviter dilatato, ambobus longe appendiculatis, 4 præcedenti haud longiori, apice intus valde et longe acuminato, extus rotundato. Antennæ validæ, crassæ, compactæ, cylindricæ, absque clava, et apice vix crassiores, articulis 1 magno, 2 subquadrato, 3 paulo longiori, subobconico, 4 latitudine sua vix longiori, 5 paulo longiori, 6 fere transverso, 7 quadrato, 8 transverso, 9 præcedenti duplo longiori, 10 nono paulo longiori, 11 longiori, basi truncato, paulo crassiori, obtuse acuminato. Prothorax grosse sed non confluenter punctatus, capite et latitudine sua longior, antice sat abrupte attenuatus, lateribus ante medium rotundatus, dein ad apicem sinuato-angustatus, disco pone medium gibbosus et obtuse tuberculatus, ante basin fovea subelongata, foveis lateralibus magnis. Elytra grosse sed disperse punctata, latitudine sua longiora et basi leviter attenuata, humeris obliquis, elevatis, et obtuse carinatis, disco utrinque carina, valida, recta et subintegra. Abdomen densius punctatum, elytris longius, lateribus late et alte marginatum, segmentis duobus primis dorsalibus utrinque obtuse carinatis 1° apice medio obtuse prominenti, 2 medio tuberculo magno, ovato, obtuso, 3 apice obtuse producto. Pedes validi, tibiis anticis leviter incurvis et medio extus crassioribus, intermediis subrectis, apice leviter incurvis, posticis vix incurvis.

♂. Metasternum late et profunde impressum utrinque longitudinaliter obtuse carinatum, segmentis ventralibus 2 medio valde et profunde impresso, margine postica emarginata, 3 toto medio ovatim impresso, postice leviter emarginato, 4 et 5 transversim minus impressis, 6 medio longitudinaliter et leviter impresso, istis impressionibus fundo lævibus; tibiis intermediis apice magis incurvis.

Genus HAMOTOPSIS, n g.

Crassus. Antennæ basi distantes, crassæ, clava conspicua, triarticulata, articulo ultimo magno. Palpi maxillares magni, articulis 1° minuto, attamen conspicuo, 2 multo majori et crassiori, subcylindrico, intus medio subemarginato, 3 minutissimo, subtriangulari, 4 maximo, irregulariter subgloboso, basi oblique truncato, extus rotundato, intus subrecto, simplici, haud canaliculato, tantummodo depresso, apice intus foveola rotundata appendicem membranaceum ferenti. Prothorax foveis tribus sulco transverso junctis ornatus. Elytra brevia. Abdomen subelongatum, late marginatum, segmentis subæqualibus. Coxis intermediis haud contiguis, posticis distantibus; trochanteribus anticis brevibus, intermediis elongatis, posticis subelongatis; tarsorum articulis tribus, 1° minutissimo, 2 subconico 3 præcedenti fere duplo longiori, unguibus binis æqualibus.

This genus is very closely allied to *Hamotus* from America, and more especially to the subgenus *Pseudohamotus*, from which it differs only by the maxillary palpi; in *Hamotus* the 1st joint is invisible, the 2nd is long, more or less conical, the 4th one is more or less sulcate inwardly. In *Hamotopsis* the first joint of the maxillary palpi is rather long and subconical; the 2nd is abruptly thick and cylindrical but a little bent with the internal margin sinuate; the 3rd one is still smaller than in *Hamotus;* the 4th one has no groove, but is only flattened inwardly, and at the apex there is inwardly a fovea in which the membranaceous appendage is inserted.

HAMOTOPSIS AUSTRALASIÆ, n.sp.

(Plate x., figs. 7 and 8.)

Oblongus, ferrugineus, elytris castaneis, palpis rufis, sat dense sed breviter fulvo-setosus, disperse et obsoletissime punctatus. Caput deplanatum, latitudine sua longius, ante oculos sat abrupte attenuatum, inter oculos anterius foveis duabus validis, suboblongis et in fronte medio fovea altera minori, sulciformi.

Antennæ crassæ, validæ, articulis 1 magno, cylindrico, 2 paulo minori, quadrato, 3-8 transversis, 9-10 majoribus, vix transversis, 11 magno, duobus præcedentibus longiori et latitudine sua maxima paulo longiori, irregulariter pyriformi, extus oblique truncato et apice obtuso. Prothorax capite latior et longitudine subæqualis, antice attenuatus, latitudine maxima ante medium sita, lateribus subrectis, foveis tribus subæqualibus sulco transverso parum profundo unitis, basi ipsa carinula brevi longitudinali prædita. Elytra longitudine sua paulo latiora, basi attenuata, humeris obliquis et notatis, basi foveis duabus magnis, stria suturali integra, sulco dorsali brevissimo et diffuso. Abdomen elytris multo longius, segmentis inter se æqualibus, 1 et 2 basi medio longitudinaliter carinatis. Metasternum medio longitudinaliter depressum. Pedes validi, femoribus parum incrassatis; tibiis vix incurvis. ♀. Long. 2·50 mm.

Clarence River, N.S.W. (Mr. A. M. Lea).

Genus Hamotulus, Schaufuss.

Tijds. v. Ent. xxx. p. 108.

This genus has been created by Dr. Schaufuss for his *Bryaxis chamæleon* from Australia: it includes *Tyrus mutandus*, Sharp, from New Zealand, and also another new species from Australia.

It resembles *Hamotus*, but it differs much in the sha e of the

deplanatum et vertice transversim convexum, inter oculos anterius foveis duabus magis inter se quam ab oculis distantibus, juxta tuberculum frontalem tuberculis duobus minutis, parum distantibus, tuberculo frontali leviter transverso et obsolete diviso. Oculi magni. Antennæ validæ elongatæ, basi distantes, articulis 1° majori, 2° subcylindrico, latitudine sua longiori et cæteris fortiori, 3 obconico, 4-6 latitudine sua paululum longioribus, 7 quadrato, 8 subtransverso, clava magna triarticulata, 9 leviter transverso et paululum obliquo, 10 majori, subquadrato, apice obliquo et angulo apicali externo producto, 11 præcedenti angustiori, ovato. Prothorax nonnihil transversus, capite latior, antice valde attenuatus, postice lateribus subrectis, foveis tribus quarum media multo minori sulco transverso junctis. Elytra subquadrata, convexa, basi vix attenuata, humeris subrotundatis et notatis, basi foveis duabus magnis, stria suturali integra, sulco dorsali medio abbreviato. Abdomen breve, convexum, marginatum, segmento 1° dorsali paulo majori et basi transversim profunde impresso. Metasternum altum, subcordatum, late deplanatum. Segmento ventrali ultimo, apice profunde bisinuato, oblonge foveato. Pedes validi, femoribus sat incrassatis, tibiis omnibus ad apicem leviter incurvis et apice minutissime calcaratis. ♂. Long. 1·90 mm.

This species differs much from *chamæleon*, Schfs., which has the antennæ simple, the head without tubercle, the elytra much more attenuate at base, with a well marked and much longer dorsal stria.

Forest Reefs, N.S.W. (Mr. A. M. Lea).

Genus Leasymus, n.g.

Oblongus, tuberculo antennario lato, antennæ basi distantes, elongatæ, clava triarticulata; maxillarum cardine prominenti, angulo inferiori externo longe spinoso; palpi maxillares magni, articulis 1 minuto, 2 magno, securiformi, in angulo inferiori processu longo, incurvo et apice penicillato, processu altero medio, breviori, recto, penicillato, prædito, 3 ad angulum secundi supernum inserto, basi pyriformi et processu apice penicillato, longissimo, prædito, 4 in secundum flexo, transverso, apice valde subulato et acuminato, infra medio angulato et processu longo penicillato prædito. Prothorax ovatus. Abdomen late marginatum, segmento 1° dorsali cæteris multo majori, ventralibus 7 ♂ 6 ♀ instructum, 2 ventrali cæteris parum majori (♂ sexto apice profunde angulato-emarginato et 7° minuto, rhomboidali). Coxæ posticæ valde distantes. Prosterni lateribus infra (episternis) anterius productis. Trochanteribus anticis et posticis parum elongatis, intermediis elongatis, tarsis gracilibus, elongatis; unguibus binis mediocris, æqualibus.

This new genus resembles *Didimoprora*, but is very different; the construction of the maxillary palpi is most extraordinary, and bears some analogy to the same organ in another Australian insect, *Schistodactylus phantasma*, Raffr., which in every other respect is exceedingly distinct.

LEANYMUS PALPALIS, n.sp.

(Plate x., figs. 5-6.)

Totus ferrugineus, antennis ad apicem nigrescentibus, palpis testaceis, nitidus, parce nigro-pubescens. Caput latitudine sua

fundo oblonge foveatum, juxta coxas intermedias transversim carinato-elevatum, ista carina utrinque in cornu divaricata; segmentis ventralibus sexto basi impresso apice angulatim emarginato, 7 minuto, transversim rhomboidali; pedium anticorum trochanteribus basi et femoribus basi minute sed longe spinosis, tibiis apice leviter incurvis, femoribus incrassatis, intermediorum femoribus incrassatis et infra medio obtuse angulatis, tibiis leviter incurvis et apice minute calcaratis, posticorum femoribus leviter incrassatis, tibiis elongatis et subrectis.

♀. Prosterni episternis obtuse productis; metasternum simplex; pedium anticorum trochanteribus et femoribus validius spinosis, intermediorum trochanteribus basi et femoribus basi spinosis. Long. 2·40-2·70 mm.

Clarence and Tweed Rivers, N.S.W. (Mr. A. M. Lea).

Genus DIDIMOPRORA, Raffray.

Rev. d'Ent. 1890, p. 148.

I had established this new genus for *Tyrus Victoriæ*, King, and it is very different from *Tyrus*, which has no representative in Australia. There are several new species of these fine insects.

A 2. Prothorax smooth, hardly punctate.

B 2. Posterior tibiæ ♂ with a strong ante-apical spur.

C 2. Spur very strongly compressed; joints of the antennæ 8-10 transverse in the ♂ ... . *Victoriæ*, King

C 1. Spur hardly compressed; joints of the antennæ 8-10 square in the ♂....... .... *armata*, n.sp.

B 1. Posterior tibiæ without ante-apical spur, but provided with a strong basal tooth and another median one. ♂.. .... *Leana*, n.sp.

A 1. Prothorax strongly punctured and even rugose.

B 2. Joints of the antennæ 9-10 square; punctuation of the prothorax strong, but simple.. .... *puncticollis*, n.sp.

B 1. Joints of the antennæ 9-10 transverse; punctuation of the prothorax very strong, confluent and rugose . ... *dimidiata*, n.sp.

DIDIMOPRORA VICTORIÆ, King

Tyrus Victoriæ, King, Trans. Ent. Soc N.S.W. i 1865, p. 168.
(Plate x., fig. 39.)

Oblonga, nigro-picea, antennis pedibusque castaneis, elytris rubro-castaneis (in ♂ basi infuscatis), palpis testaceis, pube castanea. Caput latitudine sua longius, tuberculo antennario elongato et valde sulcato, basi foveato, foveis duabus alteris magis inter se quam ab oculis distantibus. Antennæ validæ, articulis 1° majori, 2 quadrato, 3 subobconico, 4-5 elongato-quadratis, 6 simili sed angustiori, 7 subquadrato, cæteris in utroque sexu diversis. Prothorax oblongus, antice posticeque angustatus, lateribus rotundatus, foveis lateralibus magnis, media ante basali oblonga. Elytra latitudine sua longiora, ad basin attenuata, humeris obliquis et notatis, sulco dorsali obsoleto et brevissimo. Abdomen elytris longius, convexum.

♂. Antennarum articulis 8 transverso, 9 transverso, octo duplo latiori, 10 transverso majori, angulo interno apicali producto, ambobus infra deplanatis, 11 magno, ovato, basi truncato, infra basi oblique impresso. Metasternum convexum, late nec profunde longitudinaliter impressum. Segmentis ventralibus 4 medio depresso, 5 impresso, 6 foveato. Trochanteribus intermediis

DIDIMOPRORA ARMATA, n.sp.

Praecedenti simillima, differt elytris brevioribus et basi minus attenuatis, abdomine castaneo.

♂. Antennæ minores, articulis 4-7 subcylindricis, latitudine sua longioribus, 8 et 9 quadratis, 9 præcedenti duplo majori, 10 majori, vix transverso, 11 ovato. Metasternum medio irregulariter impresso, juxta coxas intermedias medio tuberculo aurantiaco fasciculato, utrinque fovea magna et tuberculis duobus. Segmentis ventralibus 5 medio transversim impresso et utrinque tuberculato, 6 medio depresso, utrinque profunde foveato et valde convexo, 7 rhomboidali, magno, ultimo dorsali arcuatim carinato; trochanteribus intermediis medio dente breviori compresso, acuto, posticis basi dente brevi leviter recurvo præditis; tibiis intermediis apice minute calcaratis, posticis ante apicem calcare longo, vix compresso armatis. Long. 2·50 mm.

The ♀ is not known and will very likely prove to be very closely similar to the ♀ of *Victoriæ*.

Clyde River, N.S.W.

DIDIMOPRORA LEANA, n.sp.

(Plate x., fig. 38.)

Oblonga, capite prothoraceque nigris, abdomine antennis pedibusque ferrugineis, elytris rubris, pube brunnea. Caput latitudine sua longius, tuberculo antennario lato et valde sulcato, inter oculos foveis duabus inter se et ab oculis subæqualiter distantibus, postice in vertice medio foveola minuta singula. Antennæ valide, articulis 1° magno, 2 quadrato, 3 subconico, 4 quadrato, 5 6 subcylindricis et paulo longioribus, 7 paulo breviori, 8 leviter transverso, cæteris in utroque sexu diversis. Prothorax disperse et vix perspicue punctatus, oblongus, antice plus postice minus attenuatus, lateribus medio rotundatis, foveis lateralibus magnis, media oblonga. Elytra latitudine sua longiora, basi leviter attenuata, humeris rotundatis et notatis, sulco dorsali obsoleto et brevi. Abdomen elytris longius, convexum.

♂. Antennarum articulis 5 paulo latiori, 7 subobconico, 8 leviter transverso et intus apice leviter producto, 9 magis transverso et producto, 10 transverso, minus producto, 11 oblongo, infra basi fovea magna. Metasternum late depressum, segmentis ventralibus 5 vix impresso, 6 convexo, 7 ovali. Trochanteribus intermediis dente maximo, compresso et acuto armatis, coxis posticis et trochanteribus basi obtuse productis; tibiis intermediis apice minute calcaratis, posticis, basi intus dente maximo, lato, compresso, laminiformi, apice oblique truncato et ad medium dente altero apice truncato et intus fasciculato præditis.

♀. Antennarum articulis simplicibus, 9-10 transversis, 11 ovato; segmentis ventralibus 5 arcuatim apice emarginato, 6 transverso, ultimo dorsali producto, triangulari, parte inferiori deplanata. Long. ♂. 3·40-3·60; ♀. 2·90-3·20 mm.

This species differs from the preceding ones by the antennal tubercle broader and shorter, and the scattered, hardly visible punctures of the prothorax.

Clarence, Tweed, and Richmond Rivers, N.S.W. (Mr. A. M. Lea).

DIDIMOPRORA PUNCTICOLLIS, n.sp.

(Plate x., fig. 37.)

♂. Metasternum juxta coxas intermedias, antice, profunde bifoveatum, istis foveis carina transversa antice limitatis et inter istam carinam et coxas intermedias penicillis duobus recurvis aurantiacis, postice leviter impressum et utrinque setosum; segmentis ventralibus 5-6 medio impressis, 7 suborbiculari; femoribus anticis, basi infra oblique impressis et carinulatis; trochanteribus valde angulatim dilatatis et compressis; tibiis posticis pone medium intus calcaratis et setosis, dein ad apicem leviter emarginatis et apice minute calcaratis. Long. 3·50 mm.

The ♀ is not known.

Tamworth, N.S.W. (Mr. A. M. Lea),

DIDIMOPRORA DIMIDIATA, n.sp.

Magis ovata et anterius attenuata, rubro-castanea, capite et prothorace dense confluenter et valde rugoso-punctatis, obscurioribus, pube longa, obscura. Caput elongatum, antice valde attenuatum, tuberculo antennario elongato et angustato, obsolete sulcato, foveis obsoletis. Antennæ elongatæ, articulis 1° valde elongato, cylindrico, 2 quadrato, 3-5 oblongis, subcylindricis, 6 quadrato, 7-8 transversis, 9-10 multo majoribus, transversis, 11 ovato. Prothorax subhexagonus, subconvexus, foveis lateralibus oblongis, media obsoleta. Elytra latitudine sua vix longiora, basi attenuata, humeris obliquis et notatis, sulco dorsali parum profundo et brevi. Abdomen leviter ampliatum. Metasternum convexum, obsolete sulcatum; segmentis ventralibus 5 arcuatim emarginato, 6 transverso, ultimo dorsali abrupte deflexo haud producto. ♀. Long. 2·70-2·80 mm.

This species is very distinct, the body being shorter and more attenuate in front; the first joint of the antennæ is unusually long and the punctuation on the head and the prothorax is rugose and confluent. The ♂ is not known.

Bridgetown, W.A. (Mr. A. M. Lea).

Genus SPILORHOMBUS, n.g.

Caput elongatum, antice parum attenuatum, altum, lateribus et postice retusum, supra planum et plurifoveatum. Palpi maxillari mediocres, articulis 2° basi gracili apice sat abrupte ovaliter incrassato, 3 obconico et latitudine sua longiori, 4 præcedenti duplo longiori, multo crassiori, ovato et leviter intus securiformi et breviter pubescenti. Oculi magni, postice siti. Antennæ crassæ, basi leviter distantes. Prothorax cordatus, antice sat abrupte attenuatus. Elytra sat magna, basi valde bifoveata et sulcata. Abdomen elytris paulo longius, late marginatum; segmentis tribus primis dorsalibus longitudine leviter crescentibus, ventralibus 2-5 subæqualibus. Coxis intermediis approximatis, posticis parum distantibus; trochanteribus intermediis elongatis, anticis et posticis brevibus; pedes elongati et graciles, unguibus binis æqualibus.

This genus differs from *Didimoprora* by the palpi, the last joint of which is much thicker and somewhat securiform; by the head which is raised, but flattened, without antennal tubercle; and by the prothorax which is more cordate.

SPILORHOMBUS HIRTUS, n.sp.

Oblongus, cinnamomeus, longe et dense hirtus, pubescentia subaurea pilis obscuris intermixta. Caput latitudine sua multo longius, postice et lateraliter abruptum, lateribus leviter obliquis, angulis posticis fere acutis et dense obscure ciliatis, foveis duabus notatis, basi foveis duabus magnis aureo pubescentibus, stria depressa et sulcis dorsalibus subintegris. Metasternum subconvexum, postice vix impressum. Femoribus parum crassis, intermediis leviter incurvis, tibiis intermediis et posticis leviter sinuatis. Abdomen infra medio leviter depressum, segmentis ventralibus 6° transversim apice emarginato, 7 minuto, transversim triangulari. ♂. Long. 2·50 mm.

Swan River, W.A. (Mr. Brewer).

Genus TYROMORPHUS, Raffray.

Rev. d'Ent. 1883, p. 240.

This genus is very different from *Didimoprora*. The head has no more an antennal tubercle: the palpi differ much—the first joint is very small; the second one is slender at base and inflated at the apex, which is more or less obliquely truncate; the third one is rather large, but not so long as the adjoining ones, it is more or less angulate in the middle inwardly; the fourth one is large, ovate, more or less obliquely truncate at the base, generally sulcate or at least strongly attenuate inwardly, obliquely truncate outwardly at the apex, this truncature bearing the setiform appendix: the intermediate trochanters are comparatively short, the metasternum is broad and short: the posterior coxæ very distant; and the 2nd ventral segment larger than the others.

This genus includes *nitidus*, Raffr., type of the genus, *comes*, Schfs., and *Tyrus spinosus* and *humeralis*, Westw., which I do not know, besides several new species.

Here follows a synopsis of this genus :—

A 2. Entirely punctate.
B 3. Head short, rounded; three last joints of the antennæ black; 3rd joint of the maxillary palpi ovate, elongate.... .... *humeralis*, Westw.
B 2. Head square; antennæ black, except the three first joints; third joint of the maxillary palpi short, subtriangular.................. .... *nigricornis*, n.sp.
B 1. Head rather long, attenuate in front.
C 2. Antennæ bicolorous.

D 2. Joints of the antennæ 3-10 black, 11th testaceous; 3rd joint of the palpi ovate; impression on the first dorsal segment small and without carinæ... *cribratus*, n.sp.

D 1. Antennæ castaneous, 11th joint rufous; 3rd joint of the palpi short, subtriangular; impression of the first dorsal segment obliquely sulcate and carinate on each side.................................... *Mastersi*, Macl.

C 1. Antennæ unicolorous.

D 2. Impressions of the first dorsal segment small, without carinæ; 3rd joint of the palpi briefly ovate.. *comes*, Schfs.

D 1. Impressions of the first dorsal segment very long, sulciform, divergent and carinate; 3rd joint of the palpi short and triangular......... .... *dispar*, n.sp.

A 1. Entirely smooth.

B 2. Black, elytra red; joints of the antennæ 7-10 black. *spinosus*, Westw.

B 1. Body unicolorous; antennæ little darker towards the apex.

C 2. 3rd joint of the palpi strongly angulate inwardly at the middle; penultimate dorsal segment produced above the pygidium............................ *nitidus*, Raffr.

C 1. 3rd joint of the antennæ ovate, rounded inwardly; penultimate dorsal segment simple............. ... *lævis*, n.sp.

TYROMORPHUS NIGRICORNIS, n.sp.

Oblongus, castaneus, elytris pedibusque dilutioribus, antennarum articulis 1-3 castaneis, cæteris nigris, palpis rufis, totus crebre et

1° dorsali basi transversim depresso et utrinque longitudinaliter et parum oblique impresso et minute carinato.

♂. Corpore angustiori et graciliori. Elytra basi minus attenuata; trochanteribus anticis basi spina minuta recurva, intermediis basi dente valido, recurvo præditis; tibiis anticis apice minute calcaratis, femoribus anticis minutis. Metasternum late concavum et medio sulcatum; segmento ultimo ventrali utrinque sinuato, medio longe lobato et producto, impresso.

♀. Corpore latiori et robustiori. Elytra basi magis attenuata; trochanteribus anticis basi spina elongata et femoribus anticis basi spina breviori armatis. Metasternum medio sulcatum; segmento ultimo ventrali utrinque leviter sinuato, medio obtuse lobato; ultimo dorsali abrupte declivi, haud producto. Long. 2·10-2·30 mm.

This species is likely very closely allied to *humeralis*, Westw.; but according to the description and figure, the head is more elongate, the third joint of the maxillary palpi is much shorter, and the last joints of the antennæ much more transverse.

Clarence and Tweed Rivers, N.S.W. (Mr. A. M. Lea).

TYROMORPHUS CRIBRATUS, n.sp.

Oblongus, castaneus, elytris pedibusque dilutioribus, antennarum articulis 3-10 nigris, ultimo testaceo, totus crebre et valde punctatus, breviter obscure et sublente pubescens. Caput latitudine sua multo longius, antice attenuatum, deplanatum, inter oculos foveis duabus, fronte medio impresso et foveola minuta utrinque in angulo externo antico. Palpi validi, pallide testacei, articulis 2 apice valde clavato, 3 ovato, intus rotundato-ampliato, 4 magno, ovato, basi oblique truncato, intus late sulcato. Antennæ elongatæ, articulis 1° longo, subcylindrico, 2 oblongo, 3-7 oblongis et decrescentibus, 8 fere quadrato, 9 magno, subobconico, latitudine sua longiori. 10 subquadrato, 11 ovato. Prothorax cordatus, fovea media ante basali, vix perspicua. Elytra latitudine sua longiora, basi attenuata, humeris valde obliquis, parum notatis, sulco dorsali valido, profundo, dimidiam partem disci attingenti.

Abdomen elongatum, segmento 1° dorsali basi transversim impresso, foveis duabus minutis absque carinis. Metasternum late deplanatum, sulcatum.

♂. Segmento ultimo ventrali utrinque sulcato, medio lobato et producto, ultimo dorsali convexo, medio apice angulatim emarginato.

♀. Segmento ventrali ultimo brevi, medio sinuato, utrinque latiori; ultimo dorsali abrupte declivi et concavo. Long. 2·80-3·20 mm.

Clarence River, Tamworth, N.S.W. (Mr. A. M. Lea).

This species differs from *nigricornis* by the head more elongate, the antennæ longer and the impressions of the 1st dorsal segment small and without carinæ.

One specimen (♀) from Tamworth has the abdomen piceous-black and shorter, the elytra less attenuate at the base and the shoulders less oblique, the head shorter, and the joints of the antennæ 9-10 a little longer; the body is stouter. Long. 2·50 mm. The ♂ being unknown, I dare not consider this insect as a distinct species.

TYROMORPHUS MASTERSI, Macleay.

Trans. Ent. Soc. N.S.W. 1871. ii. p. 152.

ratis, notatis, sulco dorsali medium superanti. Abdomen minus elongatum, basi transversim impressum, utrinque sulco obliquo divergenti et carinato. Metasternum sulcatum; segmento ultimo ventrali rhomboidali, apice haud sinuato, medio subgibboso; trochanteribus anticis breviter sed acute spinosis. ♂. Long. 2·40 mm.

This species is stouter and shorter than the preceding one, the elytra are more square, and the impressions of the 1st dorsal segment very different.

Gayndah, Q.

Tyromorphus comes, Schaufuss.

Tijds. v. Ent. xxix. 1886, p. 285.

The type of Mr. Schaufuss is a ♂. The intermediate trochanters are compressed, with an angular and obtuse dilatation; the metasternum is concave and sulcate, the ultimate ventral segment is sinuate in each side and produced in the middle, the last dorsal one is convex and emarginate; the palpi and antennæ are very much the same as in *cribratus*, but the antennæ are unicolorous and joints 9-10 are more elongate and more ovate. The body is shorter and stouter. Long. 2·40 mm.

Rockhampton, Q.

Tyromorphus dispar, n.sp.

Subovatus et postice ampliatus, totus rufo-castaneus, totus disperse punctatus, brevissime, sublente et pallide pubescens. Caput latitudine sua multo longius, antice leviter attenuatum, fronte medio impresso, inter oculos foveis duabus. Palpi validi, articulis 1° valde clavato, 3 brevi, intus angulato, 4 magno, ovato, basi truncato, intus leviter sulcato. Antennæ elongatæ, articulis 1 elongato, cylindrico, 2-7 oblongis, leviter decrescentibus, 8 obconico breviori, 9 magno, breviter ovato, 10 haud longiori, sed latiori, 11 ovato. Prothorax cordatus, foveola basali inconspicua. Elytra latitudine sua multo longiora, ad basin sensim attenuata, humeris obliquis et notatis, sulco dorsali fere integro, arcuato. Abdomen elytris multo brevius, segmentis dorsalibus 1° basi transversim valde impresso, sulcis duobus validis, carinatis, divergentibus et inter eos disco convexo, 2° apice medio producto, cæteris valde declivis, ultimo leviter concavo. Metasternum sulcatum; pedium anticorum trochanteribus medio et femoribus basi acute spinosis. ♀. Long. 2·20 mm.

Tamworth, N.S.W. (Mr. A. M. Lea).

This species differs from all the others by the short abdomen, the deep and oblique sulci of the first dorsal segment The ♂ is not known.

TYROMORPHUS LÆVIS, n.sp.

Oblongus, ferrugineus, elytris dilutioribus, nitidus, lævis, parce sublente pubescens, antennis, articulo primo excepto, infuscatis, palpis testaceis, pedibus rufo-testaceis. Caput latitudine sua longius, fronte medio impresso, inter oculos foveis duabus. Palpi validi, articulis 2 clavato, intus subangulato, 3 ovato, intus rotundato, 4 mediocri, ovato, paulo breviori, intus breviter sulcato. Antennæ elongatæ, graciles, articulis 1° elongato, cylindrico, 2-7 oblongis, leviter decrescentibus, 8 præcedenti multo breviori, 9

Genus Abascantus, Schaufuss.

Tijds. v. Ent. xxix. p. 258.

I shall only mention this genus, referring the student to Dr. Schaufuss's description, and the complementary observations I gave in the Rev. d'Ent. (1890, p. 148, tab. iii. figs. 15, 15'). It is closely allied to *Tyromorphus*, Raffr., but the last joint of the maxillary palpi is comparatively much larger and strongly truncate at apex; this truncate part is carinate all round with the internal angle sticking out and has also a longitudinal carina on the upper surface, the third joint is irregularly ovate and much smaller than the last one; the first dorsal segment is also much larger than the others.

It includes only one species, *A. sannio*, Schfs., (*loc. cit.*) and Raffray, Rev. d'Ent. (1890, Pl. iii. figs. 15, 15').

King George's Sound, W.A.

Genus Durbos, Sharp.

Trans. Ent. Soc. Lond. 1874, iv. p. 495.

This genus has been characterised in a masterly manner by Dr. D. Sharp. It comes close to *Gerallus* and *Schaufussia*, and agrees with the latter in having the eyes situated near the posterior angles of the head; the maxillary palpi are also long, the joints being slender at the base and clavate at apex, the three last ones do not differ much in size, the 4th has also a small clivage at apex. It differs from *Schaufussia* by the antennal tubercle, which is as broad as the head, and the latter is not angularly constricted in front; the elytra are very short. Both *Durbos* and *Schaufussia* differ from *Gerallus* by the position of the eyes, which are in the median part in *Gerallus*; and by the large size of the first dorsal segment.

This genus includes only two species, *priscus*, Sharp, (*loc. cit.*) from Champion Bay, W.A., and *interruptus*, Schfs., (Tijds. v. Ent. xxix p. 291) from Sydney. I received the latter of these from Mr. Masters under the name of *Rhytus punctatus*, King.

Genus SCHAUFUSSIA, Raffray.

Rev. d'Ent. 1883, p. 238; 1890, pp. 149, 159 and 160, Pl. iii. fig. 11 : Ann. Soc. Ent. Fr. 1896, p. 132.

I have discussed already (Soc. Ent. Fr. 1896, p. 132) the rather intricate synonymy of this genus, and I will reproduce it.

♀. *formosa*, King, (*Tyrus formosus*) Trans. Ent. Soc. N.S.W. 1863, p. 41, Pl. 16, fig. 1.—Parramatta.

♂♀. *brevis*, Schfs., (*Bryaxis brevis*) Nunq. Otios. iii. 1880, p. 498; *Schaufussia brevis*, Raffr., Rev. d'Ent. 1883, p. 239, Pl. iv. fig. 15.—Tasmania.

♀. *angustior*, Raffr., (*Schaufussia angustior*) Rev. d'Ent. 1883, p. 239, Pl. iv. fig. 16.—New South Wales.

♀. *intermedia*, Schfs., (*Durbos intermedius*) Tijds. v. Ent. xxix. p. 292.—Australia.

♀. *affinis*, Schfs., (*Durbos affinis*) Tijds. v. Ent. xxix. 1886, p. 291. —Tasmania.

♂. *constrictinasus*, Schfs., (*Tyromorphus constrictinasus*) Tijds. v. Ent. xxix. p. 285.—Wide Bay, Q.; Tasmania.

All these different descriptions refer to the same species, which according to seniority must retain the name of *formosa*, King.

or less ovate with the shoulders very little marked; they have a strong dorsal stria.

This genus is rather numerous in Australia; the species are very closely allied to each other, and I think it will be useful to give a synopsis of the species I possess.

A 2. Prothorax smooth or hardly punctate; elytra more or less punctate.

B 3. Prothorax entirely smooth; elytra with a moderate punctuation. ♂. Anterior femora with a small sharp spine beneath not far from the base; anterior tibiæ with a small tooth in the middle; intermediate trochanters with a compressed and recurved tooth at apex; posterior femora with an impression near the middle of the external side; ♀, intermediate trochanters with a very small, sharp and curved spine.—Rockhampton; Wide Bay; Clarence River... .. ♂. *punctipennis*, Schfs.* ♀. *globulicornis*, Schfs.†

B 2. Prothorax with a very fine and scattered punctuation; elytra with a very strong and coarse punctuation.—Swan River........... *dimidiatus*, n.sp.

B 1. Prothorax obsoletely punctate; elytra with a moderate punctuation.—West Australia.... *nanus*, Sharp.‡

A 1. Prothorax and elytra both more or less strongly punctate.

B 2. Punctures on the head and prothorax strong but rather scattered, stronger and denser on the elytra.—Tasmania..................... *modestus*, Schfs §

B 1. Punctures subequal on head, prothorax and elytra.

C 2. Punctures rather small; prothorax broad and subcordate, broadest in front, with the sides oblique behind; elytra with the sides oblique and the shoulders well marked. ♂. Anterior

* Nunq. Otios. iii. p. 509.

† Tijds. v. Ent. xxix. 1886, p. 295.

‡ Trans. Ent. Soc. Lond. 1874, iv. p. 494.

§ Tijds. v. Ent. xxix. p. 295.

trochanters with a strong, sharp, straight and subapical spine, intermediate ones simply carinate; intermediate femora strongly emarginate at base outwardly, thickened in the middle, emarginate inwardly before the end.—Australia.......... *longipes*, Schfs.*

C 1. Punctures very strong.

D 2. Prothorax regularly ovate; elytra, sides and shoulders rounded; palpi nearly as long as the antennæ and slender, joints 3, 4 subequal to each other and hardly longer together than the second one; head nearly flat, sides parallel. ♂. Intermediate trochanters with a very strong, basal and compressed tooth, obliquely truncate and minutely penicillate at apex, intermediate femora strongly emarginate at base outwardly, tibiæ with a sharp, long apical spur.—Parramatta; Clarence River; Tasmania.. *palpalis*, King.†

D 1. Prothorax a little cordate, broadest before the middle; head narrow and depressed in front of the eyes, vertex rather convex; palpi much shorter than the antennæ, each joint much thicker; punctures very strong but a little scattered. ♀.—Clyde River.... *cribratipennis*, Schfs.§

tatus, subcordatus, convexus, capite latior. Elytra grosse et rude sed non confluenter punctata, basi attenuata, humeris nullis, lateribus leviter rotundatis, sulco dorsali parum profundo et medio evanescenti. Abdomen impunctatum, elytris multo brevius. Pedes elongati, femoribus pone medium incrassatis, posticis basi confertim minute strigosis.

♂. Antennarum clava majori. Metasternum grosse punctatum, late nec profunde punctatum.

♀. Metasternum impunctatum minus impressum. Long. 1·60 mm.

Swan River, W.A. (Mr. A. M. Lea).

This species is closely allied to *punctipennis*, but the antennæ are longer, the prothorax is more cordate and the punctuation on the elytra is much stronger.

Genus Rytus, King.

Trans. Ent. Soc. N.S.W. 1866, p. 303.

This genus resembles much *Gerallus*, but the palpi are very different: the last joint instead of being pedunculate at the base and clavate at the apex is clavate at the base and thin and slender (subulate) from the middle to the apex, which is generally a trifle thicker, with a very small section on which is inserted an exceedingly small and short membranaceous appendage. The ♂ has always the head more or less irregular and the penultimate joint of the antennæ longer. There is a good deal of confusion in the synonymy on account of these sexual differences. The species are found exclusively in Australia.

Rytus convexus, King.

Trans. Ent. Soc. N.S.W. 1865, p. 167; [illegible], [illegible]. [illegible] iii. p. [illegible].

I do not know *convexus*, King, and the description of the head is rather incomplete, but I am pretty certain that it is [illegible] [illegible]. The head [illegible] [illegible] [illegible] [illegible] the frontal one is trapezoidal and a little convex, bounded behind by a transverse groove and bearing two strong and short brushes of yellow hairs; the posterior part is much raised all round, but entirely hollowed, the raised sides ending abruptly and horn-like on each side, in front of the eyes; the head is punctate except at the bottom of the cavity; there are few scattered punctures on the body; the maxillary palpi are long, the basal club of the fourth joint is thick, briefly ovate and evidently shorter than the apical subulate part, which is very slender; the antennal joints 3-5 are about twice longer than broad, 6-7 quadrate, 8 nearly transverse, 9 obconic, not longer than broad, and 10 transverse.

The above description is drawn up from two typical specimens of *porcellus*, Schfs. If *corniger*, King, should not agree with this description it must be considered as a different species.

Clyde River, N.S.W.

RYTUS PROCURATOR, Schaufuss.

♂. *procurator*, Schfs., Tijds. v. Ent. xxix. p. 286; ♀. *orientalis*, Schfs., *loc. cit.* p. 287.

The body is shining, smooth, with very scattered but distinct punctures; the head is rather strongly and densely punctate.

♂. Head nearly bipartite, the posterior part strongly punctate,

RYTUS EMARGINATUS, King.

♂. *emarginatus*, King, *(Tyrus)* Trans. Ent. Soc. N.S.W. 1865, p. 303; ? ♀. *Victoriæ*, King, *(Rytus) l.c.* 1865, p. 304.

I have not seen *emarginatus*, King, but I have received from Mr. Masters a specimen labelled *Victoriæ*, King, which is a ♂; and certainly King's description of *Victoriæ* refers to a ♀, and that of *emarginatus*, King, to a ♂. Both have the same size and a very obsolete punctuation. It is most likely that both ♂ and ♀ belong to the same species.

R. emarginatus has certainly the head very much like the head of *procurator*, Schfs., but the strong punctuation on the head of *procurator* would have been certainly mentioned by King, and for this reason I consider it as a different species.

As regards *Victoriæ*, King, (the supposed ♀ of *emarginatus*, King), it is certainly different from *orientalis*, Schfs., (the ♀ of *procurator*, Schfs.); in *procurator (orientalis*, ♀) the punctuation is much scattered, but rather strong and distinct; in *Victoriæ* the punctuation is exceedingly obsolete, hardly visible and very much scattered; the head is longer, more attenuate in front; the maxillary palpi are shorter, the basal club of the 4th joint is thicker and a trifle longer than the apical subulate part, which is not quite so slender as in *procurator*; the 9th joint of the antennæ is at least as long as broad, and much longer than the tenth, which is transverse; the elytra are more rounded and more convex

Parramatta (*emarginatus*, ex King); New South Wales (*Victoriæ*, ex Masters).

RYTUS SUBULATUS, King.

Tyrus subulatus, King, Trans. Ent. Soc. N.S.W. 1864, p. 103: *Rytus punctatus*, King, *l.c.* 1866, p. 303, Pl. v. fig. 6; ♂. *subasper*, Schfs., Nunq. Otios. iii. p. 510.

The identity of *subulatus* and *punctatus*, King, is evident, as it was only a substitution of names; *subulatus*, being the older, must be retained. *R. subulatus* is evidently similar to *subasper*, Schfs. In the collection of Schaufuss I found a specimen labelled "*punctatus*, King; type, Parramatta"; this insect is a ♀ and the types of *subasper*, Schfs., from Tasmania and New South Wales, are all ♂'s.

It is very likely that King had seen both ♂ and ♀ without noticing it; he says "the head is free from other markings except (in some specimens) a slight depression hardly amounting to a fovea." Those specimens with the depression are ♂. It agrees with *subasper*, and the two species are certainly identical.

The head and prothorax are very coarsely and densely punctured; on the elytra the punctuation is equally strong, but much more scattered; the maxillary palpi are moderately long, the basal club of the 4th joint is strong, ovate, as long as the apical subulate part.

♂. Head nearly flattened, a little attenuate in the anterior part; in front is a small transverse groove, abbreviated on each side; exactly behind it is a small oblong depression bounded in front by a transverse tubercle; the 3rd joint of the antennæ longer than the others, 4-6 a little longer than broad, 7 square, 8 a little transverse, 9 trapezoidal, as long as broad, 10 transverse.

♀. Head nearly flat in front with two very shallow foveæ hardly visible, the front hardly divided, the antennal tubercles

♂. The head resembles much the head of *procurator*, Schfs.; the vertex is much raised, trapezoidal and a little transverse, very abruptly emarginate in front, this truncature being ciliate, more especially on the sides and bearing in the middle two small brushes of whitish hairs, the anterior part is much lower, excavated, smooth and shining; in the middle of this excavation is a transverse and a little convex elevation, the extreme edge of the front is narrow, raised, rugose, and above each antenna there is a rugose tubercle with a small brush of hairs; the antennæ are long, joints 3-5 nearly three times and 6 only twice as long as broad, 7 about as long as broad, 10 trapezoidal and very little transverse.

♀. Vertex a little convex, in front of the eyes on each side is a large but shallow fovea, the frontal part square, rather deeply sulcate. The antennæ are shorter, joints 4 and 5 not more than twice as long as broad, 6 only a little longer than broad, 7 square, 8 a little transverse, 9 hardly longer than broad, and 10th decidedly transverse.

Rockhampton, Q. (♂., type of Schaufuss); Clarence River, N.S.W. (♀., Mr. A. M. Lea).

Subfamily CLAVIGERIDÆ.

Genus ARTICERUS, Hope.

Trans. Ent. Soc. Lond. iv. 1845, p. 106.

ARTICERUS HAMATIPES, n.sp.

(Pl. x., fig. 40.)

Subelongatus, ferrugineus, nitidus, sat longe et disperse fulvopubescens Caput disperse varioloso-punctatum, elongatum, ante oculos attenuatum et apice subtriangulare. Antennæ breves, capite paulo longiores et crassiores, regulariter valde clavatæ, apice nonnihil attenuatæ. Prothorax obsolete varioloso-punctatus latitudine sua paulo longior, antice plus postice minus attenuatus, lateribus leviter rotundatis, fovea media ante basali parum profunda, subelongata. Elytra disperse et aspere punctata, latitudine sua longiora. Abdomen læve, nitidum, elytris longitudine æquale, apice valde obtusum, cava maxima, latitudine sua longiori, fundo angulatim plicatula, lateribus late et longe carinatis, subrectis, carinula externa basi crenatula, interna subparallela, fasciculata. Metasternum gibbosum. Segmento 2° ventrali transversim declivi, apice medio foveato. Femoribus simplicibus, tibiis anticis et posticis leviter clavatis et mediocriter compressis, intermediis valde compressis, extus medio maxime dilatatis et acute hamatis, postice abrupte angustatis, intus leviter angulatis. ♂. Long. 1·40 mm.

The abdomen is deeply and nearly entirely excavate, the sides are straight, strongly carinate; this carina is flattened, its outer and inner margins are nearly parallel; the strong hook of the dilatated intermediate tibiæ, the short and nearly regularly conical antennæ lead at once to the identification of this species.

Swan River, W.A. (Mr. A. M. Lea).

ARTICERUS BIPARTITUS, n.sp.

Rufo-castaneus, nitidus, breviter sat dense pubescens. Caput minute subconfluenter, aspere punctatum, latum, subdeplanatum, apice obtusum, lateribus subparallelis. Antennæ sat elongatæ,

The ♂ of this species is not known; two specimens have been sent to me by Mr. Lea, together with the preceding species, as being ♂ and ♀ of the same species, which, after all, is not quite impossible; but without mentioning the antennæ, which are known to vary immensely in the ♂ and ♀ of the same species, there are so many and so important other differences that it seems to me difficult and improbable that the two belong to the same species; the head is much broader and shorter with a fine rugose punctuation; the prothorax is short and broad with a shallow longitudinal sulcus; the abdomen is broader, shorter and rounded behind, in the sides, before the middle being abruptly and angularly constricted, so as to appear divided into two parts; the carina of the base is much shorter, and its sides are oblique and convergent so as to form an elongated triangle; the excavation of the abdomen is large, deep, but transverse.

Swan River, W.A. (Mr. A. M. Lea).

ARTICERUS CULTRIPES, n.sp.

(Plate x., fig. 23.)

Brevis, ferrugineus, nitidus, brevissime et vix perspicue setosus, totus, abdomine excepto, in capite densius, aspere punctatus. Caput sat latum et mediocriter elongatum, apice obtusum. Antennæ elongatæ, prothoracis medium attingentes, basi attenuatæ, medio subcylindricæ, apice leviter incrassatæ, infra leviter curvatæ. Prothorax leviter transversus, antice plus, postice minus attenuatus, lateribus rotundatis, basi media foveola parum profunda et sulco longitudinali lato, sed parum profundo. Elytra latitudine sua parum longiora. Abdomen latum, elytris nec longius, postice rotundatum, cava maxima profunda, carina laterali marginibus subrectis, ante medium sulco transverso abrupta, vix fasciculata. Metasternum gibbosum. Femoribus sat incrassatis, intermediis crassioribus; tibiis anticis valde compressis et dilatatis, extus medio acute angulatis, intermediis compressis extus rotundatis, præsertim ad apicem incurvis, posticis minus compressis, leviter sinuatis. ♂. Long. 1·70 mm.

This species much resembles *bipartitus* in the general shape of the body; however, the antennæ are more cylindrical, the punctuation is more apparent on the prothorax and the elytra, the abdomen is not angularly contracted on the sides, but the lateral carina is abruptly ended by a transverse groove, and its two margins are nearly parallel. The ♀ is not known.

Bridgetown, W.A. (Mr. A. M. Lea).

ARTICERUS FALCATUS, n sp.

(Plate x., fig. 25.)

This species is very similar to *curvicornis*, Westw.; and it was looked upon by Dr. Schaufuss as the ♂ of *curvicornis*, but of this latter I have both sexes, and this species is evidently different. The shape of the body is practically the same, but the punctuation in *falcatus* is much stronger and more ocellate. The antennæ are very different; they are a little shorter, as broad as the head, compressed, very slender at base, rather abruptly ampliated and rotund outwardly and a little curved inwardly, the widest part being at the apex; it is obsoletely and longitudinally impressed above, and has beneath a rather deep and oblong impression; the feet are alike, but the internal apical spur of the intermediate tibiæ is not so strong.

lateribus paullo tantum rotundatis, antice vix magis convergentibus quam postice; processus clypeo imminens apice truncatus, foramine magno, aperto; prope foramen utrinque punctum nigrum prominens. Antennæ 15 articulatæ, capitis longitudine, crassæ, ab articulo 2° usque ad medium sensim incrassatæ, dein apicem versus sensim attenuatæ, articulis præsertim mediis subglobosis. Labrum lanceolatum, acutum, mandibulorum medium haud attingens. Mandibulæ nigræ, capite triente breviores, subrectæ, apice tantum curvatæ, haud dentatæ. Prothorax capite fere duplo angustior, basin versus valde angustatus, subtriangularis, margine antico et postico truncatis et in medio paullo incisis, lateribus prothoracis subrectis, angulis anticis rotundatis.

It does not agree with any of the Australian species described by Mr. W. W. Froggatt.

It belongs to the genus *Termes*, L., subgenus *Coptotermes*, Wasm. It is nearly allied to the Indian *Coptot. Gestroi*, Wasm., (Ann. Mus. Civ. Gen. [2] xvi. p. 628), and *travians*, Havil., (Journ. Linn. Soc. Lond. Zool. Vol. xxvi. p. 391), but differs specifically by the longer, cylindrical-oval head; by the antennæ thickened in the middle, with nearly globose joints; by the truncate hind margin of the prothorax, whose sides are also much more narrowed towards the base.

EXPLANATION OF PLATE X.

Fig. 1.—*Hamotulus dispar.*
Fig. 2.——— ————, palpus.
Fig. 3.—*Gnesion rufulum.*
Fig. 4.——————, palpus.
Fig. 5.—*Leanymus palpalis.*
Fig. 6.——————, palpus and cardo.
Fig. 7.—*Homatopsis Australasiæ.*
Fig. 8.——————, palpus.
Fig. 9.—*Eupines diversicolor*, antenna (♂).
Fig. 10.—*Eupinoda diversicornis*, antenna (♂).
Fig. 11.—*Eupines nigriceps*, antenna (♂).
Fig. 12.—*Eupinoda amplipes*, posterior leg (♂).
Fig. 13.——————, anterior leg (♂).

Fig. 14.—*Eupines globulifer*, antenna (♂).
Fig. 15.—*Eupinoda Leana*, antenna (♂).
Fig. 16.—*Eupines biclavata*, antenna (♂).
Fig. 17.——— *triangulata*, „ (♂), last joints.
Fig. 18.——— *nodosa*, „ (♂) „
Fig. 19.——— *compressinoda*, antenna (♂), last joints.
Fig. 20.——— *concolor*, „ (♂), „
Fig. 21.——— *sternalis*, „ (♂), „
Fig. 22.——— *lævifrons*, „ (♂), „
Fig. 23.—*Articerus cultripes* (♂).
Fig. 24.——— *curvicornis*, antenna (♂).
Fig. 25.——— *falcatus*, antenna (♂).
Fig. 26.—*Batrisus asperulus*.
Fig. 27.——— *ursinus*, head (♂).
Fig. 28.——— *cyclops*, head (♂).
Fig. 29.—*Batrisodes tibialis*, anterior leg (♂).
Fig. 30.——— *Mastersi* (♂).
Fig. 31.—*Briara frontalis* (♂).
Fig. 32.—*Mesoplatus tuberculatus*.
Fig. 33.—*Paraplectus biplagiatus*.
Fig. 34.—*Briara basalis*, head and antenna (♂).
Fig. 35.——— *capitata*, head and antenna.
Fig. 36.—*Ctenisophus laticollis*, palpus.
Fig. 37.—*Didimoprora puncticollis*, posterior leg (♂).
Fig. 38.——— *Leana*, „ „
Fig. 39.——— *Victoriæ*, „ „
Fig. 40.—*Articerus hamatipes*, intermediate leg (♂).

INDEX

OF THE GENERA AND SPECIES MENTIONED AND DESCRIBED.

(Names in Italics are synonyms).

AUSTRALIAN *PSYLLIDÆ*.

By Walter W. Froggatt, F.L.S.

(Plates xi.-xiv.)

Introduction.

The insects of this family of the Homoptera, like those of several other groups of the order, are very interesting because of the remarkable larval and pupal transformations they exhibit in the course of the metamorphosis, as well as for the curious protective coverings—lerps or scales—which many of the tiny larvæ commence to fabricate as soon as they emerge from the egg. The lerp or scale is added to as the insect increases in size up to that of the full-grown pupa; then crawling from beneath it, it casts the final pupal shell, and emerges as the perfect insect.

The best known Australian species are those that form the leaf manna upon the foliage of Eucalypts. As far back as 1849 Anderson (1) described the chemical composition of some "manna" (lerp) that had been obtained in the Mallee-gum scrubs in the

Beveridge (5) in a paper on the Aborigines of the Lower Murray and Darling Rivers has also referred to it under the heading of "Laarp," which, he says, "is the excrement of a small green beetle wherein the larva thereof is deposited." He gives a very remarkable account of how the natives collected and fed upon the lerp-scales during the summer months: and he adds that it is so plentiful "that a native can easily gather from 40 to 60 pounds weight of it in a day." But this must be a slip, for old residents of the Wimmera, where it was very plentiful before the Mallee scrub was cleared off, have informed me that 2-3 lbs. was quite as much as any one could obtain in a day; and that the blacks used to gather it for food in winter, rolling it up in bark and hiding it in the trees: when they wanted to eat it they first moistened it with water.

Many species form regular galls and blisters upon leaves, chiefly those of Eucalypts. These first appear as little pits, which swell into either bubble-like excrescences or thickened rounded masses enclosing the larva. This emerges from an opening either on the upper or under surface of the leaf.

Others again hide under loose bark on the trunk or branchlets of a tree, enveloping themselves in a mass of flocculent matter, which exudes and forms white spots dotting the trunk all over. These species are so diligently looked after by several kinds of ants which sometimes form galleries over them that it is difficult to collect specimens.

Most of the naked species are more common upon Acacias and other scrub trees than upon Eucalypts, and swarm in such numbers on the under surface of the leaves or over the young branchlets, as at first sight to be easily mistaken for aphides.

Some of the true lerp-producing species present very curious examples of insect architecture. The lerp-scales are sometimes like little cockle shells with delicately crenulated edges, semi-transparent or opaque, black or richly tinted with yellow or red; some are smooth and flattened, others convex and covered with fine hairs, sometimes they are closely attached to the leaf, but the more delicate ones are simply fixed to it by a hinge at the apex, the larva being free to crawl in or out.

All the lerp-scales are fabricated by the larvæ and pupæ from the excess of sap or juice sucked up through their sharp bills from the food-plant. This is ejected in small globules from the anus, but it is quite different from the excrement. It is another form of honey-dew, which when drawn out into fine threads by the feet and spun into the net-like sugar lerps, solidifies and hardens in the sun. In the naked species the larvæ expel the sap which forms a real honey-dew as in the case of the European pear-tree Psylla (*P. pyricola*), in which it is so excessive that the whole foliage and trunk of the tree become smothered with the exudation; this in turn is attacked with a fungus which covers it with a smutty black coat (fumagine) seriously injuring the tree.

The Australian fauna is very rich in species of Psyllidæ, but from their small size and active habits in a country so rich in larger and more conspicuous insects they have been naturally rather overlooked, for they are seldom to be met with except by sweeping among the brush with a net, or by breeding them from the larvæ and pupæ upon infested foliage. In the British Museum Catalogue of the Homoptera (1850-51), Walker records five species, all from Tasmania. Another species also from Tasmania was added to the list by the same author in his Descriptive Catalogue of

In 1895 Tryon (9) gave a brief description of the eggs and larvæ of a species which is very plentiful in the spring and autumn upon the foliage of the "Moreton Bay Fig" (*Ficus macrophylla*). The larvæ prick the leaves with their sharp rostra, and live in social groups protected by masses of coagulated sap.

In 1898 a paper by Maskell was published shortly after his death in which three species from Australia were described (9).

The last paper dealing with Australian species is that by S. A. Schwarz (10), who redescribed Dobson's *Psylla eucalypti*, placing it in Signoret's genus *Spondyliaspis*; and added a new species.

The above is a summary of all that has been written on Australian *Psyllidæ;* but the foreign species have had many admirers, and have been well worked out.

In a group of Micro-Homoptera like the *Psyllidæ*, where the largest are hardly over two lines in length, there is a very great difference in examining carded or mounted specimens, sometimes many years old, in Museum Collections, and series of living insects of the same species, with a knowledge of their life-histories. Therefore, to describe them properly, there is no doubt that the correct mode of studying them is upon the spot and in relation to their food-plants. Fortunately they are very easily bred out, and in most cases when the food-plant is obtained specimens in all stages of development, from the egg to the perfect insect, may be found and worked out upon the same branch of foliage, the whole life-history being under review at once.

I have to tender my thanks to the following gentlemen for their kindness in sending me specimens and notes:—Messrs. C. French and C. French, Jr., of Victoria, Mr. H. Tryon of Brisbane, and Mr. A. M. Lea of Tasmania. To Dr. Horváth of Hungary and Dr. Howard of America I am indebted for papers dealing with the family, and to Mr. J. H. Maiden of Sydney for notes and papers upon the subject of manna.

Classification.

The *Psyllidæ* form a very well defined family of the suborder Homoptera, closely allied to the *Aphidæ* in habits and form. I have one species bred from the Kurrajong (*Sterculia heterophylla*) that, on a superficial examination, is wonderfully like a green-winged aphid. In their larval habits they are much more coccid-like, and some of the lerp-scales produced might very easily be mistaken for a coccus. They seem also to have some relationship, particularly in the larval stages, with the *Aleurodidæ*. Most of the larvæ of *Aleurodes* form rounded disc-like shells or tests under which they pupate, but those of some of the Australian species pit the leaves of Eucalpyts exactly like some Psyllids; however in the adult stage *Aleurodes* shows a much closer relationship to the Coccids, and might be described as a higher type of this group, with both sexes winged, but with a very primitive venation. In the *Psyllidæ* the venation of the wings is very well-defined, and characteristic of the species. Sharp (11) places the *Psyllidæ* in the sixth family of the Homoptera after the *Jassidæ* and before the *Aphidæ*, a very natural position. In this he follows most of the later systematic entomologists who have placed them from their external characters.

A great deal has been written about the classification of the *Psyllidæ*. Réaumur named them Faux Pucerons from their relation to the *Aphidæ*. Linnæus placed them in the genus *Chermes* (now restricted to a group of the *Coccidæ*). Geoffroy, considering that the later name had been improperly used by

Most subsequent writers have followed Low's classification. In the same year Scott (18) formed a fifth subfamily for Walker's two genera *Carsidara* and *Tyora*.

The members of the family have a wide geographical range. Europe is particularly rich in numbers: Southern Asia and the Northern parts of Africa have a fair supply. Riley and some of the earlier writers have described the North American species, and Scott several from South America. Buckton has dealt with three from India, and Schwarz with one from Japan. They appear to be more numerous in temperate climates, or dry semi-desert lands, than in the tropics, and the dry open ranges of Australia covered with stunted Eucalyptus shrubs are very rich in species.

The eggs are either bright yellow or brown, sometimes scattered all over the foliage: at others, particularly in the case of the lerp-producing species, clustered together in irregular rows from ten to forty in number. Their form is generally elongate-oval, sometimes pointed at the extremities: and sometimes there is a slight keel down the centre of the dorsal surface. The shell splits down the centre when the tiny larva emerges.

The larva is generally pale yellow, elongate in form, with the head and the abdomen rounded at the extremities; the antennæ short, stout, and pointed at the tip, the eyes red, small, and irregular in form: the legs stout and long, with the digitules of the tarsi long. During its successive moults it may change its colour several times: a bright yellow larva frequently changes to bright green or red before its final moult into the full grown pupa; the indications of the fuscous-brown or black dorsal marks become more distinct and defined at each ecdysis, so that these graduations of colour become very interesting to the observer. Several writers have noticed this variation of colour. Reaumur observed it as far back as 1773 (17), when studying species. The larva at a very early stage of development shows two rounded pairs of projections upon the sides of the thoracic segments, which afterwards form the wing-covers of the pupa, the transformations being so gradual that it is difficult to define the borderland between the larva and pupa, or where one stage ends and the next commences. In dealing with their development, when all stages have been obtainable, I have taken the smallest form to be found as the larva, and defined the pupa from the largest, frequently just as it is ready to cast its skin and emerge as the perfect insect. Though the changes are gradual, each moult brings some alteration; first, the abdominal segments show the line of separation from the thorax, next the line between the base of the head and thorax, and the enlargement of the wing-covers. The antennæ, though not increasing much in length, show more joints; in the earlier stages the 3rd joint is very long, and the additional joints, until the normal number of nine is reached, appear to divide off from the apical portion of the elongated third joint, which, however, is generally the longest in the perfect Psylla. Many of the larvæ and pupæ are covered with fine hairs; most of the species that are naked (not forming lerps or galls) have the hairs upon the dorsal surface covered with tiny particles of white sugary secretion, with those on the sides converted into white filaments sometimes of considerable length.

The perfect insect might be compared to a minute cicada in general form, but there the actual resemblance ends. The head is generally broader than long, sometimes deflected, with large eyes; the ocelli three in number, the lateral ones placed on the

scutellum rounded. The wings are fully developed in both sexes; the elytra or forewings longer than the abdomen, with stout nervures, the costal with or without a stigma; the venation is simple and constant, and of both generic and specific value. From the primary vein run the radius parallel with the costal, the cubitus which branches into an upper and lower arm, each again bifurcated into a more or less regular cell at the apex; the cubitus with or without a petiole; the hindwings simple. The legs are formed for jumping; the coxæ of the hind pair armed with a rounded spine; the apex of the tibiæ of the hind legs armed with a fringe of fine spines; the tarsi two, with large double claws. The abdomen is composed of six segments. The genitalia of the male comprise an upper and lower valve, two curved processes known as the forceps, and an enclosed penis. The female genitalia consist of two more or less elongate valves enclosing the ovipositor. The form and structure of the genital organs are also of use for specific characters, and are constant in each species.

The colouration and size are not constant. Löw says, with reference to the European species (18), that not only are there differences in the colour according to the age of the insects, but the same species on a different food-plant varies; and that the successive generations change with the seasons in both particulars. In Australian species so far I do not find any perceptible difference in the successive broods (many of which live all through the year), probably on account of the comparatively uniform climate; but the colouration is very variable in some species, particularly in those hiding under bark or crawling upon the foliage in their larval state, while the males are often the smallest. Though some do not vary in the least, being distinctly marked when they emerge from the pupæ, others often take several days to attain the dark markings upon the wings, and these vary and fade away to the faintest tinge of brown in individuals of the same brood; pale yellows become ochreous, and browns black. After a time many of the richest colours fade or change into darker tints.

All the species that I have examined appear to fall very naturally into the subfamilies defined chiefly upon the structure of the wings by F. Löw. Those forming lerp-scales or hiding under bark are referable either to the *Liviinæ* or *Aphalarinæ;* most of those living among flocculent matter upon foliage or forming rudimentary lerps to the *Psyllinæ;* and all the true gall-producing species (with a few exceptions) to the *Triozinæ*.

I have not proposed new genera for any species that I could at all reasonably refer to genera already well-defined; and if I have erred on the side of caution, specialists will be able to rectify this defect. This course appears to be preferable to forming new genera on such scanty material, as has been done in the past; one of Walker's genera, for instance, being based on the examination of a single specimen minus the head.

In a large series of specimens one frequently meets with examples possessed of an extra cell or cross nervure in the wings; such, if examined alone, would certainly not fit the genus to which the species belongs. It also appears to me that some of the latest genera established by Riley and others are so minutely defined that they can only take in the single species upon which they are founded, whereas if they had received more general treatment they might have included all allied forms.

In Schwarz's paper (10) he discusses the position of the genus

lower branch of cubitus; upper furcation very long and narrow; upper fork of lower branch of cubitus very long, more than twice as long as lower fork.

Genus i.—LIVIA, Latr., Hist. Nat. Ins. Vol. xii. p. 374, 1804 (*Diraphia*, Illig.).
ii.—CREIIS, Scott, Trans Ent. Soc. Lond. p. 462, 1882.
iii.—LASIOPSYLLA, n.g.

Genus CREIIS, Scott.

Head: crown down the centre more than half the breadth between the eyes. In front of each eye a short angular tooth, front margin convex. Front lobes long, stout, vesicate. Antennæ long, slender. Eyes viewed from above hemispherical, placed on the side of the head.

Thorax: pronotum narrow, widest at lateral margins, within which is a small fovea; dorsulum moderately convex; mesonotum of an irregular hexagonal shape. Elytra elongate, rounded at apex; radius joining marginal nerve before apex, furcations of cubitus elongate.

Type *Livia longipennis*, Walker.

C. LONGIPENNIS, Walker.

(Plates xi., fig. 1; xii., fig. 17; xiv., fig. 8).

Livia longipennis, Walk., B.M. Cat. (Homoptera) p 910, 1851; *Psylla livioides*, Walk., Ins. Saunders. Homop. pt. iii. p. 111; *C. longipennis*, Scott, Trans. Ent. Soc. Lond. 1882, pt. 3, p. 463.

Lerp formed upon the leaves of *Eucalyptus* sp., not more than one or two upon a leaf, rich canary-yellow, attached to the leaf by a regular flange from which it swells out, lobed on either side to a broadly rounded apex, convex, the edges in contact with the leaf somewhat flattened; formed of opaque threads running from the flange in a crescent pattern so close as to give it a granulated appearance, but the whole of the upper surface thickly clothed with long hair-like filaments, giving it a very beautiful appearance; the outer hairs easily abraded, and many specimens are more or less denuded of the outer covering. Diameter through centre 5, across 6, height above surface of leaf 1 line.

Larva and *pupa* unknown.

Imago.—Length ·27 inch, antennæ ·13 inch.

General colour red, eyes dark brown, apex of central abdominal segments on dorsal surface banded with black, lower portion of genitalia yellow; wings pale brown, semiopaque, coriaceous and very finely wrinkled, nervures red. Head small, deeply cleft in front, with a median suture and shallow fovea on sides, truncate behind base of antennæ; a small angular tooth in front of eyes, arcuate behind. Face lobes large, rounded at apex and clothed with grey hairs. Antennæ very long, cylindrical, 1st and 2nd joints short and rounded, 3rd long, 4-5th shorter, 6th-8th longer, 9th short, 10th very short and rounded at tip. Eyes rounded on outer margins: ocelli—central ocellus large, oval, at apex of median cleft; lateral ocelli large. Thorax: pronotum narrow, depressed on sides; dorsulum large, rounded on both sides, coming to a point on sides; mesonotum very large, depressed in centre, swelling out on sides, with the scutellum almost cordiform. Legs short and stout. Wings thrice as long as broad, rounded at apex; primary stalk rather short, stalk of subcosta short; costal cell elongate, with a second or false costal nervure running to base of stigma in a line with the outer one; stigma small, running out

This is not a common species, and individuals are generally found singly upon leaves. Walker's two specimens in the British Museum come from Tasmania. Mr. Lea has lately sent me several worn lerp-scales from the neighbourhood of Hobart. When fresh the lerp is one of the most beautiful Australian forms, but the thick covering of filaments soon wears off when exposed to the weather.

Genus LASIOPSYLLA, n.g.

Head similar to that of *Creiis*, with a short tooth on sides of eye, and long slender antennæ. Thorax large, convex; pronotum short, sharply rounded in front; dorsulum large, arcuate in front, broadly rounded on sides. Wings nearly thrice as long as broad, with an inner or false costal nervure running close and parallel to costal nervure, merging into it at shoulder; stigma wanting; radius short, nearly straight, stalk of subcosta a little shorter than stalk of cubitus; upper fork of lower cubitus very long, curving in at centre. Apex of hind tibiæ dilated, and bearing three fine black spines on the edge.

LASIOPSYLLA ROTUNDIPENNIS, n.sp.

(Plates xi., fig. 2; xii., fig. 4; xiv., fig. 11).

Lerp.—Large, flattened, thin white scales, up to nearly ½ inch in diameter, irregularly rounded, arcuate at the hinge, attached to the leaf by a small hinge from which the scale grows in concentric rings, giving it a reniform shape; convex in centre, with outer margin pressed close to the leaf; on the leaves of *Eucalyptus melliodora*, *E. polyanthema* and several other allied species; sometimes single or half a dozen overlapping each other.

Larva.—General colour dull yellow, tinged with pink on abdomen and legs; antennæ barred with black; two large blotches on head, and a double row of impressed black spots running to tip of abdomen where they come to a v-shaped point; anal tubercle black. Head truncate, slightly rounded in front, forming with thorax a solid piece widest at base of abdomen; eyes very small;

antennæ very slender, short, mottled with black. Abdomen flat, swelling out from base, broadest in centre, outer edges of segments rounded; anal tip pointed.

Pupa.—General colour pale sea-green with blotches on head, two smaller ones in front; legs, antennæ, two spots at hind margin of head, and those on the thorax and abdomen as in larva black. Head large, lobed in front, arcuate behind; eyes swelling out, rounded behind; antennæ very long, curving round and tapering to tips, 1st-2nd joints short, stout, 3rd long, slightly elbowed; eyes projecting, slightly rounded. Thorax long, wing-cases large; legs long and stout. Abdomen as in larva.

Imago.—Length ·23 inch, antennæ ·09 inch.

General colour yellowish-brown deeply tinted with pink; antennæ reddish, with the apex of each of the last 8 joints dark brown; face red, shaded on the outer margin with chestnut; eyes dark reddish-brown: segments of thorax dull red shaded with pale brown, segmental divisions black, the red replaced in male both on head and thorax with black: legs yellow, tarsi brown; wings semiopaque, coriaceous, the apical portion in female thickly clouded or mottled with fuscous, nervures red; abdomen beautifully barred with red, edged on either side with green and a fine black band below. Head large, truncate in front, with a deep

back to scutellum, truncate at apex; scutellum large, truncate in front, rounded behind. Legs long, femora of forelegs very stout and thickened; tarsi long. Wings large, very long, over thrice as long as broad; costal nervure rounded at base, slightly hollow in centre and curving round at tip to hind margin of wing, which is nearly straight; primary stalk long and stout; stalk of subcosta shorter than stalk of cubitus; costal nervure double at base, but without any true subcostal cell or stigma; the costal cell long and slender, tapering out into a tail between costal and subcostal nervures; radius long, turning up at tip of wing; stalk of cubitus longer than stalk of radius, upper branch short, bifurcated about centre of wing, upper and lower forks of equal length, running out at tip of wing and forming a very elongate slender cell; lower branch of cubitus shorter than upper, upper fork very long, curving down before reaching tip of wing; lower fork transverse, running out at a sharp point; clavus very thick and short, clavical suture running through centre of the long, slender, anal cell. Abdomen stout and rounded to tip. Genitalia (♂) short and broad; lower genital plate short, angular; forceps oval; penis hidden: upper genital plate long, slender: (♀) upper and lower genital plates short and pointed, clothed with fine hairs.

Hab.—Melbourne (on *E. melliodora*; Mr. C. French, Junr.), and Bendigo, Vic. (on *E. polyanthema*; W. W. Froggatt); Hobart, Tas., (lerp only, on *E.* sp.; Mr. A M. Lea); Bathurst, and Tumut, N.S.W. (on *E. melliodora*, and *E. polyanthema*; W. W. Froggatt); Brisbane, Q., (lerp only, on *E.* sp.; Mr. H. Tryon).

This is a very common species where the particular species of Eucalypts enumerated grow, and has a very wide range over the eastern portion of Australia. I have seen bushes about Bathurst covered with the white lerp-scales in the early summer. I have another form of the lerp collected on the foliage of a low scrub Eucalypt growing on the river flats near Bourke, Darling River, that has puzzled me very much, for though I can see no specific differences in the pupa or perfect Psyllid, yet the lerp-scale, while of the same colour and shape as the more common coastal species, is quite different in its structure. Instead of being formed in concentric rings, it is made up of close delicate parallel bars, which run from the base to the outer margins, giving it a beautifully striated appearance.

LASIOPSYLLA BULLATA, n.sp.

(Plates xi., fig. 3; xii., fig. 16; xiv., fig. 15).

Lerp.—Thin bubble-like galls or excrescences upon the upper surface of the leaves, produced by the attacks of the larvæ on the under surface of the leaves of several species of Eucalypts; narrow and constricted at base, but swelling out in an elongate-oval or rounded gall, from 4-8 lines in height and 4-5 in diameter; very variable in shape, and from 1-12 on a single leaf; when fully developed the general green colour is frequently brightly tinted with red and yellow. Basal orifice large, but closed with a circular cake of saccharine matter; the small larva crawls about in its spacious chamber until nearly ready to pupate; then it is enveloped in a mass of white flocculent wool, though in its earlier stages quite free from such filaments.

Larva bright canary-yellow, legs and antennæ semitransparent, eyes bright red, the dorsal surface showing traces of the fuscous marking of the pupa. Head short, broad, rounded in front, antennæ very long and stout, clothed with coarse hairs. Thorax

General colour light chestnut and bright yellow; wings coriaceous, light brown, with reddish-brown nervures; clavical suture bright pink, giving it a very distinctive appearance. Head small, curving down in front, truncate at base, deeply cleft in front, rounded to eyes, with a median ridge and large shallow depression on either side. Face lobes short and broad, rounded and clothed with long grey hairs. Antennæ long, slender, springing from below inner margin of eye; 1st-2nd joints very short, 3rd very long, 4th-9th slender, uniform in length, 10th short and slightly thickened at tip. Eyes large, reddish-brown, not projecting: ocelli large, central ocellus at the apex of frontal cleft, lateral ocelli close to hind margin of eyes. Thorax: pronotum very small, convex in front; dorsulum hexagonal, convex on summit, rounded at apex; mesonotum deeply arcuate in front, large, rounded on sides and behind. Legs short and thick. Wings thrice as long as broad, rounded to tip, but sharply turned down and somewhat straight on hind margin; primary stalk straight, rather short; stalk of subcosta shorter than stalk of cubitus; a distinct false or second costal vein running from base to apex of subcosta, forming a thickened costal band tapering to tip; stigma wanting; radius slightly curved upwards, not reaching extreme tip of wing; upper branch of cubitus short; upper and lower furcations very long, of equal length, forming a narrow uniform cell, and slightly curved up at tips, lower branch nearly as long as upper, upper fork very long, curving down in centre, rounded at tip; lower fork long, sloping inwards; clavus stout, clavical suture very distinct, bright red in sunlight. Genitalia (♂): lower valve short and angular; forceps arcuate on outer edge, coming to a point at apex; upper genital plate large, flask-shaped, swollen and rounded, apex nipple-shaped.

Hab.—Sydney (on *E. capitellata*) and Mittagong, N.S.W. (on *E. dives;* W. W. Froggatt).

This is a very remarkable Psyllid, allied to *Creiis longipennis*, but with the wing of a distinctly different shape. The larvæ living in the leaf-galls are quite different from the shield-shaped

Trioza larvæ found in various galls; and are much more like the sugar-lerp-forming Psyllids.

ii. Subfamily APHALARINÆ, F. Loew.

Front of head either swollen or prolonged into two conical processes, or roughly rugged; eyes prominent. Stalk of cubitus as long as, or longer than stalk of subcosta.

Genus i.—Euphyllura, Först, Rheinl. u. Westphal. Verh., 1848.
ii.—Rhinocola, Först., *loc. cit.*
iii.—Aphalara, Först., *loc. cit.*
iv.—Psyllopsis, F. Löw, Verh. Zool.-Bot. Gesell. Wien. xxviii., 1879.
v.—Thea, Scott, Trans. Ent. Soc. London, 1882.
vi.—Phytolyma, Scott, *loc. cit.*
vii.—Phyllolyma, Scott, *loc. cit.*
viii.—Spondyliaspis, Sign., Ann. Soc. Ent. Franc. 1879.
ix.—Cardiaspis, Schwarz, Proc. Ent. Soc. Washington, iv., 1896.

Genus Rhinocola, Förster.

Head with eyes prominent; front of head prolonged into two conical processes; central ocellus on inner border of vertex some-

General colour pale yellow, with the tips of the antennæ, a broad patch on either side of the dorsal surface of the head, a double row of smaller markings on the thorax and basal abdominal segment, wing-covers and apical portion of abdomen slate-grey; eyes reddish-brown. Head almost globular, eyes prominent, well round on the sides of the head; antennæ thick, standing out on either side; wing-covers oval, large. Abdomen long, narrow, rounded at the tip, finely edged with short hairs. Dorsal surface flattened; under surface pale yellow.

Maskell has described and figured this species in detail in his paper (6), so that I need not again go over the same ground. He found this small, greyish-brown psyllid common upon the foliage of blue gums (*E. globulus*) growing in New Zealand; and he suggested that as this is an introduced tree in that country the insects might be Australian. My specimens were obtained in considerable numbers upon young seedlings in Purchase's Nursery, Parramatta, and also in the Botanic Gardens, Sydney.

Rhinocola revoluta, n.sp.

(Plates xi., fig. 12; xii., fig. 8; xiv., figs. 19-19*a*).

The larvæ form a very remarkable, double-valved, opaque white lerp like a rounded, flattened oyster shell, about 2 lines in diameter, placed along the edges of the young leaves of several different Eucalypts; they are thin when first formed upon the surface of the leaves, but as the actions of the feeding larvæ cause the leaf to become discoloured and curl up, the lerp-scales assume a horizontal position and become packed in rows side by side, rolled up in the enveloping leaf, sometimes as many as twenty-four in a row; when few the lerp-scales are generally much larger. When the larvæ are ready to emerge the valves open at the apex.

Pupa short and broad, head thorax and abdomen yellow tinged with green, the last darkest and banded with interrupted black bars; legs and antennæ yellow; wing-covers chocolate-brown. Head broad, truncate in front; antennæ rather short, standing out on either side, slightly hairy, pointed at the tips; eyes small. Thorax broad, rounded in front; wing-covers small, elongated, covered with fine white hairs. Legs short and thick. Abdomen broad, rounded, swelling out at the base, and rounded to the apex.

Imago.—Length 0·15, antennæ 0·045 inch.

General colour bright green, thorax mottled with reddish-yellow, tarsi brown; wings semi-opaque, finely granulated, tinged with pale dull yellow, darkest along hindmargin, the tips clouded with fuscous and black running into the cells; nervures light brown. Head not as broad as the thorax, deeply cleft in front, with a slight median suture and shallow fovea on either side, sloping down on either side behind the antennæ; arcuate at hindmargin; face lobes large, broad and rounded, thickly clothed with hairs; antennæ in front of eyes moderately long, basal joints thickest, each clouded with fuscous at apex, 3rd and 4th joints longest, 5th-6th shortest, 7th-8th long and slender, 9th shorter and broadest at apex, 10th oval, short and broad. Eyes very large, rounded, and contracted on inner edge: central ocellus small, pink; lateral ocelli large, vitreous. Thorax with pronotum very narrow, curving round; dorsulum very broad, convex, round in front, the sides produced into slender points, slightly arcuate on the sides to rounded apex; mesonotum large, clavus long and stout, clavical suture long and distinct. Abdomen stout, tapering to tip.

♂. Smaller than female; with the wings semitransparent, but without fuscous markings. Genitalia (♂) short and broad; lower genital plate short and rather angular; forceps short, thick, and turned downward; upper genital plate long, thick, and rounded at tip.

Hab.—Bendigo, Vic. (on *E. leucoxylon*); and Tumut, N.S.W. (on *E. macrorrhyncha*, and *E. hemiphloia*; W. W. Froggatt).

RHINOCOLA ASSIMILIS, n.sp.

(Plate xi., fig. 13).

Closely allied to the preceding species. The habits of both larvæ and pupæ are identical, except that the lerp-coverings within the curled leaf are very much more irregular in form, and somewhat larger. The minute larvæ do not at once commence to form a lerp-scale, but crawl about enveloped in white filaments, and take shelter in the first curl of the leaf. General colour pale yellow.

Pupa dull brown, with darker markings on the head, legs and antennæ; a row of spots down each side of thorax and abdomen black.

Imago.—Length 0·15, antennæ 0·03 inch.

General colour ochreous, eyes reddish-brown; wings coriaceous, light brown, nervures reddish-brown. Head narrow, curved down in front, arcuate behind, with front margin arcuate in centre and on either side at base of antennæ, with a short median suture and small rounded fovea on either side; face lobes small, hidden from above; antennæ short, apex of terminal joints clouded with chestnut, 1st and 2nd joints very stout and short, 3rd joint cylindrical, twice the length of the two preceding combined, 4th-9th shorter, 10th short, oval at the tip. Eyes not projecting, angular on the inner margin: central ocellus very small, lateral ocelli small, close to the eye. Thorax: pronotum angular in front, broadest in centre, with an impressed fovea, narrow and rounded at extremities; dorsulum spindle-shaped; mesonotum large, almost truncate in front, rounded almost to a point on the sides, convex behind, scutellum small and angulated in front. Legs short, stout. Wings as in the preceding species but without the fuscous markings on the tips. Genitalia (♂) short and broad, details unknown.

Hab.—Cooma, N.S.W., (on *E. viminalis;* W. W. Froggatt).

It is very curious that this species constructs such a similar lerp in the earlier stages, and has the wings so exactly similar in shape and venation. The colours of the larva and perfect insect, however, are different. The head and thorax also are very different in their form and structure. A difference in colour might perhaps be due to seasonal variation; but the differences in shape of the different parts of the thorax are both distinct and pronounced.

RHINOCOLA CORNICULATA, n.sp.

(Plates xi., fig. 11; xiv., fig. 13).

Lerp common upon the slender leaves of *E. gracilis*, generally singly, but sometimes two or three close together; pale horn-colour to yellowish-brown, opaque and transversely striated; length 6, diameter at apex 2½ lines; commencing at a slender point, tapering to a broad rounded apex, convex on upper surface,

barred with black; wings light brown, coriaceous, semiopaque; nervures ochreous. Head very short, turned down in front, not as broad as thorax, deeply cleft in front, with a median suture and elongate fovea on either side, arcuate behind. Face lobes very large, broad and round, fringed with long hairs. Antennæ with first joint very short and broad, 2nd short, rounded, 3rd slightly bent at base, cylindrical, 4th-8th increasing slightly in length towards apex, 9th much shorter, 10th short, oval. Eyes large, angulated on inner margins: central ocellus small, hidden from above; lateral ocelli large, close to inner edge of eyes. Thorax: pronotum almost angular in front, deeply arcuate behind, short, not reaching to outer margins of eyes; dorsulum convex in front, pointed at extremities, sloping on either side of hind margin to centre which is truncate; mesonotum very large, with both sides of front margin lobed, sides produced into a point, hind margin sloping round to the centre where it is truncate; scutellum large, rounded behind; the entire dorsal surface of thorax finely shagreened. Legs stout, tarsal claws black. Wings twice as long as broad, long and narrow, broadest and rounded towards tip; primary stalk moderately long; stalk of subcosta short, costal cell elongate, rounded; stigma moderately long, angular at base; radius short, running out on upper edge of wing; stalk of cubitus longer than subcosta; upper branch of cubitus long; upper fork longest, curving down just below tip of wing, lower fork curving inwards; lower branch of cubitus long, upper fork long, curving upwards, lower fork long, sloping outward; clavus stout, rounded, clavical suture very slight. Abdomen stout and thick set. Genitalia (♂) unknown; (♀) short and broad, valves rounded on sides

Hab.—Bendigo, Vic. (on *E. gracilis*) and Wagga, N.S.W. (also on *E. gracilis;* W. W. Froggatt).

I have specimens of two lerp-scales which may be identical with those of this species, but I know nothing about the insects. The first came from Mr. J. H. Maiden, who received them upon botanical specimens of *E. largiflorens* growing near Bourke; these lerps are in clusters upon the leaves, and are of a darker yellow tint, opaque and more robust in form. The second were sent to the late Mr. A. S. Olliff from Western Australia, covering the leaves of *E. rudis* in thousands; these are lighter-coloured than the former, and somewhat smaller. This is the lerp for which Mr. Olliff proposed to form a new genus *Xylolyma*, but he merely exhibited the specimens at one of the Society's meetings (Proceedings, 1894, p. 740), and did not publish anything in the shape of description, if he had the adult insects.

Rhinocola ostreata, n.sp.

(Plates xi., fig. 14; xiv., fig. 20).

Larvæ producing rounded lerp-scales upon the leaves of *E. gracilis; lerp* 2 lines in diameter, convex, commencing at an elongate flange at the base, swelling out and circular at apex, attached to the lerp by the flange, but the encircling edge also fitting close to the leaf-surface; semi-opaque, dull horn-colour, with a smoky tint produced in concentric rings from the flange, giving it a regular circular formation.

Pupa.—General colour bright red; antennæ, legs, wing-covers and markings on the dorsal surface of head, thorax, and abdomen black; eyes dark yellow. Head broad, rounded in front, eyes well back, on sides of head; antennæ short, stout, and pointed at tips. Thorax broad, flattened; wing-covers small, rounded at

length towards tip, 9th-10th short and stout, forming a slight club. Eyes large, circular, flattened, not projecting: central ocellus small, silvery at apex of median suture; lateral ocelli vitreous, large, standing out from hindmargin of eyes. Thorax: pronotum narrow, curving round, raised in centre, arcuate behind, rugose behind eyes; dorsulum convex, rounded in front, coming to an obtuse point on either side, and broadly rounded behind; mesonotum broad, very much raised, convex, arcuate in front, rounded on outer margins into a short blunt point on either extremity, truncate behind; scutellum arcuate in front, convex, rounded behind. Legs stout, tarsi black. Wings coriaceous, thrice as long as broad, rounded at tip; primary stalk short and stout, stalk of subcosta a little shorter than stalk of cubitus; costal cell large, stigma long, slender; radius not reaching tip of wing, turning upward, then downward, and again curving upward at extreme tip; upper branch of cubitus long, upper and lower forks of nearly equal length forming a long slender cell, and both curving outwards at tip; lower branch of cubitus long, upper fork curving upwards, rounded; lower fork short, turning downward; clavus very stout, clavical suture very distinct. Abdomen short and stout, genitalia (♀) short, coming to a cylindrical point, clothed with fine silvery hairs.

Hab.—Bendigo, Vic. (on *E. gracilis;* W. W. Froggatt).

Rhinocola pinnæformis, n.sp.

(Plates xi., fig. 8; xiv., fig. 12).

Lerp-scales generally in clusters of three or four upon the surface of the leaves of *E.* sp., mature specimens and others just forming side by side; 2 lines in diameter, a little longer than broad, light brown, opaque and very convex on the dorsal surface, undersurface white; attached by a flange at base to the leaves, irregularly rounded with the free edge produced into slender fingers forming a fringe right round. In immature lerp-scales these fingers appear to be the ends of the transverse ribs used in the construction of the scale, but in the perfect scales these marks are obliterated and the upper surface is perfectly smooth and rounded.

Larva and *pupa* unknown.

Imago.—Length 0·16, antennæ 0·025 inch.

General colour chestnut-brown marbled with ochreous-yellow; wings hyaline, nervures dark brown. Head short, curving down in front, slightly rounded in front, with median suture and dark brown fovea on either side, rounded on sides and arcuate behind. Face lobes hidden from above, very short and broad, clothed with fine hairs. Antennæ short and slender, yellow, apical joints black, 1st-2nd short and broad, 3rd-8th slender, 9th-10th shorter and thickened. Eyes large but not projecting: central ocellus at apex of median suture; lateral ocelli small, on hindmargin of head. Thorax: pronotum convex in front, broadest in centre, arcuate behind, extremities curved up behind eyes; dorsulum broad and angulated, extremities of front margin pointed, hindmargin with an angulated point on either side; mesonotum large, arcuate in front, rounded on sides and hindmargin; scutellum elongate, arcuate in front, rounded behind. Legs short, stout. Wings large and broad, twice as long as broad, primary stalk moderately long, straight; stalks of subcosta and cubitus of equal length, costal cell broad; stigma distinct, stout, angular; radius short, curving upward and coming out above tip of wing; upper branch of cubitus long and straight; upper fork long, curving

Lerp shaped like the valve of a mussel shell, attached to the leaf by a scale-like hinge; 2½ lines in diameter, round, convex. When freshly formed the colours and outline of the larva can be distinguished through the lerp, but later on it becomes browner and opaque.

Pupa variable in colour; when full grown, pale greenish-yellow with the wing-covers dark brown, upper surface of the thorax blotched with bright red; eyes, centre of the head, and abdomen of a similar colour. Two large blotches between eyes, blotches on lower half of thorax, and interrupted bars across abdominal segments black.

Imago.—Length 0·09, antennæ 0·03 inch.

General colour ochreous, antennæ banded with brown; front of thorax, a patch on either side behind wings, and scutellum black; abdomen pink, with a slender black band on apical edge of each segment; wings light brown, thickly mottled with irregular brown spots; nervures light brown. Head narrow, convex behind, rounded on the sides, with a deep median suture and fovea on either side. Face lobes broad, short, hairy. Antennæ short and slender, 1st joint very short, thick, 2nd slightly longer than 1st, 3rd longest, 4th much shorter, 5th-9th joints of about the same length, 10th short and rounded at the tip. Eyes very large, rounded on outer margin: central ocellus small, at apex of median suture; lateral ocelli large, circular, bright red, close to hindmargin of eyes. Thorax: pronotum narrow, convex in front; dorsulum large, rounded in front, coming to a flange at extremities, sloping down and truncate behind; mesonotum large, arcuate in front, rounded behind, sides armed with a rounded spine in a line with forewings; scutellum arcuate in front, rounded behind. Legs short and stout; apex of thighs, base of tibiæ, and tarsi brown. Wings short and broad, rounded at tips, slightly more than twice as long as broad, broadly rounded from base to tip and slightly arcuate in centre of hindmargin; primary stalk short, thickened at apex, stalk of subcosta short, costal cell large, with a stout transverse nervure between it and the long slender stigma; radius short, curving upward to front margin of wing; stalk of cubitus longer than subcosta; upper branch of cubitus short, upper and lower forks coming out on either side of tip of wing, and forming a long slender cell; lower branch of cubitus moderately long, upper fork long, curving round, lower fork transverse, short; clavus stout and thickened, clavical suture very distinct. Abdomen short, stout, wedge-shaped. Genitalia (♂) viewed from above forming a pair of pincer-like projections.

Hab.—Sydney (Botanical Gardens), Botany, and Tumut, N.S.W. (in each case on *E. robusta*; W. W. Froggatt).

RHINOCOLA VIRIDIS, n.sp.

(Plates xi., fig. 9; xiv., fig. 17).

Larvæ forming a lerp-scale upon the foliage of *E. robusta*; a rare species, the scales always single.

Lerp 2½ lines in length, about 2 in diameter at apex; glassy, opaque, semitransparent, the structure indistinct but apparently transversely striated; commencing at a finger-like hinge, from which the lerp rapidly swells out into a rounded, somewhat fan-shaped convex scale, not unlike the scale of the adult female of several typical coccids of the genus *Chionaspis*.

Larva pale yellow, with a broad red mark in centre and at tip of abdomen.

stout and broad, 2nd much smaller, 3rd longest, 4th-9th of uniform length, 10th short, elongate-oval. Eyes round, large, projecting: central ocellus at apex of median suture; lateral ocelli small, vitreous. Thorax: pronotum narrow, forming a rounded knob at extremities behind eyes; dorsulum large, rounded in front, slightly angulate on sides, rounded behind; mesonotum large, arcuate in front, rounded behind; scutellum arcuate, rounded behind. Legs short, stout; tarsi broad, rounded. Wings not quite thrice as long as broad, slender, rounded at tip; primary stalk stout, long; stalk of subcosta short, cross-nervure forming a long slender stigma nearly reaching tip of wing; radius long, not reaching extreme tip of wing, curving upward slightly; stalk of cubitus long, upper branch of cubitus short, upper fork long, emerging in centre of tip of wing; lower fork shorter, curving down, forming a moderately long slender cell; lower branch of cubitus long, not as long as upper, upper fork long, turning downward, lower fork short, curving in at extremity; clavus stout. Abdomen large, coming to a regular point. Genitalia (♂) short and broad at base; lower genital plate short, truncate at tip; forceps short, broad, and rounded at tip; upper genital plate long and slender; penis hidden.

Hab.—Botany, N.S.W., (rare; two specimens on *E. robusta;* W. W. Froggatt).

RHINOCOLA MARMORATA, n.sp.

(Plate xii., fig. 3).

Early stages and life-history unknown; imago caught by sweeping a scrub of *Leptospermum* bushes.

Imago.—Length 0·14, antennæ 0·125 inch.

General colour: head pale green, basal joints of antennæ and legs yellow, apical portion of former dark brown; thorax reddish-brown and yellow tinted with green; abdominal segments black with apical edge bright red; genitalia red; under surface reddish-yellow tinted with green; wings hyaline, richly clouded with dark brown at tips and with yellow on hindmargin; nervures reddish-brown. Head very broad between eyes, rounded and lobed in front, truncate at base of antennæ, with a distinct median suture and no fovea on sides, deeply arcuate behind. Face lobes short and broad, thickly clothed with fine hairs extending over the face. Antennæ very long and slender, clothed with fine hairs, standing out in front of eye; 1st joint very short and stout, 2nd short, cylindrical, 3rd longest, 4th-9th shorter than 3rd, 10th elongate-oval, not half length of preceding ones. Eyes very large, reniform, angulated on hindmargins; projecting: central ocellus very small, well up from base of median suture; lateral ocelli red, small, close to hindmargin of eyes. Thórax: pronotum small, narrow, terminating in a rounded knob at extremities at inner margin of eyes; dorsulum short, broad, round in front and behind, tuberculate on sides; mesonotum large, arcuate in front, rounded behind; scutellum small. Legs and undersurface lightly clothed with hairs; tibiæ long and slender, tarsi large. Wings long, slender, nearly thrice as long as broad, deeply curved at base of costa, rounded at tip; primary stalk long; stalk of subcosta short; costal cell short, elongate-oval, slightly angular at apex; stigma long and slender; radius long, curving upward and then downward at tip; stalk of cubitus longer than subcosta; upper branch long, curving upward, upper fork longer than lower,

[illegible] down below ti[illegible] forming a short ang[illegible]

The pupæ of this species do not protect themselves with any kind of lerp-scale, but are found in a naked state, more or less covered with filaments, upon the twigs and foliage of *Fuchsia excorticata*, a New Zealand shrub.

The adult insect is pale yellow, with the head and dorsal surface mottled with black. Wings hyaline.

Genus APHALARA, Förster.

Head either swollen or produced into conical processes. Thorax broad, wings rounded at tip, membranous; stigma wanting; radius curved; stalk of cubitus in the forewing longer than stalk of subcosta. Genitalia: male genital valves prolonged into two slender processes encircling the penis.

APHALARA TECTA, Maskell.

Trans. Roy. Soc. S. Australia, 1898, p. 6, pl. ii., figs. 5-10.

This species was described from specimens from Victoria, the exact locality not being given. I am unable to identify it, though I have a number of Victorian species. Maskell's figures of the lerp or "pupal shield" are unlike any specimens in my collection; and the lerp-scales may be peculiar to *E. stuartiana*, the food plant.

"General colour yellow with the dorsal surface of the thorax and abdomen marked with a few black patches; antennæ yellow with brown tips: wings hyaline." Maskell figures the head with long face lobes like those of *Spondyliaspis*, and gives the wing with a distinct stigma and a subcostal cell. The latter character should remove it from this genus, which Forster defined as without a stigma. The lerp-scale is narrow at the flange or hinge, swelling out and rounded at the apex, transversely striated; yellow and opaque.

APHALARA CARINATA, n sp.

(Plates xii., fig. 7; xiv., fig. 16).

The larvæ attack the extreme tips of the leaves of *E. capitellata*, forming half rounded galls through the tips of the leaves swelling out and curving round.

Lerp gall as much as 4½ lines in diameter, red to brown in colour, general form rounded, turning up, and narrow at base; under side curving upward to the leaf, being open but covered with an opaque horn-coloured lerp, 2 lines in diameter, curled upward with a rim on the lower edge so that it is half gall, half lerp.

Pupa.—General colour: head pale green, thorax and abdomen deeper green; antennæ, legs, front of head, and centre of thorax yellow; two elongate spots between eyes, wing-covers, and a double row of spots on abdominal segments black.

Imago.—Length 0·20 inch, antennæ (?)

General colour pale green, legs yellow, antennæ and thorax marked with yellow; scutellum black; wings semiopaque, horn-colour, nervures ochreous. Head small, not as broad as thorax, lobed in front, and sloping down to eyes, median suture ridged with a very large fovea on either side occupying the whole of the forehead, arcuate behind. Face lobes very large and broad, slightly truncate at tip. Antennæ slender, 1st joint stout, cylindrical, 2nd joint half as long as first, 3rd longest, 4th-9th uniform in length, 10th shorter, swelling out and rounded to a pointed apex. Eyes large, hemispherical, projecting on the sides: central ocellus at apex of median suture; lateral ocelli large, well up on hind margin of head. Thorax: pronotum narrow,

very distinct, long and slender. Genitalia (♂) short, stout [too much damaged for details].

Hab.—Mosman's Bay, near Sydney (on *E. capitellata*: W. W. Froggatt).

Genus PHYLLOLYMA, Scott.

Head with crown broad, length down centre about equal to half width between eyes. Face lobes narrow, ribbon-shaped. Antennæ short. Eyes moderately large, on side of head, their inner margins separated from lateral margin of crown on its ower half by a lunate or cuneate plate. Thorax: pronotum narrow, convex, lateral margins convex, scarcely reaching beyond middle of posterior margin of eyes; mesonotum across insertion of elytra not wider than head and eyes together: dorsulum transverse, semihexagonal. Elytra rhomboidal: stigma wide at mouth; radius terminating in upper apical angle; cubitus petiole shorter than upper arm, longer than lower one.

Type *Psylla fracticosta*, Walker.

PHYLLOLYMA FRACTICOSTA, Walker.

Psylla fracticosta, Walk., B.M. Cat. (Homoptera) p. 275, 1850-1; *Phyllolyma fracticosta*, Scott, Trans. Ent. Soc. Lond. 1882, 457, pl. xviii., figs. 5-5*e*.

This species was described from Tasmania by Walker; and was again described by Scott after examining the specimens in the British Museum collections. The latter author figures the elytron, a front view of the head, the antennæ, and the genitalia.

It is a large species, measuring 1½ lines in length: fuscous-brown in colour, with broad rounded wings clouded with brown; a remarkable broad angular pale blotch in the centre of elytra, and three similarly coloured marks on the tips. These characters should render the species easily recognisable. I have not so far seen it.

Genus CARDIASPIS, Schwarz.

Head emarginate posteriorly, vertex flat, narrowing to base of antennæ; frontal processes sharply separated from vertex:

antennæ slightly longer than width of head; eyes large, globular, projecting, occupying the whole side of head. Thorax: pronotum greatly convex transversely, lateral impressions large and deep; dorsulum less than twice the length of pronotum, sharply rounded in front; mesonotum very little convex in front; side-pieces of pro- and mesosternum prominent. Wings elongate-oval; veins fine, none of them curving much; stalk of cubitus longer than that of subcosta; a distinct large pterostigma; tip of wing at termination of fourth furcal. Epimera of mesosternum greatly developed, transverse; spiniform processes very small, vertical. Legs short and robust, hind tibiæ without a broad tooth; tarsi normal.

Type *Cardiaspis artifex*, Schwarz.

Cardiaspis artifex, Schwarz.

(Plates xi., fig. 10; xii., fig. 9; xiv., fig. 14).

C. artifex, Schwarz, Proc. Ent. Soc. Washington. Vol. iv. p. 72, 1896.

Lerp convex, 3 lines in diameter; singly upon the leaves of *E. robusta*, attached to the leaf by a brown hinge from which comes an opaque-whitish angular piece, rounded at the apex, convex and curving upward, from which on all sides radiate a number of

either side; abdomen dark brown to blackish, with the apical edges of each segment barred with reddish-yellow; wings hyaline, finely granulated, nervures reddish-pink. Head short, as wide as thorax, rounded in front, with a very slight cleft; median suture and large flattened fovea on either side arcuate behind. Face lobes very large, broad, rounded at tips. Antennæ short, stout, 1st-2nd joints short and stout, 3rd longest, 4th-8th uniform, 9th-10th slightly thickened. Eyes large, prominent, rounded: central ocellus very large, situate in cleft just behind face lobes; lateral ocelli large. Thorax: pronotum very narrow, rounded in front, deeply arcuate behind, with a dark impressed fovea towards the extremities; dorsulum small, sharply rounded, and almost angulated in front, truncate behind; mesonotum very large, slightly arcuate in front, rounded to a conical point at extremities, rounded behind; scutellum long and narrow. Legs short, robust. Wings more than twice as long as wide, broad and rounded to tip; primary stalk long, straight; stalk of subcosta short, costal cell elongate; stigma angular at base, moderately long; radius short, nearly straight, coming out at upper edge of wing; stalk of cubitus long, upper branch long, upper fork long, curving downward; lower fork much shorter; lower branch of cubitus long, upper fork rounded, curving down, lower fork short curving inwards at apex; clavus stout, clavical suture very distinct. Abdomen short, stout. Genitalia: (♂) lower genital plate stout, forceps broad, rounded, upper genital plate slender.

Hab.—Manly, near Sydney, and Termiel, N.S.W. (on *E. robusta;* W. W. Froggatt); S. Australia (on *E. leucoxylon;* A. Koebele).

The lerp-scales were very numerous upon the foliage of saplings or shoots springing up where trees of swamp mahogany (*E. robusta*) had been cut down near Manly in the summer of 1892, but have never appeared there since. The extraction of the sap by the larvæ in feeding and in constructing their beautiful scales causes a large brown blotch to appear wherever one was formed, and it was the curious spotted appearance of the foliage so caused that first attracted notice.

This species was described by Schwarz from South Australian specimens. He says that, "It is distinguished from all other described genera of this tribe by its vertical head and the form of the frontal processes." The venation of the wings is similar to that of the genus *Rhinocola*, in which I had placed it before seeing Mr. Schwarz's definition of *Cardiaspis*. He supposes that the large size of the lerp-scale in comparison with the larva is accounted for by the fact that the latter is enveloped in a mass of woolly secretion. But none of our lerp-producing Psyllids are remarkable for this excretion which is unusually scanty in comparison with that present in the naked forms of larvæ.

CARDIASPIS PLICATULOIDES, n.sp.

(Plates xi., fig. 7; xii., fig. 18; xiv., fig. 9).

Lerp chocolate-brown, 1½ lines in diameter, massed together in clusters of 30-40 or sometimes singly, upon the leaves of *E. rostrata* and several other Eucalypts. General form like that of the valve of the bivalve shell, *Plicatula cristata*, Lam.; attached closely to the leaf, curving upward, banded with five distinct parallel ribs, with shorter ridges between them to the margin; the crystalline secretion between the ribs finely striated.

Pupa very small in proportion to the size of the lerp; general

3rd much longer than 4th, 4th-6th of uniform length, 7th-8th longer, 9th shorter, 10th short and rounded at tip. Eyes large and prominent: central ocellus small, lateral ocelli situated about middle of hindmargins of eyes. Thorax: pronotum very narrow; dorsulum large, rounded in front, arcuate on sides, truncate behind; mesonotum convex, broad, rounded behind; scutellum arcuate in front, rounded behind. Legs short and thick, with a row of fine black spines at apex of tarsi of hind legs. Wings nearly thrice as long as broad, rounded on front margin, somewhat pointed at tip; primary stalk short; stalk of subcosta short, costal cell separated from the slender stigma by a sloping cross-nervure; radius short, running out above tip of wing; stalk of cubitus very long, upper branch long, upper and lower fork long, of equal length, forming an elongate cell in centre of wing; lower branch of cubitus long, upper fork long, curving round; lower fork curving outward; clavus stout, clavical suture long, distinct. Abdomen short and stout. Genitalia (♂) large, upper genital plate forming two conical processes, lower genital plate small; forceps large, rounded to a conical point at tip.

Hab.—Melbourne, Vic. (on *E. rostrata*, in Botanical Gardens; C. French, Jr.); Ryde, near Sydney, Tumut, Yass, and Mittagong, N.S.W. (in each case upon *E.* sp.; W. W. Froggatt).

The little shell-like lerp is very plentiful, but it is one of the most difficult from which to obtain the perfect insects on account of the many parasitic Chalcids which infest them; out of a great number collected by Mr. French, only half-a-dozen specimens of the perfect insects were bred. I had the same difficulty with my own specimens; fully half of them when found were punctured with a small circular hole on the side by which a small wasp had emerged. In Newman's Entomologist for 1841 (Vol. i. p. 88), a woodcut of this lerp is reproduced, with the following note from Mr. A. H. Davis, of Adelaide, S.A.:—"I have now by me the leaf of a Eucalyptus, covered with little habitations, perfectly like shells, the form even of the ribs being faithfully represented as in the annexed drawing; there are a dozen on one leaf, and they are scarcely half the size here depicted; the shell is of a dirty brown colour; some species of the same family make white shells, and the shell fabricated by one species resembles that of a limpet."

One of the figures in Dr. Dobson's paper, previously noted, appears to represent the lerp of this Psyllid.

Genus COMETOPSYLLA, n.g.

Head curved sharply down in front; eyes as broad as thorax. Face lobes very short, truncate at apex, broad. Antennæ very short, slender to the tips; a rounded pad between base of antennæ and eyes. Thorax : pronotum widest in the middle, rounded at extremities, reaching to outer edge of eyes; dorsulum elongate at extremities, rounded in front; mesonotum large. Elytra short, acute at apex; costal nervure swelling out and forming a rounded knob in front of stigma; stigma large; subcostal nervure running nearly to tip of wing; radius emerging at tip of wing; stalk of cubitus much longer than stalk of subcosta; upper branch of cubitus long, both furcations short and broad. Genitalia (♂) short, broad.

COMETOPSYLLA RUFA, n.sp.

(Plates xii., figs. 6 and 21; xiv., fig. 18).

Lerp crystalline white; 1½ lines in diameter, clustering together

Eyes large. Thorax a little broader than head at junction of wing-covers. Abdomen narrow at base, with tip sharply rounded to a point.

Pupa lighter brown, with the darker markings of the larva much broader, so that the colour of the dorsal surface is much darker. General form very short and broad in proportion to length, with apical portion of abdomen showing a granulated surface; the long hairs of the larva produced into soft downy tufts. Head short, broad, slightly lobed in front, rounded to eyes, and curving in behind them, deeply arcuate behind; antennæ stout, moderately long, standing out on side of head; eyes flattened, not projecting. Thorax much broader than head, swelling out, rounded to tip of abdomen; wing-covers very broad, rounded, projecting; five black dots between wing-covers; legs short, stout. Abdomen pointed at tip.

Imago.—Length 0·125, antennæ 0·0125 inch.

General colour bright reddish-orange, marked with light brown, eyes dark brown; hindmargin of head black; antennæ mottled with brown; thoracic segments mottled with light brown, upon the mesonotum forming a dainty scroll work; abdomen marked with black; legs pale ochreous; wings pale fuscous to smoky brown, finely granulated, nervures pale brown. Head very wide, forehead very narrow, rounded in front, with a slight median suture, deeply arcuate behind. Face lobes very short, broad, hidden from above. Antennæ very short, slender, 1st joint very stout, 2nd more rounded, 3rd-10th short, rounded at both extremities. Eyes elongate, as wide as forehead: central ocellus large, situated in a fovea in centre of face, hidden from above; lateral ocelli large, at the extreme hindmargin of head. Thorax: pronotum rather large, curved in front, marked with a black spot in line with eyes, angular at tips; dorsulum irregularly rounded in front, extremities produced into two angular teeth, rounded on apical margin; mesonotum very large, arcuate in front, swelling out and rounded on sides of scutellum, the latter truncate in front, rounded behind. Legs short, thick. Wings a little more than twice as long as broad, rounded to apex of costal cell, then straight to tip, coming to a regular point at apex of marginal cell; primary stalk long, stout; stalk of subcosta short; stigma rounded in front and running to a long slender point; costal cell short, broad; radius long, curving upward, emerging at tip of wing; stalk of cubitus very long, curving downward, upper branch of cubitus long, upper and lower fork curving downward, forming an angular cell, lower branch of cubitus very short, upper fork long, curving round, lower fork short, curving in at apex; clavus stout, clavical suture thick. Abdomen short, stout. Genitalia (♂) short, broad, turned upwards; lower genital plate short, rounded; upper genital plate long, elongate, oval; forceps short, rounded at base, conical at tip; penis slender, straight: (♀) sabre-shaped, nearly as long as the rest of abdomen, fringed with stiff hairs on the underside, and regularly toothed along upper valve.

Hab.—Liverpool, Wagga, N.S.W. (on *E. melliodora;* W. W. Froggatt).

I have also found the larvæ and pupæ of this psyllid common upon the foliage of *E. hemiphloia*, and several undetermined Eucalypts in the neighbourhood of Penrith and Mittagong, but in all cases they were free upon the twigs and constructed no lerp-scales. As the structure of the lerp is very crystalline and brittle, this is probably accounted for by the fact that in damp weather the secretion will not set to form a protective covering,

branous, very long, narrow, angulated at tip, veins fine, stalk of cubitus longer than subcosta: a long but narrow pterostigma: radius long, curving downward to tip of wing: stem of second fork parallel with radius. Legs short, anterior and middle tibiæ simple, posterior tibiæ not dentate at base, but dilated at apex and produced near the outer (posterior) apical angle into a stout mucro: anterior and middle tarsi with first joint short, simple, and much longer than claw-joint: posterior tarsi with first joint as long as claw-joint and dilated beneath into a broad, flattened, membranous, cushion-like disc: metasternal epimera very large; as long as wide, without spiniform processes. Abdomen: sixth ventral segment of ♂ broadly divided for the reception of the genitalia: genital plate and forceps without lateral appendages: genitalia of ♀ not beak-shaped, the outer valve consisting of two short plates.

Type *Psylla eucalypti*, Dobson.

Spondyliaspis eucalypti, Dobson.

(Plates xii., figs. 1 and 12: xiii., fig. 5: xiv., fig. 7.)

Psylla eucalypti, Dobs., Proc. R. Soc. V. Diemen's Land, 1851, Vol. v. Pt. iii. p. 235: *Spondyliaspis eucalypti*, Schwarz, Proc. Ent. Soc. Washington, Vol. iv. p. 69, 1896.

Lerp convex, circular, opaque white, coming to a rounded point, 1½ lines in diameter, generally covering the surface of the leaves in clusters of forty or more close to but seldom overlapping each other: irregular and rugose on the outer surface, attached to the leaf right round the edge, but peeling off when dry. If enclosed in a tin the lerps liquefy to a certain extent.

Larva differing very slightly from pupa.

Pupa: general colour red: antennæ semitransparent, outer margins from sides of head to tip of abdomen pale green: two spots on head, two on pronotum, wing-covers and a double row of elongate marks coming to a point at tip of abdomen black, under-surface of head and abdomen bright green, legs, thorax, and sides of head yellow. Head short, arcuate in front, rounded behind: antennæ long, third joint bent: eyes large, rounded.

Thorax large, rounded in front, wing-covers long, narrow, rounded at tips; legs short, stout. Abdomen very broad, swelling out from base, rounded, and slightly crenulated on sides to a tuberculate anal process.

Imago.—Length 0·18, antennæ 0·055 inch.

General colour light olive-green; mesonotum, dorsulum (also marked with parallel brown bars) and some of abdominal segments bright orange-red; all the segmental divisions marked with brown, those upon abdomen forming black bands; antennæ and legs light brown, apex of joints and tip of former clouded. Wings delicate, transparent, nervure light brown. Head large, wider than pronotum, deeply angulated in centre, with deep median suture, and small fovea on either side, arcuate behind, truncate behind antennæ and eyes. Face lobes very large, long, cylindrical, projecting in front of head, covered with fine hairs, rounded at tips, with a deep cleft between them. Antennæ long, slender; 1st joint short, broad, 2nd short, truncate, 3rd very long, 4th shorter, 5th-8th uniform in length, slightly shorter towards the tip, 9th short, 10th very short, pointed. Eyes very large, somewhat reniform, prominent: central ocellus very small, at apex of median suture; lateral ocelli round, small, very close to hindmargin of eyes. Thorax: pronotum angular in centre, arcuate on sides of front margin, with a small circular fovea on

curving downward; lower fork curving downward, shorter than upper, lower branch of cubitus short, upper fork long, arching round, lower fork short, curving in at tip; clavus short, clavical suture slender. Abdomen short, stout. Genitalia (♂) long, curved over the back, shaped somewhat like a duck's head when closed, lower genital plate short, broad, rounded, with a short tubercle at tip; forceps slender at the base with a rounded knob on the inner edge, flattened and slender above, rounded at tip; penis slender, upper genital plate stout, cylindrical at base, sloping to tip on inner edge : (♀) forming a short angular hairy tip tapering to a point.

Hab.—Tasmania (on *Eucalyptus* sp.; A. M. Lea); Victoria (on *E.* sp.; C. French, Jr.); New South Wales, widely distributed (on *E.* sp., &c.; W. W. Froggatt); Townsville, Queensland (on *E.* sp.; H. Tryon).

This species is our commonest "sugar-lerp" which has a very wide range over eastern Australia It does not confine its attention to one species of Eucalypt, but is found upon *E capitellata*, *E. piperita*, *E. leucoxylon*, *E. gracilis*, and several other species.

As children we used to gather and eat the scales of this species, but it is those of the larger species that were collected and eaten by the natives in the Mallee scrubs, and which were described as "manna." In the venation of the wings this species could come under the genus *Aphalara*, but Schwarz has examined some specimens of this species and redefined Signoret's genus *Spondyliaspis* for its reception on account of the peculiar spiny structure of the hind legs, differences that have not been used very much as generic characters in this group of the Homoptera. Signoret defined the genus upon the lerp-scales only, and had never seen the perfect insects.

SPONDYLIASPIS MANNIFERA, n.sp.

(Plates xii., figs. 2 and 10; xiv., fig. 6).

Lerp white, 4 lines in diameter, circular, convex; generally singly upon the surface of the leaves of *E. polyanthema*, *E. hemi-*

phloia, and other species; attached all round the lower edge to the leaf, and thickly clothed with numbers of looped filaments giving it a woolly appearance.

Pupa light yellow, darker on abdomen; legs and antennæ semitransparent, except the tip of the latter, which is blackish: a brown mark on either side of head, a broad stripe on either side of thorax, a number of irregular spots, and inner edges of wing-covers brown, sides of abdomen clouded with brown, with a double row of spots as in the former species. Head large, lobed in front, straight on sides, swelling out behind eyes, arcuate at base; antennæ rather stout, standing out on sides; eyes rounded, rather prominent; thorax a little broader than head, longer than broad; wing-covers slender, oval, pointed at apex; legs short, stout; abdomen large, globular, constricted at base, swelling out and rounded to an anal tubercule, with an angular projection on either side. The whole insect is fringed with fine hairs which are covered with minute particles of manna.

Imago.—Length 0·16, antennæ 0·07 inch.

General colour bright yellow marked with light brown, eyes reddish-brown, legs and antennæ fuscous, darker markings on thorax and apical segments of abdomen, wings semitransparent, outer nervures light brown, inner nervures yellow. Head as broad as thorax, short, deeply cleft in front, with a rounded lobe

slender. Wings long, slender, more than thrice as long as broad: primary stalk long; stalk of subcosta short, costal cell long and slender, with costal nervure thickened, no true stigma; radius very long, slender, curving round at tip; stalk of cubitus longer than subcosta; upper branch of cubitus long, turning upward; upper fork turning down, longer than lower fork, lower branch of cubitus long, turning downward, upper fork long, curving round, lower fork curving outward forming a long cell: anal cell lanceolate, with a fine clavical suture. Abdomen stout, slender at junction with thorax. Genitalia (♂) showing a black patch on either side: lower genital plate short, rounded, coming to a point at tip; forceps large, elongate, curved inwards; penis hidden, upper genital plate elongate-oval.

Hab.—Tumut, N.S.W. (on *E. polyanthema* and *E. hemiphloia*: W. W. Froggatt); Wimmera, Vic. (on *E. gracilis*: W. W. Froggatt.)

This species has a wide range, and is closely allied to *S. eucalypti*.

Genus DASYPSYLLA, n.g.

Head short, turned down sharply in front, deeply arcuate behind: eyes large, angulated on the inner margin Face lobes large, broad, close together, truncate at apex. Antennæ very short. Thorax: pronotum very narrow; dorsulum coming to a point at extremities; mesonotum large. Elytra broad, rounded at apex, with a double costal nervure forming a narrow cell, inner nervure running into upper before reaching the wedge-shaped stigma: primary stalk short: radius short, not reaching apex of wing; stalk of subcosta not as long as stalk of cubitus: upper cubital furcation elongate, both nervures curving outward at apex. Genitalia (♂) short: lower genital plate broad, angular; forceps broad.

Type *D. brunnea*.

DASYPSYLLA BRUNNEA, n.sp.

(Plates xii., figs. 5 and 11: xiv., fig. 21).

Lerp chocolate-brown: 2 lines in diameter, formed singly upon the leaves of *E. polyanthema;* irregularly rounded, undersurface forming a thin brown shell or lower valve attached to the leaf; upper valve commencing with a small hinge, swelling out to lower edge; curving round in concentric rings, forming a convex rounded shell fringed on summit and outer margin with short filaments.

Pupa dark red; legs, two spots on head, wing-covers and the spots between them, and two rows of spots down centre of abdomen black; ventral surface of head and thorax light red. Head short, broad, rounded on sides; antennæ very long, slender, curving round to centre of wing-covers; eyes small. Thorax short, wing-covers short, rounded; legs short, thick. Abdomen short, constricted at base, swelling out, rounded to tip, broader than long.

Imago.—Length 0·185, antennæ 0·045 inch.

General colour ochreous, with creamy-white and pink tints; head and thorax finely shagreened, thickly covered with small black spots; abdominal segments dark brown to black, with lower margins ochreous; wings thickly mottled with fine spots and larger blotches of black and brown, nervures white at base, black in clouded portions, pink in clearer portions. Head as wide as thorax, short, face turned downward, slightly cleft in centre, arcuate behind antennæ and basal margin. Face lobes large, broad, rounded, thickly fringed with white hairs. Antennæ very short, 1st joint short, black; 2nd short, cylindrical; 3rd slender,

on front margin, broad at tip, costal nervure very stout at base forming a double or secondary costal nervure to stigma; primary stalk short; stalk of subcosta not as long as stalk of cubitus; costal cell short; stigma short, angular; radius short, curving upward, emerging on upper edge of wing; stalk of cubitus long; upper branch straight, upper and lower forks long, curving in at centre and turning outward at tips; lower branch of cubitus long, straight, upper fork long, curving round, lower turning outward at apex; clavus stout, clavical suture slender. Abdomen short, stout. Genitalia (♂) long, stout; lower genital plate short, angular at apex; forceps long, formed of two broad-tipped processes closing down over the penis; upper genital plate slender, conical, truncate at apex, projecting beyond forceps.

Hab.—Tumut, N.S.W. (on *E. polyanthema*; W. W. Froggatt).

Genus Thea, Scott.

Head twice as broad between eyes as length down centre; crown with posterior margin almost straight, sides straight from base to front of eyes, where they are produced into a short triangular tooth. Face lobes narrow, ribbon-shaped. Antennæ long. Eyes on sides of head projecting nearly their whole width beyond pronotum. Thorax: pronotum narrow; mesonotum convex, as wide as head and eyes at the insertion of forewings. Elytra elongate, apex acute; stigma short; radius long, terminating at apex; cubitus upper furcation longer than arm, petiole a little longer than lower arm.

Type *Psylla triguttata*, Walker.

Thea formicosa, n.sp.

(Plate xi, fig. 4).

Larva dull white, eyes pink, faintly marked with light brown; apex of abdomen darkest, apical segment fringed with hairs. *Second stage:* Head, thorax, and base of abdomen creamy white; legs, antennæ, and wing-covers light brown; eyes dark brown, two large blotches on head dark brown; small, black, angular spots forming a regular pattern on thorax; basal segments of abdomen marked with fine black interrupted bars; apical portion light chocolate-brown, rugose, flattened, and marked with irregular punctures. Legs, antennæ, and outer edge of abdominal segments fringed with coarse hairs.

Pupa with white colour of larva changing into opaline-blue, brightly spotted with red, bands upon abdominal segments thickened. Head truncate between antennæ, sloping down to eyes, arcuate behind; eyes very large; antennæ long, thick, very stout at base. Thorax broad, legs thick, stout; wing-covers small, pointed at apex. Abdomen large, flattened at apex, and irregularly rounded to tip.

Imago.—(♀) Length 0·21, antennæ 0·08 inch.

General colour pale opaline-blue; ocelli, edges round eyes, spots on head, and outer margins of abdominal segments bright red; first and second joints of antennæ black; terminal joints, legs, and ventral surface ochreous; thorax barred with light chestnut; abdominal segments black, with apical margins banded with blue and red, genitalia red; wings semi-opaque, the transverse nervure at base of subcostal and outer edge of wing blotched with black, nervures bright reddish-brown. Head short, as broad as thorax, slightly lobed in front, rounded, arcuate at base of antennæ, with shallow median suture, rounded on sides, arcuate behind. Face thrice as long as broad, rounded in front to a sharply curved tip, broadest at second cubital cell, primary stalk long; stalk of subcosta short; costal cell short, broad; stigma short, angular, rounded in front: radius long, curving downward, emerging at tip of wing; stalk of cubitus longer than subcosta, upper branch of cubitus curving upward, upper and lower forks forming a short broad cell, lower branch of cubitus very short, upper fork curving upward, lower fork very short, turning sharply downward, forming a very short, broad cell: clavus stout, clavical suture very distinct. Abdomen short, stout. Genitalia (♀) very long, stout, sabre-shaped, rounded at base, upper valve longest, slightly truncate at tip.

♂. Smaller than ♀, with no black markings upon wings, and of a uniform brown colour with pale reddish tints. Abdomen very short. Genitalia (♂) very large, lower genital plate elongate, rounded at apex; forceps large, curved upward; upper genital plate longer and more slender than forceps.

Hab.—Thornleigh, Botany, Mittagong, N.S.W. (on *E. piperita*: W. W. Froggatt).

The larvæ form no lerp but hide under loose bits of bark on the trunks of several white-stemmed gums, thickly enveloped in white flocculent matter, thickest round the abdomen, which exudes from beneath their shelter and reveals their hiding place, and when abundant dots the trunks all over with white blotches. Other colonies are found congregated on the stems of small trees, where they are frequently covered by ants with a thick felted sheath of woody débris, sometimes extending for four or five feet from the ground and completely sheltering them. The ants, *Iridomyrmex nitidus*, Mayr, swarm over them in this covered gallery, and evidently protect them for the sake of the honey-dew that is secreted.

THEA OPACA, n.sp.

(Plates xi., fig. 5; xii., fig. 14; xiii., figs. 1-4).

Larva similar in appearance to that of *T. formicosa*, but of a more uniform ochreous colour; thoracic markings similar.

Pupa.—Head broad, slightly lobed in front; antennæ long, slender, tapering to tips; wing-covers short, broad, darker-coloured than legs; abdomen tufted with hair on either side.

Imago.—Length 0·24, antennæ 0·07 inch.

General colour reddish-pink, marbled with brown and black. Head in form as in the former species. Pronotum a little broader; dorsulum more elongate at extremities, with a deep impressed fovea on either side towards front margin; mesonotum larger, more rounded at extremities. Legs similar. Wings of same shape and colouration; primary stalk longer and straighter; stigma not rounded in front, but transverse: forks of upper branch of cubitus of equal length, forming a long slender cell; lower branch of cubitus moderate, upper fork longest, curving round, lower fork curving in at apex. Abdomen short, stout. Genitalia (♂): lower genital plate short, broad, rounded beneath; forceps short, broad, slightly pointed at apex; upper genital plate very long, pointed at tip; penis hidden.

Hab.—Canterbury, Croydon, Sydney, N.S.W. (on *E.* sp.; W. W. Froggatt).

This is not so plentiful as *E. formicosa*, but is found in similar localities, and is always covered with ants.

THEA LEAI, n.sp.

increasing in length towards tip, 10th short, elongate-oval. Eyes rounded, projecting very slightly: central ocellus small, at extreme apex of median suture; lateral ocelli small, about centre of hind-margin of eyes. Thorax: pronotum broad, rounded in front, arcuate behind, a small fovea on either side; dorsulum convex, rounded in front, pointed at tips, sloping to a truncate apex at junction with mesonotum, the latter broad, arcuate in front, running to a blunt point on sides, rounded behind, broad and nearly square. Legs long, stout. Wings more than twice as long as broad, rounded in front, pointed at apex, rounded to base; primary stalk long; stalk of subcosta short; stigma short, rounded at base, pointed at apex; radius long, curving upward, then turning down above tip of wing; stalk of cubitus longer than subcosta, upper branch of cubitus long, upper and lower forks nearly equal in length, forming a long slender cell, lower one curving downward; lower branch of cubitus moderately long, upper fork short, curving round, lower fork very short, turning outward, forming a short, broad cell; clavus stout, clavical suture long, slender. Abdomen short, stout. Genitalia: (♂) ochreous, broad, short; lower genital plate round, broad; forceps short, broad, rounded at tip; upper genital plate long, slender: (♀) sabre-shaped, long, slender.

Hab.—Tamworth, N.S.W. (on *E.* sp.; A. M. Lea).

Four specimens in the collection of the Department of Agriculture, Sydney, pinned and labelled "Tamworth, collected by Mr. Lea."

LITERATURE.

1.—Anderson, T. On a new Species of Manna from New South Wales: Edinb. New Phil. Journ. xlvii. 1849, p. 132.

2.—Dobson, T. On Laap, or Lerp, the Cup-like Coverings of Psyllidæ found on the leaves of certain Eucalypti: P. & Proc. R. Soc. Van Diemen's Land. Vol. i. Pt. 3, 1851, p. 235.

3.—Wooster, W. W. How the Lerp Crystal Palace is built: Journ. Micros. Soc. Victoria. Vol. i. 1882, p. 91.

4.—Tepper, J. G. O. Remarks on the "Manna" or Lerp Insect of South Australia: Journ. Linn. Soc. Lond. Zool. Vol. xvii. 1883, p. 109.

5.—Beveridge, P. On the Aborigines inhabiting the Great Lacustrine and Riverine Depression of the Lower Murrumbidgee, L. Lachlan and L. Darling: Journ. & Proc. R. Soc. N.S. Wales. Vol xvii. 1883, p. 63.

6.—Maskell, W. M. On some Species of *Psyllidæ* in New Zealand: Trans N.Z. Inst. Vol. xxii. 1890, p. 157, pl. x.-xii.

7.———— On some Australian Insects of the Family *Psyllidæ*: Trans. Roy. Soc. S.A. Vol. xxii. 1898, p. 4, pl. i.-iii.

8.—Scott, J. On certain Genera and Species of the Group of *Psyllidæ* in the Collection of the British Museum: Trans. Ent. Soc. Lond. 1882, p. 463.

9.—Tryon, H. Three Undescribed Insects whose Food-plant is the Moreton Bay Fig, &c.: Trans. Nat. Hist. Soc. Queensland. Vol. i. 1892-94 [1895], p. 60.

10.—Schwarz, E. A. Notes on the Lerp Insects (*Psyllidæ*) of Australia: Proc. Ent. Soc. Washington. Vol. iv. 1897, p. 66.

11.—Sharp, D. Homoptera, Family *Psyllidæ*: Cambr. Nat. Hist. Insects. Part ii. 1899, p. 578.

12.—Foerster, A. Uebersicht der Gattungen und Arten in der Familie der Psylloden: Rheinl. u. Westphal. Verhand. 1848, pp. 65-98

13.—Loew, F. Zur Biologie u. Charakteristik der Psylloden, &c.: Verh. K. K. Zool.-bot. Gesell. in Wien. xxvi. Bd. 1876, p. 187.

14.———— Zur Systematik der Psylloden: *op. cit.* xxviii. 1879, p. 585.

15.———— Revision der paläarktischen Psylloden, &c.: *op. cit.* xxxii. Bd. 1883, p. 227.

16.———— Katalog der Psylloden: Wiener Entom. Zeitung. Heft ix. 1882, p. 209.

17.—Reaumur, R. A. F. de. Mémoires pour servir à l'Histoire des Insectes. T. iii. 1737.

Fig. 10.—*Cardiaspis artifex*, Schw.; elytron.
Fig. 11.—*Rhinocola corniculata*, n.sp. „
Fig. 12.——— *revoluta*, n.sp. „
Fig. 13.——— *assimilis*, n.sp. „
Fig. 14.——— *ostreata*, n.sp. „

Plate xii.

Fig. 1.—*Spondyliaspis eucalypti*, Dobs.; elytron.
Fig. 2.——— *mannifera*, n.sp. „
Fig. 3.—*Rhinocola marmorata*, n.sp. „
Fig. 4.——— *liturata*, n.sp.
Fig. 5.—*Dasypsylla brunnea*, n.sp.
Fig. 6.—*Cometopsylla rufa*, n.sp.
Fig. 7.—*Aphalara carinata*, n.sp. „
Fig. 8.—*Rhinocola revoluta*, n.sp.; genitalia (♂).
Fig. 9.—*Cardiaspis artifex*, Schw. „
Fig. 10.—*Spondyliaspis mannifera*, n.sp. „
Fig. 11.—*Dasypsylla brunnea*, n.sp. „
Fig. 12.—*Spondyliaspis eucalypti*, Dobs. „
Fig. 13.—*Thea formicosa*, n.sp.
Fig. 14.——— *opaca*, n.sp. „
Fig. 15.—*Lasiopsylla rotundipennis*, n.sp. „
Fig. 16.——— *bullata*, n.sp. „
Fig. 17.—*Creiis longipennis*, Walk.
Fig. 18.—*Cardiaspis plicatuloides*, n.sp. „
Fig. 19.—*Rhinocola liturata*, n.sp.
Fig. 20.—*Thea Leai*, n.sp.
Fig. 21.—*Cometopsylla rufa*, n.sp. „

Plate xiii., Figs. 1-4.—*Thea opaca*, n.sp.

Fig. 1.—Pupa enlarged.

Fig. 2.—Imago (♀). 1*a*. Face lobes. 2*a*. Prothorax. 3*a*. Mesonotum. 4*a*. Dorsulum. 5*a*. Scutellum.

Forewing (Elytron). 1*b*. Costal nervure. 2*b*. Primary stalk. 3*b*. Clavus. 4*b*. Clavical suture. 5*b*. Stalk of subcosta. 6*b*. Stalk of cubitus. 7*b*. Subcosta. 8*b*. Lower branch of cubitus. 9*b*. Upper branch of cubitus. 10*b*. Lower fork of lower cubitus. 11*b*. Stigma. 12*b*. Upper fork of lower branch of cubitus. 13*b*. Radius. 14*b*. Lower fork of upper cubitus. 15*b*. Upper fork of upper cubitus.

Fig. 3.—Genitalia (♂). 1. Apex of abdomen. 2. Lower genital valve. 3. Forceps. 4. Penis hidden. 5. Upper genital valve.

Fig. 4.—Genitalia (♀). 1. Upper valve enclosing ovipositor. 2. Lower valve.

Plate xiii., Figs. 5, 5*a*.—*Spondyliaspis eucalypti*, Dobs.

Fig. 5 —Head; 5*a*. face lobes.

Plate xiv.

Fig. 6.—*Spondyliaspis mannifera*, n.sp.; lerp-scales.
Fig. 7.——— ——— *eucalypti*, Dobs. ,,
Fig. 8.—*Creiis longipennis*, Walk.
Fig. 9. – *Cardiaspis plicatuloides*, n.sp.
Fig. 10. –*Rhinocola liturata*, n.sp.
Fig. 11.—*Lasiopsylla rotundipennis*, n.sp. ,,
Fig. 12.—*Rhinocola pinnæformis*, n.sp.
Fig. 13.——— *corniculata*, n.sp.
Fig. 14.—*Cardiaspis artifex*, Schwarz.
Fig. 15. –*Lasiopsylla bullata*, n.sp.
Fig. 16.—*Aphalara carinata*, n.sp.
Fig. 17. –*Rhinocola viridis*, n.sp.
Fig. 18.—*Cometopsylla* (?) *rufa*, n.sp.
Fig. 19.—*Rhinocola revoluta*, n.sp.
Fig. 19*a*.——— ——— (open leaf) ,,
Fig. 20.——— *ostreata*, n.sp. ,,
Fig. 21.—*Dasypsylla brunnea*, n.sp. ,,

(Natural size or enlarged).

ON SOME NEW SPECIES OF EUCALYPTUS.

By R. T. Baker, F.L.S., Curator, Technological Museum, Sydney.

(Plates xv.-xix.)

Eucalyptus vitrea, sp.nov.

"White Top Messmate."

(Plate xv.)

A tall tree with a roughish bark similar to *E. amygdalina*, Labill., the extremities of the branches being smooth.

Sucker leaves alternate or opposite, with a short petiole or sessile, ovate-lanceolate, acuminate, lateral veins diverging from below the middle of the midrib, prominent on both sides, intra-marginal vein removed from the edge, not shining; under 6 inches long, 1½ inches broad. Mature leaves narrow-lanceolate, about 6 inches long, and 6 to 9 lines wide, petiole short; *shining on both sides*, a dull green when fresh but drying a light slate colour; lateral veins few and almost parallel to the midrib, two generally commencing at the base of the midrib and running the whole length of the leaf almost parallel to the midrib. Oil glands very numerous.

Peduncles axillary, short, 2-3 lines, bearing generally from 5-8 flowers. Buds from 2½ to 4 lines long, operculum hemispherical, shortly acuminate. Ovary flat-topped. Anthers kidney-shaped, connective prominent.

Fruit hemispherical, about 3 lines in diameter, rim thick, red, slightly convex, shining, pedicel about 1 line long.

Hab.—Crookwell (J. J. Hook), Moss Vale (S. Farrell), mountains north of Marulan (R. H. Cambage), Bungendore (W. Bäuerlen).

Timber.—A hard, close-grained timber, full of shakes and gum veins, and apparently of little economic value. It possesses none of the good qualities of *E. amygdalina*, Labill., which is fissile, soft and easily worked.

Oil.—Leaves obtained from Crookwell, New South Wales, 16th June, 1900. Average yield of oil for three distillations = 1·48 per cent. The crude oil is almost colourless. It contains much phellandrene, over 20 per cent. eucalyptol, and citral is probably present in the higher boiling portions, the lemon odour being very marked and the aldehyde reactions readily obtained The constituent having an odour of peppermint, present in the oil of *E. coriacea*, appears to be absent in this oil, and it thus approaches more closely the oil of *E. amygdalina*. Less than 2 per cent. distilled below 173° C.; between 173° C. and 183° C. 78 per cent. distilled (corrected).

Specific gravity crude oil at 15° C. = 0·886.

Specific gravity fraction 173-183° C. at 15° C. = 0·8792.

Specific rotation crude oil = $[a]_D$ —33·92°.

Specific rotation first fraction = $[a]_D$ —37·76°.

Eucalyptol in fraction 173-182° C. = 26 per cent. (H. G. Smith).

This tree is known locally as "Silver Top Messmate," "Pepper-

The venation of the leaves resembles that of *E. amygdalina* and *E. coriacea*, but more particularly the latter species. The immature fruits are difficult to distinguish from those of *E. amygdalina*, whilst the mature ones bear a strong likeness to those of *E. coriacea*. The bark is almost identical with that of *E. amygdalina*, but the timber is quite distinct, resembling more closely that of *E. dives*, from which species, however, it differs in the shape and venation of both sucker and mature leaves, fruits and constituents of the oil.

From *E. coriacea* it differs in the nature of its timber, bark, oil and leaves. Summarised, this species has (*a*) a bark similar to that of *E. amygdalina*; (*b*) timber similar to that of *E. dives*; (*c*) leaves and venation similar to those of *E. coriacea*; (*d*) fruits approaching in form to those of *E. amygdalina*; and (*e*) sucker leaves differing from those of any of the species above enumerated. It is most closely allied to *E. coriacea* and *E. dives*, but yet very distinct from both.

Its specific name alludes to the glossy surface of the leaves.

Eucalyptus Delegatensis, sp. nov.

"White Ash," "Silver-Topped Mountain Ash."

(Plate xvi.)

A very tall tree occurring on the top of mountain ranges in the south-east corner of the colony. Bark stringy, reddish, extending well up the trunk.

Sucker leaves large, broadly lanceolate, oblique, venation prominent, spreading, intramarginal vein removed from the edge. Mature leaves comparatively large, often 9 inches long and 2 broad, lanceolate, acuminate; venation prominent, lateral veins spreading, intramarginal vein removed from the edge. Oil glands numerous.

Peduncles axillary, about 6 lines long, slightly compressed, bearing from 6-10 flowers. Buds clavate, 6-7 lines long, calyx short, merging into a pedicel 3-4 lines long; operculum hemispherical, obtuse. Ovary flat-topped. Anthers kidney-shaped; stamens all fertile.

Fruit pyriform, about 4 lines long and 3 broad, rim thick, truncate or countersunk.

Hab.—Delegate Mountain, N.S.W. (W. Bäuerlen).

Timber.—Pale-coloured, very fissile; used for general indoor purposes in the above locality.

Oil.—Leaves obtained from Delegate Mountain, New South Wales, 16th February, 1899. Average yield of oil for three distillations = 1·76 per cent. The crude oil is a light lemon colour, and is but little coloured, resembling in this and other respects the oil of *E. dives*, *E. radiata*, &c. It consists largely of lævophellandrene, contains no eucalyptol, and eudesmol could not be detected. Eighty per cent. of the oil distilled between 172° and 183° C. (corrected), less than 2 per cent. distilling below 172° C.

Specific gravity of crude oil at 15° C. = 0·8602.

Specific gravity fraction 172-183° C. at 15° C. = 0·8513.

Specific rotation crude oil = $[a]_D$—68·12°.

Specific rotation fraction 172-183° C. = $[a]_D$—75·76°.

Kino.—The kino is allied to all those kinos belonging to the

as well as quite different young leaves, and its more restricted habitat on more elevated mountain situations, distinguish it from *E. obliqua.*

Mr. W. Bäuerlen, who was the first to bring this Eucalypt to notice, states that—"It is a large tree, up to 200 feet high, and 3-5 feet diam. Bark on trunk persistent, fibrous, not easy to distinguish from that of *E. obliqua*, or that of *E. fastigata*. Limbs, and even sometimes the upper part of the trunk, quite smooth, a character which distinguishes the tree from the two above-named species. Young leaves and fruit resemble those of *E. Sieberiana*, but bark and timber are quite different. The timber is highly spoken of, cut in the sawmills and used for building purposes: much preferred for splitting, for which purpose it is said to equal 'Cut Tail' (*E. fastigata*), from which species it is sufficiently removed by the foliage and quite differently shaped fruit.

"Only known, so far, from the Delegate Mountain, and there restricted to a narrow belt on the higher part of the mountain, at an elevation of from about 4000-4500 feet.

"This is not the 'White Ash,' *E. fraxinoides*, Deane & Maiden, of the Tantawanglo Mountain and Sugar Loaf Mountain, much less the 'Silver Top' from near Nimitybelle; nor is it the 'Mountain Ash' (*E. Sieberiana*), despite the similarity of the fruit, the bark being quite different.

"There is great difficulty in distinguishing the leaves of this species from those of *E. obliqua* and those of *E. Sieberiana*, so that in my opinion it will be the chemical analysis which will have to decide between these species.

"Topographically *E. obliqua* occurs on the lower part of the mountain, reaches some distance up, and is then joined by *E. fastigata*, which species ascending somewhat higher leaves *E. obliqua* behind and forms a broad belt, where it abruptly ceases, and the "White Ash" takes it place, forming a narrow belt as it ascends the mountain; then it also ceases abruptly and leaves the higher part and summit to *E. coriacea*.

"The timber-getters look upon *E. obliqua* and *E. fastigata* as the same species, and call them indiscriminately 'Stringy Bark' or 'Cut Tail.' They also strip the bark indiscriminately; but they distinguish this 'White Ash' well, and its bark is not used for stripping. The timber is lighter in colour than that of 'Cut Tail,' from which fact the tree has received its vernacular name, and not on account of the bare branches, nor on account of the leaves; in fact, the leaves are remarkably large and broad.

"I may state that instead of becoming rougher with age, as is the case with the 'Mountain Ash,' *E. Sieberiana*, and the 'White Ash,' *E. fraxinoides*, from Tantawanglo, the bark of this species becomes less and less furrowed as the trees advance in age.

"*E. Delegatensis*, notwithstanding so much resemblance in leaves, buds and fruit, is *not E Sieberiana*, and comes nearer to *E. obliqua;* however, the buds and some slight difference in the fruit, but especially the young leaves, sufficiently remove it from that species. The bark is more like that of *E. obliqua*, and quite different from that of *E. Sieberiana*, and has not the slightest resemblance to *E. Smithii*, R.T.B."

The chemical constituents are quite distinct from those of the above-named species (R.T.B.).

both surfaces, not shining: lateral veins spreading, but not prominent, and almost quite hidden: intramarginal vein close to the edge. Buds on slender pedicels from 4-6 lines long. Flowers numerous, mostly in a terminal panicle. Calyx small, pyriform. Operculum hemispherical or conical, sometimes shortly acuminate. Ovary flat-topped. Anthers all fertile, cells opening by terminal pores.

Fruit variable in shape, sometimes cylindrical, with the thin rim incurved, whilst at other times pilular in form with a constriction below the rim, 2-3 lines long as well as broad.

Hab.—Dubbo to the Darling River ("Gum"; W. Bauerlen); Nymagee, Condobolin ("Coolabah"); Mt. Hope ("Yellow Jacket" and "Gum"); Cobar ("Coolabah Gum"); Drysdale ("Coolabah"); Bodabah, 30 miles E. of Nymagee, one of the most easterly localities for this species. For these localities I am indebted to Mr. R. H. Cambage, who also informs me that the difference between "Yellow Jacket" and "Gum" is, that when rough bark goes far up the tree and gives it a yellowish appearance it is called "Yellow Jacket."

Timber.—Timber very hard and red in colour, and very interlocked, in fact so much so that it is stated to be almost impossible to split; and though having a good repute for durability, it is very little used owing to the difficulty in splitting. It should be a good timber for railway sleepers. At Eremeran Station, 30 miles south of Nymagee, and at Double Peak, Mount Hope, the timber is used for making charcoal (R. H. Cambage).

Oil.—Oil from leaves collected at Nyngan, New South Wales, December 7th, 1899.—The crude oil is brownish-orange in colour, phellandrene is not present, and no eudesmol was detected at this time. The oil contains a large quantity of dextropinene proved by its characteristic reactions. Between 165° and 170° C.,* 40 per cent. distilled, while below 165° C. only 34 per cent. had distilled. The oil does not meet the requirements we recommend to be demanded for a good Eucalyptus oil.

* Temperatures corrected to nearest whole degree.

Specific gravity crude oil at 15° C. = 0·9076.

Specific gravity fraction 165-185° C at 15° C. = 0·9016.

Specific rotation crude oil = $[a]_D + 11·8°$.

Specific rotation fraction 165-185° C = $[a]_D + 13·6°$.

Eucalyptol fraction 165-185° C. = 42 per cent.

A consignment of leaves was sent from Girilambone, 16th March, 1900, to test the constancy of the species. The oil differs from that obtained at Nyngan only in the ratio to be expected in Eucalyptus oils of the same species, and they may be considered identical oils.

Specific gravity crude oil at 15° C. = 0·9078.

Specific rotation crude oil = $[a]_D + 10·7°$.

Eucalyptol crude oil = 37·2 per cent.

The yield of oil appears to vary in these trees, but the greatest amount obtained was 0·64 per cent. The deficiency of yield and the inferior quality of the oil make this species of little use for oil distillation (H. G. Smith).

Kino.—The kino is ruby-coloured, transparent in thin pieces, somewhat tough and not easily powdered. It is exceedingly soluble in cold water to a clear ruby-coloured solution. On addition of alcohol to the aqueous solution a precipitate is formed

E. largiflorens is a "Box-tree," with the usual box bark and characteristic box timber; but this species has a smooth bark and reddish timber. The two trees are readily distinguished by the settlers.

This species differs from *E. largiflorens* (1), in the inflorescence being mostly in terminal panicles; (2), the shape and venation of the leaves, i.e., the intramarginal veins being only slightly removed from the edge, and the lateral veins being less prominent; (3) the shape and size of the fruits; (4), timber; (5), oil; and (6), bark.

The bark and timber are in colour and texture so different that they alone distinguish it from *E. largiflorens*.

The name "Coolabah" attached to this and a few other species is evidently a mistake, since the true "Coolabah" is an *Angophora* (*A. melanoxylon*, R.T.B.), which occurs at Coolabah, the town of that name on the western railway line. This Eucalypt is at Coolabah also, where it is known as "Gum" and not "Coolabah."

The meaning of the aboriginal name "Coolabah"—a gnarled, knotted tree—applies eminently more to the *Angophora* than to *E. largiflorens* or this species (W. B.).

By the general classification of Eucalypts it belongs to the Gums or smooth-barked Eucalypts, and dividing these into pale and red-coloured timbers it falls into the latter division. The fruits are quite distinct from those of any described species. The kino and timber connect it with the [illegible], but not the bark, which is entirely different.

The similarity of bark and timber and the shape and venation of the leaves lead me to place it in sequence next to *E.* [illegible], R.T.B., but the fruits and constituents of the oil differentiate it from [illegible] Gum, *E. Dunnii*, and from *E. propinqua*, Smith, although in some features it resembles this latter species, especially the anthers.

The specific name has reference to the close, interlocked timber

EUCALYPTUS MORRISII, sp.nov.

"Grey Mallee."

(Plate xviii.)

A mallee of rather dense growth, or somewhat spreading, usually about 15 feet high or somewhat higher; stems 2-3 inches in diameter; rarely growing to tree-size, about 25 or 30 feet high, and 6 to 12 inches in diameter. Stems mostly hollow. Branchlets often flattened or quadrangular. Bark grey, somewhat fibrous, or on very old trees even furrowed, approaching that of an "Ironbark."

Young leaves petiolate, generally lanceolate in form, sometimes narrower and sometimes broader than the mature ones; opposite or with a tendency to become so. Mature leaves lanceolate-acuminate, on petioles of about 1 inch long, occasionally falcate; about 6 inches long and up to 1 inch wide; not shining, venation spreading, very prominent on both sides, intramarginal vein removed from the edge. Oil glands numerous.

Peduncles axillary, not numerous, flattened and twisted, short, under 6 lines, mostly 3-4 lines long, bearing 3 to 7 shortly pedicellate or sessile buds (mostly in threes). Calyx-tube hemispherical, 3 lines in diameter. Operculum obtuse, conical, 3 lines long. Ovary domed. Anthers parallel; connective not prominent.

It attains not its greatest height, but certainly its greatest diameter, on the highest hills amongst the roughest and rockiest parts (W. Bäuerlen).

Perhaps the most remarkable specific character about it is the rim of the fruits. The word "domed" hardly expresses correctly this feature, for merging into the valves at the top it forms as it were a truncate cone resting on the hemispherical calyx base. This conformation of the rim is noticeable as soon as the stamens begin to fall off, and from this stage, until and after its full development, it gives the appearance to the fruit of a pathological affection or a monstrosity. The shape of the fruits, however, is remarkably constant throughout the extensive range of the species. It is, so to speak, the rim of *E. tereticornis*, Sm., only very much more emphasised.

This Eucalypt differs, however, from that species in the venation and texture of its leaves, shape of operculum, chemical constituents of its oil, as well as in the timber and bark.

The expanded valves are similar to those of *E. viminalis*, but this is its only connecting link with that species.

The buds, especially the operculum, resemble those of *E. santalifolia*, from which species, however, the venation of leaves, and mature fruits differentiate it.

In botanical sequence it is placed next to *E. tereticornis*, Sm.

It is a remarkably constant and well-defined species throughout the area of its distribution.

It is named after R. N. Morris, LL.D., the present Superintendent of Technical Education in New South Wales, in acknowledgment of his co-operation in our work on the economics of the genus Eucalyptus.

Timber.—A hard, close-grained, interlocked, brownish-coloured, durable timber, quite distinct from that of *E. viminalis*, Labill., and *E. tereticornis*, Sm., its allies.

Oil.—Leaves sent from Girilambone, New South Wales, 15th January, 1900. Average yield of oil from three distillations, 1·69 per cent. The crude oil is but little coloured, and has an odour of aldehydes like all good oils of this class. When rectified by distillation the oil is almost colourless, being slightly tinged yellow, also resembling oils of this class. Phellandrene is of course absent. Eudesmol was not detected at this time. There are practically but little constituents boiling at a high temperature, as 95 per cent. distilled below 185° C., consequently the rectified oil is of low specific gravity, and although containing just upon 60 per cent. of eucalyptol, yet this sample would not pass the test of specific gravity as fixed by the Pharmacopœia, as the specific gravity of the crude oil is only 0·9097 at 15° C. This is an excellent illustration of the unsatisfactory nature of this standard of specific gravity of 0·91 as fixed by the Pharmacopœia; this oil is one of the best we have distilled, yet because it is deficient in high boiling constituents, which bodies may be considered objectionable both from a medicinal and commercial point of view, it is penalised or practically condemned because it is too good. It has been proved during this research over and over again that a Eucalyptus oil must not be condemned if its specific gravity is below 0·91 at 15° C. It is again suggested that the standard of minimum specific gravity for Eucalyptus oil be reduced to 0·905 at 15° C.

This oil consists principally of dextropinene and eucalyptol. The oil commenced to distil at 166° C.,* (neglecting the water

of the first consignment: these leaves were obtained so that the constancy of the species might be tested. The oil is the same, only a little richer in eucalyptol: it varies only in the percentage amount of its constituents and in its physical properties in the ratio usually experienced in Eucalyptus oils of the same species; this slight alteration may be governed, perhaps, by situation and soil.

Average yield of oil from three distillations = 1·613 per cent.

Specific gravity crude oil at 15° C. = 0·916.

Specific rotation crude oil = $[a]_D - 4·1°$.

Eucalyptol, crude oil = 65 per cent.

This species of Eucalyptus may thus be considered to be an excellent one for distillation, as the yield of oil is large. The oil is excellent: there is a comparative absence of high boiling constituents, consequently on rectification little would be lost. The species is obtainable in any quantity. We have pleasure in bringing this species under the notice of those interested in the distillation of Eucalyptus oil. The results obtained are those that would maintain commercially the leaves and terminal branches being collected and used in the same way (Henry G. Smith).

It gives a plentiful supply of leaves, especially as the clusters would not be destroyed and would always grow up again (W. Bauerlen).

Kino.—The kino is not plentiful. It is friable, dissolves in boiling water, but becomes turbid on cooling, the turbidity being caused principally by eudesmin, but aromadendrin is also present in small quantity: a very dilute aqueous solution gives with ferric chloride a green colouration (Henry G. Smith).

———

Mr. W. Bauerlen, who was the first to collect this Eucalypt, states.—"This species, also a Mallee, grows in the same way as, and associated with Green Mallee, *E. viridis*, R.T.B.; but the leaves are so different in colour, &c., as to distinguish it at once. The bark is much the same, but persistent often right out to the branches:

at other times smooth nearly half-way down. The persistent bark is rougher and more furrowed, in the larger trees making a slight approach to the Ironbarks. In the cross-cut it is red or brown, quite different from *E. viridis*, R.T.B., and the buds, flowers, and fruit are totally different. I cannot make it agree with any of the Parallelantheræ, to which section it appears to belong, yet from its peculiar highly domed fruit one would think it might be easily placed if it is a known species. It grows on dry stony hills, extending somewhat more to the foot of the hills. It gives a plentiful supply of leaves, especially as the clusters would not be destroyed and would always grow up again. The umbels are 3-7-flowered, but there appears to be a tendency to 3-flowered umbels. This Mallee is also called "Black Mallee," as well as "Cabbage Mallee," the latter said to refer to the soft wood.

EUCALYPTUS VIRIDIS, sp.nov.

"Green Mallee," "Red Mallee," "Brown Mallee."

(Plate xix.)

E. gracilis, F.v.M., Eucalyptographia, Dec. iii. (*partim*).

A Mallee of dense growth, the stems usually 2-3 inches in diameter, though occasionally measuring 20 feet in height, but rarely growing into tree-size. Bark smooth, or only rough at the

Fruits pilular; rim thin, contracted, about 2·3 lines in diameter; on a pedicel of from 1 to 3 lines long.

Hab.—On the hills near Girilambone, N.S.W., thence across country to Cobar; also seven miles out from Coolabah on the Wilga Downs Road (W. Bäuerlen).

Timber.—A hard, close-grained, interlocked, yellowish-coloured timber. Being a Mallee, it is only rarely found in tree-form, when it has a tendency to pipe.

Oil.—Leaves obtained from Girilambone, New South Wales, 16th January, 1900.

Average yield of oil = 1·06 per cent.

The crude oil is of a light orange-brown colour, and has an odour indicating the presence of cuminic aldehyde; no eudesmol was detected. Phellandrene is not present. Eucalyptol was present, but the oil contains less than 10 per cent. of that constituent; lævopinene was also present. Cuminic aldehyde was determined in the higher boiling portion of the oil.

The oil commenced to distil at 167° C.,* between 167° and 172° C. 32 per cent. distilled, mostly lævopinene; below 183° C. 80 per cent. had distilled, and 95 per cent. was obtained below 255° C.

Specific gravity crude oil at 15° C. = 0·9006.

Specific gravity fraction 167-183° C. at 15° C. = 0·8882.

Specific rotation crude oil = $[a]_D$ —8·90.

Specific rotation fraction 167-183° C. = $[a]_D$ —8·2°.

This oil at present has no commercial value (H. G. Smith).

Kino.—Friable; when its aqueous solution is extracted with ether it is found that eudesmin exists alone, aromadendrin being entirely absent. A very dilute aqueous solution gives a green colouration with one drop of ferric chloride (H. G. Smith).

This species, along with another, is figured on the same plate by Baron von Mueller in his Eucalyptographia (Dec. iii.) under the name of *E. gracilis*, F.v.M. As by a natural classification it

* These temperatures are corrected to the nearest whole degree.

can be shown that two species have been included under one name, I propose as in a former paper (Proc. Linn. Soc. N.S.W. June, 1898) to give this species specific rank, leaving Mueller's name of *E. gracilis* to apply to the Victorian and South Australian Mallee. Almost all the specimens in the National Herbarium, Melbourne, are referable to it, whilst only a few specimens to Green Mallee, *E. viridis*. I endeavoured to restore it under the name of *E. fruticorum*, but that species is in such inextricable confusion that I think science would be better served if it were given specific rank under the name of *E. viridis*.

The anthers, stamens, section of buds and fruiting twig in the lower left side of the plate (*loc. cit.*) illustrate the Victorian Mallee.

The stamens of the New South Wales Mallee *are all fertile*, whilst in the southern species the outer ones are sterile. The anthers of the southern species are attached by the connective to the stamens at the back by an attenuated point of the filament, and the cells open by terminal pores. The fruit of the southern species conforms to Bentham's description (B.Fl. iii. p. 211) and and also to Mueller's (Eucalyptographia, Dec. iii.), and is figured by the Baron in the lower left hand corner of his plate.

The anthers and fruits of the Green Mallee, *E. viridis*, are, as

however, by no means constant, as the bark is mostly of an ashen-grey colour, in fact, is red or brown, chiefly for some time only after decortication, when the colours are indeed very striking."

"The name 'Green Mallee' refers to the vivid lustreless green of the leaves, so different from other Eucalypts; where this species occurs on and around hills it imparts quite a feature to the landscape, especially as the individual trees grow densely massed together. This characteristic is constant, and is the very one by which the species is at once most readily distinguished in the field."

"The maximum dimensions, as far as seen, are 40 feet in height and 1 foot in diameter. The trunk is almost always hollow, leaving only a few inches of solid wood. Branches, twigs and leaves have rather a stiff upright appearance. The bark is very curious on account of a rich yellow tinge right through the texture, not merely yellow in the inner layer as in some of the Stringybarks. This species must be reckoned amongst the smooth-barked Eucalypts, though a roughish persistent bark runs up sometimes to the height of 6 feet or so."

"It grows generally around the foot of rocky or stony hills in gravelly not purely sandy soils, sometimes found growing densely over patches several miles in extent."

EXPLANATION OF PLATES.

Plate xv.—*E. vitrea*, R.T.B.

Fig. 1.—Sucker leaves.
Fig. 2.—Twig with buds, mature leaves and single fruit.
Fig. 3.—Single fruit with narrow convex rim.
Fig. 4.—Fruit with countersunk rim.
Fig. 5.—Anthers (enlarged).

Plate xvi.—*E. Delegatensis*, R.T.B.

Fig. 1.—Sucker leaf.
Fig. 2.—Buds and mature leaves.
Fig. 3.—Fruits with truncate rims.
Fig. 4.—Fruit with countersunk rim.
Fig. 5.—Anthers (enlarged).

Plate xvii.—*E. intertexta*, R.T.B.

Fig. 1.—Young leaves.
Fig. 2.—Twig with buds,
Fig. 3.—Clusters of fruits.

Plate xviii.—*E. Morrisii*, R.T.B.

Fig. 1.—Twig showing buds and flowers.
Fig. 2.—Twig with early fruit.
Fig. 3.—Section of bud (enlarged).
Fig. 4.—Anthers (enlarged).
Fig. 5.—Cluster of fruits.
Fig. 6.—Top view of fruits.

Plate xix.——*E. viridis*, R.T.B.

Fig. 1.—Flowering twigs.
Fig. 2.—Umbel of buds with acute operculum.
Fig. 3.—Anther (enlarged).
Fig. 4.—Varieties of fruits.

NOTES AND EXHIBITS.

Mr. Stead exhibited preparations of a pelagic crustacean *Galathea* sp. (from 1½ to 2 in. in length), collected by Captain W. Waller, of the s.s. Westralia, between south-eastern Australia and New Zealand; the animals are known to shipping-masters as "Whale-food"; and at times myriads of them cover so large an expanse of water as to impart to it a uniform red tinge. Also a very handsome and almost perfect living specimen of the Murrumbidgee Crayfish, (*Astacopsis serratus*, Shaw), and he drew attention to the fact that when this animal is disturbed suddenly it emits a curious hissing sound, resembling somewhat that made by a snake.* And two photographs taken at Ben Buckler, Bondi, which showed in a remarkable manner the effect produced by atmospheric erosion, aided by the vibration set up by the waves, on cliffs of the Hawkesbury Sandstone. The disintegration of the cliffs is also considerably helped by the basaltic dykes which here and there intersect the rock, and which, upon decomposing, isolate large sections.

Mr. Baker exhibited herbarium specimens and samples of the essential oils and timber of the Eucalypts described in his paper

Mr. Froggatt showed an extensive series of *Psyllidæ*, and of their lerp-structures in illustration of his paper.

Mr. Palmer described a case of hydatids in a Black-tailed Wallaby (*M. ualabatus*, Desm.), which died in captivity after a few days' illness. On dissection the chest cavity was found to contain several hundred loose pellucid cysts varying in size from that of a pea to that of a duck's egg, irrespective of a number imbedded in the lobes of the lungs.

---

* For an earlier notice of this habit, *vide* Proceedings, (2), i., 1886, p. 505—*Ed.*

Mr. Greig Smith exhibited a selection of cultures and microphotographs in illustration of his paper.

Mr. Whitelegge, on behalf of the Rev. W. W. Watts, contributed a Note on some collections of Mosses recently reported upon by Dr. Brotherus and Dr. Warnstorf, resulting in additions to the Moss-flora of the Colony of twelve new species, and four not previously recorded.

Mr. Hedley exhibited on behalf of Dr. T. H. May a series of photographs of aboriginal carvings in stone at Bingera, near Bundaberg, Queensland.

Mr. Fred. Turner exhibited the fruit, with drawings, of the true *Cydonia chinensis*, Thouin, the Chinese Quince, a new product for Australia. The fruit was egg-shaped, measuring a little more than six inches in length and fourteen inches in circumference at the widest part; and was grown by Mr. J. A. Murdoch, Wahroonga, North Sydney. The tree is only four years old and about four feet high, and this year it produced twelve fruits each of the size of that exhibited. In its native country, China, it is said to produce fruit as large as a child's head.

WEDNESDAY, JUNE 27TH, 1900.

The Ordinary Monthly Meeting of the Society was held at the Linnean Hall, Ithaca Road, Elizabeth Bay, on Wednesday evening, June 27th, 1899.

The Hon. James Norton, LL.D., M.L.C., President, in the Chair.

Messrs. HENRY W. BROELEMANN, Paris, and HORACE W. BROWN, Hunter's Hill, were duly elected Ordinary Members of the Society.

DONATIONS.

Department of Agriculture, Brisbane—Queensland Agricultural Journal. Vol. vi. Part 6 (June, 1900). *From the Secretary for Agriculture.*

Australian Museum, Sydney—Records. Vol. iii. No. 7 (June, 1900). *From the Trustees.*

Department of Mines and Agriculture, Sydney—Geological Survey: Mineral Resources. No. 7 (1900): Agricultural Gazette of New South Wales. Vol. xi. Part 6 (June, 1900). *From the Minister for Mines and Agriculture.*

Four Entomological Pamphlets (Miscellaneous Publications Nos. 358, 363, 369, from Agric. Gazette of N.S.W.). By W. W. Froggatt, F.L.S. *From the Author.*

Royal Society of New South Wales—Abstract, June 6th, 1900. *From the Society.*

Australasian Journal of Pharmacy, Melbourne. Vol. xv. No. 174 (June, 1900). *From the Editor.*

Department of Mines, Melbourne—Annual Report for the Year 1899. *From the Secretary for Mines.*

Field Naturalists' Club of Victoria, Melbourne—Victorian Naturalist. Vol. xvii. No. 2 (June, 1900). *From the Club.*

University of Melbourne—Matriculation Examination Papers, May, 1900. *From the University.*

Department of Mines, Tasmania—Progress of the Mineral Industry of Tasmania for the Quarter ending 31st March, 1900. *From the Secretary for Mines.*

Cambridge Philosophical Society — Proceedings. Vol. x. Part v. (May, 1900): Transactions. Vol. xviii. (1900). *From the Society.*

Entomological Society, London—Transactions, 1900. Part i. *From the Society.*

Linnean Society, London—Journal. *Zoology.* Vol. xxvii. No. 178 (April, 1900). *From the Society.*

Manchester Literary and Philosophical Society, Manchester—Memoirs and Proceedings. Vol. xliv. Parts 2-3 (1898-1900). *From the Society.*

Geological Survey of Canada, Ottawa—Descriptive Note on the Sydney Coal Field, Cape Breton, Nova Scotia. By H. Fletcher, B.A. (1900): Preliminary Report on the Klondike Gold Fields. By R. G McConnell, B.A. (1900). *From the Director.*

American Geographical Society, New York — Bulletin. Vol. xxxii. No. 2 (1900). *From the Society.*

American Philosophical Society, Philadelphia—Proceedings. Vol. xxxviii. No. 160 (Dec., 1899). *From the Society.*

Lloyd Library of Botany, Pharmacy and Materia Medica, Cincinnati—Bulletin. No. 1 (1900). *From the Librarian.*

Missouri Botanical Garden, St. Louis, Mo.—Eleventh Annual Report (1900). *From the Director.*

Museum of Comparative Zoology, Harvard College, Cambridge, Mass.—Bulletin. Vol. xxxv. No. 8 (May, 1900). *From the Museum.*

U.S. Geological Survey, Washington— Nineteenth Annual Report, 1897-98. Parts ii.-iii. and v. with Atlas; Twentieth Annual Report, 1898-99. Parts i., vi. and vi. contd. *From the Director.*

U.S. Department of Agriculture, Washington—*Division of Agrostology*—Bulletin. No. 22 (1900): *Division of Botany*—Bulletin. No. 24 (1900). *From the Secretary for Agriculture.*

Madras Government Museum, Madras—Bulletin. Vol. iii. No. 1 (1900). *From the Superintendent.*

Perak Government Gazette. Vol. xiii. No. 13 (May, 1900). *From the Government Secretary.*

R. Università degli Studi di Siena—Bullettino. Vol. iii. Fasc. 1 (1900). *From the University.*

Archiv für Naturgeschichte, Berlin. lxiv. Jahrgang. ii. Band. 2 Heft. 1 Hälfte (1898). *From the Editor.*

Gesellschaft für Erdkunde zu Berlin — Verhandlungen. xxvi. Band. Nos. 8-10 (1899): Zeitschrift. xxxiv. Band. No. 4 (1899). *From the Society.*

Zoologischer Anzeiger, Leipzig. xxiii. Band. Nos. 614-615 (May, 1900). *From the Editor.*

Société Scientifique de Chevtchènko à Lemberg (Autriche)—Die Chronik. No. 1 (1900). *From the Society.*

Société Hollandaise des Sciences à Harlem—Archives Néerlandaises. Série ii. Tome iii. 3e et 4e Livs. (1900). *From the Society.*

Jardin Botanique de Tiflis, Caucase—Recueil des Travaux. Livraison iv. (1899) *De la part du Représentant du Ministre d'Agriculture et des Domaines de l'Etat.*

L'Académie Impériale des Sciences de St. Pétersbourg—Annuaire du Musée Zoologique, 1899. No. 4. *From the Academy.*

Societas pro Fauna et Flora Fennica, Helsingfors—Acta. Vol. xv. (1898-9); Vol. xvii. (1898-9). *From the Society.*

Kongl. Svenska Vetenskaps-Akademie, Stockholm—Handlingar. xxxii. Band (1899-1900): Oefversigt. 56 Argången (1899): Vegetationen i Rio Grande do Sul, af C. A. M. Lindman (1900). *From the Academy.*

Zoological Museum, Copenhagen—Two Separates (1900). *From the Museum.*

South African Museum, Cape Town—Annals. Vol. ii. Part i.

NOTES ON SOME NEW ZEALAND AND AUSTRALIAN PARASITIC HYMENOPTERA, WITH DESCRIPTIONS OF NEW GENERA AND NEW SPECIES.

By William H. Ashmead, Assistant Curator, Division of Insects, U.S. National Museum.

(Communicated by W. W. Froggatt, F.L.S.)

The following notes and descriptions of new genera and new species of parasitic Hymenoptera are based upon a small but most interesting collection of these insects, sent me last summer by Mr. Walter W. Froggatt, Government Entomologist of New South Wales, or upon specimens collected by Mr. Albert Koebele, formerly an Assistant Entomologist in the U.S. Department of Agriculture, but at present Government Entomologist in the Hawaiian Islands.

Superfamily iii.—ISSPOIDEA.

Family xxxiii.—BETHYLIDÆ.

Genus Ateleopterus, Forster.

(1) Ateleopterus longiceps, n.sp.

♀.—Length 4·5 mm. Very elongate, black and shining; head anteriorly finely alutaceous, metathorax delicately shagreened; tibiæ ferruginous, tarsi yellowish; antennæ, except the scape basally, with the first four or five joints of the flagellum, yellowish; scape basally and the other joints fuscous or dark brown. Wings, except the basal third which is hyaline, are fuscous; subcostal vein—the only vein present—brown. Head large, oblong and slightly wider than the elongate thorax, nearly thrice as long as wide, the sides being parallel. Thorax about once and one-half as long as the head, rounded anteriorly and slightly narrowed posteriorly, metathorax about as long as scutellum, mesonotum and pronotum united. Abdomen conic-ovate, as long as the head and thorax united and much wider than the thorax, with the apical margins of dorsal segments 2-4 slightly sinuate medially.

Type—No. 4870, U.S.N.M.

Hab.—Rose Bay, near Sydney, N.S.W.

Described from four ♀ specimens, bred by Mr. Froggatt, March 6, 1892, from "hollow stem of *Acacia discolor*." The species will be found to be parasitic on some wood-boring coleopterous larva inhabiting the stem.

Genus Sierola, Cameron.

(2) Sierola antipoda, n.sp.

♀.—Length 1·5 to 1·8 mm. Black, shining; scape brown, flagellum yellowish, shading off into brown at tip; legs dark brownish-piceous, all trochanters, anterior and middle tibiæ, tips of hind tibiæ and all tarsi yellowish; wings hyaline, subcostal vein, prostigma and stigma brown, the other veins pallid: one closed discal cell and a closed marginal cell.

(3) SIERROLA WEBSTERI, n.sp.

♀.—Length 2·4 mm. Black; all tibiæ and tarsi rufo-testaceous; wings hyaline, tegulæ, parastigma and stigma piceous-black; veins brownish-yellow.

Head and thorax above finely coriaceous, the former with a few minute punctures scattered over its surface. Antennæ black, except joints 2-5, the first joint or the scape subglobose, the others small, moniliform.

Type—No. 4872, U.S.N.M.

Hab.—New South Wales (received from Prof. F. M. Webster).

Superfamily vi.—CYNIPOIDEA

Family lviii.—FIGITIDÆ.

Subfamily iii.—ANACHARINÆ.

Genus ANACHARIS, Dalman.

(4) ANACHARIS ZEALANDICA, n.sp.

♀.—Length 1·6 mm. Polished black; antennæ, except first two joints and base of the third, and legs, except coxæ, trochanters and most of femora, dark honey-yellow; wings hyaline, venation dark brown.

Head and thorax clothed with a sparse, silvery white pubescence, dense on metapleura; first joint of flagellum the longest, being as long as the scape and the pedicel united; parapsidal furrows distinct, crenate; scutellum with two large oblique foveæ at base, its disc coarsely reticulate; the mesopleura anteriorly finely shagreened, the mesepipleura being perpendicularly aciculated; metathorax rugose. Abdomen longly petiolate, the petiole smooth and as long as the hind femur, body ovate, subcompressed and highly polished.

Type—No. 4873, U.S.N.M.

Hab—New Zealand (A. Koebele, collector).

(5) ANACHARIS AUSTRALIENSIS, n.sp.

♂.—Length 1·8 mm. Agrees closely with *A. zealandica*, except as follows:—The antennæ from the apex of the second joint and beneath are entirely brownish-yellow; the legs, except the coxæ and the hind trochanters, *above* are brownish-yellow; wings hyaline, the venation light brown; while the petiole of the abdomen is shorter, being only about four-fifths the length of the hind femur.

Type—No. 4874, U.S.N.M.

Hab.—Australia (A. Koebele, collector).

Subfamily v.—EUCOELINÆ.

Genus TRYBLIOGRAPHA, Förster.

(6) TRYBLIOGRAPHA AUSTRALIENSIS, n.sp.

♀.—Length 3 mm. Polished black; antennæ, except first joint, and legs, including all coxæ, red; scape obfuscated; wings hyaline, with a brownish-red cloud enclosing the whole of the marginal cell and a fainter cloud beneath it as far as to the basal vein.

Antennæ gradually incrassated toward apex; pedicel oval, first joint of flagellum clavate, the following gradually decreasing in length, the first being the longest, about thrice as long as thick at apex, the joints from the fifth oval-moniliform. Scutellum with two large deep foveæ at base, separated from each other only by a delicate carina, the cup broadly oval, with a deep fovea posteriorly, the sides below the cup and posteriorly

antennæ, except towards apex from fifth joint, and legs, including coxæ, yellow; antennæ from fifth joint brown; wings hyaline, ciliate, veins brown.

First joint of flagellum longest, a little more than four times as long as thick, second a little shorter, being only about four times as long as thick, joints beyond to the fifth imperceptibly shortening, the following slightly thicker and subequal. Metathorax sparsely pubescent with two parallel carinæ down the middle.

Type—No. 4876, U.S.N.M.

Hab.—Australia (A. Koebele, collector).

Superfamily vii.—CHALCIDOIDEA.

Family lxi.—TORYMIDÆ.

Subfamily ii.—TORYMINÆ.

Genus Torymus, Dalman.

(8) Torymus eucalypti, n.sp.

♀.—Length 2·8 to 3 mm.; ovipositor as long as body. Gold-green with cupreous reflections; head in front and mesopleura blue-green; mandibles rufous; palpi white; antennæ brown-black, scape and pedicel aeneous; tegulæ and legs, except coxæ and the hind femora which are metallic green, pale yellowish, tibial spurs and tarsi white; wings hyaline, stigma and marginal vein brown.

Head shagreened and punctate, ocelli red; thorax above transversely wrinkled or striate, with some sparse punctures scattered over its surface; metathorax smooth, with some short lineations at extreme base just back of scutellum; mesopleura smooth, impunctate, except some delicate lines on anterior margin; hind coxæ large, coarsely reticulated.

♂.—Length 2 to 2·5 mm. Agrees well with the female except in the usual sexual differences, and in having sometimes a brownish or metallic band on the anterior and middle femora, the extreme base of the scape being yellowish, while the flagellum is stouter, with the joints a little wider than long.

Type—No. 4877, U.S.N.M.

Hab.—Sydney, N.S.W.

Described from 3 ♀ and 5 ♂ specimens bred by Mr. Froggatt from "a soft red gall on *Eucalyptus*."

Subfamily iv.—MEGASTIGMINÆ.

Genus MEGASTIGMUS, Dalman.

(9) MEGASTIGMUS IAMENUS, Walker.

1839. Walk. Monogr. Chalcid. ii. p. 6.

Hab.—Sydney, N.S. Wales. Of this species Mr. Froggatt has sent 1 ♀, bred from galls of *Brachyscelis pileata*, and 4 ♀ specimens bred from a globular gall on *Eucalyptus*. The latter are considerably larger than the dimensions of the species as given by Walker, but otherwise seem to agree well with his description, and with the smaller form bred from *B. pileata*.

(10) MEGASTIGMUS ASTERI, n.sp.

♀.—Length 3 to 4 mm.; ovipositor as long as body. Brownish-yellow, smooth and shining, except some delicate transverse striæ on vertex of head, on mesonotum and within the femoral impression on the mesopleura; body clothed with some sparse black and white hairs, metapleura with long, white hairs; sutures of scutellum

Type—No. 4878, U.S.N.M.

Hab.—Sydney, N.S.W.

Host—Dipt. Described from 3 ♀ and 2 ♂ specimens, bred by Mr. Froggatt from a dipterous gall on Snowbush (*Aster ramulosus*).

(11) Megastigmus brachyscelidis, n.sp.

♀.—Length 2 mm. Black or dark brown, ovipositor a little shorter than body; orbits broadly, lower part of face, pronotum, lateral lobes of mesonotum, and sometimes middle lobe, axillæ, a narrow transverse line back of scutellum, mouth parts, except teeth of mandibles, and legs, except hind coxæ, pale yellowish. Abdomen paler beneath and at apex. Antennæ brown, scape pale beneath, flagellum subclavate, first joint of funicle about once and one-half as long as thick, following joints gradually shortening, seventh a little wider than long. Wings hyaline, stigma and veins brown. Vertex of head and mesonotum delicately transversely striate, scutellum and metanotum delicately shagreened.

♂.—Length 1·8 mm. Black, shining; orbits and face below antennæ, a transverse band on hind margin of pronotum, lateral lobes of mesonotum posteriorly, axillæ along their inner suture, extreme tip of abdomen, and legs, except hind coxæ, all yellowish. Antennæ brown, joints of funicle scarcely longer than wide; wings hyaline, stigma and veins brown-black.

Type—No. 4879, U.S.N.M.

Hab.—Sydney, N.S.W.

Host—Rhynch.: *Brachyscelis crispa*, Olliff (Froggatt). Described from 2 ♀ and 2 ♂ specimens.

Family lxii.—CHALCIDIDÆ.

Subfamily ii.—CHALCIDINÆ.

Genus Haltichella, Spinola.

(12) Haltichella bicolor, n.sp.

♀.—Length 3·5 mm. Head and thorax black, closely punctate, and clothed with a sparse, silvery white pubescence; scape, legs and abdomen ferruginous; flagellum long, filiform, cylindrical throughout, black.

Head concave anteriorly, vertex transversely acute, anterior depression bounded by a delicate carina, the same extending across vertex and separating front ocellus from lateral ocelli. Wings hyaline, venation brown, postmarginal vein acute, longer than stigmal vein. Hind femora much swollen and armed with numerous minute black teeth. Abdomen pointed, ovate, ovipositor subexserted.

Type—No. 4880, U.S.N.M.

Hab.—Australia (A. Koebele, collector).

Genus ENCYRTOCEPHALUS, Ashmead, n.g.

(13) ENCYRTOCEPHALUS SIMPLICIPES, n.sp.

♂.—Length 2·2 mm. Robust, ferruginous; head anteriorly shagreened; thorax above, including scutellum, closely, rugulosely punctate; club of antennæ white; scape and pedicel yellowish, funicle dark brown; abdomen rufo-piceous, obfuscated at apex; wings hyaline, with a large discoidal fuscous cloud extending across wings below marginal and stigmal veins, the cloud having, however, a clear space just beneath marginal vein.

femora much swollen but unarmed, *without* teeth or serrations; abdomen small, oval, subdepressed.

Type—No. 4881, U.S.N.M.

Hab.—Australia (A, Koebele, collector).

Family lxiii.—EURYTOMIDÆ.

Genus SYSTOLE, Walker.

(14) SYSTOLE KOEBELEI, n.sp.

♀.—Length 2·5 mm. Robust, brownish-yellow and sparsely pubescent; stemmaticum, occiput, pedicel of antennæ, except narrowly at apex, anterior margin of mesonotum and sutures dividing sclerites of thorax black. Wings hyaline, pubescent, veins brown, subcostal vein interrupted by a white bulla at its junction with marginal vein; marginal and postmarginal veins about equal, longer than stigmal. Head viewed from above subglobose, with the scrobes deep and extending nearly to front ocellus; pedicel obconical, twice and one-half longer than thick at apex; flagellum clavate, first three joints very small, together shorter than pedicel, and very much narrower, moniliform, joints beyond widening and wider than long. Thorax rugulose, except black anterior portion of mesonotum which is smoother but with delicate, transverse aciculations. Abdomen oval, finely coriaceous and sparsely pubescent, segments after second subequal in length.

Type—No. 4882, U.S.N.M.

Hab.—Australia (A. Koebele, collector).

Genus EURYTOMA, Illiger.

(15) EURYTOMA AUSTRALIENSIS, n.sp.

♀.—Length 2·4 mm. Black and similar in structure to *E. studiosa*, Say, clothed with a sparse, white pubescence. Antennæ wholly black; tegulæ, tips of femora, tibiæ, except a subfuscous median blotch *outwardly*, and all tarsi honey-yellow. Pedicel small, obconic, very little longer than thick and smaller than first joint of funicle; funicle 5-jointed, joints nearly equal in size, submoniliform; club 3-jointed, as long as last three joints of funicle united. Wings hyaline, venation brown, subcostal vein yellowish towards base. Hind coxæ outwardly towards base shagreened. Abdomen conic-ovate, smooth and polished, except the short petiole, which is rather coarsely shagreened; fourth dorsal segment longer than second and third united.

Type—No. 4883, U.S.N.M.

Hab.—Australia (A. Koebele, collector).

(16) EURYTOMA EUCALYPTI, n sp.

♀.—Length 2·6 to 3 mm. Black; anterior angles of pronotum, as seen from side, with a yellowish spot, seen from above, invisible; antennæ, except *apex* of scape and base of pedicel which are black, mandibles, tegulæ, and legs, except hind coxæ and middle of hind femora, ferruginous or brownish-yellow; wings hyaline, costal vein yellow, marginal, stigmal and postmarginal veins brown-black, the latter subpetiolate, ending in a large rounded stigma with an uncus.

Vertex of head and thorax above clothed with a rufous pubescence, that on the face, sides of thorax and metapleura white or silvery-white.

Funicle 5-jointed, joints fluted and a little wider than long. Sides of thorax and [illegible] sides *broadl[illegible]* along sutures o[illegible]ue and

thick, basal part of the following being quadrate or nearly so; abdomen with petiole four times as long as thick, shagreened, and with a grooved line down centre; body subglobose, segments subequal; hind coxæ opaque, coriaceous.

Type—No. 4884, U.S.N.M.

Hab.—Uralla, N.S.W.

Described from several specimens bred by Mr. Froggatt, from galls on *Eucalyptus*.

(17) Eurytoma binotata, n.sp.

♀.—Length 3·5 to 5 mm. Black, clothed with a whitish pubescence; pronotum with two oblong oval yellow spots, one on each anterior angle, and both distinctly visible from above; scape of antennæ, pedicel at apex, and legs, except coxæ and a blotch on middle of femora above, pale ferruginous; rest of antennæ and coxæ black. Wings hyaline, venation brown, marginal vein a little longer than postmarginal, stigmal vein normal, very nearly as long as postmarginal.

Head and thorax closely umbilicately punctate; funicle joints a little longer than thick; abdomen conic-ovate, subsessile, a little longer than head and thorax united, acutely pointed at apex, sides of segments 4-7 ciliate with white hairs, fourth segment and beyond very delicately shagreened at sides.

Type—No 4885, U.S.N.M.

Hab.—Sydney, N.S.W. "Bred from galls on the turpentine tree" (Froggatt).

Family lxiv.—PERILAMPIDÆ.

Family lxv.—EUCHARIDÆ.

Genus Metagea, Kirby.

(18) Metagea Kirbyi, n sp.

♀.—Length 4 mm. Aeneous-black, tinged with blue or purplish in certain lights. Head and thorax, except discs of parapsides which are smooth and aeneous, coarsely rugose. Legs brownish-yellow, femora, except at tips, dark brown, hind coxæ black, hind tibiæ, except tips, fuscous. Wings hyaline, venation brown. Flagellar joints, except first, submoniliform, not or very little longer than thick; first joint of flagellum as long as 2 and 3 united. Abdomen aeneous-black, petiole about four times as long as thick, smooth and impunctate.

Type—No. 4886, U.S.N.M.

Hab.—Australia, Gosford (A. Koebele, collector).

(19) METAGEA RUFIVENTRIS, n.sp.

♀.—Length 9 mm. Head and thorax bright metallic green, coarsely rugose; mouth parts, scape, pedicel and legs, except coxæ, yellow; abdomen, except petiole, second segment at base above and hypopygium wholly rufous; flagellum and second abdominal segment at base above black, petiole purplish; wings subfuscous, hyaline at base.

Flagellar joints 1 to 6 all longer than thick, terminal joints alone moniliform; first three or four joints elongate, but gradually shortening, first about five times as long as thick at apex, fourth less than two-thirds length of first, the others still shorter, 7th and 8th submoniliform. Abdomen, except as noted, red, petiole about four times as long as thick, finely microscopically punctate above.

Type—No. 4887, U.S.N.M.

flagellum subclavate, sparsely pubescent, funicle joints longer than thick. Thorax with parapsidal furrows distinct, entire; anteriorly abruptly truncate, the pronotum not being visible from above except slightly at lateral angles; scutellum large, axillæ widely separated. Legs yellow, coxæ, except at tips, metallic; trochanters, tarsi and tegulæ yellowish-white. Wings hyaline, marginal, postmarginal and stigmal veins light brown, the marginal vein being very long, fully four-fifths length of subcostal vein, postmarginal vein about half length of marginal, stigmal a little shorter than marginal and ending in a rounded knob. Abdomen conically elongate and about one-third longer than head and thorax united.

♂.—Length 1·8 mm. Blue to bluish-green; flagellum long, filiform, joints more than twice as long as thick, first and second fully thrice as long as thick; legs, except middle and hind coxæ, wholly yellow; abdomen oblong, depressed, and scarcely as long as thorax.

Type—No. 4888, U.S.N.M.

Hab.—Sydney, N.S.W.

Host—Dipt.: *Cecidomyia frauenfeldi* (W. W. Froggatt).

Family lxvii.—CLEONYMIDÆ.

Subfamily i.—CHALCEDECTINÆ.

Genus Systolomorpha, Ashmead, n.g.

(21) Systolomorpha thyridopterygis, n.sp.

♀.—Length 1·8 to 2 mm. Black, shining; flagellum brown, sutures of trochanters, tips of femora and all tibiæ and tarsi yellowish-white; wings hyaline, venation brown.

Head transverse, a little wider than thorax, about thrice as wide as thick antero-posteriorly, scrobes delicately impressed but distinct; ocelli 3, arranged in an obtuse triangle, lateral ocelli a little wider from each other than to front ocellus, surface of head distinctly coriaceous. Antennæ short, flagellum clavate, pedicel short, obconical, a little longer than thick and much larger than first two joints of funicle; funicle joints short, wider than long, all gradually widening to club. Thorax in shape similar to the Eurytomid genus *Systole*, Walker, parapsidal furrows distinct, entire, mesonotum delicately transversely aciculate; scutellum coriaceous, axillæ meeting at their inner basal angles; metathorax short, abrupt, smooth, with small, rounded spiracles. Wings hyaline, with a faint discoidal cloud, venation brown, marginal and postmarginal veins about equal, one-third the length of subcostal vein, stigmal vein a little shorter than marginal, gently curved and ending in a small knob. Abdomen subglobose, subcompressed beneath and subsessile, not longer than thorax, second segment (or first body-segment) nearly twice length of third, the following subequal. Hind femora somewhat swollen, with a slight tooth beneath before apex.

Type—No. 4889, U S.N.M.

Hab.—Adelaide, S. Australia.

Host.—Lepid.: *Thyridopteryx* sp., on *Eucalyptus* sp. Bred Oct. 22, 1886, by A. Koebele.

Genus AGAMERION, Haliday.

(22) AGAMERION COERULEIVENTRIS, n.sp.

♂.—Length 3·8 mm. Robust, metallic blue-green, thorax above bronzed-green; face from front ocellus, thorax at sides and ... legs except as noted and abdomen ...

Subfamily ii.—CLEONYMINÆ.

Genus THAUMASURA, Westwood.

(23) THAUMASURA TEREBRATOR, Westwood.

A single specimen of this remarkable and rare species is labelled "No. 125, South Australia" (W. W. Froggatt).

(24) THAUMASURA RUBROFEMORALIS, n.sp.

♀.—Length 10 mm.; to tip of ovipositor 14·5 to 15 mm. Head and thorax metallic bronzed-green, tinged with blue, and clothed with a whitish pubescence; abdomen above blue, beneath bronzed, segments bearded with white hairs at sides; flagellum brown; coxæ metallic blue-green, all femora red, rest of legs fuscous or brown-black. Wings hyaline, venation brown.

Type.—No. 4891, U.S.N.M.

Hab.—Sydney, N.S.W. (W. W. Froggatt).

Genus DINOURA, Ashmead, n.g.

(25) DINOURA AURIVENTRIS, n.sp.

♀.—Length 4·8 mm.; to tip of ovipositor nearly 7 mm. Head bluish or blue-green, with a metallic green spot on vertex enclosing ocelli; thorax above metallic green, at sides and beneath, with metathorax, blue or blue-green; femora, except tips, aeneous-black, anterior pair more or less bluish; rest of legs pale yellowish, hind tibiæ obfuscated medially. Wings hyaline, venation dark brown. Abdomen gold-green and terminating in a prominent ovipositor which is dilated into three broad leaf-like expansions, like a propeller in a naphtha launch.

♂.—Length 3·6 mm. Antennæ 9-jointed, flagellum filiform, clothed with a short, dense, felt-like pubescence, joints longer than thick; all legs aeneous-black, with white tarsi; abdomen oblong-oval, bronzed-black, metallic greenish above towards base.

Type.—No. 4892, U.S.N.M.

Hab.—Sydney, N.S.W. Bred by Mr. Froggatt from galls of *Brachyscelis pileata*.

(26) DINOURA CYANEA, n.sp.

♀.—Length 5 mm.; to tip of ovipositor 10 mm. Head and thorax blue, with a metallic green tinge on thorax above; basal half of abdomen yellowish; anterior and middle tibiæ, hind tibiæ along outer face and all tarsi ivory white Wings hyaline, as in previous species.

Type—No. 4893, U.S.N.M.

Hab.—Wellington, N.S.W. Bred Aug. 20, 1891, by Mr. Froggatt, from *Brachyscelis ovicola*, Schr.

Family lxviii.—ENCYRTIDÆ.

Subfamily i.—EUPELMINÆ.

Genus EUPELMUS, Dalman.

(27) EUPELMUS ANTIPODA, n.sp.

♀.—Length 3 mm. Head metallic green, thorax bottle-blue mesopleura with a greenish tinge; abdomen aeneous-black, ovipositor prominent, yellow for two-thirds its length; wings fuscous, except at basal third which is hyaline; all coxæ metallic green; sutures of trochanters, tips of tibiæ and tarsi yellowish-white, rest of legs aeneous-black.

Type—No. 4894, U.S.N.M.

veins longer than marginal. Ovipositor about the length of basal joint of hind tarsi, sheaths broad.

Type—No. 4895, U.S.N.M.

Hab.—Australia (A. Koebele, collector).

Family lxix.—PTEROMALIDÆ.

Subfamily i.—MERISINÆ.

Genus BRACHYSCELIDIPHAGA, Ashmead, n.g.

(29) BRACHYSCELIDIPHAGA FLAVA, n.sp.

♀.—Length 1·2 to 1·4 mm. Yellow, smooth, impunctate, or at most with the surface feebly alutaceous; eyes and ocelli black; occiput, pronotum on anterior face, a spot on middle lobe of mesonotum, a spot on inner angle of lateral lobes, a large spot on disc of scutellum, sutures surrounding scutellum and metathorax brown-black; abdomen above, except basal segment, dusty or brownish. Wings hyaline, venation dark brown; marginal and postmarginal veins about equal, one-third length of subcostal vein; stigmal vein a little shorter than marginal, terminating in a rounded knob with an uncus. Antennæ short, flagellum clavate, pedicel stout, obconical, about twice as long as thick at apex and very much longer and stouter than first three or four funicle joints united; first joint of funicle very small, about as long as thick, second also small, a little wider than long, the following gradually increasing in size and width and all wider than long.

Type—No. 4898, U.S.N.M.

Hab.—Sydney, N.S.W. Bred by Mr. Froggatt from *Brachyscelis pileata.*

Genus TEROBIELLA, Ashmead, n.g.

(30) TEROBIELLA FLAVIFRONS, n.sp.

♀.—Length 2·8 mm. Mostly black, polished and impunctate; head on vertex and posteriorly including temples black; face below front ocellus, cheeks, mandibles, palpi, antennæ and tegulæ brownish-yellow; thorax and abdomen entirely black; legs, except coxæ which are dark fuscous, yellow. Wings hyaline, venation brown, postmarginal vein one-half longer than marginal, stigmal vein slightly longer than marginal but scarcely so long as postmarginal, subclavate, with an uncus at apex. Head transverse, a little more than thrice wider than long, face above insertion of antennæ subconvex, ocelli arranged in an obtuse triangle. Antennæ inserted on, or near, middle of face, scape rather short, subcompressed, pedicel subglobose, joints of funicle short, wider than long, outer apical angles acute.

Thorax with distinct, entire parapsidal furrows, metathorax very short, abrupt and smooth medially, mesopleural ridges closely punctate or rugulose, spiracles small, rounded. Abdomen as seen from above broadly oval, shorter than thorax, dorsum flat, beneath convex, first dorsal segment occupying nearly half the whole surface, the following segments short, subequal.

Type—No. 4897, U.S.N.M.

Hab.—Sydney, N.S.W. Bred by Mr. Froggatt, from a lumpy gall on a *Eucalyptus* twig.

Genus COELOCYBA, Ashmead, n.g.

(31) COELOCYBA NIGROCINCTA, n.sp.

lobe) trapezoidal, axillæ widely separated; metathorax short but not abrupt, and smooth. Front wings with the postmarginal vein very long, much longer than marginal, stigmal vein also long, longer than marginal, but still shorter than postmarginal. Abdomen sessile, conic-ovate, a little longer than head and thorax united, flat above and beneath at base, but with ventral half compressed and carinate. Hind coxæ very large, compressed, outer face flat.

Type—No. 4898, U.S.N.M.

Hab.—Sydney, N.S.W. Bred by Mr. Froggatt from an Agromyzid gall on *Eucalyptus corymbosa*.

Subfamily ii.—PTEROMALINÆ.

Genus Pteromalus, Swederus.

(32) Pteromalus puparum, Linn.

Six ♀ and two ♂ specimens of this cosmopolitan species were bred by Mr. Froggatt from the pupa of *P*[illegible].

Subfamily iii.—SPHEGIGASTERINÆ.

Genus Ophelosia, Riley.

(33) Ophelosia crawfordi, Riley.

Hab.—Sydney, N.S.W. Twelve ♀ specimens bred by Mr. Froggatt from the adult female of *Icerya purchasi*.

Genus Tomocera, Howard.

(34) Tomocera californica, Howard.

(Syn. *Moranila testaceiceps*, Cam.)

Hab.—California; Hawaiian Islands.

A specimen of this species is in the National Museum, collected in Australia by A. Koebele. The species is parasitic on *Lecanium oleæ*, but has not before been reported from Australia.

Family lxx.—ELASMIDÆ.

Genus Euryischia, Riley & Howard.

(35) Euryischia lestophoni, Riley & Howard.

Hab.—Sydney. Ten specimens, bred by Mr. Froggatt from *Icerya rosæ*, on *Grevillea* sp.

Family lxxi.—EULOPHIDÆ.

Subfamily ii.—APHELININÆ.

Genus Pteroptrix, Westwood.

(36) Pteroptrix maskelli, n.sp.

♀.—Length 0·60 mm. Black, shining; antennæ and tegulæ light brown; tips of all femora, or knees, anterior tibiæ, except in front, extreme tips of middle and hind tibiæ, and all tarsi, except last joint, snow-white. Wings hyaline, ciliate, veins light brown.

♂.—Length about 0·54 mm. Yellow; legs ivory-white; eyes purplish-brown; a spot on pronotum anteriorly and a minute dot on hind angles, two spots on disc of mesonotum, a spot on each parapside, and the metanotum black; abdomen with dorsal segments narrowly banded with brown.

dorsal abdominal segments 3, 4, and more or less of 5, black or brown-black; first dorsal segment sometimes also dusky or slightly obfuscated; flagellum light brownish, pubescent. Wings hyaline, venation pale, nearly hyaline. Abdomen ovate. Legs pale yellowish, tibiæ and tarsi whitish.

♂.—Length about 0·6 mm. Polished black, head anteriorly brown; legs, except coxæ, yellowish-white.

Type—No. 4900, U.S.N.M.

Hab.—Hornsby, near Sydney, N.S.W.

Described from many specimens, bred by Mr. Froggatt from "a small shot gall on *Eucalyptus*."

Subfamily iv.—ELACHISTINÆ.

Genus EUPLECTRUS, Westwood.

(38) EUPLECTRUS AUSTRALIENSIS, n.sp.

♀.—Length 3 mm. Head and thorax black, clothed with long white hairs; antennæ light brownish; legs, except hind coxæ, and abdomen, extreme tip above, peach-yellow, trochanters ivory-white, hind coxæ black, polished.

Pronotum rather coarsely confluently punctate, middle and lateral lobes of mesonotum with sparse moderately large punctures anteriorly but posteriorly smooth, scutellum feebly shagreened.

Type—No. 4901, U.S.N.M.

Hab.—Australia (A. Koebele, collector).

Subfamily v.—EULOPHINÆ.

Genus DIAULOMORPHA, Ashmead, n.g.

(39) DIAULOMORPHA AUSTRALIENSIS, n.sp.

♀.—Length 2 mm. Gold-green, reticulately shagreened; antennæ black, 8-jointed, flagellum subcompressed, clothed with a short, dense pubescence, funicle 3-jointed, first and second joints longer than wide, third about as long as wide; legs, except coxæ, brownish-yellow, hind coxæ golden-green, strongly punctate;

scutellum with two delicate, parallel grooved lines, postscutellum large, half length of scutellum; metathorax short, with a median carina. Wings hyaline, veins brown, marginal vein very long, almost as long as subcostal vein, stigmal and postmarginal veins also long, more than half length of marginal. Abdomen conic-ovate, as seen from above, subdepressed.

Type—No. 4902, U.S.N.M.

Hab.—Australia (A. Koebele, collector).

Superfamily viii.—ICHNEUMONOIDEA.

Family lxxiv.—EVANIIDÆ.

Subfamily iii.—AULACINÆ.

Genus Aulacus, Jurine.

(40) Aulacus apicalis, Westwood.

Hab.—Sydney, N.S. Wales. Three ♀ specimens, bred by Mr. Froggatt from the larva of *Piesarthrius marginellus*, Hope, a longicorn beetle, feeding upon *Acacia longifolia*.

Family lxxvi.—ICHNEUMONIDÆ.

Subfamily i.—ICHNEUMONINÆ.

Genus Probolus, Wesmael.

Subfamily ii.—CRYPTINÆ.

Tribe v.—Cryptini.

Genus CHROMOCRYPTUS, Ashmead.

(44) CHROMOCRYPTUS ANTIPODIALIS, n.sp.

♀. Length 5·2 mm. Black; antennal joints 7 and 8 *above*, anterior margin of pronotum, a narrow band at apex of second dorsal segment of abdomen, and two spots on tegulæ white; metathorax, legs (except tips of hind tibiæ and their tarsi) and first segment of abdomen rufous. Wings hyaline, venation, except basal half of stigma which is white, black.

Type.—No 4903, U.S.N.M.

Hab.—Australia (A. Koebele, collector).

Subfamily iii.—PIMPLINÆ.

Tribe i.—Acoenitini.

Genus LEPTOBATOPSIS, Ashmead, n.g.

This new genus falls in next to *Leptobatus*, Grav., but may be easily separated by the following differences:—

Front wings *with* an areolet.

Areolet large, rhomboidal; ovipositor as long as or longer than the abdomen; head as in *Exetastes*; claws simple.................. *Leptobatus*, Grav.

Areolet small, petiolate; ovipositor shorter than the abdomen, the abdomen elongate, narrowed towards the base, the first segment long, slender, petioliform; claws pectinate .... *Leptobatopsis*, Ashm., n.g.

(45) LEPTOBATOPSIS AUSTRALIENSIS, n.sp.

♀.—Length 9 mm. Black, shining; two lunate spots on vertex (one near each eye), a narrow line on lower anterior orbits, clypeus, mandibles, except tips, a line anteriorly at sides of prothorax, sometimes interrupted, two triangular spots on mesonotum anteriorly (one on each side), a spot beneath tegulæ, scutellum, a spot on metopleura just above hind coxæ, first segment of abdomen, a band at base of second and third segments, a narrow band at apex of third and a large band at apex of abdomen yellow or yellowish-white; legs rufous, anterior coxæ and trochanters, middle coxæ, a moderately broad annulus near base of hind tibiæ and basal joint of hind tarsi, except apex, white; tips of hind femora, a narrow annulus at extreme base of their tibiæ, before the white annulus, the rest of tibiæ and tarsi black. Wings hyaline with a large fuscous cloud at apex, veins black.

Head narrowly transverse, smooth, impunctate; thorax rather closely punctate, *without* parapsidal furrows; metathorax closely rugoso-punctate.

Type—No. 4904, U.S.N.M.

Hab.—Australia (A. Koebele, collector).

Tribe ii.—**Labenini.**

This tribe is represented by only three genera—*Labena*, *Grotea* and *Nonnus*, Cresson. It is not yet recognised outside of the American fauna.

Tribe iii.—**Lissonotini.**

Genus EUCTENOPUS, Ashmead, n.g.

This new genus falls in next to *Phytodietus*, Grav., so far as its

(46) ECCTENOPUS NOVAZEALANDICUS, n.sp.

♀.—Length 9 mm. Head and thorax castaneous; orbits, face and clypeus, apical third of antennæ, except four or five apical joints, and legs, except parts to be noted, all lemon-yellow; middle and hind coxæ concolorous with thorax, apical third of hind tibiæ, their spurs and basal joint of their tarsi fuscous; claws black, thickly pectinate. Wings subfuscous, stigma and veins black.

Type—No. 4905, U.S.N.M.

Hab.—New Zealand (A. Koebele, collector).

Tribe [illegible].—Pimplini.

Genus ALLOTHERA, [illegible] Ashmead, n.g.

This new genus falls in a table of genera next to *Theronia* Holmgren, and the recently established genus *[illegible]*, Krieger, but may be at once separated from them by the following characters:—

Claws simple, not pectinate; metathorax [illegible] lated, but with a strong transverse apical carina, the upper hind angles [illegible] or tuberculate; hind femora [illegible] ... [illegible]

(47) ALLOTHERA [illegible], n.sp.

♀.—Length 10-11 mm. In stature and general appearance resembles somewhat *Lissonota* [illegible]. [illegible] Black, antennæ, anteriorly and [illegible] face [illegible] antennæ, clypeus, mandibles, except tips and legs except [illegible] coxæ and hind coxæ and trochanters, two [illegible] small spots at base of scutellum, a spot on metanotum [illegible] of both pairs of wings, a [illegible] metathoracic tubercles and two [illegible] small spots at apex of abdominal segments 1 to 6 white. Wings fuscous [illegible] with a black spot at apex of marginal cell, stigma and veins brown-black or black. Thorax, except a smooth median spot on mesopleura, closely, finely, rugulosely punctate, [illegible].

polished, impunctate, but with the transverse and oblique lateral furrows or impressions as in *Lissopimpla*, Kriechb.

Type—No. 4906, U.S.N.M.

Hab.—Australia.

Host—Lepid. Bred by A. Koebele from the pupa of an unknown Noctuid moth.

Subfamily iv.—TRYPHONINÆ.

Tribe v.—Bassini.

Genus B A S S U S, Gravenhorst.

(48) BASSUS LAETATORIUS, Fabr.

Hab.—Europe, North America.

Two specimens of this common European species are in the National Museum, taken by Mr. Koebele in New Zealand. Dr. Shauinsland has also taken it on the Chatham Islands.

Subfamily v.—OPHIONINÆ.

Tribe iv.—Anomalini.

Genus B A R Y L Y T A, Förster.

(49) BARYLYTA COARCTATA, n.sp.

Tribe vi.—**Paniscini.**

Genus PANISCUS, Gravenhorst.

(50) PANISCUS PRODUCTUS, Brullé.

Hab.—New Zealand (A. Koebele, collector: 8 specimens).

Family lxxvii.—ALYSIIDÆ.

Subfamily iii.—ALYSIINÆ.

Genus ASOBARA, Förster.

(51) ASOBARA ANTIPODA, Ashmead.

Hab.—Chatham Islands (Dr. Schauinsland).

Family lxxviii.—BRACONIDÆ.

Subfamily i.—APHIDIINÆ.

Genus LIPOLEXIS, Förster.

(52) LIPOLEXIS RAPÆ, Curtis.

This European species, incorrectly placed in the genus *Aphidius* by Marshall, has been taken in Australia by Mr. Koebele. It is a common species in Europe and the United States, and attacks *Aphis brassicæ.*

Subfamily iv.—METEORINÆ.

Genus METEORUS, Haliday.

(53) METEORUS ANTIPODALIS, n.sp.

♀.—Length 3·6 mm. Pale ferruginous or brownish-yellow; mouth-parts, pronotum at sides, petiole of abdomen and legs, except hind coxæ, femora and tibiæ, all pale yellowish; eyes large, purplish-brown; ocelli pale but situated on a black spot; parapsides, the crenate furrow at base of scutellum and metathorax black or blackish. Wings hyaline, stigma and veins brown; submedian cell longer than median, recurrent nervure received by second cubital cell near its base. Hind coxæ, femora and tibiæ ferruginous, petiole of abdomen with two black spots at apex.

Type—No. 4908, U.S.N.M.

Hab.—Australia (A. Koebele, collector).

Subfamily ix.—CHELONINÆ.

Genus PHANEROTOMA, Wesmael.

(54) PHANEROTOMA AUSTRALIENSIS, n.sp.

♀.—Length 4 mm. Uniformly brownish-yellow, the surface finely rugulose; eyes broadly oval, brown-black; ocelli close together on a black spot; teeth of mandibles black, outer very long, acute. Wings hyaline, stigma and most of veins brown, median and submedian veins yellowish; first abscissa of radius and second transverse cubitus short, first transverse cubitus unusually long, being fully thrice as long as second transverse cubitus, thus giving the second cubital an unusual shape, which will at once distinguish the species. Abdomen 3-segmented, rugulose.

Type—No. 4909, U.S.N.M.

Hab.—Australia (A. Koebele, collector).

(55) Orgilonecura antipoda, n.sp.

♀.—Length 5 mm.; ovipositor longer than abdomen. Head, except eyes, and anterior legs, except base of femora, trochanters and coxæ, and middle legs, except femora, trochanters and coxæ, ferruginous; antennæ, thorax, rest of legs and abdomen, except first two segments, which are pale yellowish except a discal blotch on second, black. Wings fuscous, stigma and veins dark brown, almost black.

Type—No. 4910, U.S.N.M.

Hab.—Australia (A. Koebele, collector).

Subfamily xii.—MICROGASTERINÆ.

Genus Acoelius, Haliday.

(56) Acoelius australiensis, n.sp.

♀.—Length 1·2 mm. Body including legs wholly black; wings hyaline, with a large brown band across middle and occupying all but apical and basal thirds of wing. Antennæ very long, about twice length of body, tapering off to a point at apex, broadest a little beyond middle. Head and thorax finely, closely punctate, opaque, and clothed with a dull sericeous pubescence; metathorax shining, areolated. Abdomen oblong-oval, shining, impunctate, ovipositor subexserted.

Type—No. 4911, U.S.N.M.

Hab.—Australia (A. Koebele, collector).

Genus Apanteles, Förster.

(57) Apanteles antipoda, n.sp.

♀.—Length 2·1 to 2·4 mm. Black: mandibles ferruginous: palpi white: legs brownish-yellow, hind coxæ, except apex, black, finely rugulose, apex of hind femora, apex of hind tibiæ and their tarsi, more or less fuscous: ventral abdominal segments 1-4 brownish-yellow.

Head smooth, polished, impunctate, except indistinctly on face below antennæ. Antennæ, except apex of pedicel, wholly black,

18-jointed, and a little longer than body. Mesonotum and scutellum closely punctate, opaque or subopaque, lateral hind angles of the former shining and less closely punctate; mesopleura with a large, smooth, shining, impunctate disc, but the surface surrounding it, as well as the mesosternum, closely punctate; metathorax rugulose, *without* a median carina. Wings hyaline, stigma and veins light brown, costal veins, in both wings, to near their apex pale yellowish; areolet open behind, the inner side, or what represents the transverse cubitus, being as long as first abscissa of radius. Abdomen, except as noted, black polished, first two segments sculptured or irregularly longitudinally rugulose; first segment trapezoidal, a little longer than wide at apex, one-half longer than second, second and third about equal, fourth, fifth and sixth subequal, shorter than third; ovipositor with its tip alone exposed.

♂.—Length 1·5 mm. Differs from the female only in its smaller size, in having longer antennæ, smaller abdomen, and by the second dorsal segment of abdomen being less distinctly sculptured and a little shorter than third.

Type—No. 4912, U.S.N.M.

Hab.—Sydney, N.S. Wales.

Host—Lepid.: *Agrotis* sp. Bred by Mr. Froggatt.

Subfamily xv.—BRACONINÆ.

This subfamily I have separated into three tribes, as follows:—

Submedian cell in front wings much shorter than the median; eyes very large, extending clear to the base of mandibles, the face very narrow........................	Tribe i.—*Aphrastobraconini*.
Submedian and median cells in front wings equal; eyes non-extending to base of mandibles..	Tribe ii.—*Braconini*.
Submedian cell very distinctly longer than the median; eyes normal	Tribe iii.—*Euurobraconini*.

(The type of *Euurobracon*, Ashm., is *Bracon penetrator*, Smith, described from Japan).

Tribe ii.—Braconini.

Genus CALLIBRACON, Ashmead, n.g.

This new genus falls in between *Melanobracon*, Ashm., and *Coeliodes*, Wesmael, agreeing, however, more nearly with the former, especially in regard to the venation of the front wings. The two genera may be separated as follows:—

Scape rather long, cylindrical, truncate at apex, the pedicel much shorter than the first joint of flagellum, the third flagellar joint shorter than either the first or second; second dorsal

Genus Microbracon, Ashmead.

(61) Microbracon thalpocharis, n.sp.

♀.—Length 3 to 3·5 mm. Head, antennæ, legs and abdomen black, segments of abdomen narrowly banded with white at apex; thorax orange or brownish-yellow; wings subfuscous, stigma and veins brown; ovipositor a little longer than abdomen.

♂.—Differs from ♀ in having the tips of middle femora and their tibiæ and tarsi brownish-yellow.

Type—No. 4915, U.S.N.M.

Hab.—Australia; California. Bred by A. Koebele and D. W. Coquillett, from *Thalpochares coccophaga*.

(62) Microbracon tricolor, n.sp.

♀ – Length 2·6 mm. Head and thorax (except mesopleura and metathorax), anterior legs and knees of middle legs brownish-yellow; eyes and ocelli brown-black; antennæ, mesopleura and metathorax, middle and hind legs, and abdomen black, extreme lateral margin of segments 1-3, extreme apical margin of third dorsal segment, and venter white; ovipositor as long as abdomen. Wings subfuliginous, stigma and veins dark brown, almost black.

Type—No. 4916, U.S.N.M.

Hab.—Australia (A. Koebele, collector).

Genus Iphiaulax, Förster.

(63) Iphiaulax trinotata, n.sp.

♂.—Length 3·5 mm. Elongate; head, thorax and legs honey-yellow; stemmaticum, eyes, antennæ, a spot on parapsides near tegulæ, a central blotch on metathorax, mesosternum and dorsum of abdomen black or brown-black. Wings subhyaline, stigma, except outer margin and veins, brown; outer margin of stigma and costal veins black.

Type—No. 4917, U.S.N.M.

Hab.—Australia (A. Koebele, collector).

(64) IPHIAULAX AUSTRALIENSIS, n.sp.

♂.—Length 10 mm. Black, except as follows: face from front ocellus, occiput, lower part of temples, cheeks, palpi, pronotum, except hind angles, anterior and middle legs, basal half of wings, stigma and a blotch beneath, yellow; first three segments of abdomen orange-red, apical margin of sixth and the following white.

First and second dorsal segments rugose, first tricarinate, middle keel highly elevated; second segment with broad depressions laterally, lateral and apical margins elevated, third segment with a broad, crenate furrow across base and a transverse punctate, subapical line, following segments to sixth with transverse impressed lines.

Type—No. 4918, U.S.N.M.

Hab.—Australia (A. Koebele, collector).

ON THE CARENIDES (FAM. *CARABIDÆ*).

No. IV.

By Thomas G. Sloane.

In this paper are included some new species sent to me for description by Mr. C. French, of Melbourne (the types being in his possession); also a synoptic table of the groups of species into which the Genus *Carenum* may be divided, with some notes on synonymy, &c.

Genus Laccopterum.

Laccopterum humeralis, n.sp.

Elongate-oval; prothorax lobate, lobe wide and roundly truncate; elytra with base deeply emarginate, humeral angles prominent, a longitudinal row of three or four large foveiform punctures along middle of each elytron; anterior tibiæ tridentate. Black, nitid, with subviolaceous tints at basal foveæ of prothorax and on inflexed margin of elytra.

Head subquadrate (2·5 × 4 mm.), convex, transversely impressed behind frontal spaces, these convex; frontal sulci deep, parallel; clypeus with median part truncate, intermediate angles short, triangular; eyes not prominent; two supraorbital punctures on each side. Prothorax lightly transverse (3·3 × 4·3 mm.), convex, parallel on sides, widely rounded at posterior angles, strongly sinuate on each side of basal lobe; anterior margin truncate with angles shortly but decidedly advanced; border reflexed, a little wider behind posterior angles; marginal channel narrow on sides, tripunctate; basal area defined by a strong curved transverse line; a well marked rotundate foveiform impression at each basal sinuosity; median line strongly impressed. Elytra truncate-oval (8 × 4·3 mm.), narrowed to base, roundly and abruptly declivous on sides, deeply and abruptly declivous at apex; suture deeply impressed; humeral angles upturned in a strong short obtuse prominence; border reflexed; margin wide on apical curve; four punctures on base of each elytron near humeral angle; a closely placed row of punctures along margin; anterior discoidal puncture near basal fourth, about equidistant between humeral angle and second puncture, two posterior punctures not far apart near apical third; inflexed margin wide, narrowed posteriorly, vertical at apex. Length 15·5, breadth 4·3 mm.

Hab.—North-West Australia, between Port Darwin and the Lennard River (Coll. French).

The only previously described species to which *L. humeralis* seems closely allied is *Carenum multiimpressum*, Casteln., with the description of which it agrees generally, except in regard to the large punctures of the disc of the elytra. Castelnau, in his description of the elytra of *C. multiimpressum*, says, "they present on the middle of their breadth a longitudinal series of

Genus Carenum.

Seven years ago I attempted to arrange the species of the central genus *Carenum* into groups as an aid in the identification of species, at the same time enumerating the species that I believed to belong to each group.* Further knowledge shows me that the system of groups then adopted was faulty, and that a number of species were referred to groups which were not their natural place. I now offer a new arrangement of the species into groups, following the lines of my former classification, but presenting alterations where such have been necessary to bring the tabulation more into accord with what seems to me a natural system, the chief alteration being the removal of the *C. anthracinum* group from Division "ii." to Division "i."

The plan of using the names of species for distinguishing groups of species in large genera seems to me decidedly better than using numbers. It allows for the interpolation of new groups in their natural position, if required, without altering the designation of every subsequent group, as would be necessary with a numeral system; it also permits of a different arrangement of the groups without causing any confusion if subsequent research shows that the first order adopted is not the most natural; in the case of numbered groups a rearrangement may alter the number of several of the groups, with the result that all subsequent references must state which system of groups is meant. The species first described should always be used as the name-species of a group.

Table of Groups in the Genus Carenum.

I. Penultimate joint of labial palpi not swollen towards apex (usually narrow and longer than apical joint). Suborbital antennal scrobes straight, single. Inflexed margin of elytra wide behind first ventral segment.

* P.L.S.N.S.W. 1893, viii. (2), pp. 462-465.

A. Reflexed border of elytra extending past humeral angle on to base, narrowly reflexed and not forming a thickened projection at shoulder .... C. brevicolle Group.

AA. Reflexed border of elytra ending at humeral angle and forming a thickened upturned humeral projection.

B. Elytra impunctate.

C. Anterior tibiæ tridentate.

d. Prothorax with posterior angles prominent and strongly marked......... C. transversicolle Group.

dd. Prothorax with posterior angles rounded.......... .... C. Macleayi Group.

CC. Anterior tibiæ bidentate.

e. Prothorax with posterior angles prominent, subrectangular. .... C. reflexum Group.

ee. Prothorax with posterior angles rounded, not marked.

f. Prothorax without a basal lobe.... C. lævipenne Group.

ff. Prothorax with base widely lobate in middle, a strong sinuosity on each side of lobe, angles of lobe prominent .... C. perplexum Group.

BB. Elytra bipunctate.

g. Prothorax with posterior angles prominent and strongly marked; the posterior marginal punctures placed on

k. Head with a longitudinal supraorbital sulcus behind each eye........................ C. planipenne Group.

kk. Head not longitudinally impressed on each side of occiput behind eyes......... C. marginatum Group.

JJ. Elytra quadripunctate.... .... C. anthracinum Group.

II. Penultimate joint of labial palpi short, thick, swollen. Suborbital antennal scrobes divided longitudinally in middle by an oblique ridge. Inflexed margin of elytra narrow behind first ventral segment.

L. Inflexed margin of elytra becoming gradually narrowed to apex. (Elytra quadripunctate) C. bonellii Group.

LL. Inflexed margin of elytra very narrow behind first ventral segment. (Facies elongate.)

m. Elytra quadripunctate.

n. Frontal sulci hardly diverging backwards.................. .. .. C. scaritioides Group.

nn. Frontal sulci deep and diverging strongly backwards......... .... C. 4-punctatum Group.

mm. Elytra bipunctate on apical third...... C. subplanatum Group.

The following is a list of species of the genus *Carenum* arranged in groups according to the table given above. In my list of seven years ago a number of species that were unknown to me in nature were placed in wrong groups: therefore, in order that any such errors in the present list may cause as little confusion and trouble as possible, I have printed the names of all species definitely known to me in Roman letters, and have used italics for those species that are not at present known to me in nature. I now only assume responsibility for the accuracy of position of those species printed in Roman letters, though I have endeavoured to place the others in their proper places as far as descriptions will allow.* Species of which the position seems doubtful are referred to in notes after the list.

* It may be noted that although I may know a species in nature, it should not be assumed that I, in every case recognise its validity as a species, for I believe there is some synonymy yet to be dealt with, but it is a matter only to be treated of when one's knowledge is sufficient.

C. brevicolle Group.

C. brevicolle, Sl.

C. transversicolle Group.

C. frenchi, Sl.
C. occidentale, Sl.
C. transversicolle, Ch.

C. macleayi Group.

C. macleayi, Blkb

C. reflexum Group.

C. reflexum, Sl.

C. lævipenne Group.

C. cordipenne, Sl.
C. ineditum, Macl.
C. lævipenne, Macl.
C. politum, Westw.

C. perplexum Group.

C. perplexum, Westw.

C. rectangulare Group.

C. rectangulare, Macl.
C. tibiale, Sl.

C. smaragdulum Group.

C. breviforme, Bates.
C. nickleri, Ancey.
C. odewahnii, Cast.
C. optimum, Sl.
C. opulens, Sl.
C. ovale, Sl.
C. porphyreum, Bates.
C. rufipes, Macl.
C. serratipes, Sl.
C. smaragdulum, Westw.
C. speciosum, Sl.
C. splendens, Cast.
C. tumidipes, Sl.
C. virescens, Sl.

C. subcyaneum Group.

C. rugatum, Blkb.
C. subcyaneum, Macl.
C. sulcaticeps, Sl.

C. planipenne Group.

C. planipenne, Macl.
C. purpureum, Sl.
C. vicinum, Sl.

C. marginatum Group.

C. amplicolle, Sl.
C. convexum, Ch.
C. decorum, Sl.
C. fugitivum, Blkb.

C. devastator, Cast., the largest species of *Carenum*, is unknown to me in nature. The description seems that of a species which should be placed in my Division "ii.," and leads me to suppose that it is a species with the prothorax of *Carenidium*, and elytra resembling those of *C. regulare*. No mention is made of the anterior tibiæ, but I should expect them to be bidentate.

C. nitescens, Macl., is unknown to me in nature, but the description gives ample evidence that it is not allied to *C. marginatum*. The elongate shape of the elytra indicates that their inflexed margins will prove to be narrow, and suggests that its place is near to *C. subplanatum*.

CARENUM PUSILLUM, Macl.

A careful examination of specimens sent to me by Mr. French labelled "Cape York," which I identify as *C. pusillum*, Macl., shows this to be an isolated species which cannot be placed in any of the groups into which I have divided *Carenum;* its affinity seems rather towards *Laccopterum cyaneum*, Fabr., (also a species of doubtful position), than to any other described species. It requires a separate group, but as I am not prepared to suggest its relative position towards the other groups, it has not been worked into my table. The following description gives some features not alluded to in the original description :—

bidentate with a small projection just above base of second tooth. Upper surface cyaneous. Length 8·5, breadth 2·9 mm.

Carenum rectangulare, Macl.

Specimens of *C. rectangulare*, Macl., have been sent to me by Mr. French as having been found between Alice Springs and Charlotte Waters, Central Australia. These specimens I have compared with the types of *C. rectangulare*, Macl., with which they proved conspecific. They are intermediate between the typical forms of *C. rectangulare* and *C. tibiale*, Sl., and indicate that *C. tibiale* should probably be placed under *C. rectangulare* as a variety—the only differences being the less truncate base of the prothorax and the more cordiform elytra of *C. tibiale*. The colour of *C. rectangulare* is variable, one of Mr. French's specimens having the elytra of a rich purple colour narrowly margined with greenish-blue, and the prothorax purple near the sides, with the margin coloured like that of the elytra.

Table of Species of the C. smaragdulum Group.

A. Head with one supraorbital puncture, prothorax with two marginal punctures on each side.*
  B. Prothorax with posterior angles strongly marked.
    c. Elytra with shoulders prominent, the base emarginate between them............ *C. emarginatum*, Sl.
    cc. Elytra with shoulders rounded, the base hardly emarginate .... *C. interioris*, Sl.
  BB. Prothorax with posterior angles rounded.
    D. Prosternum with intercoxal part widely longitudinally channelled, truncate at base.
      e. Anterior angles of prothorax strongly advanced.
        f. Border of elytra hardly folded over at humeral angles—not raised or prominent (intermediate tibiæ with a short external spur at apex)...... *C. ovale*, Sl.

* Excepting *C. tumidipes*, Sl., which has three marginal punctures on each side.

ff. Border of elytra decidedly thickened at humeral angles and forming a raised prominence.

G. Anterior femora not swollen in middle of lower side.

h. Suborbital antennal scrobe long, its lower edge forming a ridge.

i. Frontal sulci short, parallel.

j. Prothorax without a well-defined basal area.

k. Prothorax and elytra with cupreous margins......... { *C. cupreomarginatum*, Blk. / *C. opulens*, Sl. }

kk. Prothorax and elytra purple-black, with greenish margins......... *C. dispar*, Macl.

jj. Prothorax strongly lobate behind and with well-defined basal area......... *C. virescens*, Sl.

ii. Frontal sulci divergent...... *C. froggatti*, Sl.

hh. Suborbital antennal scrobe short, lower side not forming a ridge. (Posterior tarsi with basal joint short—shorter than two succeeding joints together).

l. Prothorax and elytra with

nn. Posterior tibiæ light, prothorax with three marginal punctures on each side..... *C. tumidipes*, Sl.

ee. Anterior angles of prothorax very little advanced (not prominent)..... *C. iridescens*, Sl.

DD. Prosternum with intercoxal part obliquely rounded at base, the middle forming a subtuberculiform prominence. (Prothorax and elytra with cupreous margins)............... *C. lepidum*, Sl.

AA. Head with two supraorbital punctures, prothorax with three marginal punctures on each side. (Frontal sulci strongly divergent backwards).

o. Posterior tibiæ thick, incrassate........... { *C. odewahni*, Cast. / *C. speciosum*, Sl. }

oo. Posterior tibiæ slender.

p. Prothorax with marginal channel and reflexed border wide .... { *C. elegans*, Macl. / *C. splendens*, Cast. / *C. rufipes*, Macl. / *C. cognatum*, Sl. / *C. optimum*, Sl. }

pp. Prothorax with marginal channel and reflexed border narrow........... *C. distinctum*, Macl.

I offer the following notes on some species of the *C. smaragdulum* Group, including the species I have omitted from the table given above :—

C. porphyreum, Bates, is evidently closely allied to *C. cupreomarginatum*, Blkb., and *C. opulens*, Sl., but should be readily distinguishable from them by the green margins of the prothorax and elytra. It is described as having the prothorax with "the hind margin not lobed and distinctly trisinuate."

C. smaragdulum, Westw., must present an affinity to *C. virescens*, Sl. Its shining green colour distinguishes it from allied described species excepting *C. virescens*; however, from Westwood's figure the humeral angles of the elytra seem more rounded with the border less erect than in *C. virescens*. I believe, judging from Westwood's figure, that *C. smaragdulum* resembles *C. habile*, Sl., in facies.

C. habile, Sl., (type destroyed), appears to be closely allied to *C. smaragdulum*, Westw., from which only the colour seems to readily differentiate it.

C. coruscum, Macl., is unknown to me in nature. Sir William Macleay apparently regarded it as allied to *C. elegans*, Macl.

C. splendens, Casteln., is closely allied to *C. elegans*, Macl.; *C. cognatum*, Sl., is a form connecting them. It seems to me probable that with further knowledge these three forms will be united under *C. elegans*, but more data than I possess are necessary for a confident opinion on the subject.

C. rufipes, Macl.—I have examined the type of *C. rufipes* in the collection of the Australian Museum, Sydney; it is an immature specimen with slender posterior tibiæ. A specimen from Norseman, S.W. Australia, has been given to me by Mr. C. French, and a comparison with the type of *C. rufipes* shows it to be that species. This Norseman specimen only seems to differ from *C. elegans*, Macl., by having the elytra wholly of a steel-blue colour; to me it seems conspecific with *C. elegans*, but might, perhaps, be regarded as the south-western form of that species and given rank as a variety.

C. optimum, Sl.—This form, from the Murchison District of Western Australia, seems to be the north-western representative of *C. elegans*, Macl.; further knowledge may show it to be a

Black, shining; elytra dark purple, brighter towards sides, marginal channel greenish-blue; lateral channel of prothorax of a bluish tinge.

Head smooth, subquadrate (3·5 x 5·2 mm.), convex across occiput; front depressed; facial sulci short, lightly divergent posteriorly; median frontal space not divided from occiput; lateral frontal spaces abruptly declivous externally, lightly raised at base above sides of occiput; clypeus with median part truncate, intermediate angles obtusely dentiform, narrow at base and projecting sharply from margin of head on outer side; preocular sulcus distinct, short: preocular process narrow, convex on outer edge: eyes depressed; one supraorbital puncture on each side. Antennæ setaceous, lightly compressed. Prothorax transverse (4·3 x 6·8 mm.), wider between anterior than between posterior angles; sides lightly rounded; anterior angles shortly advanced, widely obtuse; posterior angles rounded, not marked; base shortly lobate, obliquely truncate towards sides; basal lobe wide, rounded; sinuosity on each side of base wide; border reflexed, wide on sides, very widely reflexed at posterior angles, narrower on basal lobe; marginal channel wide; median line strongly impressed; two marginal punctures on each side. Elytra a little wider than prothorax (10·5 x 7·1 mm.), rather depressed; sides lightly rounded, widening very gently behind shoulders: base emarginate, strongly (not roundly) declivous; humeral angles prominent; border reflexed, reaching peduncle, strongly upturned and prominent (not dentiform at humeral angles; margin wide, especially on apical curve; a catenulate series of evenly placed fine punctures along sides; three punctures on each side of basal declivity. Prosternum with intercoxal part longitudinally impressed, base truncate and unipunctate at each angle. Posterior legs light, posterior coxæ impunctate. Length 20, breadth 7·1 mm.

Hab.—Nickol Bay, W.A. (Coll. French).

A very distinct species; it differs from *C. interioris*, Sl., by the elytra being longer, less strongly rounded on sides, and with prominent humeral angles. The posterior marginal puncture of the prothorax is placed a little before the posterior angle, which is an unusual position in *Carenum*.

Carenum ovale, n.sp.

Elliptical-oval, robust, convex, lævigate; head large, with parallel frontal sulci; prothorax transverse, posterior angles rounded; elytra oval, bipunctate posteriorly, humeral angles not marked; anterior tibiæ tridentate. Black, prothorax and elytra with narrow violaceous margins.

Head transversely subquadrate (4·6 × 6·8 mm.), convex; frontal sulci short, parallel; clypeus lightly declivous to labrum, median part wide, truncate, intermediate angles short, obtuse; eyes convex. Prothorax transverse (5 × 8·5 mm.), convex; sides lightly rounded, subparallel in middle, rounded at posterior angles; basal curve short, lightly bisinuate; anterior margin truncate between anterior angles—these shortly advanced, obtuse; border thick, reflexed, more widely upturned at posterior angles; two marginal punctures on each side. Elytra oval (12·6 × 9·2 mm.), convex, roundly declivous to humeral angles; sides evenly rounded; base roundly truncate; humeral angles rounded; border reflexed, hardly folded over (not thickened or upturned) at humeral angles;

prosternum not bordered along anterior margin; basal joint of posterior tarsi shorter, &c. This species was sent to Mr. French as having been found on a journey from Port Darwin to the Lennard River near Derby.

CARENUM SERRATIPES, n.sp.

Oblong-ovate, form rather light; prothorax with anterior angles porrect, posterior angles rounded, base lobate, marginal channel wide, bipunctate; elytra oval, bipunctate on apical third; anterior tibiæ tridentate, outer edge denticulate above three large teeth. Black, shining, elytra faintly suffused with violet on lateral declivities, lateral channels of elytra and prothorax cyaneous-purple.

Head transverse (3·5 x 5 mm.), smooth, strongly declivous on sides, not longitudinally impressed on sides of occiput; front rather depressed; frontal sulci deep, elongate, lightly divergent posteriorly; clypeus deeply emarginate-truncate in middle, intermediate angles triangular, obtuse; preocular sulcus narrow, deep; preocular process narrow; eyes lightly convex, not prominent, deeply set in orbits; one supraorbital puncture on each side; submentum strongly raised from gulæ, divided from genæ on each side by a longitudinal sulcus. Prothorax transverse (3·7 x 6·2 mm.),

impression; a short wide shallow impression near each basal sinuosity. Elytra oval (12·2 × 8·8 mm.), lightly emarginate, obliquely declivous between humeral angles, convex, evenly rounded on sides; border thick, prominent and upturned at humeral angles: punctures of base (5 or 6) in two transverse rows (the punctures of posterior row smaller than those of anterior row). Prosternum with intercoxal part widely channelled, truncate at apex, with a single puncture at each side. Anterior tibiæ tridentate, two small prominences above larger teeth. Length 25, breadth 8·8 mm.

Hab.—North-West Australia (Colls. French and Sloane). Sent to me by Mr. French, who reports it as having been found by one of his friends on a journey from Port Darwin to the Lennard River.

Its position is between *C. habitans*, Sl., and *C. elegans*, Macl., to both of which it has affinities, and which it proves to belong to one group of species. Differences between it and *C. habitans* to which attention may be directed are, the anterior tibiæ more swollen in middle of lower side, posterior tibiæ much more slender and less incrassate to apex; the prothorax with three marginal punctures on each side. The larger and heavier head with one supraorbital puncture on each side differentiates it from *C. elegans* and allied species. The divergent frontal sulci readily separate it from *C. virescens*, Sl., and other species with short parallel

p. 201) seems to be *C. subcyaneum*. I think it is quite likely that further investigations with specimens from many different localities will show that *C. rugatum*, Blkb., and *C. sulcaticeps*, Sl., will have to be placed under *C. subcyaneum*, perhaps as varieties.

Carenum subporcatulum, Macl.

C. subporcatulum, Macl., Trans. Ent. Soc. N.S.W. 1865, i. p. 184; *C. politulum*, Macl., *l.c.*, 1873, ii. p. 98.

A comparison of the types of these species in the Australian Museum convinces me of the correctness of the above synonymy.

Carenum anthracinum, Macl.

C. anthracinum, Macl., Trans. Ent. Soc. 1864, i. p. 135; *C. ebeninum*, Casteln., Trans. Roy. Soc. Victoria, viii. p. 134.

C. ebeninum, Casteln., should, from the description, be a synonym of *C. anthracinum*, Macl.; specimens are in the Macleay Museum labelled "*C. ebeninum*, Casteln.," which are conspecific with *C. anthracinum*. Mr. Blackburn has noted the variability in colour of *C. anthracinum* (Trans. R. Soc. S. Aust. 1887, x. p. 58); specimens taken by Mr. C. French, Junr., in the North-west of Victoria, vary in colour of the elytra from black on the disc with purple reflections on the sides to black only near the middle of the disc with the lateral declivities bright metallic purple: these specimens cannot be separated from *C. anthracinum*, and seem also (though a little larger) conspecific with *C. gracile*, Sl., which in view of their evidence cannot be regarded as more than a variety of *C. anthracinum*. I do not think *C. cyanipenne*, Macl., is a species distinct from *C. anthracinum* (*vide* Blackburn, *l.c.*)

Carenum scaritioides, Westw.

C. scaritioides, Westw., Arcan. Ent. i. p. 192; *C. intermedium*, Westw., Trans. Ent. Soc. 1849, v. p. 203; *C. atronitens*, Macl., Trans. Ent. Soc. N.S.W. 1863, i. p. 137; *C. oblongum*, Macl., *l.c.*, p. 138; *C. nigerrimum*, Macl., *l.c.*, p. 176; *C. ambiguum*, Macl., *l.c.*, p. 177; *C. subquadratum*, Macl., *l.c.*, p. 177; *C. striato-punctulatum*, Macl., *l.c.*, p. 178; *C. ignotum*, Sl., P.L.S.N.S.W. 1891, (2), vi. p. 427.

Most of the synonymy given above has already been published by me (P.L.S.N.S.W. 1897, xxii p 211), and I have now to add *C. ambiguum*, Macl., after examination of the type. *C. ignotum*, Sl., is, from comparison of types, conspecific with *C. ambiguum*. After seeing the type of *C. subquadratum*, Macl., I can find no decided difference, beyond its somewhat more depressed form, between it and *C. scaritioides*, Westw., of which I regard it as a synonym.

Carenum regulare, n.sp.

♂. Elongate, robust, parallel; head large, deeply bisulcate, lightly constricted behind eyes; prothorax a little broader than long, subparallel on sides, obliquely angustate to base; elytra twice as long as broad, subdepressed, strongly declivous on sides and apex, quadripunctate, border thickened and upturned at humeral angles, inflexed margin very narrow behind first ventral segment; anterior tibiæ bidentate. Under surface, legs, head (generally), and disc of prothorax black, some greenish reflections on sides of occiput, below eyes and on sides of prosternum; margins of prothorax viridescent; elytra purple-black, violaceous

mentum and genæ very strongly and abruptly raised from gulæ. Antennæ stout; four basal joints cylindrical, six succeeding ones compressed, apical joint elongate, oval. Prothorax very little wider than head (6·6 x 7·5 mm.), convex, lightly declivous to base; basal area defined by a transverse impression; sides subparallel, rounded at posterior angles, hardly sinuate on each side of base; anterior margin truncate; anterior angles widely rounded and hardly at all advanced; base wide, rounded; border narrow, more strongly reflexed behind posterior angles, thick on base; lateral marginal channel narrow, not continued across base; a distinct wide shallow impression in front of each basal angle; median line linear, deep (only the anterior marginal puncture present in specimen before me). Elytra about same width as prothorax (15 x 7·7 mm.), very little rounded on sides; apical curve wide, short, base truncate, vertical above peduncle; disc subdepressed; sides and apex roundly and strongly declivous; border narrow, thickened towards apex (the edge not reflexed on apical curve), decidedly thickened and upturned (not dentate) at humeral angles; four ocellate punctures in a slight depression on base near each humeral angle, a fifth puncture at a little distance from these; lateral row of punctures rather widely placed, except near humeral angles. Suture between second and third ventral segments entire; second ventral segment with a shallow concave impression in middle between posterior trochanters; reflexed border of apical ventral segment not foveolate on each side of anus. Legs long; anterior femora thick, not dilatate in middle or with a notch on lower side before apex, unipunctate near lower margin of inner side towards apex; anterior tibiæ bidentate, apical plate without a dentiform projection below tarsus; intermediate tibiæ with a strong triangular external tooth at apex; posterior legs long, light, trochanters pyriform, triangular and pointed at apex. Length 27, breadth 7·7 mm.

Hab. Barrow Creek, Central Australia (Coll. French; unique).

Closely allied to *C. acutipes*, Sl., with which it agrees in every feature of importance, but differing in colour, larger size, basal sinuosities of the prothorax almost obsolete, posterior trochanters with the apex less drawn out into an acuminate point It seems impossible to unite the two unique specimens of *C. regulare* and *C. acutipes* before me under one specific name, though subsequent observations may show them to be varieties of a widely spread and variable species. The anterior discoidal punctures of the elytra are placed close behind the humeral angles (2 mm. distant), the posterior about the apical fifth. I have placed *C. acutipes* and *C. regulare* in the same group as *C. quadripunctatum*, Macl., though they seem to have but little affinity to that species.

Carenum subplanatum, Bates.

I identify as *C. subplanatum*, Bates, a species in Mr. C. French's collection, of which the following is a description :—

Head black with purple reflections near posterior extremities of frontal sulci; prothorax black, violaceous towards margins; elytra bright violaceous; under surface black. Head transverse (2·5 × 3·7 mm.), convex, transversely impressed across occiput; frontal sulci deep, diverging lightly backwards and meeting posterior transverse impression at full depth; frontal spaces convex, the lateral ones abruptly acclivous and bipunctate above eyes; clypeus with median part truncate, intermediate angles prominent, short, acute; eyes deeply set in orbits, not prominent or convex; suborbital channels to receive antennæ wide, short,

Carenum (Paliscaphus) felix, Sl., resembles *C. subplanatum*, Bates, but a specimen is not now available to me for comparison. It seems to me that *Paliscaphus*, though closely allied to *Eutoma*, cannot be united with it, and is as much entitled to generic rank; however, I prefer to merge it in *Carenum*.

Genus EUTOMA.

I offer the following grouping of the species of *Eutoma* in the hope that it may help to make the identification of species in that genus a little less difficult than at present. *Eutoma aberrans*, Sl., should be referred to *Conopterum*.

I. Elytra bipunctate.

A. A single large puncture on the base of each elytron near shoulder.

- *E. cavipenne*, Bates.
- *E. cupripenne*, Macl.
- *E. gratiosum*, Sl.
- *E. splendidum*, Macl.
- *E. violaceum*, Macl.

AA. Several punctures in a cluster at base of each elytron.

- *E. adelaidæ*, Blkb.
- *E. episcopale*, Cast.
- *E. filiforme*, Cast.
- *E. glaberrimum*, Macl.
- *E. levissime*, Sl.
- *E. Mastersi*, Macl.
- *E. purpuratum*, Cast.
- *E. subrugosulum*, Macl.
- *E. substriatulum*, Macl.
- *E. tinctillatum*, Newm.
- *E. undulatum*, Macl.

II. Elytra quadripunctate.

- *E. brevipenne*, Macl.
- *E. digglesi*, Macl.
- *E. magnificum*, Macl.
- *E. viridicolor*, Sl.

III. Elytra sex- or octopunctate.

- *E. frenchi*, Sl.
- *E. punctipenne*, Macl.

The *E. splendidum* Group belongs to South-West Australia, with the exception of *E. gratiosum*, Sl., which is said to be from the Mallee District of Victoria, a locality which I cannot help thinking may have been ascribed to it in error.

E. tinctillatum, Newm., seems to be the only species of the genus found east of a line drawn from Moreton Bay to Port Phillip. I have no doubt but that *E. bipunctatum*, Macl., *E. punctulatum*, Macl., *E. newmani*, Cast., *E. læve*, Cast., *E. loddo-*

nense, Cast., are synonyms of *E. tinctillatum*. I have not examined Sir William Macleay's types carefully, but have great doubt whether his species from Eastern New South Wales, viz., *E. glaberrimum*, *E. mastersi*, *E. subrugosulum*, *E. substriatulum* and *E. undulatum* are different from *E. tinctillatum*.

EUTOMA LÆVISSIME, n.sp.

Elongate, subdepressed, lævigate; head wide across eyes, decidedly narrowed behind eyes; prothorax a little broader than long, widely lobate at base; elytra truncate-oval, bipunctate, base with three punctures in a depression near each angle, marginal row of punctures wide apart in middle of sides (only three or four). Polished black, prothorax narrowly margined with purple, elytra becoming purple on lateral declivities.

Head wide (2·8 × 3·5 mm.), transversely impressed behind vertex, frontal sulci deep, diverging posteriorly; space between sulci convex, with an oblique foveiform impression on each side at narrowest part; spaces between sulci and eyes convex, roundly prominent and convex before eyes; preocular sulcus obsolete; eyes convex, more prominent than supra-antennal plates, deeply enclosed in orbits at base; postocular processes sloping gently to head; two supraorbital punctures on each side. Prothorax a little broader than long (3·3 × 3·5 mm.), subdepressed, lightly

The legs are as usual in *Eutoma;* the frontal sulci are shallower posteriorly and do not extend as far backwards as in *E. tinctillatum*, Newm. In the specimen before me, just behind the extremity of each sulcus there is a small punctiform impression, which may represent the usual end of the sulcus in the species, and the presence of which in the type specimen may be caused by a slight interruption of the ordinary course of the sulcus.

Allied to *E. tinctillatum* but more depressed; the head wider across the eyes and more strongly narrowed posteriorly, eyes more prominent and convex; prothorax proportionately wider, less convex and less declivous to the sides (the sides are decidedly narrowed to the apex from the anterior marginal puncture); elytra shorter, the marginal punctures fewer in number and placed at wider intervals on middle of sides; posterior trochanters shorter and more oval (more widely rounded at apex). The genæ rise abruptly from the gulæ at the base, instead of by a gentle slope as in *E. tinctillatum*, and are not divided from the submentum by a sulciform impression. Its exact habitat is doubtful, but I believe it is either from Cape York or the North-west Coast.

E. PUNCTIPENNE, Macl.

Three specimens of this species have been given to me by Mr. C. French as from Cape York, Queensland. The colour is metallic-purple (including under surface), each elytron has four large punctures—wide apart—placed longitudinally along the middle (one specimen has five punctures on the right elytron). The measurements of the largest specimen are—head 2·8 × 3·2mm., prothorax 3·5 × 3·3 mm., elytra 7 × 3·3 mm. Length 11·5-13·5, breadth 2·8-3·3 mm.

E. frenchi, Sl., is very closely allied to *C. punctipenne*, Macl., but seems to me specifically distinct; it differs by its green colour, the elytra with only three punctures on each, the anterior angles of the prothorax a little more prominent, &c. It should be noted that by some oversight I described *E. frenchi* as having the elytra quadripunctate, whereas the type, which is in my possession, has three punctures on each elytron, the posterior puncture being near the apex on the apical declivity.

Genus CARENIDIUM.

CARENIDIUM SAPPHIRINUM, Bates.

A species of *Carenidium* from the North-west Coast is in the possession of Mr. C. French, which I identify as *C. sapphirinum*, Bates. The following is a brief description:—

Upper surface dark violet becoming green on sides of prothorax, elytra and head, also in the frontal sulci; under surface black with bluish metallic tints on gulæ; prosternum, inflexed margins of elytra, sides of abdomen and legs black. Head large (4·7 x 7 mm.), upper surface shagreened; labrum emarginate in a regular curve; clypeus sinuate-emarginate behind labrum; intermediate angles obtuse, wide at base, projecting lightly. Mandibles thick, the left with the tooth in middle thick, short, slightly elevated. Prothorax convex, transverse (6 x 8·7 mm.); sides rounded; border widely reflexed. Elytra oval (14 x 9·5 mm.), subdepressed; base abrupt behind peduncle, lightly emarginate at suture; four or five obliquely placed punctures on each side of base. Length 29, breadth 9·5 mm.

From *C. gagatinum*, Macl., and *C. superbum*, Casteln., two

obtuse, sinuosities on each side of basal lobe stronger; elytra more convex, narrower at the base, more roundly ampliate behind the shoulders, lateral declivities more rounded—almost hiding the border in the middle when viewed from above. From *C. sapphirinum*, Bates, it differs by facies; the labrum smaller and more deeply emarginate; clypeus more deeply and evenly emarginate; anterior margin of head, outwards from intermediate angles of clypeus, forming an even oblique curve (not curving decidedly forward in middle as in *C. sapphirinum*). The mandibles are flat on the upper surface as in *C. superbum*.

Genus MONOCENTRUM.

MONOCENTRUM LONGICEPS, Chaud.

I have found a species of *Monocentrum* in New South Wales, near Grenfell and Urana, which seems to be *M. longiceps*, Chaud. The late Mr. Geo. Barnard found *M. longiceps* at Coomooboolaroo, Dawson River, Queensland.

DESCRIPTIONS OF TWO NEW SPECIES OF DIPTERA FROM WESTERN AUSTRALIA.

By D. W. Coquillett.

(Communicated by Arthur M. Lea.)

Phytomyza betæ, n.sp.

Yellow, an ocellar dot, upper half of occiput except along the margin, mesonotum except lateral margins and hind angles, a spot on each side of the scutellum, the metanotum and abdomen except sides and hind margins of the segments of the latter, blackish-brown, tibiæ yellowish-brown, tarsi dark brown, antennal arista and hairs and bristles of entire insect black; mesonotum polished, destitute of short, bristly hairs except in front of the suture, bearing three pairs of dorsocentral bristles, the anterior pair shortest and situated slightly in front of the suture; wings greyish-hyaline, hind cross-vein wanting. Length 1 mm.

Hab.—West Australia; three specimens bred by Mr. Arthur M. Lea, from larvæ mining the leaves of the beet.

Closely related to the European *flavoscutellata*, but the latter is twice as large, has the third antennal joint black, mesonotum opaque, grey-pruinose and bearing four pairs of dorsocentral bristles, etc. This is the only species of *Phytomyza* known to me to attack the beet. It is very probable, however, that it originally fed upon some other plant, one that is indigenous to Australia.* No species of this genus was described from Australia by the older authors, nor is any species of *Phytomyza*

* This surmise is quite correct, as I have reared it from several poisonous plants of the genus *Anthocercis*; and in Tasmania from both beet and mangels.—A.M.L.

mentioned by Van der Wulp in his recent Catalogue of the Diptera of South Asia.

MYIOPHASIA FLAVA, n.sp.

Yellow, the hairs and bristles chiefly black, a medio-dorsal brown vitta on the abdomen; eyes of male separated about the width of the lowest ocellus, antennæ reaching five-sixths of distance to the oral margin, the third joint slightly more than thrice as long as the second, the lower front corner produced in the form of a blunt tooth, arista thickened on the basal two-fifths; mesonotum somewhat polished, marked with five grey pruinose vittæ, three pairs of postsutural dorsocentral macrochætæ, two marginal pairs on the scutellum, sternopleura bearing two macrochætæ, abdomen subopaque, bearing only marginal macrochætæ; wings hyaline, third vein bristly about half-way to the small cross-vein, hind cross-vein nearer to the latter than to bend of fourth vein, first posterior cell petiolate, the petiole about half as long as the hind cross-vein. Length 5 mm.

Hab.—West Australia; two males bred from adults of the Scarabæid, *Anoplostethus opalinus*, Burm.

Although aberrant in its colouring, this interesting species possesses practically all the structural characters of the type species of the genus *Myiophasia;* the two heretofore known

the middle very distinctly rugulose. Two basal segments of *abdomen* with moderately large punctures. Length (including rostrum) 2, width ⅓ mm.

Hab.—Geraldton, W.A.

Differs from *cæcus* in being smaller, thinner, and paler; the hair shorter, the abdomen more distinctly punctate, and the anterior tibial hooks shorter and thinner. It is possible that it should be regarded as a variety only of *H. cæcus*, but it is at least a variety worthy of a name. A number of specimens were obtained from the "outer beach" at Geraldton burrowing at the roots of a small species of salt-bush *(Atriplex)*.

Halorhynchus cæcus, Woll.

This species I have repeatedly searched for under seaweed and beach-growing plants at and about Fremantle (the original locality), but never succeeded in obtaining more than two specimens. These were taken from about four inches below the surface at the roots of a species of spinifex grass growing close to Cottesloe Beach.

Tasmanica, n.g.

Head not distinctly separated from the rostrum, their outline slightly incurved at middle, combined length equal to about two

THE DOUBLE STAINING OF SPORES AND BACILLI.

By R. Greig Smith, M.Sc., Macleay Bacteriologist to the Society.

Until quite recently the methods recommended by various authors for staining the spores of bacilli produced results that were far from being satisfactory. Fortunately this state of affairs has been removed by Klein*, of Amsterdam, who has published a method which, with a little improvement, is eminently successful.

The methods in general use consist in the first place in preparing a film of the material to be stained by mixing a portion of an agar or potato culture of the bacillus with a drop of water upon a clean cover-glass and allowing the suspension to dry. The dried film is then fixed by passing it three times through the bunsen-flame, a process which has for its object the coagulation of the bacterial protoplasm and the attachment of the bacteria to

noted that it was much easier to kill bacteria and spores when they were in the moist condition than when they had been dried, and from this he reasoned that it would be much easier for the stain to get through the moist than through the dried spore-case. Accordingly he stained first and fixed afterwards.

His procedure is as follows:—A platinum loop is filled with a culture of the bacillus grown upon potato for 24 hours (at 37° C.), and this is introduced into a small quantity—say, two or three drops—of normal saline contained in a watch-glass. The culture is stirred in the saline until a homogeneous suspension is obtained. An equal number of drops of carbol-fuchsin is added, and the two fluids are thoroughly mixed. The watch-glass is placed high over a micro-chemical burner so that the heat applied is just sufficient to cause a slight vapour to hover over the surface of the fluid. A larger watch-glass is placed over the first to keep out the dust and to enable one to judge the intensity of the heating. There should be just a slight film of condensed water upon the covering glass. At the end of *six* minutes the watch-glass is taken from the burner and allowed to cool for a few moments. The bacilli and spores which have precipitated more or less are again distributed in the stain by imparting a rotatory motion to the cover-glass. A loop of the suspension is spread over a clean cover-glass and allowed to dry in the air. The film is fixed by passing it twice through the flame. The bacilli are then decolorised by immersing the cover-glass in 1% sulphuric acid for one or two seconds, after which the acid is washed off with water and the bacilli counter-stained in dilute watery-alcoholic methylene-blue for three or four minutes. The film is washed in water, dried and mounted in balsam. After this treatment the spores appear red, the bacilli blue.

The method as recommended by Klein is excellent as far as the principle is concerned, but the details might be altered with advantage. Some spores, instead of staining red, show only a pink margin. Klein does not push the staining process far enough, and indeed it is hardly possible to do so when a watch-glass is used to hold the fluid. There are spores, as, for example, *Bac. lactis XII.* (Flügge), which stain quite readily with carbol-fuchsin in the usual manner after fixation by heat, while others, as a *Bac. leptosporus* sp., stain but faintly when Klein's method is employed. The refractory spores are stained a deep red by the following method:—Four drops of normal saline are pipetted into a small test-tube, and the spore-bearing material is rubbed up with this until a homogeneous suspension is obtained. Four drops of fresh carbol-fuchsin are pipetted into the tube and the mixture shaken. A plug of cotton wool is inserted and the tube placed into a beaker of boiling water. The water is boiled for a quarter of an hour, when the tube is taken out and shaken. A loopful of the bacterial suspension is withdrawn and spread uniformly over a cover-glass which is dried either in the air or high over a bunsen-flame. The film is next fixed by passing the cover-glass three times through the flame in the usual manner. The bacilli are decolorised in methylated spirit containing 1·5% (by volume) of concentrated hydrochloric acid. When the film appears colourless, the cover-glass is withdrawn and moved about in water to remove the alcohol, after which the film is stained with carbol-methylene-blue in the ordinary manner; it is then washed, dried and mounted.

The acidified alcohol appears to give a cleaner film than when

when the investigator might easily be led astray. Certain rod-shaped water-bacteria become vacuolated as they grow old: the protoplasm aggregates at the poles, which stain deeply, leaving the centre of the rod unstained. In these cases the central unstained vacuole, and especially when it is oval, presents an appearance similar to an unstained spore. When counter-staining is made use of, the central vacuole does not stain. It is here that perhaps the advantage of possessing a process which stains the most refractory spores is evident. One can depend upon the spore being stained. In the case of vacuoles it is possible, by limiting the decolorisation with acidified alcohol, and by counter-staining with blue, to obtain the bacteria with their dense terminal protoplasm stained red, and the vacuole pale blue.

Mr. D. G. Stead exhibited mounted preparations of various crustaceans including *Nectocarcinus integrifrons*, M.-Edw., from Port Jackson, *Cancer novæ-zelandiæ*, Jacq. & Lucas, from New Zealand, *Lithodes maia*, Leach, from Norway, and *Macrophthalmus setosus*, M.-Edw., one specimen of the last of these being distorted by the attack of a parasite (*Bopyrus* sp.).

Mr Froggatt exhibited a series of co-types of the parasitic Hymenoptera described in Mr. Ashmead's paper.

Mr. Waterhouse exhibited the sexes of the butterfly commonly known as *Papilio Erectheus*, Don.; and he raised the question of the authority for the choice of names in this and similar cases. The female was originally described and figured by Donovan in the "Insects of New Holland" (1805) as *P. Ægeus* (pl. xiv.), and the male as *P. Erectheus* in the same work (pl. xv.).

Mr. Greig Smith illustrated the technique of the double staining of spores and bacilli proposed in his paper; and he also showed mounted preparations under the microscope.

The Surveyor, Sydney. Vol. xiii. No. 6 (June, 1900). *From the Editor.*

University of Sydney—Calendar for the Year 1900. *From the Senate.*

Australasian Institute of Mining Engineers, Melbourne—Transactions. Vol. vi. (1900). *From the Institute.*

Australasian Journal of Pharmacy, Melbourne. Vol. xv. No. 175 (July, 1900). *From the Editor.*

Department of Mines, Melbourne: Geological Survey – Monthly Progress Report. New Series. Nos. 11-12 (February-March, 1900). *From the Secretary for Mines.*

Field Naturalists' Club of Victoria—Victorian Naturalist. Vol. xvii. No. 3 (July, 1900). *From the Club.*

Royal Geographical Society of Australasia, Victorian Branch, Melbourne. Vol. xviii. Part 1 (1900). *From the Society.*

Department of Agriculture, Perth, W.A.—Journal, June, 1900. *From the Secretary.*

Western Australian Year-Book for 1898-99. Vol. i. (1900). *From the Victoria Public Library, Perth.*

Royal Society of Tasmania—Papers and Proceedings for the

Royal Society, London—Proceedings. Vol. [illegible]. No. [illegible] (June, 1900). *From the Society.*

Zoological Society, London—Abstract, May [illegible], 1900. *From the Society.*

Royal Irish Academy, Dublin—Proceedings. Third Series. Vol. v. No. 4 (1900). *From the Academy.*

Geological Survey of Canada, Ottawa—Contributions to Canadian Palæontology. Vol. i. Part [illegible] ([illegible]). *From the Director.*

American Naturalist, Cambridge. Vol. [illegible]. Nos. [illegible] (May-June, 1900). *From the Editor.*

U.S. Department of Agriculture, Washington. Division of Biological Survey—Bulletin. No. [illegible] (1900). North American Fauna. No. 17 (1900). *From the Secretary for Agriculture.*

Wisconsin Natural History Society, Milwaukee—Bulletin. Vol. i. No. 2 (April, 1900). *From the Society.*

Zoological Society of Philadelphia—Twenty-eighth Annual Report, 1900. *From the Society.*

Museo Nacional de Montevideo—Anales. Tomo iii. Fasc. xiii (1900). *From the Museum.*

Museu Paraense de Historia Natural e Ethnographia, Para—Boletim. Vol. iii. No. 1 (Feb., 1900). *From the Museum.*

Perak Government Gazette, Taiping. Vol. xiii. Nos. 14-18 (June, 1900). *From the Government Secretary.*

Naturhistorischer Verein in Bonn — Verhandlungen. 56 Jahrgang. Zweite Halfte (1899): Sitzungsberichte der Niederrheinischen Gesellschaft fur Natur- und Heilkunde zu Bonn, 1899. Zweite Halfte. *From the Society.*

Verein fur Naturwissenschaftlicher Unterhaltung zu Hamburg —Verhandlungen. x. Band (1896-98). *From the Society.*

Zoologischer Anzeiger, Leipzig. Band xxiii. Nos. 616-617 (May-June, 1900). *From the Editor.*

26

Société Géologique de Belgique—Annales. Tome xxvii. 2e Liv. (May, 1900). *From the Society.*

Société Royale Linnéenne de Bruxelles—Bulletin. 25me Année. No 7 (May, 1900). *From the Society.*

Faculté des Sciences de Marseille—Annales. Tome x. (1900). *From the Faculty.*

Zoological Museum, University of Copenhagen—The Danish Ingolf Expedition. Vol. i. Part ii. (1900); Vol. ii. Part iii. (1900). *From the Museum.*

Società Entomologica Italiana, Firenze—Bullettino. Anno xxxii. Trimestre i. (May, 1900). *From the Society.*

Zoologische Station zu Neapel—Mittheilungen. xiv. Band. 1 u. 2 Heft (1900). *From the Director.*

Imperial University, Tōkyō: College of Science—Journal. Vol. xii. Part 4 (1900). *From the University.*

South African Philosophical Society, Cape Town—Transactions. Vol. xi. Part 1 (1900). *From the Society.*

Finska Vetenskaps-Societeten, Helsingfors — Bidrag. lviii. Häftet (1900): Forhandlingar. xli. Häftet (1900). *From the Society.*

Société des Naturalistes de Kiew—Mémoires.

DESCRIPTIONS OF NEW AUSTRALIAN LEPIDOPTERA.

By Oswald B. Lower, F.E.S., Lond.

BOMBYCINA.

Odonestis hilaropa, n.sp.

♂. 30 mm. Head, thorax, palpi, legs and abdomen deep mahogany-red, posterior legs paler. Antennæ ferruginous, pectinations ochreous, at greatest length 8. Forewings elongate, moderate, rather short, costa nearly straight, hindmargin bowed; mahogany-red, deeper on basal half; a broad transverse median shade, anterior edge obscure, posterior edge moderately defined, from $\frac{2}{3}$ of costa to beyond middle of inner margin, sinuate inwards on lower half; a curved transverse row of fuscous spots, from just before apex to just before anal angle; a reddish-fuscous hindmarginal line: cilia fuscous-reddish. Hindwings mahogany-red mixed with ochreous towards base; cilia as in forewings.

Cape York, Q.; one specimen, in December.

Nola vernalis, n.sp.

♂. 20 mm. Head and thorax white. Palpi 4, white, strongly infuscated on sides. Antennæ whitish on basal half, ochreous on terminal half. Abdomen whitish. Anterior and middle legs fuscous, tibiæ and tarsi obscurely ringed with whitish, posterior pair whitish. Forewings elongate-triangular, costa gently arched, hindmargin oblique, hardly rounded; whitish, irregularly suffused with fuscous; a short blackish dentate line at base of costa; a blackish line from costa at about $\frac{1}{3}$, commencing obliquely outwards on upper third, thence twice waved and terminating on inner margin at $\frac{1}{4}$; a series of blackish elongate spots, starting from costa before middle, strongly curved around and ending on inner margin in middle, somewhat obscure on lower $\frac{1}{3}$; a second row of similar, yet more elongate spots, starting from costa at $\frac{3}{4}$, and ending on inner margin at anal angle, almost parallel to previous row of spots: hindmarginal area of wing more fuscous: a row of obscure blackish hindmarginal spots : cilia fuscous, with dull whitish points at extremities of veins. Hindwings whitish, faintly infuscated around apex and upper half of hindmargin; a fine fuscous hindmarginal line; cilia whitish, fuscous-tinged.

Blackwood and Goolwa, S.A.; two specimens, in October.

GEOMETRINA.

Xanthorhoe lychnota, n.sp.

♂♀. 24-30 mm. Head, palpi and thorax light fuscous, mixed with ferruginous, palpi 2. Antennæ fuscous, in ♀ annulated with white, pectinations of ♂, 6. Abdomen greyish-ochreous, sprinkled with fine black scales. Legs dark fuscous, more or less banded with whitish-ochreous. Forewings elongate-triangular, hindmargin bowed, hardly waved; whitish-ochreous in ♀, in ♂ leaden-tinged, in both sexes strongly mixed throughout with fine

wings with hindmargin slightly waved, rounded; light fuscous, paler towards base and faintly strigulated throughout with dark fuscous. Undersurface ochreous; forewings with basal area and a broad ante-apical streak fuscous. Hindwings finely reticulated throughout with fuscous, and with a fuscous discal spot.

Broken Hill, N.S.W.; two specimens, in April and October.

An interesting and pretty species; the yellowish undersurface is a curious characteristic. The ♀ is much more gaily coloured than the ♂.

Leptomeris hypocallista, n.sp.

♂. 22 mm. Head, thorax, palpi and abdomen ochreous-ferruginous. Antennæ (broken), ciliations (?). Abdomen with a submedian band of darker fuscous. Forewings elongate-triangular, costa gently arched, hindmargin slightly rounded, oblique; ochreous-ferruginous, with fuscous markings, somewhat obscure, dentate; 1st line slightly curved outwards, from before $\frac{1}{3}$ of costa to $\frac{1}{3}$ of inner margin; median from $\frac{2}{3}$ of costa to beyond middle of inner margin, preceded by a fuscous discal dot above middle and indented above inner margin to beneath discal dot: second parallel to median, from just before $\frac{3}{4}$ of costa to $\frac{2}{3}$ of inner margin; subterminal cloudy, waved; submarginal only faintly indicated; a hindmarginal row of blackish dots: cilia reddish, terminal third fuscous. Hindwings with hindmargin rounded; colour somewhat lighter than forewings; markings as in forewings, but discal dot well marked; cilia as in forewings. Undersurface bright reddish-orange, markings of upperside reproduced, and hindmarginal dots very distinct.

Goolwa, S.A.; six specimens, in January.

In the *rubraria* group.

Diastictis epidesma, n.sp.

♀. 30 mm. Head, thorax and palpi deep chocolate. Antennæ ochreous. Abdomen fuscous, anal tuft ochreous. Legs ochreous, finely irrorated with fuscous and dark fuscous. Forewings elongate-triangular, hindmargin angularly projecting in middle, upper half slightly concave; deep chocolate; costal edge obscurely yellowish and strigulated with blackish; a row of small white dots, placed on a narrow blackish shade, from beneath ¾ of costa to ¾ inner margin, three median spots edged anteriorly with black; a row of obscure black dots along hindmargin; some minute white scales along hindmargin: cilia deep chocolate. Hindwings with hindmargin rounded, faintly waved; dark fuscous, somewhat purplish-tinged and becoming greyish on basal third; a hardly perceptible waved blackish postmedian line; some chocolate scales along inner margin; an obscure blackish hindmarginal line; cilia as in forewings.

Exeter, S.A.; one specimen, in November. (Type in *Coll. Harold Lower*).

AMELORA TETRACLADA, n.sp.

♂. 40 mm. Head and face ochreous-ferruginous, forehead with a large scoop-like horny projection. Antennæ ochreous-fuscous, pectinations 4, ochreous. Thorax shining white, anteriorly pale yellow, patagia pale yellow. Abdomen ochreous-white, beneath silvery-white. Legs ochreous-fuscous. Forewings elongate-triangular, apex round-pointed, hindmargin bowed; silvery-white, with light ochreous-fuscous markings; 4 longitudinal thick

side obscurely reproduced; a large black apical spot. Hindwings white, with a large blackish apical spot and a small blackish spot on hindmargin near anal angle.

Broken Hill, N.S.W.; one specimen at light, in May.

AMELORA PARONYCHA, n.sp.

♀ 40 mm. Head and face dull ochreous, forehead with a prominent broad scoop-like horny projection. Thorax grey, anteriorly dull ochreous, patagia dull ochreous. Abdomen ochreous-white. Antennæ ochreous. Legs ochreous, coxæ white. Forewings elongate-triangular, apex round-pointed, hindmargin bowed; silvery-white, with light fuscous markings; a thick streak along costa from base to apex, emitting from its lower edge at $\frac{1}{3}$ a thick furcate streak, upper fork terminating on middle of hindmargin, lower on hindmargin above anal angle; a strong thick oblique streak from lower edge of costal streak at $\frac{5}{6}$ to hindmargin immediately above termination of upper fork of furcate streak; a thick streak along inner margin to anal angle: cilia ochreous, (imperfect). Hindwings with hindmargin rounded, faintly waved; white, somewhat tinged with light greyish-ochreous; a moderate blackish subapical spot, indented in middle, and continued very narrowly as a hindmarginal streak to anal angle. Undersurface white; hindwings with a moderate black subapical blotch.

Broken Hill, N.S.W.; one specimen at light, in May.

Recalls *Thalaina Clara*, Walk., in general appearance and markings; it may ultimately prove to be the ♀ of *tetraclada*, although I must admit that I have no justifiable ground for considering it so, as I know of no species (and I possess the whole of the known species of both *Thalaina* and *Amelora*) in which the sexes differ to such an extent. The present and the preceding species form a valuable connecting link between the two above-mentioned genera, and at first sight either of them could very easily be mistaken for species of *Thalaina*, but the bipectinated antennæ of the ♂, and the curious frontal projection of the head unquestionably refer them to *Amelora*, their nearest ally being *leucaniata*, G'n.

PYRALIDINA.

Hypsopygia costalis, Fab.

(*Pyralis costalis*, Fab., Ent. Syst. No. 420; *P. fimbrialis*, Schiff., Wien. Verz. p. 124; *P. aurotænialis*, Chr., Bull. Mosc., 56, i. (var.); *P. hyllalis*, Walk., xvii. p. 265 (var.); *P. rubrocilialis*, Staud., Hor. Ent. Ross., 1870, p. 181.)

Cooktown and Mackay, Q.; ten specimens, in November. Not previously recorded from Australia.

Scoparia mesogramma, n.sp.

♀. 20 mm. Head, thorax, palpi and antennæ ashy-grey-whitish, thorax more whitish in middle, palpi whitish at base. Abdomen fuscous, basal segment orange. Legs ashy-grey-whitish, posterior pair whiter, tibiæ fuscous with whitish rings. Forewings elongate, moderate, slightly dilated posteriorly, costa gently arched, hindmargin obliquely rounded; ashy-grey-whitish; a well defined black streak above middle of wing, from base to end of cell, anteriorly attenuated, and somewhat curved upwards on anterior third; a fine black line beneath streak, from base to before middle, edged above by its own width of whitish; a row of elongate black spots along hindmargin, between veins: cilia ashy-grey-whitish. Hindwings rather dark fuscous [illegible]

beneath; fulvous, somewhat shining and wholly irrorated transversely with rows of minute fuscous dots, posterior half of costal edge finely dull ochreous; an elongate fuscous mark lying on inner margin at ¼: cilia bronzy-fuscous, becoming dark fuscous on basal half. Hindwings bronzy-fuscous: cilia fuscous with a greyish parting line near base.

Burnside, S.A.; two specimens, in October.

ANATROPIA PENTACOSMA, n.sp.

♂. 12 mm. Head, thorax, palpi and antennæ ochreous-whitish, tinged with ferruginous, patagia ferruginous-ochreous. Legs fuscous, tibiæ and tarsi banded with ochreous-whitish (abdomen broken). Forewings elongate, moderate, costa nearly straight, slightly arched at base, hindmargin [illegible] sinuate beneath apex: basal patch [illegible] two transverse ochreous-ferruginous fasciæ, first [illegible] second starting from middle one of 3 [illegible] outwards and ending on inner margin at [illegible] basal patch from ¼ of costa to before [illegible] with two indentations above middle and [illegible] purple-blackish transverse fascia [illegible] well defined, posterior edge [illegible] especially on lower [illegible]: a [illegible] fuscous; groundcolour between [illegible] with some scattered spots of [illegible] margin, two moderate [illegible] apical spot: followed by about 7 [illegible] of costa, thence continued [illegible] gradually diminishing [illegible] mixed with blackish and [illegible] wings with hindmargin [illegible] spotted with darker, [illegible] with darker fuscous basal and [illegible]

Blackwood, S.A.; one specimen, in [illegible]

Closely allied to *chloroxanthes*. [illegible] be a variety; but it appears distinct by the [illegible]

TORTRIX ARGYRASPIS, n.sp.

♂. 18 mm. Head, palpi, thorax and antennæ ochreous-fuscous, face lighter. Abdomen greyish. Anterior legs ochreous-fuscous, middle and posterior pair whitish-ochreous. Forewings elongate, moderate, costa rather strongly arched, hindmargin oblique, slightly sinuate beneath apex; ochreous-ferruginous with silvery markings, finely edged with blackish; costal edge finely dark fuscous from base to middle; a moderately broad streak, from base in middle to before $\frac{1}{3}$ of wing, thence obliquely deflected and ending on inner margin in middle, somewhat dilated on inner margin; a broad, somewhat suffused streak from middle of wing, gently curved downwards and ending on hindmargin above anal angle; a moderately well defined streak from just beyond middle of costa, strongly sinuate downwards and ending on apex, becoming somewhat narrowed on sinuation; two oblique silvery flecks in sinuation, just below costa, edged by fine black lines beneath: cilia reddish-ferruginous, with a silvery tooth at apex, caused by continuance of sinuate streak. Hindwings fuscous, slightly ferruginous-tinged; cilia light ferruginous-fuscous, with a darker subbasal line.

Brisbane, Q.; one specimen, in December.

In the neighbourhood of *aulacana*, Meyr.

between base and before apex, from each of which proceeds an obscure waved oblique fuscous line; a large round silvery-white spot, at ⅓ from base; a second immediately below, only separated by a fine line of groundcolour, both edged anteriorly by the second oblique line from costa; a larger similar silvery-white spot at end of cell, broadest above, edged anteriorly by 4th costal streak; a row of yellowish spots around apical fifth of costa and hindmargin: cilia yellow, strongly mixed with rosy, base with a few fuscous-purple scales. Hindwings yellowish-ochreous, faintly rosy-tinged: cilia yellow.

Mackay, Q.; one specimen, in December.

Easily recognised by the large silvery-white spots.

Hoplitica habroptera, n.sp.

♂. 16 mm. Head, thorax, palpi and antennæ reddish-ochreous, second joint of palpi whitish at base, antennæ sharply annulated with fuscous. Abdomen ochreous. Anterior legs reddish, with ochreous rings on tibiæ, middle and posterior legs yellowish. Forewings elongate, moderate, costa gently arched, hindmargin obliquely rounded, 7 to apex; reddish-ochreous, with very obscure dull fuscous-purple markings: a broad irregular fascia at base, broken in middle: a second, more oblique, from ⅓ of costa to ¼ inner margin, broken above and below middle: a third, similar fascia, from ⅔ of costa to beyond middle of inner margin, emitting a fine curved line from its posterior edge, and ending on anal angle: a fine hindmarginal line: cilia reddish-ochreous, tips ochreous-yellow. Hindwings and cilia light yellow.

Brisbane, Q.; one specimen, in March.

Eulechria sciophanes, Meyr.

Proc. Linn. Soc. N.S.W. 1883, p. 323.

Broken Hill, N.S.W.; five specimens, in March and October. Not previously recorded from New South Wales.

EULECHRIA AUTOPHYLA, Lower.

Proc. Linn. Soc. N.S.W. 1899, p. 105.

By an unfortunate error this was described as *autophylla*, a name already preoccupied by a West Australian species; it should have read *autophyla*.

EULECHRIA LEPTOMERA, n.sp.

♂. 20 mm. Head, thorax, antennæ and palpi black. Abdomen and legs dark fuscous, posterior legs whitish-fuscous. Forewings elongate, moderate, costa gently arched, apex somewhat pointed, hindmargin very oblique; blackish, sparsely irrorated with dull whitish scales, base of wing more blackish; an obscure blackish dot at end of cell: cilia dark fuscous, mixed with some dull whitish scales at base. Hindwings dull orange; cilia fuscous, mixed with ochreous at base.

Gisborne, Vic.; one specimen, in October.

PHILOBOTA MICRASTIS, n.sp.

♂. 10 mm. Head, thorax and palpi ochreous-whitish, second joint of palpi fuscous at base externally. Antennæ ochreous-fuscous, basal joint dark fuscous. Legs and abdomen ochreous-grey. Forewings elongate, moderate, rather short, costa gently arched, hindmargin obliquely rounded; ochreous-whitish, more or

mixed with whitish. Legs fuscous, posterior pair whitish. Abdomen greyish-ochreous. Forewings elongate, moderate, costa gently arched, hindmargin obliquely rounded; white, with dull ochreous-fuscous markings: costal edge narrowly fuscous, from base to $\frac{2}{3}$: a suffused spot in cell at $\frac{1}{3}$ from base; a second, larger, just below and slightly beyond; a flattened spot on costa at $\frac{3}{4}$; an ante-apical streak, continued obliquely across wing to above anal angle, where it forms a large spot, which almost touches first costal spot; a row of suffused and interrupted spots along hindmargin, upper largest, and sometimes confluent with ante-apical spot: cilia white, with fuscous median lines. Hindwings light ochreous, or ochreous-fuscous; cilia greyish-ochreous.

Semaphore, S.A.; several specimens, in November.

PHILOBOTA (?) TETRASPORA, n.sp.

♀. 30 mm. Head, palpi, and thorax dull fleshy-carmine, second joint of palpi fuscous at base beneath. Antennæ dark fuscous, basal joint fleshy-carmine. Abdomen greyish-fuscous. Anterior and middle legs dull fleshy-carmine, posterior pair ochreous, anterior and middle tarsi fuscous, with whitish rings. Forewings elongate, moderate, costa gently arched, apex rounded, hindmargin obliquely rounded, 2 from angle, 7 to hindmargin; dull fleshy-ochreous: a moderate ferruginous spot in disc just beyond $\frac{1}{3}$ from base; a second, smaller, immediately below: a third, larger, at end of cell, and a fourth, more obscure, below and beyond: a hindmarginal row of small fuscous dots, becoming obsolete on lower third: cilia fleshy-carmine, tips darker. Hindwings greyish-ochreous, somewhat fuscous-tinged: 3 and 4 short-stalked: cilia greyish-ochreous, infuscated at base.

Derby, W.A.: one specimen, in March.

Placed in the genus *Philobota* provisionally: a new genus will no doubt be required to receive it, but in the absence of the ♂ it is desirable to refrain from forming one at present.

OECOPHORA EUSEMA, n.sp.

♂♀. 14-18 mm. Head ochreous. Thorax, palpi and antennæ ochreous-fuscous, patagia light ochreous, palpi beneath light ochreous. Abdomen greyish. Legs ochreous, somewhat banded with fuscous. Forewings elongate, moderate, costa gently arched, hindmargin obliquely rounded; dull ochreous, somewhat bronzy; markings blackish; a moderately thick basal streak; a moderate spot on disc before middle; a second, similar, just below, and a much larger one at end of cell; veins towards hindmargin somewhat outlined with fuscous, but this is sometimes absent; a suffused row of hindmarginal spots: cilia ochreous-fuscous, tips greyish. Hindwings greyish-fuscous; cilia ochreous, tips greyish.

Broken Hill, N.S.W.; eight specimens, in October and November.

Superficially extremely similar to *Sphyrelata ochrophæa*, Meyr., but easily separated from that species by the stalking of veins 2 and 3 of forewings.

MACHAERITIS XERODES, n.sp.

♂♀. 9-10 mm. Head white. Thorax brownish-ochreous. Palpi, antennæ and legs fuscous, palpi internally somewhat whitish, antennæ obscurely annulated with whitish, basal joint blackish. Abdomen fuscous, basal segment yellowish. Forewings elongate, moderate, rather narrow, costa gently arched, apex somewhat pointed; whitish, minutely irrorated with fuscous and blackish scales, excepting an obscure longitudinal streak of

mixed with ochreous and whitish. Anterior legs ochreous-whitish, middle and posterior pair silvery-whitish, all tarsi banded with blackish. Forewings elongate-lanceolate; dark fuscous, with whitish-ochreous markings; a small spot at base; a moderately broad oblique fascia, from costa at ¼ to inner margin at ¼; a small triangular spot on costa about middle; a similar spot immediately below on inner margin, and just before anal angle; a small spot on inner margin just beyond; a cuneiform spot on costa at apex: cilia dark fuscous, becoming whitish on inner marginal triangular spot. Hindwings dark fuscous, somewhat bronzy; cilia fuscous, becoming ochreous at base.

Broken Hill, N.S.W.; two specimens, in October.

Macrobathra eudesma, n.sp.

♂. 10 mm. Head and palpi white, terminal joint of palpi infuscated. Antennæ fuscous-whitish, basal joint black. Thorax blackish, suffused with grey-whitish in centre, patagia white. (Abdomen broken). Legs blackish, irregularly banded with white. Forewings elongate-lanceolate; black, with white markings: a moderately broad oblique fascia, from costa at ⅕ to inner margin at ⅓; a subtriangular spot on costa beyond middle, a second, similar, beyond middle of inner margin, near ¾; a third, larger, on costa before apex, reaching more than half across wing: cilia fuscous, with some blackish scales at base. Hindwings light fuscous; cilia greyish-ochreous.

Duaringa, Q.; two specimens, in February and September.

Not unlike a dwarfed specimen of *argonota*, Meyr.; but the antennæ are not white.

Macrobathra microspora, n.sp.

♀. 8 mm. Head dull whitish. Thorax fuscous, patagia whitish. Palpi fuscous, whitish internally, with a fuscous band at extremities of joints. Antennæ fuscous, basal joint black. Abdomen greyish-fuscous. Legs blackish, irregularly banded with white. Forewings elongate-lanceolate; black, markings white; a moderate oblique fascia, from before $\frac{1}{4}$ of costa to before $\frac{1}{4}$ inner margin; a second similar fascia, from beyond middle of costa to beyond middle of inner margin; a small flattened triangular mark on costa at $\frac{5}{8}$; a small round spot on inner margin before anal angle: cilia fuscous. Hindwings very narrow; dark fuscous; cilia fuscous, lighter at base.

Mackay, Q.; one specimen, in October.

Somewhat allied to the preceding.

GELECHIADÆ.

PALTODORA SAGITTIFERA, n.sp.

♀. 10 mm. Head, thorax and palpi whitish-fuscous, apex of second joint with a broad band of fuscous beneath. Antennæ fuscous. Abdomen whitish-fuscous, whiter on sides, 2 basal segments ochreous, anal tuft whitish. Anterior legs fuscous, tibiæ and tarsi obscurely ringed with whitish; middle and posterior pair ochreous. Forewings elongate, narrow, apex pointed; dull ochreous; a fine white costal streak, from base to $\frac{2}{3}$; a fine whitish longitudinal streak from base along fold, curved up along end of cell, and continued obscurely to apex; rest of wing marked with fine blackish longitudinal lines, giving the appearance of alternate ochreous and blackish lines: cilia fuscous, with some

Forewings elongate, narrow, pointed; ochreous, with a few obscure fuscous scales; costal edge narrowly dark fuscous to middle; a fine black line along fold, ending on inner margin at 2/3; an obscure fuscous line along inner margin, from near base to 1/3; some black scales at apex: cilia ochreous, mixed with some black scales. Hindwings with apex strongly produced, termen emarginate; pale fuscous; cilia greyish-ochreous.

Broken Hill, N.S.W.; one specimen, in October.

Gelechia heliochares, n.sp.

♂. 14 mm. Head, thorax, and palpi ferruginous-fuscous, thorax with a large whitish posterior spot, palpi whitish internally, terminal joint whitish, with an ochreous subapical ring. Antennæ dark fuscous, annulated with white, dentate, shortly ciliated. Abdomen dark fuscous. Legs dark fuscous, tibiæ and tarsi irregularly ringed with whitish. Forewings elongate, moderate, narrow, costa hardly arched, nearly straight, apex somewhat pointed, hindmargin very oblique; reddish-fuscous; a narrow oblique ochreous fascia, from 1/5 costa to 1/5 inner margin, not quite reaching inner margin, anterior edge well defined, posterior edge suffused, followed by some ochreous scales in disc, on which is placed a minute black dot; a narrow inwardly oblique ochreous fascia, from 3/4 costa to 3/4 inner margin, somewhat inflated on middle, and there containing a minute black dot; a small ochreous spot on costa before apex: cilia ferruginous-fuscous, around anal angle ochreous. Hindwings with apex acute, hindmargin oblique, gently sinuate beneath apex; fuscous; cilia ochreous.

Parkside and Semaphore, S.A.; five specimens, in December.

Gelechia heliopa, n.sp.

♂♀. 8-11 mm. Head, thorax, palpi and antennæ ochreous, fuscous-tinged, second joint of palpi grooved. Antennæ faintly annulated with fuscous. Abdomen greyish-ochreous, middle segments orange-yellow, segmental margins silvery-white. Legs whitish, tibiæ and tarsi fuscous, tarsi with whitish rings. Forewings elongate, moderate, costa gently arched, apex round-pointed; ochreous, slightly fuscous-tinged, with darker irregular spots throughout, obscure and sometimes hardly traceable, not forming definite markings: cilia ochreous. Hindwings with apex pointed, hindmargin sinuate beneath apex; light fuscous: cilia light fuscous, with an ochreous basal line.

Broken Hill, N.S.W.; four specimens, in October.

Ypsolophus chloranthes, n.sp.

♂♀. 8 mm. Head, thorax, antennæ and palpi pale ochreous, terminal $\frac{2}{3}$ of antennæ fuscous, second joint of palpi beneath somewhat fuscous. Abdomen fuscous. Legs ochreous-whitish, anterior pair fuscous-tinged. Forewings elongate, moderate, costa nearly straight, apex somewhat pointed, hindmargin oblique: pale ochreous, along costal edge somewhat whitish; extreme base of costa obscurely blackish; a fine black dot in middle, above anal angle: cilia pale ochreous. Hindwings with apex pointed, termen sinuate; light fuscous; cilia light ochreous.

Broken Hill, N.S.W.; five specimens, in October and November.

Ypsolophus macrosemus, n.sp.

♂♀. 8-10 mm. Head and palpi whitish, second joint of

Broken Hill, N.S.W.; ten specimens, at light, in October.

Allied to *tetrachrous* and its allies, but distinguished from those species by the well defined elongate black mark along cell.

ELACHISTIDÆ.

BATRACHEDRA CRYPSINEURA, n.sp.

♂. 9 mm. Head, thorax, palpi and antennæ white, antennæ obscurely annulated with fuscous, second joint of palpi infuscated at apex. Legs whitish, anterior pair infuscated. Abdomen dull fuscous, strongly mixed with whitish. Forewings elongate, narrow, pointed; dull ochreous-whitish, mixed with light fuscous, veins more or less obscurely outlined with whitish; a fine white line along costa, from base to beyond middle, partially edged below with a fine black line; a second similar white line in middle of disc; a third along fold; second and third lines dusted with a few minute blackish dots: cilia grey-whitish. Hindwings linear-lanceolate; whitish; cilia 5, whitish.

Broken Hill, N.S.W.; three specimens, in October.

PYRODERCES THERMOPHILA, n sp.

♂♀. 7-9 mm. Head, antennæ and palpi ochreous-whitish, antennæ somewhat annulated with fuscous, terminal joint of palpi fuscous. Thorax fuscous, patagia shining whitish-ochreous. Abdomen and legs greyish-ochreous, somewhat shining. Forewings lanceolate; dull ochreous-whitish; four irregular, moderately straight-edged dark fuscous transverse fasciæ; first basal, outer edge somewhat angulated outward, from $\frac{1}{6}$ costa to $\frac{1}{6}$ inner margin; second oblique, broken in middle, from before middle of costa to about middle of inner margin; third broad, from $\frac{3}{4}$ of costa to before anal angle; fourth apical: cilia greyish, with a blackish spot at apex. Hindwings linear-lanceolate; light fuscous; cilia light fuscous, becoming ochreous at base.

Broken Hill, N.S.W.; six specimens, at light, in March.

SYNTOMACTIS OXYPTERA, n.sp.

♂. 8 mm. Head, palpi and thorax ashy-grey-whitish, terminal joint of palpi with two white rings, below and above middle. Legs whitish. Abdomen dark fuscous, with greyish segmental rings. Forewings elongate-lanceolate; ashy-grey-whitish; markings ill-defined; a fuscous spot on inner margin before middle, edged anteriorly by a clear streak of whitish; a black spot above anal angle, and two or three along costa towards apex, where the groundcolour becomes fuscous-tinged: cilia greyish, with a fuscous tooth at apex. Hindwings lanceolate, narrow; greyish; cilia 3, greyish, ochreous-tinged at base.

♀. 8-10 mm. Head, thorax, antennæ, palpi, abdomen and legs dark fuscous, palpi with two whitish rings on terminal joint, above and below middle, legs mixed with whitish, abdomen with a patch of fine metallic scales in middle, beneath whitish. Forewings formed as in ♂; dark fuscous; all markings lost, excepting the oblique white streak preceding the fuscous spot on inner margin: cilia fuscous. Hindwings as in ♂.

Broken Hill, N.S.W.; nine specimens, in October.

SYNTOMACTIS XENONYMPHA, n.sp.

Broken Hill, N.S.W.; not uncommon, during April and October.

The curious hindwing of the ♂ is a special characteristic.

Syntomactis chionomera, n.sp.

♂♀. 8-10 mm. Head, palpi, antennæ, thorax, legs and abdomen dull white, palpi beneath light ochreous-fuscous, thorax fuscous in middle, abdomen beneath white, in ♀ the abdomen has a metallic patch of scales in middle. Forewings narrow, elongate-lanceolate, apex pointed: dull whitish, in ♀ faintly ochreous in disc; a few scattered black scales below middle, not forming definite markings; the underside of forewings of ♀ has a large distinct black spot, just before apex: cilia ochreous-whitish. Hindwings lanceolate-linear; whitish; cilia 3½, whitish.

Broken Hill, N.S W.; several specimens, in October.

Although an obscurely coloured species, it is easily recognised, especially the ♀, by the curious black spot on underside of forewings.

TINEIDÆ.

Zelleria cremnospila, n.sp.

♂. 10 mm. Head, thorax, antennæ, palpi, legs and abdomen dull whitish, anterior legs fuscous. Forewings elongate, narrow, costa gently arched, apex pointed; pale slaty-grey-whitish, with some fine scattered black scales, not forming definite markings, but more developed above inner margin, beneath costa, and around apex; a larger well defined black spot above anal angle: cilia silvery-grey, with black scales at base. Hindwings elongate-lanceolate; greyish; cilia whitish.

Port Victor, S.A.; three specimens, in November.

Zelleria hemixipha, n.sp.

♀. 16 mm. Head ashy-grey-whitish. Thorax fuscous, patagia whitish. Palpi white, second joint fuscous on sides, terminal joint infuscated. Antennæ white, annulated with fuscous, basal joint white, blackish above. Abdomen greyish-ochreous. Legs fuscous-whitish, anterior pair darker. Forewings elongate, narrow, costa moderately arched, apex acute, somewhat produced, hindmargin and inner margin continuous, nearly straight; white, suffusedly irrorated with ashy-grey, so as to appear ashy-grey-whitish, except a streak along costa, from $\frac{1}{5}$ to $\frac{3}{4}$, posteriorly suffused into ground-colour; a suffused dark fuscous line along upper margin of cell, more pronounced in middle; a similar line along fold; an obscure fuscous dot at end of cell, and a larger spot on fold at $\frac{1}{3}$ from base: cilia greyish-fuscous, at apex mixed with ashy-grey-whitish; and an obscure whitish streak. Hindwings lanceolate; grey; cilia nearly $2\frac{1}{2}$, ochreous-grey.

Highbury, S.A.; two specimens, in November.

ZELLERIA LEUCOMORPHA, n.sp.

♀. 10 mm. Head, thorax, palpi, antennæ and legs white, palpi slightly infuscated. Abdomen pale whitish-ochreous. Forewings elongate, narrow, costa gently arched, apex acute, hindmargin and inner margin continuous, nearly straight; whitish, faintly irrorated with light fuscous throughout, except on margins: cilia whitish-ochreous, mixed with fine black points around apical portion. Hindwings narrow, lanceolate; whitish-ochreous; cilia

black spots along its lower extremity; a small blackish mark on costa at extremity of streak, and another, less distinct, at $\frac{5}{8}$, groundcolour between, obscurely ochreous; an obscure ochreous tooth at apex; an obscure row of elongate blackish marks along hindmargin, sometimes continued along veins: cilia ochreous-fuscous, narrowly chequered with black. Hindwings greyish; cilia greyish-ochreous.

Parkside and Exeter, S.A.; three specimens, in November and December.

ON *DIDYMORCHIS*, A RHABDOCOELE TURBELLARIAN INHABITING THE BRANCHIAL CAVITIES OF NEW ZEALAND CRAYFISHES.

By Professor William A. Haswell, M.A., D.Sc., F.R.S.

(Plates xx.-xxi.)

The Rhabdocoele Turbellarian with which this paper deals was found by me in the branchial chambers of *Paranephrops neo-zelanicus* from streams in the Province of Otago, New Zealand. It is, so far as I have had the opportunity of observing, practically an invariable companion of the Crayfish, having been found, though never occurring in large numbers, in all the specimens examined; and it appears to pass its entire existence in the interior of the branchial chambers of its host. My attention was first directed to it owing to some resemblances to the Temnocephaleæ, for allies of which I was then searching; and some of these resemblances are of sufficient importance to justify the conclusion that we have here probably the nearest known Rhabdocoele relative

A remarkable feature is that cilia are developed only in a portion of the ventral surface, and are entirely absent round the margin and on the dorsal surface. Along the anterior and posterior margins open the ducts of numerous rhabdite-forming glands the viscid secretion of which, with the contained rhabdites, passes out from them, sometimes in considerable quantity.

Locomotion is effected sometimes by the action of the cilia of the ventral surface producing a slow gliding movement; at other times the animal "loops" along actively, the adhesive anterior and posterior regions being alternately fixed to and liberated from the surface of the substratum after the fashion of suckers.

The mouth has the appearance of a small transverse slit situated on the ventral surface nearly on a level with the eyes. The common genital aperture is a much smaller slit situated in the middle of the ventral surface about a fourth of the total length from the posterior margin.

Alimentary System.

The pharynx (*ph.*) is of the type termed by Von Graff "pharynx doliiformis." Its length is about a sixth of that of the entire body; its breadth somewhat less. In the substance of its wall between the radial fibres are numerous large unicellular glands, the ducts of which are arranged regularly and open into the internal cavity close to the anterior margin of the pharynx. The intestine (*int.*) has a thick wall composed of large vacuolated cells full of granules which are collected into spherical masses, some of which are of relatively large size.

Nervous System.

The nervous system does not present any very noteworthy features. The brain is a transverse band of fibrous material situated on the dorsal side immediately in front of the anterior margin of the pharynx, and of groups of nerve-cells connected with this in front and at the sides. From the brain a single pair of longitudinal nerve cords passes backwards close to the lateral margins of the body.

The eyes are of very simple structure and resemble those of Rhabdocoeles in general.* Each consists of a cup of pigment, the granules of which are few, large and of regularly rounded form, and an enclosed structureless body not divided into cells

Excretory System.

The excretory system opens on the exterior on each side on the ventral surface near the lateral margin almost opposite the posterior extremity of the pharynx. The opening (*exc.*) leads into a very narrow canal which enters a rounded mass of granular material. In this the canal winds about. Eventually two main canals are given off, one (*a.v.*) passing forwards, the other (*p.v.*) backwards. The former soon bifurcates. Of its two branches the inner soon also divides into an outer and an inner branch. The outer branch of the main canal runs forwards to the region in front of the eyes, where it divides into two branches, of which one is continued forwards towards the anterior margin, where it runs inwards and perhaps anastomoses with the corresponding vessel of the opposite side, while the other runs almost directly inwards and unites with the outer branch of the inner division of the main anterior canal to form a short single transverse vessel which passes into the corresponding vessel of the opposite side. The posterior main canal soon divides into two vessels

d the middle of its length. The right vas deferens across the middle line to join the left. The common rmed, a wide thin-walled tube, winds in a close spiral e extremity to the base of the penis, where it opens vesicula seminalis, from which the ejaculatory duct gh the penis to open into a median chamber, the a. Into the vesicula seminalis open also two sets of of the granule glands, the granules from which sses in the interior of the base of the penis. The ., p., and Pl. xxi., figs. 1 and 3) is a chitinous body tle to the left of the middle line, with its long axis quely, the proximal end anterior and external, the ior and internal. It consists of a tubular or rather asal portion, and a distal portion composed of a ines. The basal portion is very wide at the proximal atest width being nearly equal to the length. The n or twelve in number, some straight and dagger-s curved and tapering.

organs consist of a single ovary (Pl. xx. and Pl. xxi., ov.), oviduct and uterus (ootype), vitelline glands, ts, and a receptaculum vitelli. The ovary is a body situated to the right of the middle line. The en off from the narrower left end of the ovary; it is liately joined by the wide short duct of the recepta-. The common duct thus formed varies greatly in erent specimens. In some it presents a sharp bend in the lumen of this part of its course a mass of

where the short wide duct runs backwards from it, open the two vitelline ducts. The vitelline glands are a pair of narrow, irregu-larly lobed bodies which extend on either side between the intestine and the lateral border. There is no bursa copulatrix.

The eggs were found attached separately to the epipodites the Crayfish. Each is enclosed in a chitinoid shell, having slender pencil-like process projecting from it. Before it is hatched the young animal has completely assumed the characteristics the adult, except that the reproductive organs are not developed.

The Rhabdocoele above described apparently approaches the whole nearer to the *Vorticida* than to any of the other groups. The form of the pharynx is essentially the same, the general arrangement of the reproductive apparatus corresp fairly closely. Perhaps also there may be some resemblance regard to the excretory system, since in the genus *Vortex* vessels seem to open on the exterior in the neighbourhood of pharynx.

In the reproductive system the chief difference, in addition the posterior position of the testes and the peculiar form of penis, seems to consist in the absence of a bursa copulatrix. integument of the *Vorticida* also appears to differ from that the new form, the epidermis in that family assuming the

owing appear to be the leading characteristic features w form, for which I propose the name *Didymorchis opis*:—Pharynx a "pharynx doliiformis." Excretory ening on the ventral surface by two apertures situated om the mouth and pharynx. A single reproductive A single compact ovary and two elongated vitelline No bursa copulatrix. A pair of compact testes situated in close contact with one another. A complex chiti consisting of a tubular basal part and a distal system

---

EXPLANATION OF PLATES XX-XXI.

Plate xx.

iew of the external features and internal organisation of *Didy* n the dorsal aspect, (× 250).
rior main vessel of excretory system. *ex.*, opening of excretory a., genital opening. *int.*, intestine. *oot.*, ootype [illegible] *p.*, penis. *ph.*, pharynx. *p.v.*, posterior main vessel of excretory r., vitelline receptacle. *t.t.*, testes. *vit.*, vitelline glands.

Plate xxi.

ntral view of the reproductive apparatus. *gr.d.*, granule duct. *r.d.*, right vas deferens. *te.*, *te.*, testes. Other letters as in Plate xx.
tline of female reproductive organs of another specimen dorsal aspect. ♀ opening of vagina. *sp.*, mass of sperms in oviduct. Other letters as in Plate xx.
rsal view of chitinous parts of penis.

The appearance which he figures is, in fact, presented of my preparations of entire specimens of Australi and it is only with difficulty that it can be clearly der in these that the true interpretation is as I have state

Monticelli† states (*l.c.* p. 87) that the sac which is by other authors termed receptaculum seminis and whi called receptaculum vitelli, is in *T. brevicornis* always with yolk but with spermatozoa. In the Australian Zealand species, on the other hand, when any sperms to be detected in the interior of the receptacle, they present only in small number, the bulk of the content usually, the entire contents, consisting of the finely vitelline matter. This sac, in fact, acts as the receptacl the vitelline matter collects in anticipation of the discl mature ovum from the ovary; when this discharge takes vitelline matter is found to have become transferred to in which the completed egg becomes formed—the rec being now empty or nearly so. It is the anterior p oviduct that performs the function of retaining the sper proper designation of the receptacle is thus, in the Au species at least, not receptaculum seminis, but rece vitelli. ‡

* Compare this with fig. 1 of Plate xv. in my "Monograp

EXPLANATION OF PLATE XXII.

e., ejaculatory sac. *l.v.d.*, left vas deferens. *r.v.d.*, right vas deferens. vesicula seminalis.

1.—*Temnocephala tasmanica*: outline of tentacles; from a living specimen.

2.—*T. tasmanica*: penis and neighbouring parts; from dorsal side.

3.—*T. aurantiaca*; penis and neighbouring parts; ventral aspect.

4.—*T. cæca*: penis and neighbouring parts; from the dorsal side.

5.—*T. fasciata*: portion of a horizontal section in the plane of the duct (*dt.*) connecting the anterior (*a.t.*) with the posterior testis (*p.t.*).

CONTRIBUTION TO THE BACTERIAL FL[illegible] THE SYDNEY WATER SUPPLY, I.

By R. Greig Smith, M.Sc., Macleay Bacteriolo[illegible] Society.

As far as I am aware, the only paper that has bee[illegible] dealing with the bacterial flora of the Sydney water [illegible] contributed by Katz to these Proceedings in 1886[illegible] time bacteriology was beginning to perfect its meth[illegible] nique, and it is to be regretted that the five orga[illegible] paper are not sufficiently described to enable them to [illegible] with any degree of certainty. It therefore seemed t[illegible] that a paper or a series of papers upon the bacterial [illegible] Sydney water might be of interest to the Society.

The bacteria which will be hereafter described we[illegible] from the tap-water in the Society's Bacteriological [illegible] which is supplied directly from one of the city r[illegible] Centennial Park, into which it is pumped from [illegible] pumping station and reservoir in Crown Street. T[illegible] originally taken from the Nepean, Cordeaux and Cat[illegible] the catchment area of which covers 345 square miles. [illegible] rivers it is conducted by a series of tunnels and op[illegible] Prospect reservoir, thence by open canal and pipes to

the number of bacteria that are consumed in water is nportance than the kinds of bacteria, and especially ls happen to be of an injurious nature.

t wherein there is reason for the numerical test as he purity or otherwise of water is that the numbers ortional to the food material, whether it be saline or l in less degree to the temperature. It is obvious ganic matter will represent inversely the degree of ation to which the water has been subjected. Surface h have undergone little or no filtration will contain c matter than spring water which has had the bulk iic matter and micro-organisms removed by natural rough successive layers of soil and sand. However · may be naturally filtered, it is never sterile, and the iherent in it will multiply to an extent proportionate supply and to the temperature. Thus the number of n the water may be taken to represent the organic therefore indirectly the degree of natural filtration. tural filtration that the bacteria which inhabit the y noxious or otherwise, are removed, and in the cases, especially where the gathering grounds are in ocalities, natural filtration is not considered to be id so water is generally filtered through sand-filters supplied to towns.

s formulated a scheme for obtaining an idea respect-ity of water from the number of bacteria per cubic Pure waters vary from zero to 1,000, mediocre from)00, and impure over this number. But he further at of more importance than the absolute number of bacteria are the number of species per cubic centimetre. In a pure water he considered that this should not exceed 10.

The media generally employed for growing the colonies are either meat-peptone-gelatine or meat-peptone-agar. In Sydney water many organisms rapidly liquefy the former and utterly prevent the appearance of the more slowly growing colonies. There are others that form amœboid colonies on the surface of agar, and rapidly spreading over the plate, obliterate the fixed colonies. In the beginning of my investigation it became evident that these media were unsuited for separating the bacteria; still in working with the agar media portions of the plates could sometimes be obtained in which the amœboid activity of the colonies had been restrained. Agar also throws up the colour of the surface colonies, and this is of some assistance in picking out species. The agar was improved to a certain extent by omitting, in the preparation of the medium, the peptone and the common salt and adding 2% dextrin or gum acacia which tends to restrain the amœboid growth of the colonies.

Hesse and Niedner employ an agar medium containing agar 1·25 grm. "Nährstoff Heyden" 0·75 grm., and water 98 c.c. Abba advises a medium containing Liebig's meat extract 6 grms., gelatine 150 grms. and distilled water 1000 c.c. This is after solution made neutral to phenolphthalein and rendered alkaline

In separating the bacteria both Abba's gelatine and dextrin-meat agar were employed. The latter medium is prepared by dissolving 20 grms. of agar in 1000 c.c. of ordinary meat extract in the autoclave. After clarification with white of egg, 10 c.c. of the mixture are pipetted into warm water and neutralised to phenolphthalein with tenth-normal sodium hydrate. The calculated quantity of normal sodium hydrate is added to the bulk, together with 0·5 grm sodium carbonate; 20 grms. of dextrin or gum acacia, dissolved in a small quantity of water is added, and the medium boiled, filtered, placed in test tubes and sterilised. The water was allowed to flow from the tap for half-an-hour upon a sterilised watch-glass supported upon a tripod. From the watch-glass the water was taken up into a sterile graduated pipette and added to the previously melted and cooled gelatine (30° C.) or agar (42° C.); the tube was then shaken and the contents poured into Petri dishes, which, after setting, were inverted and incubated at the requisite temperature (15°, 18° or 22°). When the colonies had grown sufficiently, inoculations were made upon sloped agar and in gelatine (stab). When a diagnosis could not be made from these—and this was the case when the organisms were obtained for the first time—a series of gelatine plates were prepared. The colonies that developed on these plates were used both for the diagnosis of the organism and for obtaining a pure culture. Generally the cultures obtained from the primary (water) plates were impure.

Having obtained a pure culture of an organism, inoculations were made upon the ordinary media, which included peptone-meat-agar, peptone-meat-gelatine, peptone-meat-bouillon, glucose-peptone-meat-gelatine, lactose-peptone-meat-gelatine peptone-meat-bouillon with 0·5% nitrate of potassium, ordinary potato, and ordinary defatted milk. With the exception of the potato and the milk, all the media were neutralised with sodium hydrate to phenolphthalein as has been described. In observing the motility, a portion of a young agar culture was transferred to a drop of normal saline and examined. If the organisms appeared non-

appear as translucent white, raised spots, which when magnified are seen to have a wavy, erose margin and a transparent crumpled centre. Gelatine is not liquefied. In stab culture there is formed a terraced nail-head. The organism is small and oval, actively motile, and measures 0·3 : 0·6 μ. It is decolorised when treated by Gram's method. Gas is produced from glucose, and milk is not coagulated. Bouillon becomes turbid, and forms a flocculent precipitate. A slight indol reaction is obtained. On potato the growth is moist glistening, flat, spreading, and of a light stone colour, which deepens to a light drab.

This organism is decidedly small, while Wright describes it as medium-sized. The growth on potato shows a difference from *Bact. minutum.*

Bacterium No. 46, Conn.

This is a short rod-shaped bacterium with rounded ends, and measures 0·4 : 1-1·5 μ. It is actively motile, and is not stained by Gram's method. On agar it forms circular white colonies, and on gelatine translucent white colonies that sink into the medium. When the gelatine colonies are viewed under a moderate power, they are seen in the deep to be rounded or irregular, almost opaque and apparently flocose. The surface colonies are circular with fluid contents in which there are large brown floccules.

O

alkaline. Otherwise they appear to be the same organisms. The difference is perhaps too slight to justify this being called a new species.

BACTERIUM GASOFORMANS, Eisenberg.

Bubbles of gas are formed in the depth of the ordinary solid media. The cells are oval or sausage-shaped, and measure 0·5 : 1-2 μ; they are actively motile. The colonies, on gelatine plate, are circular and translucent white. Under a sixty-fold magnification the surface colonies appear rounded with a slightly erose margin; the centre is brownish and finely granular. The contents of the colony appear in motion. At a later period the contents become flocculent and the margin diffuse. In gelatine stab culture the liquefaction is crateriform, then stratiform and tubular. The stroke on agar is dirty-white, moist glistening and at the base spreading. Gas bubbles are formed in the agar. On potato there is a scanty growth which is almost indistinguishable from the medium. Milk is coagulated, the reaction being acid. Nitrates are reduced to nitrites.

BACTERIUM AURESCENS, Ravenal.

This motile organism is oval, and measures 0·3-0·4 : 1 μ. It forms threads in old cultures. In agar stroke culture the first growth is a thin yellowish-white line, which as it broadens becomes primrose-coloured, and ultimately of an old-gold shade. In gelatine stab culture there is a filiform growth; the gelatine at the upper part of the stab becomes slightly consumed, and eventually the growth around the consumed portion becomes tuberculate and yellowish-white. The colonies on gelatine are yellowish, and slowly liquefy the gelatine. Under a sixty-fold magnification they are seen to have a lacerate, transparent margin, and a pale yellowish-brown, marbled or convoluted centre. The older surface colonies have the central convolutions more marked, an annulus with coarse flakes floating in the liquefied medium, and a ciliate margin. The deep colonies are brown and rounded.

On potato the growth is orange-yellow, flat and glistening. No gas is developed in the presence of glucose, no indol is produced in bouillon, and milk is not coagulated.

The colonies on gelatine appear to differ from those described by Ravenal, but as the other characters agree this may be a variety.

BACTERIUM PULLULANS, Wright.

This is a short rod with rounded ends, and measures 0·3:0·6-1 μ; it is actively motile. On agar the colony is deep yellow and circular, with a smooth edge; the stroke is deep yellow or old-gold in colour, convex, moist glistening and restricted. The gelatine stab is at first filiform with a spreading deep yellow nail-head. The gelatine becomes slowly liquefied; the liquefied area is funicular and turbid; a deep yellow precipitate is formed. In gelatine plate culture the surface colonies have a deep yellow centre and a pale yellow margin. When magnified sixty-fold they are seen to have an irregular zonate centre and lobulated margin, from the under-surface of which rounded buds are given off. The small colonies, which have budded off, lie free around the lobular margin, and are either circular or irregular, according to their size, and brown in colour. Milk is not coagulated, and a trace of indol is produced in bouillon culture.

oon becomes shallow crateriform. There is practically no growth in the depth of the medium, and a primrose-coloured film lies on the liquefied gelatine. There is no gas developed from glucose, and milk is coagulated, the reaction being neutral. On potato the growth is canary-coloured, but changes to a pale yellow; it is raised and irregularly spread. Nitrates are not reduced. In bouillon a turbidity is formed; there is a precipitate, but no film. A strong indol reaction was obtained.

Bacterium arborescens.

Bacillus arborescens, Frankland.

On gelatine the colonies are pale yellow and circular, with a slight mycelioid appearance. The medium is liquefied round the colony. When magnified the deep colonies are seen to have an irregular centre, from which root-like fibres extend; these are closely packed together and yellowish. The surface colonies have an indefinite centre, from which loosely twisted and sharply bent comparatively wide strands apparently anastomose. The stab in gelatine is filiform, with a smooth, moist glistening, deep yellow nail-head. The gelatine is slowly liquefied in a crateriform manner, and just below the liquefied medium the filiform stab becomes expanded and diffuse. The colonies on agar are circular, raised, moist glistening and of a translucent pale buff colour. When magnified they are seen to have a granular centre, with a transparent irregular margin, beset with short irregular processes. The deep colonies are rounded, oval or lenticular and rough. The agar stroke is moist glistening, raised and spreading. The colour is a deep yellow which changes, especially in the centre of the growth, to old-gold. No gas is formed in glucose-gelatine, and nitrate is not reduced. Bouillon becomes turbid, and there is a filamentous precipitate, but no film. A slight indol reaction was obtained. Milk is not coagulated. On potato the growth is flat, spreading, glistening and deep yellow in colour. The organisms are rods with rounded ends, and may be long or short; the average size is 0·3 : 2 μ.

Bacterium miniaceum.

Bacillus miniaceus, Zimmermann.

On agar slope a brilliant vermilion, raised, spreading, moist glistening stroke is formed, from the bottom of which amœboid processes spread out and gradually cover the entire lower surface. The organism is a cocco-bacterium measuring 0·7 : 1 μ. In gelatine plate culture the colony quickly liquefies the gelatine, forming a crateriform, pink area. In gelatine stab the medium is liquefied in a stratiform manner, the fluid being very turbid from floating pink granules. The red pigment is not bleached by zinc and hydrochloric acid.

Bacterium rubefaciens.

Bacillus rubefaciens, Zimmermann.

On gelatine plate at 15° C., circular, raised, moist glistening colonies of a pale coral colour are formed. The medium is not liquefied. Under a sixty-fold magnification the surface colonies are seen to be rounded, reddish-brown and with a smooth edge; the contents are finely granular. The deep colonies are irregular and opaque. The bacteria are motile rods with rounded ends, and measure 0·45 : 1-1·5 μ; they occur singly and in twos. They are decolorised by Gram's method of staining. The agar stroke

temperature. Milk is not coagulated, nitrate is not reduced, but a trace of indol is formed in bouillon.

Zimmermann describes the organism as forming a blue-grey stroke on agar.

Bacterium salmoneum, Dyar.

The organism grows scantily in artificial media. On gelatine plate the colonies are small, hemispherical, moist glistening, and of a pale scarlet colour; when magnified the structure is seen to be finely granular and the margin smooth. The deep colonies are rounded and opaque. The stab in gelatine becomes filiform, and bears a small, rounded and raised vermilion nail-head. On potato there is practically no growth. Bouillon becomes turbid; there is a precipitate and a light reddish-coloured film. No indol is formed. On agar the stroke is narrow and restricted, at first coral-pink, then becoming light orange, and ultimately reddish-orange. The organisms are non-motile thin rods with rounded or slightly pointed ends; they stain irregularly, and measure generally 0·4 : 2 μ. Milk is not coagulated, and the reaction is unchanged. Nitrates are reduced to nitrites.

Bacterium janthinum, Zopf.

When grown upon agar the organism forms a deep violet, moist glistening, irregularly raised layer with a white margin and smooth edge; the consistency is gelatinous. The colonies in gelatine appear as very pale violet indefinite areas, the parent colony having sent out into the gelatine processes which have formed sub-colonies. Both the indefinite parent colony and the older sub-colonies are beset with processes similar to those of *Bact Zophii*. The younger colonies have a moruloid centre, and are surrounded by circular or irregular sub-colonies similar in appearance to the colonies of *Bact. pullulans*. The stab in gelatine scarcely grows in the deep; the film is violet-coloured and spreads over the surface, sending down into the medium hair-like processes which appear like a hanging veil. The gelatine begins to liquefy in two weeks, the liquefaction being shallow-stratiform. The organism is a motile rod with rounded or pointed ends, and measures 0·4-0·6 : 1-1·5 μ; involution forms in the shape of clubbed threads are soon found in the cultures. On potato there is formed a deep violet, irregularly spreading, flat, glistening layer. Bouillon becomes turbid with a flocculent precipitate and a strong violet film. The fluid becomes colourless and then violet. The indol reaction was obtained in the fluid after the elimination of the suspended bacteria by means of calcium chloride and ammonia. A control test without nitrite was made at the same time. A strong blue film formed on milk which was not coagulated; the reaction was alkaline.

Bacterium janthinum ii., n.subsp.

The stroke on agar is violet, moist glistening, and slightly raised, the margin is beset with processes which are at first white but soon become like the centre of the stroke; the consistency when the culture is touched with the needle is found to be thin. The colonies in gelatine plate are circular, crateriform, pale violet and zonate; when magnified the contents are seen to consist of large granules; the edge is diffuse. In gelatine stab the liquefaction of the medium is rapid. The liquefied area is funicular, there is a violet deposit and a thin film of the same colour;

portion of the medium. The latter becomes strongly greenish fluorescent. The organisms are oval and measure 0·4 : 1-1·5 μ. They are actively motile. The stab in gelatine is filiform with a flat, irregular, moist glistening, yellowish-white nail-head. The upper part of the stab becomes tuberculate. No fluorescence of the gelatine was observed. Milk was not coagulated and the reaction remained neutral.

Bacterium fluorescens mutabile, Wright.

i.—This is a rod-shaped bacterium with rounded ends measuring 0·5 : 0·8-1 μ and is not coloured by Gram's method of staining. It is actively motile. The stroke on agar is greenish-white and moist glistening; the growth is luxuriant and quickly widens to the sides of the glass at the bottom of the stroke. The agar becomes strongly fluorescent. The gelatine in stab culture becomes liquefied in a funicular manner: the liquefied portion is turbid with floating floccules, and there is a yellowish-white precipitate and film. There is no fluorescence of the medium. Milk is coagulated and has an alkaline reaction. On potato, a moist glistening, drab-coloured expansion covers the surface of the medium.

ii.—This organism is a trifle larger than the above and measures 0·6 : 1-1·5 μ. The stroke on agar is drier and whiter than i., that is, it is less translucent. It is also less luxuriant in its growth and not so strongly fluorescent. Otherwise the characteristics are similar.

The two organisms are closely allied to *Bact. pyocyaneum*, differing mainly in fluorescence. They are best described by Wright's designation, and one of them is probably identical with his organism.

Bacillus subtilis, Ehrenberg.

The hay bacillus forms on agar slope a dull flat membranous layer with an irregular lacerate and ciliate margin. The membrane quickly spreads over the greater part of the agar surface.

When grown at 22° C. the lower part of the growth may or may not become slightly wrinkled or folded. On potato there is formed a dry, flat, white, mealy, spreading layer which soon becomes slightly speckled. The organism is an actively motile rod with rounded ends and measures 0·8 : 2·3 μ; it forms oval, central spores. The colonies on gelatine and agar are flocose. In gelatine stab culture the growth is very rapid; the medium is liquefied in a funicular or tubular manner, but soon becomes stratiform. The liquefied gelatine, which is at first strongly turbid, becomes clearer, while a strong white film forms on the surface; the sediment is white and flocculent.

Bacillus mycoides, Flügge.

As its name implies, this organism in gelatine and agar culture forms colonies that appear mycelioid. The mycelial processes grow over the surface and penetrate into the medium. It is a rod-shaped bacillus with rounded ends, and occurs singly and in threads; the individuals measure 1 : 3 μ; oval spores are formed in the middle of the rod. According to Flügge it is non-motile, and to Zimmermann it is motile. The individuals separated from the Sydney water supply are motile. The track of the needle in gelatine stab culture becomes filiform, then villous, and finally arborescent. The gelatine is liquefied slowly, and becomes

rod with rounded ends and occurs singly and in chains. The individuals measure 0·6 : 2-3 μ and form oval, central spores. It is stained by Gram's method, and the single cells which have been grown in bouillon at 37° C. are motile. On agar at 22° there is formed a thin, white, spreading, syrupy layer. At 37° a brownish-white, flat layer rapidly spreads over the agar-surface, which soon appears as if it had been dusted with oatmeal. The upper portion of the growth is furrowed laterally and downwards. The condensed water has a strong crumpled skin. It has little similarity to *Bac. mycoides*. On potato a dry, crumpled, thin layer spreads over the surface. The colour is at first white with a fawn tinge, later it becomes reddish-brown; the potato becomes brownish. Bouillon is rendered turbid; a precipitate and a strong crumpled whitish film are formed. The indol reaction was not obtained. Milk becomes alkaline in reaction and is coagulated when warmed. Nitrates are slightly reduced to nitrites.

Frankland describes a *Bac. ramosus* which agrees with this organism fairly well except in size; he quotes it as measuring 1·7 : 7 μ. Flügge describes his bacillus as moderately large. *Bac. implexus*, Zimm., liquefies gelatine much more quickly than this organism and it measures 1·15 : 2·5 μ.

Bacillus gracilis, Zimmermann.

This is an aerobic bacillus with a terminal spore. The young rods are actively motile. In gelatine the colonies appear as dry, white specks which do not liquefy the medium. Under a sixty-fold magnification the deep colonies appear circular and opaque The amœboid surface colonies appear very granular and grained. The gelatine stab is filiform with a small nail-head, which becomes moist glistening and amœboid, while the filiform track of the needle becomes tuberculate near the top. The stroke on agar is narrow, white, moist glistening and raised; the margin is ciliate. It does not appear to grow on potato. The bacilli are long rods with rounded ends and measure 0·3-0·4 : 3-8 μ.

Bacillus circulans, Jordan.

The colonies, in gelatine, are white and sink into the medium, forming a cylindrical pit or hole which widens out at the top when the colony has grown as far as it can downwards. The medium is consumed and not liquefied. With a sixty-fold magnification the dark granular contents of the colony are seen to be in active motion, apparently circulating round the walls of the pit. The deep colonies are at first circular, irregular or amœboid, but the latter soon become round and rapidly open to the surface. In gelatine the stab assumes the form of a hollow inverted cone. No gas is produced in glucose-gelatine. The organism is a rod with rounded ends and forms a terminal round spore which becomes 1 μ in diameter. The vegetative forms measure 0·5 : 2-3 μ and are actively motile. They retain the colour when treated by Gram's method. Both on agar plate and agar slope there is formed an amœboid growth which rapidly spreads over the entire surface. In milk the casein is slowly peptonised without coagulation. Bouillon becomes turbid and there is formed a white film and a flocculent and filamentous precipitate. No indol is formed and nitrate is not reduced. A drab-coloured growth spreads over the surface of potato.

Cladothrix dichotoma, Cohn.

method of staining. No spores were observed. It grows well at 37°.

Agar plate.—Circular, white, moist glistening, raised colonies are formed. When magnified they are seen to have a smooth edge; the internal structure is slightly marbled. The deep colonies are irregular and rough.

Agar stroke.—A luxuriant, white, raised layer with smooth iridescent edge grows along the stroke. The edge becomes lacerate.

Gelatine plate.—White, circular liquefied areas are produced. Under a sixty-fold magnification, the subsurface colonies are seen to be irregular, dark brown and apparently flocose, the surface colonies have a flocose centre and a clear margin.

Gelatine stab.—At first the stab is slightly crateriform but becomes funicular. The liquefied gelatine becomes turbid and a white precipitate forms. No film appears and the turbidity persists.

Glucose-gelatine.—No gas is produced.

Bouillon.—The medium becomes turbid, a slight film is formed as well as a pale white precipitate. Traces of indol are produced.

Milk.—The casein is coagulated and the reaction becomes acid.

Potato.—There is formed at first a pale ochre moist glistening, raised growth, which becomes yellowish-brown, irregular and greasy. The layer continues to rise from the surface of the potato, and when growth stops it is mesenteric, dull, and of a light drab colour. The potato is darkened.

Nitrate-bouillon.—The nitrate is reduced to nitrite.

The affinities of this organism are with *Bact. Fairmontensis*, Wright, *Bac. aquatilis communis*, and *Bact. No. 46*, Conn. From the first named it differs in the appearance of the gelatine colonies and the growth in gelatine stab culture as well as in the growth upon potato. *Bac. aquatilis communis* is considered by Kruse to be a *Bact. fluorescens liquefaciens* which has lost its power of producing fluorescence. It therefore differs from this organism in the reaction of the milk when coagulated. *Bac. aquatilis communis* forms a yellowish-brown or reddish layer on potato and is never mesenteric. The liquefied gelatine in stab culture is at first turbid but soon clears, and further it is a medium-sized rod, 0·6 : 1·2-2·5 μ, and very motile. *Bact. No. 46*, Conn, measures 0·4 : 0·8 μ, and forms a yellowish or brownish (presumably flat) layer on potato; its colour upon agar and gelatine is yellowish-white and it coagulates milk, the reaction being alkaline.

Bacterium aerofaciens, n.sp.

Shape, etc.—This is an oval bacterium measuring 0·4 : 0·8-1 μ. It is motile and is not stained by Gram's method. It grows well at 37° C. and at 22° C. No spores were observed.

Agar plate.—There are formed translucent white, wax-like, circular, raised colonies, which when magnified are seen to be homogeneous, circular and smooth-edged. The deep colonies are irregular and rough.

Agar stroke.—The growth is raised, moist glistening, grey-white, and spreads irregularly from the inoculating line.

Gelatine plate.—The colonies are at first translucent white and slightly irregular, later they become amœboid and iridescent. When magnified the surface colonies are seen to be yellowish-brown and finely granular with a lacerate-erose margin; the deep colonies are brown or opaque and rounded with a smooth edge.

Gelatine stab.—The line of inoculation becomes filiform, with

production of gas in ordinary nutrient gelatine marks it as being new. Its closest ally appears to be *Bact. sinuosum.*

Bacterium minutum, n.sp.

Shape, etc.—A small cocco-bacterium measuring 0·4 : 0·5-0·7 μ. It is actively motile and is not stained by Gram's method. No spores were observed. It grows well at 37°.

Agar plate.—A translucent white, amœboid colony quickly grows over the surface. The amœboid processes are narrow and radiate from a central point. The structure is homogeneous and the margin smooth.

Agar stroke.—A white, moist glistening, raised growth becomes amœboid at the base. It may produce gas bubbles in the medium.

Gelatine plate.—The colonies are white and of the *B. coli* type. When magnified the deep colonies are irregular or rounded and either dark brown or opaque. The surface colonies are transparent yellowish, grained or contoured, and with an erose, rounded margin.

Gelatine stab.—The stab becomes filiform and there is formed a flat, irregularly spreading nail-head. The surface growth becomes glistening, slightly depressed and contoured. Some specimens produce gas bubbles in the medium.

Glucose-gelatine.—There is a considerable development of gas.

Bouillon.—The medium becomes turbid and there is formed a filamentous precipitate but no film. There is no indól produced.

Nitrate-bouillon.—Nitrate is strongly reduced to nitrite.

Milk—The medium is not coagulated in the cold. The reaction is acid and coagulation occurs on warming.

Potato.—There is formed an irregular, canary-coloured growth, sharply raised from the medium, which becomes dark and bluish in colour.

The organism was frequently found in the water. It has its allies in the *B. coli* and *suipestifer* groups. The potato growth is characteristic.

BACTERIUM CROCEUM, n.sp.

Shape, etc.—An actively motile rod with rounded ends, measuring 0·5-0·6 : 2-3 μ; it occurs singly, in pairs, chains and threads. It is stained by Gram's method. The organism grows best at 22° C. No spores were observed. It grows well at 37°.

Agar plate.—The colonies are circular, raised, moist glistening and buff-coloured. When magnified they are seen to have a smooth, sinuous edge and a folded or grained internal structure. The deep colonies are irregular, moruloid and slightly ciliate.

Agar stroke.—There is formed a deep yellow, moist glistening, luxuriant, spreading and slightly raised layer. The edge is at first lacerate, but becomes smooth and the colour deepens to a light orange.

Gelatine plate.—The colonies are pale buff and rounded within a zone of softened gelatine. The surface colonies when magnified (60 times) are seen to have a brownish centre and a colourless, irregularly lobed margin, within which the colony appears crumpled. The deep colonies are zonate, the zones being coloured different shades of brown; the marginal zone appears greenish.

Gelatine stab.—The growth in the deep is very slight. The surface growth is deep yellow, depressed and restricted; the medium in the neighbourhood of the film is softened. The upper

Bacterium palaeformans, n.sp.

Shape, etc.—A rod-shaped bacterium with rounded ends, measures 0·5-0·6 : 1·5-2 μ and occurs singly, in chains and threads. It is actively motile and is stained by Gram's method. It does not form spores. There is little growth at 37°.

Agar plate.—The colonies are circular, pale buff, raised and moist glistening. When magnified the structure appears finely granular and the margin erose. The deep colonies are irregular, brownish and rough.

Agar stroke.—A narrow, primrose-coloured, moist glistening, raised stroke spreads out sharply at the base, when the condensed water has been reached, until the sides of the tube are touched, thus giving the culture the appearance of a spade. The colour slowly deepens to buff. The consistency is found to be thin when the culture is touched with a needle.

Gelatine plate.—The colonies are zonate, pale buff in colour, and have sunk into the gelatine, which is slightly liquefied. When magnified sixty-fold, the deep colonies are seen to be circular, brownish and annular; the surface colonies are either circular with a corrugated dark border or lacerate-erose with a coarsely granular centre.

Gelatine stab.—There is no growth in the depth of the medium. The gelatine is liquefied in a crateriform and afterwards in a stratiform manner. The scanty deposit is deep yellow and like a crumpled film.

Glucose-gelatine.—There is no gas produced.

Milk.—The medium is not coagulated.

Bouillon.—The medium becomes slowly turbid and there is formed a thin, pale yellow film and a scanty, filamentous precipitate. The indol reaction was not obtained.

Nitrate-bouillon.—Nitrate is not reduced.

Potato.—No growth was obtained either at 22° or 30° C.

This appears to be a motile form of *Bact. dormitator*, Wright, but perhaps its closest ally is *Bact. croceum*, from which it differs in colour, luxuriance of growth, and character of the gelatine

was obtained upon agar containing 0·5 % sodium carbonate.

Bacterium subflavum teres, n.subsp.

Shape, etc.—The organisms measure 0·6 : 1·5 μ. Shorter and longer forms occur. The cells are oval and are actively motile. They are not stained by Gram's method. No spores were observed. It grows well at 37°.

Agar plate.—The colonies are circular, raised, moist glistening and buff-coloured. When magnified, the structure appears granular and the edge smooth. The deep colonies are opaque, irregular and rough.

Agar slope.—The growth is moist glistening, buff-coloured, raised and viscous; it spreads out at the base and the colour becomes deep yellow.

Gelatine plate.—The *aerogenes*-like colonies are cream-coloured and circular, moist glistening and slightly raised. When magnified, the deep colonies are seen to be circular and zonate; the surface colonies are circular, finely granular and brownish-yellow in colour.

Gelatine stab.—A scanty filiform growth is crowned by a raised nail-head, which spreads regularly. The convex growth is light buff and moist glistening.

separates it from the *coli* type of *Bac. aureus* and *Bac. subflavus*, Zimmermann. Its nearest ally appears to be Zimmermann's organism, of which it is probably a variety.

BACTERIUM ARBORESCENS AMETHYSTINUM, n.subsp.

Shape, etc.—It is a very thin rod with rounded ends, and measures 0·2-0·3 : 1·5-2·3 μ. It is not motile and is not stained by Gram's method. Threads are formed in bouillon. No spores were observed. It grows best at 22° and 30°; there is little growth at 37°.

Agar plate.—The colonies are thin and diffused, of a blue white colour and translucent. When magnified sixty-fold, the centre is seen to be granular and the margin indefinite with watery transparent processes extending outwards. The deep colonies are rounded or oval, and slightly moruloid.

Agar slope.—A translucent, moist glistening, white growth, with a decided violet tinge. The margin is indefinite and sends off watery irregular processes which have a greenish colour. The growth is luxuriant and soon spreads to the sides of the tube. The condensed water becomes pinkish-cream coloured.

Gelatine plate.—The colonies appear like delicate moulds. When magnified the structure is seen to be mycelioid or root-like. There are prominent thick main trunks, and between these delicate radial threads fill up the spaces and give the colonies a circular shape.

Gelatine stab.—The growth is fusiform and a cloudiness radiates from the top of the canal into the medium; when magnified 10 times, delicate hairs are suggested. The cloudiness hangs downwards as a veil from the margin of the buff-coloured sunken nail-head. Liquefaction slowly takes place at first in a stratiform, then a funicular manner. There is formed an orange-coloured film and a precipitate which changes from white to orange.

Glucose-gelatine.—There is no gas produced.

Bouillon.—The medium becomes uniformly turbid and there is formed a whitish precipitate. The indol reaction is decided.

Potato.—There is formed a moist glistening, old-gold expansion.

Milk.—The casein is not coagulated and the reaction is unaltered.

The organism appears to be a subspecies of *Bac. arborescens*, Frankland, from which it differs in the appearance and colour on agar stroke. Compared with the organism which I have described under that name, the colonies on gelatine are white and the processes forming the mycelioid structure are more delicate and straight.

Bacillus stellatus, n.sp.

Shape, etc.—A stout rod of variable length and breadth, measures generally 0·8-1·0 : 2-3 μ; the ends are rounded. It is non-motile and is stained by Gram's method. It forms oval central spores quickly upon potato but slowly upon agar.

Agar plate.—The colonies are dull white, circular, raised and contoured. When magnified the structure appears folded and filamentous, the margin lobed and the edge smooth. The deep colonies are irregular, opaque and warty.

Agar stroke.—The growth is dry, glistening, luxuriant, greyish-white and longitudinally ribbed or terraced, the edge is straight and rough.

Gelatine plate.—The colonies at first are white and stellate, that [illegible] radiate from a cen[illegible]

Nitrate-bouillon.—Nitrates are not reduced.

Milk.—The casein and the reaction are unaltered.

Potato.—The growth is white, dry, greasy, irregularly spreading and slightly pitted. The colour changes to a light stone or light drab.

This appears to be an ally of *Bac. verticillatus*, Rav., from which it appears to differ in the mycelioid or stellate appearance of young colonies in gelatine, the rapidity with which it forms spores on potato (2 days), the absence of a film on bouillon, and of a pellicle on the gelatine colonies, as well as in not changing the reaction of milk and in not affecting the colour of the medium in agar culture.

Two yeasts were separated, one of them pink with round to oval cells, the other a non-sporulating torula which forms in glucose-yeast-water a clear medium, with a white sediment and strong, thick pitted film. It may be Torula ii., Hansen. With the exception of *Bact. ærofaciens* and *Bact. gasoformans*, no gas was produced in lactose-gelatine by the organisms which have been described. Since these would have produced gas in gelatine in the absence of lactose, it may be said that none of these organisms produce gas from lactose.

Many of the bacteria in the Sydney water have points wherein they differ from what are probably the same organisms found in other parts of the world. Whether these differences are sufficient to warrant the formation of a new species is a matter of opinion. I consider the characters in which they differ to be practically fixed, since the cultures were made from about the fourth crop, and when the organism was diagnosed any difference from the described type was verified by a culture taken from what would be the sixth crop. For all practical purposes the types may be taken as fixed, since any work with Sydney water, sewage, etc., would be made upon cultures which had been grown in a similar manner and for no longer a time. The bacteriologist generally does not keep an unknown organism growing for several months before making a diagnosis. The characters may alter slowly after a year's artificial culture, but that is only to be expected. Any rapid change in cultural character would have occurred before the fourth crop. This was noted in a previous paper to be the case with *Bac. piscicidus bipolaris*, the colonies of which in gelatine differed greatly in the second and fourth crops. From the fourth onwards, the type was fixed. Whether the differences between the types of other authors and these organisms are sufficient to warrant the formation of a new species, a new sub-species, or merely to note the difference, is a matter of opinion, and I leave it to other bacteriologists to accept my ruling or otherwise as they think fit. I believe the formation of sub-species is to be recommended, but for this the differences must not be vital, nor yet must they be insignificant so far as the diagnostic characters are concerned. In this a hard and fast line cannot be drawn, since a character which helps the diagnosis of one class of organism may be useless for another class. As an example, the liquefaction of gelatine may be cited. This action is of immense help to the bacteriologist and is used to distinguish classes of bacteria, yet it is too well known that the power may be gained or lost under laboratory conditions, and if this is the case a similar change may be expected to occur under natural conditions.

The organisms which have been described were isolated upon

OBSERVATIONS ON THE TERTIARY FLORA OF AUSTRALIA, WITH SPECIAL REFERENCE TO ETTINGSHAUSEN'S THEORY OF THE TERTIARY COSMOPOLITAN FLORA.

By Henry Deane, M.A., F.L.S., &c.

In my Presidential Addresses to the Linnean Society of New South Wales for 1896 and 1897 I mentioned that I had devoted some attention to Australian fossil leaves, and I expressed the opinion that Baron von Ettingshausen's naming of the specimens handed over to him from Dalton and Vegetable Creek was not reliable and that his theory of the Tertiary Cosmopolitan Flora had not only not been proved but that the evidence was rather in the direction that each fossil type possessed representatives in the existing Flora of Australia. I have been urged to pursue this matter, and to aid me in my investigations have been kindly furnished with a loan of specimens by Mr. R. Etheridge, Junr., Mr. W. S. Dun and Professor Baldwin Spencer.

Before undertaking to criticise the work of so eminent a palæontologist as Baron von Ettingshausen one ought to be very sure of the ground on which one stands, but I have become so thoroughly convinced of the general correctness of the conclusions which I have indicated above, and I have received so much sympathy and encouragement from the gentlemen above mentioned and from others to whom I have explained the facts as they appear to me, that I have no hesitation in submitting to this Society a paper on the above subject.*

* See also Prof. Baldwin Spencer in the Summary of the Horn Expedition, p. 160.

The general result of the investigation so far has been to throw the greatest distrust on the theory of the "Cosmopolitan Flora," which is, briefly stated, this, that the Tertiary Floras of different countries contained the same types and were closely allied and resembled one another much more than the Tertiary Flora of any particular country resembles its existing flora. In the recent period the floras of different regions have acquired their distinctive characters, due chiefly to climatic influences, and the old types which at one time were universally distributed have disappeared in some regions and not in others. Thus with regard to Eastern Australia, the Tertiary Flora is said to contain representatives of the existing floras of all other parts of the world now absent from Australia, and in like manner the Tertiary Flora of other parts of the world, as, for example, Europe, contains, it is asserted, representatives of genera and orders such as *Eucalyptus*, *Casuarina*, *Leptomeria*, *Exocarpus*, *Proteaceæ*, &c., which are now practically confined to the Australian region, or in the case of *Proteaceæ*, to Australia and South Africa.

The theory has worked a great deal of harm. It has been published in text books as an undisputed scientific truth, and the fact is lost sight of that it exists only in the opinion of a certain school, among whom are to be counted those who, not having time

It will be my endeavour to show that it is unnecessary to seek outside Australia for the types of our fossil flora. Since writing the Presidential Addresses referred to, a further study of the subject has led to the conclusion that changes in range and distribution of the Australian flora have taken place since the Tertiary, and that consequently to some extent the character of the vegetation has been modified. This is evident from a casual inspection of the fossils from Vegetable Creek, Gunning, Wingello, Bacchus Marsh, &c. The more luxurious flora now confined to strips and patches of the coast must have formerly spread over the table land, and there seems to have been, during the period when these fossils were embedded, a moister and perhaps warmer climate, one less liable to the severe droughts that we have to endure at the present time. A difference in climate such as this would adequately account for whatever variation may have taken place in the character of the vegetation. With advancing cold from the south and droughty conditions from the west the more luxuriant flora would retreat to the warmer and damper districts, and many species, genera and perhaps orders might be completely killed out, especially in the south.

The flora of Australia has been shown to be composed of three elements which are chiefly developed if not exactly predominant in the sub-regions named by Professor Spencer Torresian (northern and eastern), Bassian (south-eastern), and Eyrean (central and western). The Torresian flora, which Mr. Hedley calls Papuan, is more or less intimately connected through New Guinea with the Malayan and South-eastern Asiatic flora; it flourishes under conditions of warmth and moisture. It is looked upon by Professor Tate as exotic in character and as belonging properly to the Oriental Region.

The conditions requisite for the Bassian element of the flora seem to be moderation of temperature, if not actual cold, accompanied by damp. This element shows affinities with certain New Zealand and South American plants, and is treated by Professor

Tate as Andean in origin and introduced. The Eyrean sub-region is one over which dry conditions largely prevail, and those conditions may now help to protect the flora from the inroads of the coast vegetation. Professor Tate treats the flora of this region as the truly indigenous one. He divides the Continent into three, the Euronotian on the east and south, which on the coast partially amalgamates with the two introduced elements of Oriental and Andean character, the Eremian in the centre and the Autochthonian of the south-west.

Referring again to Professor Spencer's subdivisions, which more nearly illustrate the views which I have to put forward, it is clear that if the range of climatic conditions varied, so that drought and moisture, heat and cold were differently distributed, the divisions between the floral sub-regions would take up different positions. Such a change seems to have taken place in the past, and the climate having been then moister it is probable that there was formerly a much larger area over which the two eastern elements, and especially the Torresian, predominated and the sway of the western element was restricted. When the centre of Australia was largely lacustrine, and hot winds and drought were less of a feature in the climate, the Torresian element of the Flora might readily be supposed to extend as far as this and there would at that time have existed a luxuriant vegetation, one of the

Department of Mines, I find four important statements summarising the author's conclusions:—

(*a*) He says (p. 3)—"The Tertiary Flora of extra-tropical Australia is, as regards character, essentially distinct from the present living flora of Australia."

(*b*) "It has also much more similarity to the Tertiary Floras at present known than to the existing flora of Australia."

(*c*) "The characteristic plants of Australia are but feebly represented."

(*d*) Then (p. 4) he says—"The genera *Myrica*, *Betula*, *Alnus*, *Quercus*, *Fagus* and *Salix* are characteristic of the European and North American floras; the *Castanopsis*, *Cinnamomum*, *Tabernæmontana*, *Premna*, *Elæocarpus* and *Dalbergia* point to East India and China; *Magnolia* especially to the warmer parts of North America; *Bombax* to tropical America; *Knightia* and *Coprosma* to Oceania."

The first questions that arise after reading (*a*), (*b*) and (*c*) are, what does Ettingshausen take to be the character of the existing flora of Australia, and what are the characteristic plants of Australia that he refers to?

There can be little doubt that in speaking of the "character of the existing flora of Australia" he is altogether ignoring the very important coastal element, which is indeed at the present time spread over a comparatively small area, but is very rich in genera and species, and that his "characteristic plants of Australia" are certain well known types of *Proteaceæ*, such as *Banksia*, *Dryandra*, *Lomatia*, *Grevillea*, *Hakea* and *Persoonia*, and a few genera belonging to other natural orders, such as *Eucalyptus*, *Casuarina*, *Leptomeria*, *Exocarpus*, &c.

If Baron von Ettingshausen had possessed any profound knowledge of the distribution of plants in Australia, would he have attributed so much importance to the results of his inquiry? He no doubt means that, because in the specimens sent to him to examine, *Eucalyptus*, *Banksia* and other peculiar Australian types do not make up the majority, a radical change is indicated. But this view is incorrect. I have pointed out above that what is now the coast climate must have extended further inland, and no doubt embraced the districts from which the fossils are obtained. If any deposit derived from the brush vegetation of the coast as it now exists could be preserved in the same way as the Vegetable Creek and Dalton deposits, we should find *Eucalyptus*, *Banksia*, &c., only in a small minority, as those types do not flourish in the "brush." Why, then, should we expect to find them abounding in the Tertiary Beds of Dalton and Vegetable Creek?

We now pass on to the consideration of proposition (*d*). Ettingshausen maintains that *Myrica*, *Betula*, *Alnus*, *Quercus*, *Fagus* and *Salix* are found and these are characteristic now of Europe and North America. It is to be remarked that *Quercus* is not only found in Europe and North America, but extends at the present time to Africa, South America, the south and east of Asia, the Philippines and Malay Archipelago, and even to New Guinea, which is in the Australian region. *Fagus* has four representatives at the present day in Eastern Australia and Tasmania, and three in New Zealand, so that the selection of *Quercus* and *Fagus* to prove the Baron's point is most unhappy. It may be mentioned that *Fagus* is also just as much at home in Patagonia. It remains to be seen whether the forms attributed to *Myrica*, *Betula*, *Alnus* and *Salix* can be only such and nothing else.

he says, points to the warmer parts of North America. It will be necessary to see whether these leaves are not equally referable to existing species of Australian plants. *Bombax* is said to point to tropical America; the fact was not known, or it was overlooked, that the genus is represented at the present day in Northern Australia.

Knightia and *Coprosma* are said to point to Oceania. In passing, it is to be remarked that the leaf named *Knightia* is a very poor fragment. *Knightia* is one of two genera of *Proteaceæ* indigenous to New Zealand, the other being *Persoonia*. Each contains one species only in New Zealand, *Knightia* being peculiar to it. *Persoonia* has many Australian representatives. The example is not a happy one. If the ancestors of *Knightia* and *Persoonia* have not been introduced fortuitously from Australia by wind or ocean currents, or by birds, which is not altogether impossible, it is probable that these *Proteaceæ* may have spread through the more or less fleeting Antarctic connections, proof of the existence of which is continually accumulating. *Coprosma* is a genus well represented in Eastern Australia at the present day; ten species are known to exist in that region.

It is to be supposed that Ettingshausen has brought his strongest examples forward, but it will be seen from the above how essentially weak the arguments adduced are; and if it can be shown, as I believe it will be, that the so-called leaves of *Myrica*, *Betula*, *Alnus*, *Salix*, *Castanopsis* and *Magnolia* do not necessarily belong to those genera, the whole fabrication created by Ettingshausen, at least as regards Australia, falls to the ground, and the reference by Heer, Ettingshausen and others of European fossil plants to Australian genera becomes also liable to doubt.

In a future paper I propose to deal with these special cases and some others suggested by the perusal of the "Contributions to the Tertiary Flora of Australia."

I wish to add a few words with reference to the reputed existence of Eucalyptus and other types, now peculiar to the Australian region, in beds of Tertiary Age in England and Europe.

Although accepted as a fact by some eminent men and adopted in the text books, the evidence is considered altogether inadequate by many others whose opinions command respect, so that it is not improbable that if the follow-my-leader practice were discarded and each writer took the opportunity of judging for himself, there would be a general acknowledgment that the assumptions rest on altogether insufficient grounds. The writers in Zittel's valuable work "Palæophytologie" throw doubt on a great many of the determinations of Ettingshausen and his school. It seems to be conceded, indeed, that the existence of *Eucalyptus*, which most of the specimens do not absolutely prove, receives strong support from the case of *E. Geinitzii* in the Cretaceous, as leaves, flowers and fruit approximating to those of *Eucalyptus* have been produced, the fruits indeed separate, but the leaves and flowers on the same stalk. Now, however, we have in Dr. Newberry's posthumous work on the Amboy Clays (Monographs U.S. Geol. Survey, Vol. xxvi.) a statement that the author has discovered Heer's fruits of *E. Geinitzii* in great abundance, that he has no doubt whatever of their being identical with Heer's specimens, and that he has proved them not to be those of any species of *Eucalyptus* at all, inasmuch as they are flattened, not round as they ought to be if of that genus, and that he has obtained them attached to a core of a cone, evidently that of a conifer (see p. 46 of the work

countries where they do not now exist, as they were then representatives of an already ancient and widespread group of Conifers. The objection raised is against the probability of there having lived in those parts the peculiar dicotyledonous types of the Australian region.

There are further considerations which point to the great improbability of the existence of *Eucalyptus* in the Cretaceous of the Northern Hemisphere. *Eucalyptus* belongs to a natural order in which the leaves are normally opposite. That the ancestral forms of that genus possessed opposite leaves is inferred from the fact of the leaves being so arranged in seedlings; in many species the change to long and alternate leaves only takes place after several years' growth; in some species, such as *E. melanophloia*, the opposite character persists throughout life. These facts seem to point to the probability of the pendent, leathery leaves alternately placed being an adaptation to conditions of drought, and in support of this supposition it has been pointed out that where species have failed to produce the vertically hanging leaves, another expedient has made itself apparent, namely, that they have not only become thick and leathery, but protected with a coating of an oily excretion giving them a glaucous appearance. The flowers themselves have lost the power of producing petals except as such may be represented in the deciduous operculum, and this gives a still stronger hint of the whole plant having become modified in the course of long ages to resist drought, whereas its closest congeners, *Tristania* and *Angophora*, which have petals, are confined respectively entirely to the coast districts or to damper situations on the eastern side of Australia, not having been able to penetrate very far into the droughty interior. These being the facts, is it likely that a genus so highly specialised to resist drought should have lived in Europe so far back as the Cretaceous? There also arises the question, if *Eucalyptus* flourished in England and Europe in the Cretaceous and Tertiary, and, if the Cosmopolitan theory is trustworthy, throughout the world in the latter age, what possible conditions could have caused its extinction everywhere else but in the Australian region? Surely it should now be growing naturally somewhere on the confines of the deserts and drier districts of Asia, Africa, or America if droughty conditions help it along. On the other hand, different species of Eucalypts have adapted themselves in Australia to all conditions of moisture and dryness, heat and cold, and there is certainly nothing in the climate of other parts of the world to show why the genus should have been killed out in every corner, especially as quite a large number of species have been tried and have been found to flourish in New Zealand, at the Cape, on the Pacific Coast and elsewhere in America, and on the Mediterranean.

The probability of the former existence in Europe of any forms now characteristic of Australia is in like manner extremely small, and it requires the very strongest evidence—not the mere opinions of men however eminent—to establish such a fact. On the other hand, the probability of their having been evolved in the region where they now most abound is very great. A glance at the geological history of the past will be found instructive.

At the end of the Devonian and beginning of the Carboniferous periods, the *Lepidodendron* flora was in full vigour and apparently distributed throughout the world. Then took place a remarkable change in the Southern Hemisphere. Arctic or glacial conditions, or such at least as produced all the phenomena attributed to

nected by land and separated from the rest of the world. What an opportunity for the developments of special forms of life! We begin with Glossopteris and Gangamopteris—we only want time for the higher orders.

The land connection continued into the lower Mesozoic when the fresh-water Karoo series of South Africa, the upper Gondwana system of India, and the Ipswich and Clarence coal measures of Australia, and the Argentine Mesozoic beds ranging from Trias to Jurassic were laid down. *Tæniopteris* and *Thinnfeldia* were widely distributed.

If South Africa and Western Australia remained connected till well into the Mesozoic Age, what a likely region for the *Proteaceæ* to have originated in! In South Africa and Australia the suborder *Nucamentaceæ*, which in Bentham's opinion show archaic characters, has its stronghold; the other great group, the *Folliculares*, including the *Banksieæ*, may be assumed to have developed in Australia after its separation from South Africa. That this combined land has been the centre of distribution seems probable. From here a few have passed northwards in Africa, and later on, by means of Tasmania and the somewhat fleeting Antarctic connection with South America, members of the suborder *Folliculares* have found their way into Patagonia and Chili.

The existence of such a centre explains perhaps better than any other theory the distribution at the present day of the two allied groups of *Ericeæ* and *Epacrideæ*, which some botanists have included in one natural order, and which, judging by their affinities, have probably originated in a common ancestor. The *Ericeæ* have spread north to the Mediterranean and beyond; the *Epacrideæ* to the islands to the north and east of Australia.

There are some peculiarities about the distribution of certain groups of the *Leguminosæ* which look like parallelism between South African and Australian development; in both these countries the *Leguminosæ* are remarkable for the number of their species; the *Rutaceæ* present a similar case, for the tribe of *Boronieæ* of Australia finds its parallel in the *Diosmeæ* of South Africa. Many other similar examples exist. The *Casuarineæ*, to judge by its present distribution, must have originated in the Australian region, as, with the exception of some endemic species in the neighbouring islands, no species exist which are not Australian and which are not likely to have been transported as seeds by ocean currents.

The capsular-fruited *Myrtaceæ*, although like the groups abovementioned specially fitted to flourish in South Africa, are nevertheless not found there. They probably, therefore, did not exist in what is now Western Australia till after the land connection with South Africa had broken up. Perhaps the most likely theory would be that they originated in Northern or North-eastern Australia, arriving at their greatest development afterwards in Western Australia, while the fleshy-fruited section of *Myrtaceæ* became differentiated at a later date (probably by natural selection) and spread northwards and westwards into Asia and Europe. In connection with the suggested origin of the Myrtaceæ in Northern Australia, it is to be remarked that as one proceeds south to Victoria and Tasmania, although certain species of Eucalypts attain an enormous development and stature, the species become fewer in number, and other genera which flourish further north disappear altogether.

When we look at the enormous development that has taken place in the zoological series since the beginning of the Tertiary,

even some of the principal natural orders, were already developed in the Jurassic Period; but have we any right, except on the most convincing evidence, to assume that existing genera had already commenced their career? There is no hesitation on the part of some palæontologists to class remains of plants of Cretaceous Age under the genera *Quercus*, *Fagus*, *Acer*, *Aralia*, *Cinnamomum*, &c., as the case may be; but is it wise to do so? It seems to me that the possibility of great changes has been underestimated. The late Baron von Mueller examined and described a considerable number of fossil fruits from the Pliocene Gold Leads of this Colony and Victoria. Not one of them corresponds with any existing fruit, and the affinities of many of them are exceedingly doubtful. Surely these facts imply change of great amount; and if so much has taken place since the Pliocene, what may not have occurred when the whole Tertiary series is taken into account?

What, therefore, I wish to suggest is that though the general character of the vegetation may have remained the same for some considerable time past, and though we are dealing in a fossil state with what are no doubt the ancestors of the existing vegetation, we have no right to assume that the ancestor of an oak had in every respect the character of the modern genus Quercus, or that the ancestor of *Cinnamomum* in the Miocene would if it could be examined be found to correspond with the description of the modern genus *Cinnamomum*. It might have been the ancestor of three or four other genera as well, and so have some of the characters of each combined. Apart from this, we have the difficulty of determining what existing plant the fossil leaf really resembles; it may resemble those of half-a-dozen plants of widely different groups, and after looking into the matter one must become convinced that it is far safer to give to a fossil a name denoting resemblance rather than to dogmatically state that such a leaf is that of *Alnus*, *Cinnamomum*, &c. If Ettingshausen had contented himself with naming his specimens *Alnites*, *Cinnamomites*, &c., the proceeding would have been free from objection, but then the cult of the Cosmopolitan Theory would have received no impetus.

Mr. D. G. Stead exhibited a specimen and described the effluvium-producing powers of the so-called "Stink-Fish," *Callionymus curvicornis*, C. & V., from Port Jackson; he also showed an undetermined snake which was found coiled up in a bunch of bananas imported from Fiji.

Mr. Waterhouse, on the invitation of the President, gave some particulars of a movement now on foot, having for its object the establishment of a local Field Naturalists' Club. Messrs. Stead, Lucas and Baker spoke in support of the movement.

Mr. Baker exhibited a model of the meteorite found at Bugaldi, near Coonabarabran, N.S.W., and recently described by him.

Mr. Hedley showed an effective French example of the application of the three-colour process to the illustration of conchological subjects.

Mr. Fred. Turner exhibited the Australian plant *Desmodium parvifolium*, DC., which has become acclimatised at Bau, in Fiji. Mr. R. L. Holmes, in forwarding the specimen exhibited for identification, says:—The trefoil appeared here some years ago, but whence I cannot say, and has spread greatly during the past wet season. It is splendid feed for stock; and cattle, sheep and horses eat it greedily. On dry, poor soils it dies off in the dry season,

WEDNESDAY, AUGUST 29TH, 1900.

The Ordinary Monthly Meeting of the Society was held at the Linnean Hall, Ithaca Road, Elizabeth Bay, on Wednesday evening, August 29th, 1900.

The Hon. James Norton, LL.D., M.L.C., President, in the Chair.

DONATIONS.

Department of Agriculture, Brisbane—Queensland Agricultural Journal. Vol. vii. Part 1, Conference Supplement (July, 1900); Part 2 (Aug,, 1900). *From the Secretary for Agriculture.*

The Queensland Flora. Part ii. *Connaraceæ* to *Cornaceæ.* (1900). By F. M. Bailey, F.L.S., Colonial Botanist. *From the Author.*

Department of Agriculture, Sydney—Agricultural Gazette of New South Wales. Vol. xi. Part 8 (Aug., 1900). *From the Hon. the Minister for Mines and Agriculture.*

Royal Society of New South Wales, Sydney—Journal and Proceedings. Vol. xxxiii. (1899). *From the Society.*

Royal Anthropological Society of Australasia, Sydney—Science of Man. Vol. iii. n.s. Nos. 1-7 (Feb.-Aug., 1900). *From the Society.*

The Surveyor, Sydney. Vol. xiii. No. 7 (July, 1900). *From the Editor.*

Australasian Journal of Pharmacy, Melbourne. Vol. xv. No. 176 (Aug., 1900). *From the Editor.*

Department of Agriculture, Melbourne—Fungus Diseases of Citrus Trees in Australia (1899). By D. McAlpine. *From the Secretary for Agriculture.*

Field Naturalists' Club of Victoria—Victorian Naturalist. Vol. xvii. No. 4 (Aug., 1900). *From the Club.*

McCoy's Prodromus of the Palæontology of Victoria. Decades i.-iii. (1874-76). *From the National Museum, Melbourne.*

Public Library, Museums, and National Gallery of Victoria, Melbourne—Report of the Trustees for the Year 1899. *From the Librarian.*

Royal Society of South Australia, Adelaide—Memoirs. Vol. i. Part ii. (1900): Transactions. Vol. xxiv. Part 1 (Aug., 1900). *From the Society.*

Department of Agriculture, Perth, W.A.—Journal. July, 1900. *From the Secretary for Agriculture.*

Department of Mines, Perth, W.A.: Geological Survey—Bulletin, No. 4 (1900). *From the Government Geologist.*

Royal Society of Tasmania—Abstract, May, 1900: Papers and Proceedings, 1897. *From the Society.*

Department of Mines, Hobart—Progress of the Mineral Industry of Tasmania for the Quarter ending 30th June, 1900.

Royal Society, London—Proceedings. Vol. lxvi. Nos. 431-432 (July, 1900): Reports to the Malaria Committee, 1899-1900. *From the Society.*

Zoological Society, London—Abstract, June 19th, 1900: Proceedings, 1900. Part i. (June). *From the Society.*

Royal Irish Academy, Dublin—Proceedings. Third Series. Vol. v. No. 5 (1900). *From the Academy.*

Geological Survey of India—General Report from 1st April, 1899, to 31st March, 1900. *From the Director.*

Indian Museum, Calcutta—Illustrations of the Shallow-Water Ophiuroidea collected by the R.I.M.S.S. "Investigator" (1900). *From the Superintendent.*

Asiatic Society of Bengal, Calcutta—Journal. Vol. lxviii. Part ii. No. 4, T.p., &c. (1899): Proceedings, 1900. Nos ii.-iv. *From the Society.*

Canadian Institute, Toronto — Proceedings. New Series. Vol. ii. Part 3 (No. 9: February, 1900). *From the Institute.*

Nova Scotian Institute of Science, Halifax—Proceedings and Transactions. Vol. x. Part 1 (1898-99). *From the Institute.*

Bernice Pauahi Bishop Museum, Honolulu—Fauna Hawaiiensis. Vol. ii. Part 3 (1900). *From the Museum.*

Academy of Natural Sciences of Philadelphia—Proceedings, 1899. Part iii. *From the Academy.*

American Academy of Arts and Sciences, Boston—Proceedings. Vol. xxxv. Nos. 10-19 (Dec., 1899-March, 1900). *From the Academy.*

American Geographical Society, New York — Bulletin. Vol. xxxii. No. 3 (1900). *From the Society.*

American Museum of Natural History, New York—Bulletin. Vol. xii. (1899); Vol. xiii. Arts. i., ix.-xi. (pp. 1-18, 95-107; April-July, 1900). *From the Museum.*

Cincinnati Society of Natural History—Journal. Vol. xix. No. 6. (June, 1900). *From the Society.*

Field Columbian Museum, Chicago—Botanical Series. Vol. ii. No. 1 (March, 1900): Geological Series. Vol. i. No. 7 (Feb., 1900): Report Series. Vol. i. No. 5 (Oct., 1899). *From the Director.*

Johns Hopkins University, Baltimore—Hospital Bulletin. Vol. x. Nos. 98-105 (May-Dec., 1899); Vol. xi. Nos. 106-107 (Jan.-Feb., 1900): University Circulars. Vol. xix. Nos. 142-143 (Dec., 1899-March, 1900): Memoirs. Vol. iv. Nos. 3-4 (1899-1900). *From the University.*

New York Academy of Sciences—Memoirs. Vol. ii. Part 1 (1899). *From the Academy.*

Rochester Academy of Science—Proceedings. Vol. iii. Part 2 (March, 1900). *From the Academy.*

U.S. Department of Agriculture, Washington: Division of Vegetable Physiology and Pathology—Bulletin. No. 19 (1900). *From the Secretary of Agriculture.*

U.S. Geological Survey, Washington—Bulletin. Nos. 150-162 (1898-99): Monographs. Vols. xxii. Part ii., xxiii.-xxiv., xxvi.-xxviii. (1899). *From the Director.*

Naturwissenschaftlicher Verein für Steiermark—Mittheilungen. Jahrgang, 1899. *From the Society.*

Zoologischer Anzeiger, Leipzig. xxiii. Band. Nos. 618-620 (June-July, 1900). *From the Editor.*

Société Belge de Microscopie, Bruxelles—Annales. Tome xviii. 1er Fasc. (1894); T. xix. 2me Fasc. (1895); T. xxi. (1897): Bulletin. Tomes xvii.-xxiv. (1890-99). *From the Society.*

Société Entomologique de Belgique—Annales. Tome xliii. (1899). *From the Society.*

Société Royale de Botanique de Belgique — Bulletin. Tome xxxviii. (1899). *From the Society.*

Société Royale Malacologique de Belgique—Annales. T. xxxi. Fasc. 2 (1896); T. xxxiii. (1898); T. xxxiv., pp. xcvii.-cxxviii. (1897). *From the Society.*

Musée d'Histoire Naturelle de Marseille—Annales. Série ii. Bulletin. Tome i. Supplément (1899). *From the Museum.*

Muséum d'Histoire Naturelle, Paris—Bulletin. Année, 1899. Nos. 6-8; 1900. No. 1. *From the Museum.*

Société des Sciences Naturelles de l'Ouest de la France, Nantes—Bulletin. Tome ix. 4me Trimestre (1899). *From the Society.*

Société Zoologique de France—Bulletin. Tome xxiv. (1899): Mémoires. Tome xii. (1899). *From the Society.*

Comité Géologique de St. Pétersbourg—Bulletin. Vol. xviii. Nos. 3-10 (1899-1900): Mémoires. Vol. vii. Nos. 3-4 (1899); Vol. ix. No. 5 (1899); Vol. xv. No. 3 (1899). *From the Committee.*

Societas Entomologica Rossica—Horæ. T. xxxiii. Nos. 1-2 (1900): Index, 1859-1894. *From the Society.*

Finska Vetenskaps-Societeten, Helsingfors—Oefversigt. Vol. xl. (1897-98). *From the Society.*

L'Académie Royale des Sciences et des Lettres de Danemark, Copenhague—Bulletin, 1900. Nos. 2-3. *From the Academy.*

Istituto Botanico dell' Università di Pavia—Atti. ii. Serie. Vols. i., iii., iv., v. (1888-99). *From the Institute.*

La Nuova Notarisia, Padova. Serie xi. Luglio, 1900. *From the Editor, Dr. G. B. De Toni.*

DESCRIPTIONS OF SOME NEW ARANEIDÆ OF NEW SOUTH WALES. No. 9.

By W. J. Rainbow, F.L.S.,

(Entomologist to the Australian Museum).

Plates xxiii.-xxiv.

The present paper contains descriptions of several interesting as well as typical forms.

The first, *Dysdera australiensis*, constitutes a new generic record for Australia. Simon, in his masterly work, "Histoire Naturelle des Araignées,"* defines the geographical range of *Dysdera* as: "Europa et regio mediterranea; Africa sept. et max. austr.: ins. Atlanticæ; Asia centra.; America sept. et austr." For a genus, the species of which are so scattered, it does not, to me, appear remarkable that it should be found to occur in Australia; indeed, it seems surprising that it has not been recorded before.

The species described hereunder is a typical form, and calls for no special comment. All the members of its family (*Dysderidæ*) live by hunting, rushing out upon their prey from under stones, cracks of walls, or dark, damp, mossy situations.

The species constituting the family *Hersiliidæ* are all striking forms, and may be easily distinguished by their spinnerets, which are characteristic. The superior spinners (two) are very long; the basal joint is robust, cylindrical, and exceeds the entire length of the other spinnerets, whilst the terminal joint is tapering, and in some genera (*Hersilia*, Aud. *in* Sav., and *Tama*, E. Simon) very long and attenuated.

* Deuxième Fasicule, Tome 1er., p. 318.

All the individuals are exceedingly active and difficult to capture. Simon says*: "Les uns (*Hersilia*) se tiennent sur les troncs d'arbres et les vieilles murailles où leur coloration, grise ou blanchâtre, leur permet de se dissimuler; elles ne filent aucune toile; les autres (*Tama*, *Hersiliola*) habitent sous les pierres ou dans les fissures de rochers; elle se tiennent toujours sous la voûte formée par la pierre où elles tendent quelques fils très irréguliers, à la manière des *Pholcus*."

Only two members of the family *Hersiliidæ* have previously been recorded from Australia (*Tama novæ-hollandiæ* and *T. pickerti*), and these were described by L. Koch under the generic name *Chalinura* (1876). Simon, however, in his work already quoted† assigns them to the genus *Tama*, E. Sim.,‡ (1882), the former being a synonym. It will be seen from the above quotation that spiders of the genus *Tama* should be sought for under stones or in the fissures of rocks. The specimens captured by me (2 ♂ and 2 ♀), and for which I propose the name *Tama eucalypti*, were taken from the trunks of two Eucalyptus trees at Balmoral, Sydney, the grey bark of which they closely simulated. I saw none of these spiders on any of the neighbouring rocks, nor did I observe any webs constructed by them for the capture of prey. Indeed, from their movements it appeared to me that they relied upon their dexterity for food. At the foot of his

added a short note to the effect that *C. novæ-hollandiæ* was taken at Sydney by Mr. Bradley, but that he found no web; the specimens collected by Mr. Bradley were captured from shrubs and plants. [*C. novæ-hollandiæ* wurde von Mr. Bradley bei Sydney gefunden; nach den Notizen desselben verweilte das Thierchen bewegungslos an einen Zweige und hatte kein Gewebe; Mr. Bradley erhalte es auch durch Klopfen von verschiedenen Sträucher und Pflanzen.] It will be seen, therefore, that, according to the observations of Mr. Bradley and myself, the Australian forms of this genus, at any rate, are of arboreal habits.

Of the *Argiopidæ*, two new species of *Araneus* are described; of these, *A. singularis* is a very remarkable and interesting form; the other (*A. parvulus*) is normal. Lastly, a variety of Cambridge's *Dicrostichus furcatus* (var. *distinctus*) is recorded and described.

Family DYSDERIDÆ.

Subfamily DYSDERINÆ.

Genus Dysdera, Latr.

Dysdera australiensis, sp.nov.

(Plate xxiii., figs. 1, 1*a*.)

♀. Cephalothorax 5 mm. long, 4 mm. wide; abdomen 7·3 mm. long, 4 mm. wide.

Cephalothorax longer than wide, ovate, dark mahogany-brown, glabrous. *Pars cephalica* high, convex, obtusely truncated in front. *Pars thoracica* convex, sloping posteriorly, impressed transversely at centre, lateral radiating markings faint. *Marginal band* narrow.

Eyes six, compactly grouped, arranged in two rows, and in the form of a small, transverse oval; the pair comprising the anterior row are somewhat the largest, and are separated from each other by a space equal to about twice their individual diameter; the second row (four) is strongly procurved, and of these the median pair are not only the closest together, but are also slightly larger than their lateral neighbours.

Legs reddish-yellow, moderately long, hairy, the tibiæ and metatarsi of the third and fourth pairs armed with short black spines. Relative lengths 1, 2, 4, 3.

Palpi concolorous, hairy.

Falces mahogany-brown, long, conical, porrected; superior surfaces sparingly hairy, inner margins thickly fringed with long hairs.

Maxillæ concolorous, long, straight, convex, pointed at tips, and enlarged at base where the palpi are inserted.

Labium concolorous also, long, narrow, convex, very slightly tapering towards apex; the latter truncated and hollowed out.

Sternum mahogany-brown, oval, convex, impressed laterally, attenuated and truncated in front, obtusely pointed posteriorly; near the haunches there are a few long hairs.

Abdomen pale yellowish, pubescent, slightly projecting over base of cephalothorax.

Hab.—Sydney.

Family HERSILIIDÆ.

deep; there is also a broad, shallow, lateral groove commencing at base of cephalic eminence, and terminating near posterior angle.

Eyes eight, poised upon two tubercles; anterior pair much the smallest of the group, elevated upon a small tubercle, and separated from each other by a space equal to about twice their individual diameter; median pair of the second series (four) large, and separated from each other by a space scarcely equal to their individual diameter; lateral eyes of median group small, oval, and seated below posterior eyes; the latter are widely separated from each other, and are equally as large as the median pair of the second series.

Legs yellowish-grey, long, tapering, hairy, armed with long, fine spines. Relative lengths 1 = 2, 4, 3; of these the third pair are much the smallest.

Palpi concolorous, moderately long, robust, similar in clothing and armature to legs. Copulatory organ as in figure.

Falces yellow, cylindrical, vertical, sparingly clothed with coarse hairs or bristles; apices divergent.

Maxillæ concolorous, glabrous, arched, strongly inclining inwards; apices obtuse.

Labium concolorous also, glabrous, moderately convex, broader than long at base, rounded off at apex.

Sternum yellow, glabrous, cordate, sparingly clothed with yellowish pubescence, truncated in front.

Abdomen oblong, grey, hairy, slightly projecting over base of cephalothorax; anterior angle truncated, and deeply indented at centre; at about one-fourth its length from anterior margin there are two large, deep, round depressions or pits, widely separated from each other; and at a distance rather more than half the length there are two more, but these are much larger than the latter, are oval, seated obliquely, and rather wider apart; near the posterior angle there are again two more, and these are the smallest of the series, and are placed much closer together; commencing in front, and running between the depressions herein described, there is a broad, dark grey band, of uneven width, and irregular in outline; when dry the latter is barely visible, but after immersion in alcohol it is clear and distinct; inferior surface light grey, hairy; at about one-third its length from the front there are two large depressions or pits seated rather closely together, and below these two others, smaller, and rather wider apart; commencing near the centre and running towards base of spinnerets, there are two lateral rows of small punctures; the rows converge sharply inwards (but do not nearly meet) for about one-third their length, and then continue in nearly straight lines to their termination. *Superior spinnerets* long, grey, tapering; the terminal joint of each is annulated with dark grey near the base, and again just below the centre; tips dark grey; each of these spinnerets is clothed with hairs and armed on the inner margin with two rows of spindles of not less than 20 each.

♀. Cephalothorax 2 mm. long, 2 mm. broad; abdomen 3 mm long, 2·4 mm. wide.

The *cephalothorax*, except that it is a little larger, agrees in every detail with the ♂; the *eyes* are similarly arranged, and the only point of difference (making allowance for size) in the legs is that they are tinged with small reddish markings; the *palpi* are long, and similar in colour and armature to the legs.

Family ARGIOPIDÆ.

Subfamily ARGIOPINÆ.

Genus ARANEUS, Clerck (= *Epeira*, auct.).

ARANEUS PARVULUS, sp.nov.

(Plate xxiv., figs. 3, 3*a*, 3*b*, 3*c*.)

♂. Cephalothorax 1·5 mm. long, 1·4 mm. broad; abdomen 2 mm. long, 1·5 mm. broad; total length 3·2 mm.

Cephalothorax obovate, longer than broad, arched, sparingly hairy. *Pars cephalica* yellow, high, arched, somewhat attenuated in front, obtusely truncated. *Pars thoracica* broad, brown, glabrous, arched, deeply depressed at centre, radial grooves indistinct.

Eyes in three series of 2, 4, 2; the median group forms a square and are seated upon a tubercular eminence; of these the anterior pair are slightly the smallest, and are separated from each other by a space equal to three times their individual diameter; the posterior pair are separated from each other by a space equal to twice their individual diameter; lateral eyes minute, oblique, contiguous, and elevated upon small tubercular projections.

Legs long, yellow, annulated with broad brown rings, and armed with long, dark brown spines. Relative lengths 1, 2, 4, 3.

Palpi yellow, short, similar in clothing and armature to legs. *Copulatory organ* large, broad, and exceedingly complicated.

Maxillæ and *labium* yellow, normal.

Sternum cordate, dark brown, glabrous.

Abdomen ovate, projecting over base of cephalothorax, convex, pubescent, dark brown, with a large median patch of chrome yellow, the latter being intersected with a short, dark brown, irregular, longitudinal and transverse line, somewhat in the form of a †; upon the upper surface there are four large punctures, arranged in the form of a trapezium, and of these the anterior pair are much the closest together.

♀. Cephalothorax 2 mm. long, 1·8 mm. broad; abdomen 2·4 mm. long, 2·4 mm. broad; total length 3·3 mm.

Cephalothorax obovate, orange-yellow, clothed with long, coarse, hoary hairs. *Pars cephalica* high, arched, obtusely truncated. *Pars thoracica* broad, arched, median and radial grooves distinct.

Eyes: central group forming a trapezium, elevated upon a small protuberance; anterior median pair separated from each other by a space equal to fully twice their individual diameter; posterior pair somewhat larger than the preceding, and these are also separated from each other by a space equal to twice their individual diameter; lateral eyes minute, oblique, contiguous, and elevated upon small tubercular projections.

Legs moderately long, yellow with brown annulations, hairy, and armed with long, black spines Relative lengths 1, 2, 4, 3.

Palpi similar in colour and armature to legs.

Falces yellow, glabrous, moderately long, arched, hairy, apices divergent.

Maxillæ and *labium* yellow, normal.

Sternum cordate, glabrous, yellow, hairy.

Abdomen broad, ovate, strongly arched, boldly projecting over base of cephalothorax, very finely pubescent, dark reddish-brown in front and laterally; from near anterior margin, and extending for about one-third the length, there is a large patch of chrome yellow reticulated with fine, reddish-brown lines; from thence backwards dark reddish-brown mottled with chrome yellow;

Cephalothorax obovate, longer than broad, yellow-brown, clothed with hoary hairs. *Pars cephalica* high, arched, attenuated, obtusely pointed in front. *Pars thoracica* broad, arched, strongly depressed at centre, radial grooves distinct. *Marginal band* narrow.

Eyes arranged in three series of 2, 4, 2; the four comprising the median group form a trapezium; and the pair comprising the anterior row are separated from each other by a space equal to about three times their individual diameter; the posterior pair are larger than their anterior neighbours, and are also separated from each other by a space equal to fully three times their individual diameter; lateral eyes sessile, minute, oblong, seated well back, and close to marginal band, nearly contiguous, lower eyes slightly the smallest.

Legs moderately long and strong, hairy, spined, dark brown, with straw-yellow annulation. Relative length 1, 2, 4, [illegible].

Palpi similar in colour and armature to legs.

Falces reddish-brown, moderately long, strong, apices divergent.

Maxillæ yellowish, glabrous, arched, broad, divergent.

Labium concolorous, glabrous, arched, broader than long, rounded off at apex.

Sternum cordate, convex, dark brown, moderately clothed with long hoary bristles.

Abdomen globose, of a greyish hue, overhanging base of cephalothorax, pubescent; upon the upper surface there are four large punctures arranged in the form of a trapezium, and of these the anterior pair are much the closest together; inferior surface pubescent, grey, streaked laterally with long irregular dark brown lines, the latter converging towards spinnerets.

Epigyne a large, high, glabrous, dark brown eminence, rounded in front, lobed laterally, and terminating in an obtuse point.

Hab.—Bungendore.

Genus Dicrostichus, E. Simon.

Dicrostichus furcatus, Camb., *var.* distinctus, var.nov.

Cyrtarachne furcata, Cambridge.

This species was described and figured by Cambridge in 1877,* from a specimen collected at Parramatta. Some time ago, Master Clark, son of Dr. Clark, North Sydney, collected a specimen, which is unquestionably a variety of *D. furcatus*, an example sufficiently differentiated from the typical form to warrant description.

In the form under consideration all the parts except the abdomen agree with Cambridge's description. In Cambridge's specimen, the abdomen is described as having on the upper side "some not very large, bluntish, conical protuberances; two of them are in a straight transverse line, wide apart towards the fore margins; the rest, eight in number, are arranged in a somewhat circular group at the posterior extremity."

Upon the upper surface of the form collected by Master Clark there is near, and in front of, the two anterior protuberances, a large, broad, transverse depression, which is much the deepest laterally; the tubercles are located near the centre, straw-yellowish at the base, and the apices mahogany-brown; between the protuberances there is a short transverse ridge, the sides of which

six smaller individuals; of the latter the anterior pair are the smallest, the lateral pair much the broadest at the base, and encircled with a fine black line; the intervening space between the anterior and posterior tubercles is very uneven; at the base of the three protuberances forming the median group there are both in front and behind two rather deep punctures or pits, and of these the anterior pair are the widest apart.

Again, in his description, Cambridge says:—"The upper side is of a dull sooty hue, mapped out into rather roundish-angled patches, of various forms and sizes, which are divided from each other by clear and intersecting straw-yellow stripes; most of these patches have a central blackish spot on the fore part." From this description, however, the form under study differs in that the roundish-angled patches are moss-green; the intersecting straw-coloured stripes are present as described by Cambridge, but whereas, according to this writer, "the patches above described are almost obsolete on the middle and hinder part of the upper side, which are of a plain straw-yellow colour," in this example these patches are distinct and of a yellowish-grey hue; furthermore, there is present at the posterior extremity two distinct dark brown lines, connected in front by a gently procurved transverse bar; in addition there are also two disconnected, lateral, faintly visible brown lines; these curve gently outwards to the posterior angle, and are, like their more distinct neighbours, continued on the underside, and converge towards the spinnerets. Cambridge also described his example as having a number of long, black, lanceolate bristles upon the anterior part of the abdomen and sides (*vide* description and figure), but these are altogether wanting in the form before me. The inferior surface, sides and hinder slope agree with Cambridge's description. The abdomen, both above, laterally, and below, is clothed with fine yellowish pubescence, but these hairs are not nearly so long as those figured by Cambridge.

Hab.—North Sydney.

In a former paper, "Descriptions of Some New Araneidæ of New South Wales," No. 8, published in the P.L.S.N.S.W., 1897,

Part 3, I described and figured (p. 517, pl. xvii., figs. 3, 3*a*, 3*b*) a species of *Araneus*, for which I proposed the name *Epeira variabilis*. This specific name, I find, was preoccupied, having been used for an American species of that genus; therefore I now propose to alter the name to *Araneus concinnus*.

DESCRIPTION OF PLATES.

(Plate xxiii.)

Fig. 1 .—*Dysdera australiensis* (♀).
Fig. 1*a*.— „ „ maxillæ and lip.
Fig. 2 .—*Tama eucalypti* (♀).
Fig. 2*a*.— „ „ eyes.
Fig. 2*b*.— „ „ (♂) palpus.
Fig. 2*c*.— „ „ (♂) copulatory organ.

(Plate xxiv.)

Fig. 3 .—*Araneus parvulus* (♀).
Fig. 3*a*.— „ „ (♂) palpus, viewed from the side.
Fig. 3*b*.— „ „ (♂) copulatory organ.
Fig. 3*c*.— „ „ (♀) epigyne, viewed from the side.
Fig. 4 .—*Araneus singularis* (♀).
Fig. 4*a*.— „ „ position of eyes, viewed from the side.
Fig. 4*b*.— „ „ cephalothorax and falces, viewed from the front.

STUDIES ON AUSTRALIAN MOLLUSCA.

Part II.

(Continued from p. 100.)

By C. Hedley, F.L.S.

(Plates xxv.-xxvi.)

Chlamys blanda, Reeve.

Reeve, Conch. Icon. viii. 1853, *Pecten*, pl. xxxiv., ff. 162*a*, *b*.

Though this species was reported by Tapparone-Canefri as having been collected by the "Magenta" in Sydney Harbour,* it has never been entered in local lists. I have lately seen a specimen dredged by Mr. J. Brazier in eight fathoms off the Bottle and Glass Rocks near Sydney Heads.

Chlamys bednalli, Tate.

Tate, Trans. Roy. Soc. S.A. ix. 1887, p. 73, pl. iv., f. 3.

(Plate xxv., figs. 10, 11, 12, 13.)

Specimens of this species from Sydney Harbour so differed in sculpture from the type described by Prof. Tate that I considered it a new species and prepared the following description and figures. Since doing so I have, through the kindness of Mr. W. T. Bednall, enjoyed the opportunity of studying the actual type of *Pecten bednalli*. Our form has fewer and stronger ribs, not so markedly disposed in twos and threes as the South Australian shell, but so resembles it in other characters that I have thought it best to publish the figures and description, to withdraw the

* Viaggio Magenta, Zoologia, 1878, p. 253.

name I had proposed to bestow, and leave the specific status to be decided by one commanding a larger series.

Shell small, much higher than long and very shallow. Colour reddish-brown, passing into yellow at the apex, irregularly splashed with dark brown and opaque white. Anterior auricles largely, posterior moderately developed, anterior with six, posterior with four strong radiating, nodulose ribs. The radial ribs of the valve number about sixteen; they increase either by interpolation or by an even splitting of the primary ribs; their interstitial grooves are broad and deep. This sculpture scallops the margin and prints the interior. The whole external surface is covered by minute, frail, imbricating scales; arched in the grooves and flat on the ribs (fig. 12). The ctenolium has five well developed and one rudimentary tooth, crowded together. Hinge plate broad. One faint cardinal rib on either side. Resilium small. Chondrophore well within the margin. Height 24, length 20·5, breadth of conjoined valves 7 mm.

Hab.—Off Green Point, Port Jackson; two specimens dredged by Mr. J. Brazier, 10th July, 1886, in 8 fathoms, on a bottom of stones and sand, with broken valves of *Trigonia*.

Type to be presented to the Australian Museum.

The delicate microscopic sculpture of this species should serve to distinguish it readily; at first sight one might think the valves

in Salamander Bay, Port Stephens. The specimen, which is slightly smaller than the type, was, he tells me, dredged alive, enclosed in a nodule of hard mud. This curious habit is like that of *Choristodon rubiginosum.**

Labiosa meridionalis, Tate.

Raeta meridionalis, Tate, Trans. Roy. Soc. S. Aust., xi. 1889, p. 61, pl. xi., f. 3.

(Plate xxv., figs. 5, 6, 7, 8, 9.)

This species has hitherto been known from a single valve found on the beach of Aldinga Bay, South Australia. This year I have taken a whole shell containing part of the animal, and on another occasion a broken valve on "Chinaman's Beach," Middle Harbour. Prof. Tate, to whom one valve was submitted, kindly informs me that there is no essential difference between it and the type of *meridionalis*. My specimens are smaller, being 28 mm. in length and 21 mm. in height. Being perfect, I have utilised my example to draw the valves in apposition and other details not obtainable from the single valve hitherto known.

Though disagreeing by vermiculate sculpture, the species seems to me nearer to the subgenus *Raetella*, Dall. than to any other division of *Labiosa*.

Mylitta gemmata, Tate.

Dr. W. H. Dall has recently described† a South Australian shell as *Mylitta inæqualis*, "immediately separable from any of the other species of Mylitta by its form and inæquilateral, feebly sculptured valves."

Since these are exactly the characters of the species described by Prof. Tate as *Pythina gemmata*,‡ it seems that Dall was not aware of that diagnosis and has re-named the shell.

* *Vide* P.Z.S. 1867, p. 942.

† The Nautilus, xii., Aug. 1898, p. 41.

‡ Proc. Phil. Soc. S. Aust., ii. 1878-9, p. 132, pl. v., f. 8.

It was suggested by Smith* that *P. gemmata* should be transferred to *Mylitta;* and this classification was accepted by the author of the species.†

In arriving at the conclusion that *M. inæqualis*, Dall, should be reduced to a synonym of *M. gemmata*, Tate, I have been aided by material kindly forwarded by Mr. E. H. Matthews, the finder of *M. inæqualis.*

VENUS SCABRA, Hanley.

Reeve, Conch. Icon. xiv., *Venus*, pl. xxi., sp. 97, 1863.

This species appears not to have been recorded from N.S. Wales. I have found it rather common from Sydney northwards, as dead shells on beaches facing the ocean.

GLYCYMERIS FLAMMEA, Reeve.

Reeve, Conch. Icon. i., *Pectunculus*, pl. ii., sp. 7, 1843.

On the beach at Boydtown, Twofold Bay, N.S. Wales, I found recently a valve of this species Not only is an addition thus made to our local fauna, but the real habitat of the shell is now, I believe, first announced. On doubtful grounds this species has been attributed to New Zealand. Hutton reduces it to a synonym of (*P.*) *laticostatus*,‡ a classification at variance with Reeve's figures of the two shells. Reeve has quoted the species as

CARDITA BEDDOMEI, Smith.

Smith, Chall. Rep. Zool. xiii. 1885, p. 211, pl. xv., f. 5.

A valve of the unusual dimensions of 21 mm. long and 22 mm. high, occurred to me at Twofold Bay. Like the foregoing stragglers from the south, it is an addition to our fauna.

CADULUS LÆVIS, Brazier.

Dentalium læve, Brazier, (not *D. læve*, Schlotheim), Proc. Linn. Soc. N.S. Wales, ii. 1877, p. 59; *Cadulus lævis*, Pilsbry & Sharp, Man. of Conch. xvii. p. 195.

(Plate xxvi., figs. 8, 9, 10.)

The author of this species has supplied me with co-types from Darnley I., Torres Straits. The shell selected for illustration measures directly from end to end, that is along the chord of the arc, 26 mm. It is exceptional in having the small end bifid; most are circular and simple. A submedian constriction (marking a rest point in growth?) noted in the original description is present in but few examples and varies in position. The degree of curvature varies, young shells being more bent. All under the lens are concentrically wrinkled throughout their length.

Considering the meagre details at their disposal, Messrs. Pilsbry and Sharp estimated with remarkable accuracy the systematic position of the species.

PUNCTURELLA KESTEVENI, n.sp.

(Plate xxv., figs. 15, 16, 17.)

Shell the smallest of the genus known, thin but opaque, elevated, about as high as broad, the summit posterior and excavated by the slit, sides steep, the posterior slope surmounted by a conical projection, being the stump of the apex, which almost overhangs the basal margin. Colour pale brown. Sculpture: uniformly finely shagreened. Aperture oval, broader anteriorly. Slit on the summit, lanceolate, twice as long as broad. The interior not visible from above through the oblique passage, divided into two chambers by the septum. Septum advanced to a third of the shell's length into the cavity, projecting a median sinus, recurved upwards and forwards on each side to join the shell wall. Margin expanded, externally marked off by a furrow. Length 2·32, breadth 1·48, height 1·28 mm.

Hab.—La Perouse, N.S. Wales (H. L. Kesteven); one specimen, in shell sand.

Type to be presented to the Australian Museum.

This species is named in honour of one of my most successful pupils, Mr. H. L. Kesteven, a gentleman to whose talents both as a student and a collector I have been frequently indebted for assistance. Only two species of this genus have hitherto been recorded from Australian seas. From *P. harrisoni*, Beddome, (=*P. henniana*, Brazier)* and from all named species the hump-backed shape and great development of the septum amply distinguish the novelty. But it is probably allied to, and possibly identical with, an unnamed species noted by Watson† from Torres Straits.

CROSSEIA LABIATA, Tenison-Woods.

Ten.-Woods, Proc. Roy. Soc. Tasm. 1875, p. 151.

(Plate xxvi., fig. 18.)

of the Challenger Expedition," p. 133, examined the nomenclature and decided that it should bear the name of *Nerita punctata* because Quoy & Gaimard "distinctly say [*N. punctata*] is a species from New Holland." Those writers made no such statement; their actual words are: "Habite l'Ile de France, sur le rivage, vis-à-vis l'habitation de M. Charles Telfair, au Mapou." This locality, so precisely given, leaves us no doubt that whatever else the Australian shell may be called, it cannot be *N. punctata*.

Von Martens* chooses for our shell the name of *Nerita nigra*, Gray, but Gray used this name as of Quoy & Gaimard,† and neither author introduced it into literature with sufficient formality to render its use acceptable. In any case, as Tryon observes, there is a prior *N. nigra* of Chemnitz.

Another solution of the problem is offered by Tryon,‡ who selects *Nerita atrata*, of Reeve, arguing that as Chemnitz was not a binomial author his preoccupation of the name *Nerita atrata* should not count. This argument does not cover all the ground, for the name "Nerita atrata" is used binomially, and with a description, by authors intermediate in time between Reeve and Chemnitz. As for example, by Deshayes in the 2nd edition (1838) of the Animaux sans Vertèbres, viii., p. 603.

The fact that the only road out of the confusion was to bestow a fresh name on our species occurred simultaneously to two writers. In these Proceedings (Vol. ix., 1884, pt. 2, p. 354) Hutton redescribed the species as *Nerita saturata*. And E. A. Smith in the "Zoological Collections of H.M.S Alert," p. 69, proposed for it the name *Nerita melanotragus*. Von Martens, in the Zoological Record for 1884, observed that these two names clashed, but was unable to decide which had priority. The "Alert" Volume is in the preface dated June 20th, 1884, and is reviewed in "Nature" of September 18th. The Part of these Proceedings in question was issued August 19th. At my request Mr. E. R. Sykes kindly

* Conch. Cab. Nerita, 1888, p. 100.

† In Dieffenbach's Travels in New Zealand, ii., 1842, p. 240.

‡ Man. Conch. x., 1888, p. 26.

ascertained the date of publication of Smith's name to be August 1st, 1884. Therefore precedence must be given to *Nerita melanotragus*, Smith.

In the last Volume of these Proceedings (p. 239) Mr. W. R. Harper drew attention to the use made of this shell by an extinct tribe of aboriginals.

LIOTIA ROSTRATA, n.sp.

(Plate xxvi., figs. 4, 5, 6, 7.)

Shell small, rostrate, subdiscoidal, spire sunk, base widely excavate, thin and translucent. Whorls three and a half, separated by a deeply channelled suture, rapidly increasing, last keeled. Parallel to the suture, along the periphery and around the umbilicus run three solid opaque ridges; from the sutural band to the periphery and from that again to the umbilical border, radiate a dozen connecting bars. This sculpture may be otherwise expressed as a dozen tongue-shaped spaces excavated out of the substance of the shell above and below the periphery; the hollows translucent, the elevations opaque Crossing ridges and furrows alike are minute, close lines, which on higher magnification (fig. 7) resolve themselves into strings of oval pearls. Base wide and deeply excavate. Aperture oblique externally

pale yellow. Whorls five, separated by an impressed suture. Sculpture most elaborate; numerous small spiral keels, amounting to thirteen behind the aperture, encircle the shell, wider apart and more prominent at the periphery and growing weaker and closer as they retreat from it. Crossing the spirals, so as to form rhomboidal meshes, are longitudinal ribs of the same calibre, amounting on the last whorl to thirty-two. At the point of intersection the latter rise into small vaulted prickles. This sculpture gradually fades away on the earlier whorls, the longitudinals outlasting the spirals. The first three whorls are smooth. At the edge of the umbilicus the cancellate sculpture ceases abruptly. The base is rounded, about one-fifth of its diameter being occupied by an open, funnel-shaped umbilicus. Aperture oblique, oval, angled above. Outside the incrassate lip is less massive than in southern species of the genus, on the base it is not thickened at all. A smooth callous ridge is spread on the preceding whorl. Major diameter 5, minor 4, height 3 mm.

Hab.—Off Cape York, Queensland; one specimen, dredged by Mr. J. Brazier.

Type to be presented to the Australian Museum.

The novelty is related to *L. calliglypta*, Melvill, from the same district, but is of less height, greater breadth and has more developed spiral sculpture. The slight development of the thickened lip agrees with the latter and with *Cyclostrema cingulifera*, A. Ad.

CYCLOSTREMA ANGELI, Tenison-Woods.

Ten.-Woods, Proc. Roy. Soc. Tasmania, 1876, p. 153.

(Plate xxv., fig. 14.)

It was pointed out last year by Prof. Tate* that under this name Tryon had described and figured another species. As the real *Rissoa angeli* has yet been left unfigured, I now give a drawing of an example received from the late Mr. C. E. Beddome. This specimen measured ·8 mm. long; ·54 mm. broad.

* Trans. Roy. Soc. S.A. xxiii., 1899, p. 219.

Mr. J. Brazier has handed to me examples of both *C. angeli*, Ten.-Woods, and *C. crebrisculptum*, Tate, which he washed out of sandy mud contained in an old bottle, obtained May 22nd, 1886, in 8 fathoms, off the Bottle and Glass Rocks, Sydney Harbour.

Elusa subulata, A. Adams.

Pyramidella subulata, A. Ad., Thes. Conch. ii. 1855, p. 815, pl. clxxii. f. 13; P.Z.S. 1853, p. 177, pl. xx., f. 6.

(Plate xxv., figs. 19, 20, 21.)

Specimens of this shell dredged by Mr. J. Brazier off the Queensland coast—viz., in 12 fathoms off Cape Grenville and in 20 fathoms off Darnley Island—add another species to the Australian fauna. Beyond our limits it is reported by Tryon from the Philippines, Red Sea and Japan.

Whether it adds a genus also is a matter of opinion. Adams* separated this and others from *Pyramidella*, as a new genus. Subsequent authors have, however, held the group to be merely of sectional value. Tenison-Woods reported† a Tasmanian species of *Elusa*, but that is now unanimously considered to be a *Turbonilla*.

As the published drawings give scant details I now tender

RISSOA MACCOYI, Tenison-Woods.

Ten.-Woods, Proc. Roy. Soc. Tasm. 1876, p. 154; Tate, Trans. Roy. Soc. S.A. 1899, p. 234.

(Plate xxvi., fig. 11.)

This species was recorded from Sydney Harbour by Mr. A. U. Henn.* I have not, however, been fortunate enough to find it. Having lately obtained an example collected in the Derwent, Tasmania, by the late Mr. C. E. Beddome, I now take the opportunity of figuring the species that others may more easily recognise it. The original of my figure is 1·76 mm. long and ·4 mm. broad, and judging from T. Woods' description is probably not mature. Dr. Verco's researches have extended the range of the species to South Australia.

RISSOA TENISONI, Tate.

R. tenisoni, Tate, Trans. Roy. Soc. S. Aust. xxiii. p. 233 = *R. australis*, Ten.-Woods, Proc. Roy. Soc. Tasm. 1877, p. 146.

(Plate xxv., fig. 4.)

A figure of this well known Tasmanian species is now for the first time presented.

CHILEUTOMIA ANCEPS, Hedley.

Menon anceps, ante, p. 90.

Prof. Tate kindly points out to me that my genus *Menon* is synonymous with *Chileutomia*, Tate and Cossmann,† proposed for an Eocene fossil from Muddy Creek, Victoria. To assure me of this he forwarded an example of the type species *C. subvaricosa*, T. and C.

Having examined this with care I am quite satisfied to withdraw my name. Nevertheless I maintain that the genus in question ought to be included, where I placed it, in Eulimidæ,

* Proc. Linn. Soc. N.S. Wales, xxi. 1896, p. 500.

† Proc. Roy. Soc. N.S. Wales, xxxi. 1897, p. 403.

not in the Rissoidæ, to which *Chileutomia* was allotted. My oversight of *Chileutomia* was due to this displacement of the family. Prof. Tate now approves of the affinity to the Eulimidæ, and remarks that thus another item is added to the continuance of Eocene types in our recent fauna.

COUTHOUYIA GRACILIS, Brazier.

Vanikoro gracilis, Braz., Proc. Linn. Soc. N.S. Wales (2) ix., 1894, p. 169, pl. xiv. fig. 4.

(Plate xxvi., fig. 13.)

From the time I drew the figure quoted, I have doubted the generic position assigned to this species, but without being able to improve it. It was noted after the description that the type "is evidently young." Mrs. C. T. Starkey has kindly placed in my hands an adult example of this rare shell, which I now figure.

The fact that the last whorl is finally free from the body of the shell at once suggests *Couthouyia* for its reception, and the remaining characters agree satisfactorily with that genus. This specimen, which is from Middle Harbour, Sydney, is 3·8 mm. in length and 2·2 mm. in breadth.

Scalenostoma striatum, n.sp.

(Plate xxvi., figs. 15, 16, 17.)

Shell conical, rather glossy, opaque and solid. Colour dead white. No trace of epidermis. Whorls six, including two which are apical, smooth and glossy. The initial one is as tall but narrower than its follower, thus projecting as a style. The four subsequent whorls are parted by an impressed suture, below which they are slightly shelved, thence moderately rounded. The last is abruptly angled, rather than keeled, at the periphery, and rounded on the base. Sculpture: the adult whorls are closely girt by numerous fine spiral threads, between which are grooves of equal breadth and corresponding depth. These are obliquely and irregularly crossed by fine and coarse growth lines. An obscure varix occurs three-quarters of a whorl behind the aperture. The aperture is oblique; in profile the lip is shown to have a slight, much shallower than in the type, sinus below the suture, thence it curves forward slightly to the periphery, whence it slopes backward to the base. Aperture ovate, rounded below, angled above, lip sharp; columella thickened, a little reflected, deeply entering. Length 6·5, breadth 3·5 mm.—[No operculum. J.B.]

Hab.—Several specimens "found round the anus of a species of *Goniocidaris*, dredged in 10 fathoms, sandy mud, Port Molle, Queensland"; one specimen dredged in 25-30 fathoms off Darnley I., Torres Straits (J. Brazier).

Type to be presented to the Australian Museum.

In this species another genus is added to the Australian fauna. From all co-generic forms its few whorls, comparatively greater breadth and sculpture, amply distinguish it. Another point of interest is that it adds a genus to the list of parasites.

The parasitic Gasteropoda are not many, and it is remarkable that they have chosen their hosts from one class only, the Echinodermata. No distinction was made by earlier writers between parasites proper and mere commensals. Reckoning *Bathyscia-dium*, *Conchiolepis*, *Ovula*, *Thyca*, &c., among the latter, I can only find the following described as true parasites:—*Stilifer*, *Eulima*, *Styliferina*, *Entoconcha*, *Entocolax* and *Robillardia*.

MEGALATRACTUS ARUANUS, Linn.

(Plate xxv., fig. 18.)

In my note on this species (*ante*, p. 99) I remarked that though certain writers had seen the nidamental capsules of this mollusc they had refrained from giving any information about them.

Dr. T. H. May, writing from Bundaberg, Queensland, tells me that he sometimes finds these egg-capsules there, occasionally with embryo shells in situ. He most kindly forwarded me a dry and empty capsule which I now figure and describe.

The mass before me weighs three-quarters of an ounce, measures —length 5, breadth 2¾, and thickness 1½ inches. The shape is oblong-reniform; the concave side appears to have been attached to some foreign body which passed through an orifice at one end. The capsule is transversely divided into a dozen compartments, hinged together at the back, that is the attached, concave side. Each compartment, now empty, may be supposed to have contained several embryos; each fits into its predecessor and receives

CLATHURELLA LEGRANDI, Beddome.

Drillia legrandi, Beddome, Proc. Roy. Soc. Tasm. 1882 (1883), p. 167; *Clathurella legrandi*, Pritchard & Gatliff, Proc. Roy. Soc. Vic. N.S. xii. 1900, p. 178.

(Plate xxv., figs. 1, 2, 3.)

This species has not yet been figured. This opportunity is, therefore, taken of publishing an illustration of an authentic Tasmanian specimen received from the author of the species. The individual drawn measured 6 mm. in length and 2·5 mm. in breadth. Attention may be directed to the peculiar apex, of which the describer took no notice.

TEREBRA FICTILIS, Hinds (?).

Hinds, Thes. Conch. i. 1847, p. 183, pl. xlv. ff. 109, 110.

(Plate xxvi., fig. 14.)

This species, vaguely assigned to "Australia"[?] and not very definitely described, has never been recognised by Australian Conchologists. Tryon considers it identical with *T. bicolor*, Angas, a determination denied by Pritchard and Gatliff. There is a species of *Terebra*, locally known as *T. assimilis*, Angas, which I have collected at Manly Beach, and seen from other points of our coast. This I now figure and describe with the suggestion that it is probably Hinds' long-lost species. Though closely allied to *T. bicolor*, it is separable by form, stronger fewer ribs, and different colour-pattern. It appears to be a deeper water species than *T. bicolor*.

Shell small, rather stout, turreted, glossy. Colour of diverse patterns and shades. One before me is entirely dull white, another wholly a rich chestnut, others are cream variously streaked with purple-brown. Whorls ten, separated by an impressed suture. Nucleus dark brown, smooth, of two whorls. The next two are more elongate with incipient longitudinal ribbing. A third of each of the succeeding six whorls is occupied by a smooth constricted post-sutural band. These whorls have a dozen apiece of broad low ribs, more opaque than their interstices, on the earlier whorls strongly and on the later weakly developed. Each rib projects at the border of the constriction as a nodule and vanishes before reaching the suture. In one example the nodules coalesce into a continuous ridge. Under the lens the entire surface is seen to be sculptured by fine, close, spiral striæ, decussated by equally fine growth-lines. Base rounded, canal short. Columella edged with a callous ridge. Length 15, breadth 4 mm.

SCAPHANDER MULTISTRIATUS, Brazier.

Brazier, Proc. Linn. Soc. N.S. Wales, ii. 1877, p. 84; Pilsbry, Man. Conch. xv. 1893, p. 252.

(Plate xxvi., fig. 12.)

Having been favoured by Mr. Brazier with a specimen from Darnley Island, one of the original lot of this species, I am enabled to illustrate it. This individual is 11 mm. long, and 3½ mm. broad.

LEUCOTINA HELVA, n.sp.

(Plate xxvi., figs. 19, 20, 21, 22.)

SALINATOR, nom.nov.

Ampullarina, auctorum.

If any conchological textbook be consulted as to the status of *Ampullarina*, it will be found given as a subgenus of *Amphibola*, and ascribed to Sowerby, under date 1842, with *A. fragilis* as type.

Scudder, in the Supplemental List of the Nomenclator Zoologicus, 1882, p. 18, treats the name thus :—" Ampullarina, ———. Teste Sowerby, Conch. Man. ed. 2, p. 64. (Err. typ.? pro Ampullacera). 1842. Moll."

The second edition is inaccessible to me, but on consulting the third (1846) edition of Sowerby's Manual, the suspicions aroused by Scudder are amply justified. *Ampullarina* is doubtless *Ampullacera* wrongly copied from a MS. label. The name is introduced not as of Sowerby, but as of an unknown author. It is thus defined :—" A genus formed for the reception of *Ampullaria avellana*, f. 58. From Australia." The figure quoted represents the New Zealand species, *avellana*, not the Australian *fragilis*. On p. 312, in the explanation of plates, Sowerby actually notes that the genus is that called *Thallicera* by Swainson.

Authors appear to have assumed that because Australia was named as the habitat of the type, that *fragilis* not *avellana* was indicated. Such assumption is quite unjustifiable in the face of the facts that—(1) Sowerby names *avellana*, (2) that he figures it, and (3) that he regards *Ampullarina* as synonymous with *Thallicera*.

The literary history of *Ampullarina* offers a singular parallel to that of *Pelicaria* proposed for a New Zealand species and wrongly referred to an Australian shell.*

As the group typified by *Ampullaria fragilis* of Lamarck is now shown to be nameless, it devolves on me to suggest a name

* G. F. Harris, Cat. Tert. Moll. B.M. Pt. i., 1897, p. 218.

for it. I, therefore, propose to replace *Ampullarina* of authors, not Sowerby, by *Salinator*.

Diplommatina oreadis, n.sp.

(Plate xxv., fig. 22.)

This shell much resembles *D. obesa*, Hedley,† from New Caledonia, and can best be described by comparison with it. The Queensland shell is without the auricular expansion of the periphery, has twice as many lamellæ, and is narrower than the New Caledonian species. *D. oreadis* is 1·8 mm. long and ·9 mm. broad. The only specimen I have seen was collected by the late C. E. Beddome, 20 miles inland from Cardwell, Queensland; it will be preserved in the Australian Museum.

EXPLANATION OF PLATES.

Plate xxv.

Figs. 1-3.—*Drillia legrandi*, Beddome; shell, apex and aperture.

Fig. 4.—*Rissoa tenisoni*, Tate.

Figs. 5-9.—*Labiosa meridionalis*, Tate; valves from different aspects, with details of hinge and sculpture.

Figs. 10-13.—*Chlamys bednalli*, Tate; valves from different aspects, with details of hinge and sculpture.

Fig. 14.—*Cyclostrema angeli*, Tenison-Woods.

Figs. 15-17.—*Puncturella kesteveni*, Hedley; different aspects of shell.

Fig. 13.—*Couthouyia gracilis*, Brazier.

Fig. 14.—*Terebra fictilis*, Hinds.

Figs. 15-17—*Scalenostoma striatum*, Hedley; shell, with details of apex and aperture.

Fig. 18.—*Crosseia labiata*, Tenison-Woods.

Figs. 19-22.—*Leucotina helva*, Hedley; shell, with details of apex, sculpture and aperture.

All enlarged and to various proportions.

TWO NEW SPECIES OF PHYTOPHAGOUS HYMENOPTERA BELONGING TO THE FAMILIES *ORYSSIDÆ* AND *TENTHREDINIDÆ*, WITH NOTES ON OTHER SAWFLIES.

By Gilbert Turner.

The first of these insects agrees in all particulars with Latreille's genus *Oryssus*, the type of which is such a peculiar insect that Dalman suggested that it should be separated from the *Uroceridæ* and raised to the rank of a family; this has since been done, and it is now placed before the *Uroceridæ*. This family contains only about 20 species in the single genus *Oryssus*, which are very rare but have a wide geographical range, though the species presently to be described is the first recorded from Australia. The second belongs to the genus *Clarissa* of the family *Tenthredinidæ*, the type of which was described by Kirby from a specimen collected and forwarded by me from this district (Mackay, Q).

yellow hyaline colour; nervures black except on the lighter parts of the wings, where they are yellowish; hindwings iridescent hyaline.

The whole insect is covered with fine punctures except a small round spot, which is quite smooth, situated on each side of the upper surface of the first abdominal segment, near its base. Head almost hemispherical, a little broader than long viewed from above; fovea forming a raised ridge or horse-shoe-shaped mark on summit of head; labrum rugose, flattened; jaws short; antennæ hidden at base, composed of the usual ten joints in the female, being very irregular and variable in form, the terminal one very slender. Thorax rounded in front; pronotum forming a regular collar; front of mesonotum slightly keeled in centre, slightly constricted behind at junction with abdomen. Legs stout; tibiæ of hind legs dilated and serrate along upper margin, terminated with a single spine at apex; fore tarsi three-jointed, middle and hind tarsi five-jointed, the penultimate joint in all being the smallest and not always easy to distinguish. Ovipositor slender, spine-shaped, exserted in one specimen but not visible externally in the other.

Mackay, Q.; in March and August.

Described from two specimens, one being considerably worn and showing only traces of the pile on the body.

The specimen taken in August last year was found on a gum fence post and I captured it easily in my fingers; the other was caught by my brother in a net on a fallen log in the scrub where each of us afterwards on different occasions saw another which unfortunately we failed to secure. On attempting to capture it in my fingers it ran quickly along the log for a short distance and on being approached closely it took to the wing, first making a jump off the log, in a manner similar to that of many Chalcids; it again alighted on the log and I then tried to catch it in the net but only succeeded in frightening it away, and never saw it again.

The genus *Oryssus* is remarkable on account of the structure of the ovipositor, the paucity of veins in the wings and the insertion of the antennæ beneath the clypeus close to the mouth. Dr. Sharp* gives an illustration of an American species (*O. Sayi*) and a diagram of the head showing the peculiar structure.

TENTHREDINIDÆ.

LOPHYRIDINÆ.

Clarissa divergens, Kirby.

Mr. W. F. Kirby† founded the genus *Clarissa* on a single ♀ specimen which I sent him; I have since then been fortunate enough to capture four more specimens of what I consider to be without doubt the same species; they all, however, differ from Kirby's description in having only 11 joints in the antennæ instead of 12, one of the terminal joints being absent; three of the specimens also have the first three joints of the antennæ more or less rufo-testaceous. The front legs vary from rufo-testaceous to whitish, passing to black on the femora and basal portion of the tibiæ of one specimen; the base of the 1st joint of the hind tarsi is in some rufo-testaceous; and the lighter parts of the legs are of a lighter shade than the general colour of the insect. There is also in all the specimens a small creamy white spot on

hind legs except the coxæ, femora and in some the tarsi, whitish. Labrum whitish. Wings iridescent-hyaline with blackish nervures. Antennæ as in *C. divergens*, Kirby, (♀) except that one of the terminal joints is absent.

♀. Long. corp. 6-7 mm.; exp. al. 14-15 mm.

Shiny black; with a small white spot on each side of the 2nd abdominal segment. Four front tibiæ whitish at the base, passing to brown at the apex, tarsi brown; hind legs with trochanters and basal two-thirds of tibiæ whitish; tarsi black. Labrum whitish. Wings and antennæ as in ♂.

Mackay, Q.; in January and February; one pair taken in copula.

Described from five ♂ and seven ♀ specimens, all taken on the same flowering shrubs as *C. divergens*.

Eurys inconspicua, Kirby.

Mr. Kirby in his description of this insect (*l.c.* p. 47) omits all mention of the colour of the abdomen, which is luteous with the tip black.

Besides the species already mentioned, I have collected the following in the Mackay district:—

TENTHREDINIDÆ.

CIMBICINÆ.

Perga glabra, Kirby.
Gravenhorstii, Westw. (?)
polita, Leach.
univittata, Kirby.
Brullei, Westw.

HYLOTOMINÆ.

Hylotoma apicale, Kirby.

PTERYGOPHORINÆ.

Pterygophorus insignis, Kirby.
interruptus, Klug.
Leachii, Kirby.
uniformis, Kirby.

All the species of *Perga* are found in the forest country only and are very rarely met with. *Pterygophorus Leachii* was fairly abundant in October, 1893, also in the forest, most of the specimens being found on dead saplings; at other times I have found it decidedly rare, having only occasionally come across it, sometimes on blossom. *P. insignis* and *interruptus* and *Eurys inconspicua* are so uncommon that I am unable to say whether they belong to the forest or scrub, nearly all the specimens of these species that I have captured having been found on blossom, usually not far from scrub. The two species of *Clarissa*, *Hylotoma apicale*, and *Pterygophorus uniformis* are undoubtedly scrub insects; the last named, which is the only sawfly that I have found in large numbers, being sometimes very abundant in March, April and May on a vine which grows up after scrub has been felled, the leaves of which, I have very little doubt, form the food of the larvæ.

My thanks are due to Mr. W. W. Froggatt for assistance in preparing this paper, especially for confirming my opinion as to the generic identity of *Oryssus Queenslandensis*, Letreille's definition of the genus not being accessible to me.

CONTRIBUTIONS TO THE MORPHOLOGY AND DEVELOPMENT OF THE FEMALE UROGENITAL ORGANS IN THE MARSUPIALIA.

By Jas. P. Hill, B.Sc. (Edin.), F.L.S., Demonstrator of Biology in the University of Sydney; George Heriot Research Fellow, Edinburgh University.

(Plates xxvii.-xxix.)

II. On the Female Urogenital Organs of *Myrmecobius fasciatus*.

(Pl. xxvii., fig. 1; Pl. xxviii., figs. 2-4).

The following account is based on a macroscopic and microscopic examination of the genital organs of a young (about half-grown) *Myrmecobius* measuring 11·5 cms. (snout to root of tail). For the opportunity of examining the genital organs of this interesting and somewhat rare form, I have to thank Mr. Geo. Masters, Curator of the Macleay Museum.

The only reference to the genital organs of *Myrmecobius* with which I am acquainted is contained in a paper* by W. Leche, which deals, however, mainly with the muscular anatomy, the description of the female genital organs being quite short and unaccompanied by figures. No apology is, therefore, needed for the present communication.

In Pl. xxvii., fig. 1 the genital organs are shown as seen from the dorsal aspect. The ovaries (*o.v.*) in this young specimen are smooth oval bodies measuring 3 mm. in length by 2 mm. in breadth, and situated ventrally to the anterior extremities of the uteri. The

---

* Leche, W. "Beiträge zur Anatomie des Myrmecobius fasciatus." Verhandlungen des Biologischen Vereins in Stockholm, 1891.

Fallopian tubes (*f.t.*) are thin and convoluted, not sharply marked off from the uteri, and with fimbriated openings partially investing the ovaries. The uteri are of characteristic form and disposition. Each consists as in *Perameles* of two portions—body and neck, but these two portions are here more sharply differentiated from each other than they are in that form. The body of the uterus (*ut.b.*) is somewhat fusiform in shape, twice as broad as long, and with its long axis directed almost transversely to the long axis of the animal. They are connected by the anterior free portions of the two broad ligaments which are united in front in the median line as in *Perameles*. Anteriorly the uterine bodies contract to pass over into the Fallopian tubes, while posteriorly they similarly narrow to continue on as the uterine necks (*ut.n.*). These latter pass back parallel with each other, and almost at right angles to the uterine bodies, to become imbedded dorsally in the common connective tissue mass, which also encloses the neck of the bladder ventrally and the forwardly directed portions of the vaginal canals (fig. 2). The uterine necks increase slightly in transverse breadth posteriorly, and are so closely united as to appear externally as a single structure which terminates behind in a bulbous enlargement (fig. 1, *ut.n.*). The posterior portion of this enlargement is occupied by the two median vaginæ which have the form of quite short and small, completely separated cul-

Thiele der beiden Vaginæ, in der Mittellinie eng vereinigt, bis hinab zum Sinus urogenitalis ohne denselben einzumünden." From the median vaginæ there pass off anteriorly and ventrally the vaginal canals. The morphologically anterior portions of these have essentially the same relations as the corresponding parts in *Perameles*—*i.e.*, they run directly forwards parallel with each other, in the connective tissue just ventral to the uterine necks (fig. 2, *a.vag.c*). About the middle of the extent of the latter, the two canals leave the connective tissue mass and bend outwards at right angles to their former longitudinal course to form free tubes which, as Leche notes, are of considerable length and characteristically coiled (fig. 1, *l.vag.c.*). After forming one or two turns on each side the vaginal canals again pass inwards towards the median line and just behind the median vaginæ come to lie parallel with each other and with the urethra. The three canals are imbedded in a longitudinal strand of connective tissue (fig. 3, *l.vag.c.*, *ureth.*), representing the urogenital strand of *Perameles*, and as in that form the part of the tissue between the vaginal canals lies in the direct continuation of the median vaginæ. After a short course (about 3 mm.) in the same, the two vaginal canals and the urethra all open by separate apertures at the same level into the urogenital sinus.

The urogenital sinus is, unlike that of *Perameles*, a long narrow canal, having in this small female an approximate length of about 15 mm.

As Leche has pointed out, a distinct cloaca is present, into which the rectum opens dorsally and the urogenital sinus ventrally. A pair of large "anal glands" are situated dorso-laterally in the circular cloacal musculature (fig. 1, *a.g.*), and open by fine ducts just within the margin of the cloacal opening. There are also present in the wall of the cloaca, as in *Perameles*, numbers of tubular glands which open partly into the ducts of the anal glands and partly directly into the cloaca.

The clitoris in my specimen is attached throughout its length, and, as Leche notes, is simple. Apically it is divided into two united halves by a septum.

From the foregoing account it will be seen that while the female genital organs of *Myrmecobius* agree in certain important respects with those of *Perameles*, as described in Contribution I., they also exhibit certain well marked differences. The two forms, *Myrmecobius* and *Perameles*, agree—(1) in the possession of a relatively small median vaginal apparatus, consisting in virginal animals of two quite separate cul-de-sacs, and completely imbedded in the tissue of the genital cord. (2) In the absence of any well marked separation between the uterine necks and the median vaginæ, the former passing over directly into the latter without the intervention of distinct ora. (3) In the relations of the morphologically anterior, forwardly directed portions of the lateral vaginæ which remain permanently imbedded, with the uterine necks and the median vaginæ, in the tissue of the genital cord, and thus retain the position and course presented by the Müllerian ducts in the fœtus. And (4) in the possession of a distinct cloaca. In these four respects the genital organs of *Myrmecobius* exhibit what I have regarded as primitive features.

As regards the points of difference in the organs of the two forms, two are worthy of remark—(1) the freedom in *Myrmecobius* of the middle portion of the lateral vaginæ and their coiled character, and (2) the presence in the same of a long urogenital sinus.

III. On the Female Genital Organs of *Tarsipes rostratus*.

(Pl. xxvii., fig. 5; Pl. xxviii., figs. 6-8).

The female genital organs of *Tarsipes* have not hitherto been described. I again owe the material for the present description to Mr. Masters, who kindly allowed me to remove the genital organs from two females collected by him near King George's Sound, W.A., many years ago. Considering the age of the specimens and the fact that they were simply put into spirit entire, the organs proved to be remarkably well preserved. Both sets were examined in serial sections.

One of the females had four young in the pouch, measuring g.l. 8 mm., and h.l. 4 mm.; the other slightly smaller specimen had a distinct pouch, with, however, very small teats. Both sets of organs presented essentially the same structural features. The drawings and measurements given refer mainly to the organs of the first-mentioned female.

The genital organs are shown from the ventral aspect in Pl. xxvii., fig. 5. The ovaries (*ov*) are smooth ovalish bodies, measuring 1·25 mm. in length by ·75 mm. in breadth, lying in contact with the dorso-mesial borders of the uteri. The Fallopian tubes are sharply marked off from the uteri; they are slightly convoluted and of no great length. The bodies of the uteri are somewhat ovalish, dorso-ventrally compressed structures, with their long axes directed transversely and measuring in the first female 4 mm. in length by 3 mm. in breadth, and in the second 3·5 mm. by 1·5 mm. Posteriorly the body of each uterus contracts to form the uterine neck which, as in *Myrmecobius*, passes back almost at right angles to the long axis of the body. Histologically the necks are distinguished as in *Myrmecobius* by the absence of uterine glands.

The two uterine necks continue back side by side, surrounded by a common muscular layer and quite free from any adjacent structure, to open into the median vagina. They form at their posterior ends a prominent papilla which projects for some distance into the cavity of the vagina and at whose apex the two ora are situated.

The median vaginal apparatus in *Tarsipes* consists of a single short canal, without any trace of division into two. In front the median vagina (fig. 6, *m.v.c.*) possesses a fairly large lumen, but it rapidly narrows behind and passes back as a dorso-ventrally compressed canal, which directly opens at its hinder end into the anterior extremity of the urogenital sinus (fig. 8, *m.v.c.*). It lies in the hinder part of its extent in the connective tissue of the urogenital strand, between the two lateral vaginal canals above and the urethra below (figs. 7 and 8, *m.v.c.*). *Tarsipes* thus agrees, as I pointed out in Contribution I., with certain species of the family *Macropodidæ* in the possession of a direct and, after the first parturition, permanently open median passage for the birth of the young. In the second set of organs at my disposal, the direct communication is also present, from which fact I conclude from analogy with Macropods that this female had also bred at some previous time. From the anterior end of the median vagina there arise the lateral vaginal canals. These have a quite short free course and are only very slightly curved outwards around the inpassing ureters. Immediately below the level of the latter, the vaginal canals approximate and eventually run back parallel with each other and side by side dorsally to the

an internal septum. A distinct cloaca is present. The cloacal opening is of characteristic form, being long, narrow and spout-like (fig. 5). A cloacal sphincter muscle is not present. A pair of large anal glands open far back into the cloaca.

IV. Notes on the Female Urogenital Organs of *Acrobates pygmæus* and *Petaurus breviceps*.

The female organs of *Acrobates* have already been shortly described by Owen,* whose observations I in the main confirm, and extend. The following notes are derived from an examination of a series of serial sections through the organs of a pregnant female, with shrivelled blastodermic vesicles in the uteri and with three 2 cm. young in the pouch.

The two uteri pass back side by side and posteriorly gradually contract to form the uterine necks, which after a quite short free course enter the connective tissue between the forwardly projecting portions of the lateral vaginal canals, to open each by a medianly situated aperture on a slight papilla, into the corresponding median vagina. As Owen described, two median vaginæ are present. They are separated by a common partition wall over by far the greater portion of their extent, but in my specimen, unlike Owen's, the septum disappears posteriorly, so that there is here a quite short common median vagina recalling the condition in the multiparous *Perameles*. This ends blindly in the connective tissue between the lateral vaginæ about ·6 mm. above the anterior end of the urogenital sinus. Although remains of a pseudo-vaginal passage are not certainly recognisable, I am inclined to believe that such will be found to occur at parturition. The lateral vaginal canals, as Owen describes, pass forwards alongside, and external to, the uterine necks and then bend outwards, "forming a curve like the handles of a vase." They then converge and pass back with the urethra and median vaginæ in

* Owen, R. "On the Generation of the Marsupial Animals, &c." Phil. Trans. 1834.

the common mass of connective tissue, to open by a single aperture into the urogenital sinus. The sinus has a length of about 5 mm. It opens with the rectum into a distinct cloaca. The clitoris, situated ventrally at the junction of sinus and cloaca, is attached throughout its length; it is deeply grooved dorsally and markedly bifid at its apex. Two pairs of anal glands are present. The cloacal opening is narrow, somewhat spout-like, and ventrally directed.

In *Petaurus breviceps* the uteri and lateral vaginæ are related very much as in *Acrobates*. The median vaginal apparatus, however, consists of a single and quite short undivided canal, which ends blindly in the connective tissue, at a considerable distance from the anterior end of the urogenital sinus. According to Forbes* in *P. sciureus* "there are apparently two small cul-de-sacs; but the specimen examined does not allow me to say whether or no they unite."

V. On the Existence at Parturition of a Pseudo-Vaginal Passage in *Trichosurus vulpecula*.

(Pl. xxix, fig. 9).

In Contribution I.,† while discussing the general significance of the occurrence of a direct median passage for the birth of the

unexpected occurrence of the direct median passage in *Dasyurus* impressed on me the necessity of testing the above assumption by the examination of serial sections of the median vaginal apparatus in *Trichosurus*, with the result that a pseudo-vaginal passage was also found to occur in this form.

The female genital organs of *Trichosurus* were described and figured by Brass* in 1880, and in the following year W. A. Forbes,† in a paper on the Koala, very shortly, in a footnote, refers to the uteri and median vaginal apparatus without alluding to Brass's account.

According to the latter author, the uteri open "in die, von Vaginæ gebildeten, colossalen Blindsäcke. *Dieselben sind in der Mittellinie vollständig durch ein Septum von einander geschieden* und bis hinab zum Sinus urogenitalis verlängert ohne jedoch in denselben einzumünden" (p. 13. *Italics mine*). His fig. 1 of Taf. ii., representing the organs from the dorsal side, shows the median vaginal apparatus ending abruptly in contact with the approximated posterior portions of the lateral vaginal canals, while his fig. 2, representing a horizontal section, shows the same part terminating freely behind and unconnected in any way with the lateral canals. I would again point out that such representations of the relations of these parts are wholly inaccurate. The posterior section of the median vaginal apparatus of *Trichosurus* passes down to enter, and lies imbedded in, the connective tissue, which also encloses the posterior portions of the lateral vaginæ and the urethra.

According to Forbes, "each os tincæ projects as a prominent and quite free papilla into *the common vaginal chamber*, formed by the coalescence and fusion of the two diverticula present in *Phascolomys* and *Phascolarctos*. This chamber is capacious and has only a very slight indication of a median partition left" (p. 190. *Italics mine*).

---

* Brass, A. "Beiträge zur Kenntniss des weiblichen Urogenitalsystems der Marsupialen." Inaug. Diss. Leipzig, 1880.

† *Loc. cit. (ante)*, p. 190. Footnote.

From these two quotations it will be seen that, according to Brass, the median vaginal apparatus consists of two completely separated halves; according to Forbes, of a single chamber with only slight traces of a median partition. My own observations show that both conditions occur. In the organs of undoubted virgin animals, the apparatus consists, as Brass describes, of two cul-de-sacs completely separated by a thin, often semi-transparent partition. The two cul-de-sacs terminate in the connective tissue ventral to the converging posterior portions of the lateral vaginæ, between the latter and the neck of the bladder and some distance in front of the anterior end of the long urogenital sinus. In the organs of females which have given birth to young, on the other hand, I find that the median partition has over by far the greater portion of its extent disappeared, thus placing the two cul-de-sacs in wide and open communication. In such females, remnants of the septum are present, usually in the form of dorsal and ventral median folds of slight though varying width. Occasionally the two folds may meet in front and behind, and exceptionally I have found the two cul-de-sacs in communication through a large aperture in the posterior portion of the septum, here largely persistent.

The question now arises, at what period does the vaginal septum break down? What little definite evidence I possess

analogy with *Perameles* where I have shown that the common median vagina is actually formed as the result of parturition, we may reasonably conclude that in *Trichosurus* the same also holds good.

Pseudo-vaginal passage.—The genital organs (*a*, *b*, and *c*) of three females have been examined in serial sections, commencing at the level of the hinder portion of the median vagina and extending backwards. The first two sets of organs (*a* and *b*) were unaccompanied by young, but, as will be pointed out, had obviously been taken from females in which the young had recently been born. In the third set a young one was sent by my collector along with the organs.

In (*a*) both uteri were enlarged, the left having been the pregnant one. Its body measured 27 mm. in length by 16 mm. in breadth. In a female with a young one just ready to be born, the body of the pregnant right uterus measures 26 by 18 mm. As before mentioned, the vaginal septum presented the appearance of having been recently ruptured. From these facts, and from the condition of the pseudo-vaginal passage, I conclude that parturition had been completed only a few hours previously. In section the posterior portion of the median vagina is seen to lie in the connective tissue, between the lateral vaginæ, now running parallel with each other and the neck of the bladder. Posteriorly as the sections are traced back, the lumen of the epithelially lined vagina is found to be directly continued back, after the disappearance of the lining as a large quite irregular cleft—the pseudo-vaginal passage—in the connective tissue ventral to the lateral vaginæ. As in *Perameles* and *Dasyurus*, the passage is bounded solely by connective tissue, in which indeed it appears as a mere tear, presenting as it does every appearance of having been caused by mere mechanical rupture. In outline it is quite irregular owing to the presence of inwardly projecting shreds of connective tissue, while fragments of the same occur free in the lumen. The formation of the passage has been accompanied by a considerable extravasation of blood, large and small clots occurring in the tissue both in and around the passage (Pl. xxix.). Such clots, however, are limited mainly to the more anteriorly situated portion of the passage. In the connective tissue leucocytes are also present in considerable numbers. Posteriorly the passage becomes much reduced in size, appearing in section as a transversely extended narrow cleft. Through an unfortunate accident I was unable to see the hinder opening of the passage in this set of organs.

In (*b*) the body of the left uterus measured 23 by 15 mm. I consider the female from which this set of organs was taken had given birth to the young one within the previous twenty-four hours. Both in this set of organs and in the next (*c*) the vaginal septum is represented by low dorsal and ventral folds with uniform free margins.

The sectional appearances presented are essentially the same as in the preceding set of organs. The pseudo-vaginal passage continues back from the median vagina as a quite irregular space in the connective tissue. It lies at first ventrally to the two lateral vaginæ, but posteriorly it is situated between the ventral halves of the same. In front the passage has the same irregular outline as in (*a*). The connective tissue projects in the same irregular fashion into its lumen, and presents the same torn and ragged appearance. Extravasated blood is present in the connective tissue around the anterior portion of the passage, but in

In the sections practically the only trace of the pseudo-vaginal passage now present is a slight condensation in the connective tissue in the direct line of continuation of the median vagina, and distinguishable by its more deeply staining qualities. From the condition in this set of organs it would be quite impossible to foretell the existence of a pseudo-vaginal passage in *Trichosurus*. In this connection it may be noted that in *Dasyurus*, as I have elsewhere described, on the third day after parturition, the pseudo-vaginal passage has completely healed up, and practically all trace of it has disappeared. In these two forms then, as in *Perameles*, the pseudo-vaginal passage must be reformed anew at each succeeding act of parturition.

Trichosurus is the first Diprotodont genus in which a pseudo-vaginal passage has certainly been found to exist, and its occurrence in a member of this suborder supports, it seems to me, the view put forward in Contribution I, of the origin of the direct post-partum communication between the median vagina and the urogenital sinus as it exists in the majority of the *Macropodidæ*, and it further renders the suggested occurrence of a pseudo-vaginal passage in *M. major* all the more probable.

ERRATUM.

In Contribution I (Proceedings, 1899, Part i.), on page 49, third line from bottom, and on page 51, third line from top—*for* fundus *read* neck.

EXPLANATION OF PLATES.

Reference letters.

a.g. Anal gland. *a.vag.c.* Anterior forwardly directed portion of lateral vaginal canal. *bl.* Bladder. *cl.o.* Cloacal opening. *fim.* Fimbriated opening of Fallopian tube. *f.t.* Fallopian tube. *l.vag.c.* Lateral vaginal canal. *m.v.c.* Median vaginal canal. *ov.* Ovary. *rect.* Rectum. *ur.* Ureter. *ureth.* Urethra. *ut.* Body of uterus. *ut.n.* Uterine neck.

(Plate xxvii.)

Fig. 1.—Urogenital organs of *Myrmecobius fasciatus*, from the dorsal aspect (×4).

Fig. 5.—Urogenital organs of *Tarsipes rostratus*, from the ventral aspect (×2).

(Plate xxviii.)

Figs. 2, 3 and 4.—Trans. sections through the genital organs of *Mrymecobius fasciatus*. See text. (× $18\frac{2}{3}$).

Figs. 6, 7 and 8.—Trans. sections through the preceding genital organs of *Tarsipes rostratus*. See text. (×24).

(Plate xxix.)

Fig. 9.—Reproduction from photo-micrograph of trans. section, urogenital organs, *T. vulpecula*, showing the pseudo-vaginal passage. The dark mass to one side of the passage is formed by extravasated blood (× about 55).

THE MEASUREMENT OF BACTERIA.

By R. Greig Smith, M.Sc., Macleay Bacteriologist to the Society.

Bacteria are generally measured by means of the micrometer eyepiece, which contains a scale graduated into divisions. The values of these divisions are actually determined for the various objectives used by the microscopist by focussing the scale of a stage-micrometer and noting the number of ocular divisions included in a certain number of $\frac{1}{100}$ millimetre divisions of the stage-micrometer scale. The value of a single division is then calculated, and is thus a known constant for the objective with a certain tube length.

In measuring bacteria it is usual to employ the micrometer eyepiece and the $\frac{1}{12}$ oil-immersion objective, with a tube length advised by the maker of the objective. On no account should the value of a micrometer division be assumed or accepted without personal confirmation. For instance, the values of the divisions with a Leitz $\frac{1}{12}$ oil-immersion, micrometer eyepiece ii. and tube length 170 mm., by actual determination was found to be equal to 1·5 μ; according to Leitz's price list, it is 1·8 μ.

With a micrometer eyepiece the unit of measurement of which equals say 1·5 μ, the measurement of bacteria is uncertain unless the boundaries of the organisms coincide with the divisional lines. Fractions of the unit (1·5 μ) necessitate an estimation, and it is here that the uncertainty occurs, for the eye cannot divide a small space into 10 or 15 equal parts. Errors of measurement frequently happen. As far as the length is concerned, this is of little consequence, because on a film the bacteria are found in lengths varying between the normal length of a mature cell and twice that length when the cell is about to divide into two individuals. Small bacteria may even appear less than the normal when they are lying in the film at an angle with the coverglass. It is of greater consequence to have a correct estimation of the breadth of bacteria, and especially the breadth in relation to the normal length, for then we can have a true picture of the organism. Small differences in breadth influence the general appearance of the cells to a greater degree than small differences in length. This can be clearly seen by comparing the two figures, in one of which (fig. 1) the diagrammatic bacteria have a constant breadth and varying length, and in the other (fig. 2) they have a constant length and varying breadth.

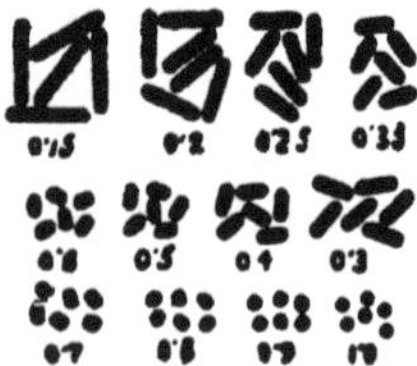

FIG. 1.

FIG. 2.

Since, owing to the method of reproduction of the fission fungi,

double printing they obtain an image of the bacteria upon a network of squares. This method is good for recording the measurements in a pictorial manner, and for actually measuring the breadth it is better than the micrometer eyepiece, inasmuch as the unit of size becomes 1 μ instead of 1·5 or 1·8 μ. For exact measurement, it would be easy, once the bacteria were photographed, to place the negative under a low power objective and measure the breadth with the micrometer eyepiece, or to project the image upon a screen by means of the projection lantern and measure with a centimetre or millimetre rule. This of necessity involves photographing the organism, a process which is not always desired.

The method I employ in determining the breadth of an organism is to fix upon a bacterium in the microscopic field and measure its length. Then I compare the organism with a series of diagrams representing bacteria, the breadths of which have been accurately measured in terms of the length. From this series one group that appears identical with the organism fixed upon is noted, and the number of this group is multiplied by the length of the organism. The result is the breadth. The breadth of another organism in the same film may be calculated in a similar manner, and the second result will generally be identical with the first. For example, the organism is a short rod measuring 1·5 μ, and on comparison with the diagrammatic table it appears identical with the group whose type number ($\frac{\text{breadth}}{\text{length}}$) is 0·4. On multiplying 1·5 μ by 0·4, the breadth 0·6 μ is obtained This result will be more exact than that obtained by estimating the breadth with the micrometer eyepiece. Since this estimation of the breadth when done from a longer and a shorter form in the same film necessitates two different calculations, these, when they agree, are more likely to be correct than when several are estimated by a similar mental estimation as obtains when the eyepiece micrometer alone is used. It, however, goes without saying that it is advisable to check the one method against the other.

A large diagram of types may be employed, but perhaps a better idea is obtained when the types are reduced to sizes approximating those observed with the oil-immersion. These are given in the accompanying figure (fig. 3), where the long rods measure three millimetres and the shorter rods 1·2 mm.

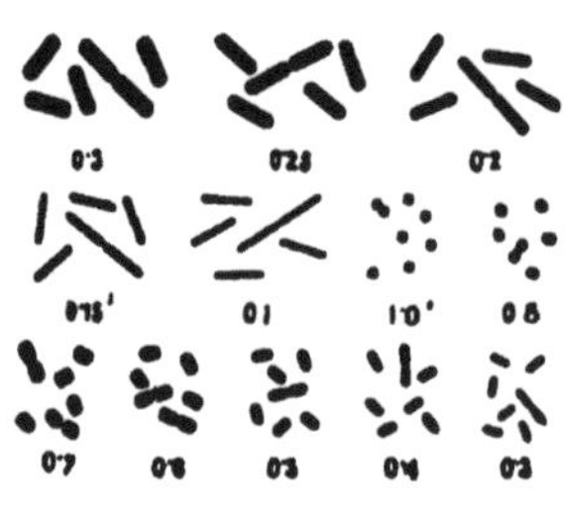

FIG. 3.

The diameters of micrococci, streptothrix and other forms might be confirmed after micrometer measurement by comparing the coccus, etc., relative to the micrometric scale division lines, with lines ruled at intervals of 1·2 mm. (the length of the smaller diagrammatic organisms) upon a coverglass which is superposed over the shorter diameter of the diagrammatic types. If the coccus or streptothrix occupies a space in the divisions of the eyepiece similar to that occupied by one of the types when viewed under such a ruled coverglass, it is obvious that the diameter of the coccus, etc., will be the type number multiplied by the value of the micrometer divisions. Such rulings can be made upon a coverglass by dipping the latter into a dilute solution of gelatine (0·5 %) and ruling the lines with Indian ink upon the thin dry gelatine film.

NOTES TO ACCOMPANY FIGURES OF BOISDUVAL'S TYPES OF SIX SPECIES OF AUSTRALIAN *CURCULIONIDÆ* BASED UPON OBSERVATIONS AND SKETCHES BY M. P. LESNE.

By Arthur M. Lea, F.E.S.

(Plate xxx.)

Some years ago I wrote to my valued correspondent, Monsieur P. Lesne, of the Paris Museum, asking him if Dr. Boisduval's types of Australian Coleoptera were in that Museum. In reply, he informed me that they were not there but in the Brussels Museum.

As it is impossible to recognise many of Boisduval's species from his descriptions,* and as I was desirous of knowing more about his Australian *Cryptorhynchides*, Monsieur Lesne very kindly promised that, when an opportunity offered, he would examine such of the types as remain; and would write to me about them.

This he has now done, not only supplying notes on a number of species, but also sending sketches of *Cryptorhynchus ephippiger*, *C. lithodermus*, *C. fuliginosus*, *C. dromedarius*, *Gonipterus notographus* and *G. reticulatus*.

These sketches are in pencil, with explanatory notes in French. They were intended to be working drawings, and not finished illustrations. But even so, it seems desirable that the information which M. Lesne has brought together with some trouble should be published for the information and guidance of other Australian entomologists. For process reproduction I have

* Voyage de l'Astrolabe. Entomologie, 2me Partie (1835).

therefore retraced them, and in place of the notes on the drawings I have affixed letters. Owing to lack of space on the plate, sketches of a portion of *Gonipterus notographus*, of *Cryptorhynchus fuliginosus* and of *C. ephippiger* could not be given. The notes written on the original sketches will be found in the explanation of the plate; those contained in Monsieur Lesne's letter are given under my own notes.

CRYPTORHYNCHUS DROMEDARIUS, Boisd. (Pl. xxx. figs. 1-2).

This is (as already noted in Masters' Catalogue) *Protopalus Stephensi*, Boheman. The synonymy of several species of *Protopalus* is somewhat complicated, and will be dealt with by me later on.* The length of the type is 22 mm The original description is—"Major, niger; thorace granulifero in medio cristato; elytris striato-punctatis basi tuberculo prominulo, sutura elevata, gibbosa, humerisque acutis."

CRYPTORHYNCHUS LITHODERMUS, Boisd. (Pl. xxx. figs. 3-4).

The sketches and notes of this insect leave no doubt whatever in my mind that it is *Poropterus varicosus*, Pascoe. The length is given as 11½ mm. The original description is—"Griseus, tomentosus; thorace antice prominulo, subbituberculato; elytris punctis impressis, cristulis sparsis, seriatis."

dorsalis, but narrower, with the prothorax longer and the sides less rounded." From the sketches and notes he supplies, however, I have no doubt that it is only a slight variety of *dorsalis*. The original description is—"Subelongatus, cinereus; thorace nigro subvariegato, coleopteris punctato-striatis plaga media, communi, quadrata nigro-fusca."

GONIPTERUS RETICULATUS, Boisd. (Pl. xxx. figs. 11-14).

I have previously* expressed the opinion that this species was possibly Boheman's *Oxyops cancellata*. The notes and sketches of Monsieur Lesne leave no doubt whatever in my mind that such is the case. He gives the length of the type as 11½ mm., (my own specimens vary from 11 to 12 mm.) and he remarks of it—"Between the eyes there is a deep impression, in front of which extends a thin carina to the anterior third of the rostrum. Segments 3-4 of the abdomen are equal in length. The mesosternum is prominent and pointed in front. The body is quite black above." *Oxyops cancellata*, or, as it should henceforth be called, *O. reticulata*, is an exceptionally distinct species. The original description is—"Niger, thorace rugoso, elytris cancellatis."

GONIPTERUS NOTOGRAPHUS, Boisd. (Pl. xxx. figs. 15-16).

I have not been enabled to recognise this species in my collection. Monsieur Lesne's notes and sketches represent quite an ordinary *Gonipterus*. He gives the length as 8 mm., and he remarks—"The under parts of the body are thickly clothed with grey scales. On the disc of the elytra these scales are smaller than on the sides and posterior declivity; on each side of apex they form a small conical tuft. There is a subhumeral callus on each side. The mesosternum is not prominent or pointed." The original description is—"Velutinus, nigro-cinereus; elytris oblongo-triangularibus amplis, punctis impressis striatim digestis"

* P.L.S.N.S.W., 1897, p. 612.

EXPLANATION OF PLATE XXX.

Figs. 1-2.—*Cryptorhynchus dromedarius*, Boisd.

Fig. 1.—Side view.
Fig. 2.—Seen from above.

A. Projecting carina, smooth and shining at summit. B. Large shining granules covering disc and sides of prothorax. C. Large tubercle, very shining at summit. D. Raised interstice. E. Large punctures. F. Clusters of brown squamose hairs. G. Sutural swelling, unequal on the surface and with stiff brown hairs, scale-like in front, setiform and denser behind.

Figs. 3-4.—*Cryptorhynchus lithodermus*, Boisd.

Fig. 3.—Seen from above.

A. Carina. B. Tubercle. C. Sutural swelling. D. Velvety-black and slightly raised fascicles situated parallel with the sides.

Fig. 4.—Side view.

A. Sutural swelling. B. Fascicles formed of blackish scales on the raised interstices.

Figs. 5-7.—*Crypterhynchus fuliginosus*, Boisd.

Fig. 5.—Seen from above.

A. Slightly raised and costiform interstice.

Fig. 6.—Side view.
Fig. 7.—Side view of head and prothorax.

A. Groove. B. Punctures, each containing a scale.

Figs. 9-10.—*Cryptorhynchus e hippiger*, Boisd.

Fig. 12.—Side view.

A. Mesosternal projection. B. Metasternal episternum, of which the anterior superior angle corresponds to the notch in the elytra.

Fig. 13.—Side view.

Fig. 14.—Antenna.

Figs. 15-16.—*Gonipterus notographus*, Boisd.

Fig. 15.—Side view. The flanks and under parts of the body are clothed with white scales, which are dense and long. The upper scales are grey or ferruginous.

A. Marginal stria. B. Lateral tubercle. C. Metasternal episternum. D. In consequence of the insect being badly prepared, the abdomen is slightly displaced.

Fig. 16.—Seen from above.

A. Punctures of the elytra in strong series on the disc, effaced towards the sides and posteriorly. B. Elytra clothed, especially behind, with white scales, moderately seriate in arrangement and rather long. C. Prominent cluster of scales.

NOTES AND EXHIBITS.

Mr. D. G. Stead exhibited an interesting preparation of the fruit of *Barringtonia cupania*, which he had recently found stranded on Maroubra Beach. The husk was covered with barnacles (*Lepas australis* ?) and, judging by their size, the specimen had been floating about for a considerable time. Besides the barnacles, there was on one side of the husk a colony of Bryozoa, and in one corner a hole had been excavated in which was safely ensconced a female specimen of the interesting little "Wanderer Crab," *Planes minutus*. In the same cavity was a polychætous annelid, and in parts where the husk was beginning to disintegrate were numbers of a small brownish-black amphipod of an undetermined species. Although the fruit had obviously been immersed for some considerable period, the kernel was apparently quite uninjured.

Mr. R. T. Baker exhibited a section of portion of an old pump made of White Ironbark, *Eucalyptus paniculata*, Sm., recently taken from a well found in making excavations in Elizabeth-

Mr. Palmer showed a stone-axe of a remarkable character, from Lawson, but the circumstances under which it came into notice were not such as to preclude the possibility of its being of non-Australian manufacture.

Mr. Greig Smith exhibited a series of thirty-five photomicrographs of Bacteria from Sydney water, in illustration of a paper read at last meeting.

The Superintendent of the Botanical Gardens, on behalf of Mr. Maiden, exhibited two well grown pot-plants in flower, viz., *Æchmea Mariæ-reginæ*, Wendl., (N. O. Bromeliaceæ) from Costa Rica, and *Cypripedium hirsutissimum*, Lindl., from Java.

Mr. Fletcher invited the attention of members to specimens of the two forms of the common Pittosporum (*P. undulatum*, And.) as usually met with about Sydney—namely, the form with rudimentary stamens which fruits abundantly, and that with perfect stamens and, as far as early appearances go, a satisfactory pistil and a viscid stigma, but which sets no fruit. As the Pittosporums are just now coming into bloom, a request was made for evidence of any intermediate condition, if still to be met with; or of the occurrence of the two kinds of flowers on the same tree.

WEDNESDAY, SEPTEMBER 26TH, 1900.

The Ordinary Monthly Meeting of the Society was held at the Linnean Hall, Ithaca Road, Elizabeth Bay, on Wednesday evening, September 26th, 1900.

The Hon. James Norton, LL.D., M.L.C., President, in the Chair.

DONATIONS.

Department of Agriculture, Brisbane—Queensland Agricultural Journal. Vol. vii. Part 3 (Sept., 1900). *From the Secretary for Agriculture.*

Two Separates (Queensland Agric. Journ. Vol. vi., 1900). By F. M. Bailey, F.L.S. *From the Author.*

Department of Mines and Agriculture, Sydney: Geological

Eleven Separates (Agric. Gazette of N.S.W., Miscellaneous Publications. Nos. 220, 224, 239, 260, 269, 300, 307, 308, 312, 333, 334); One Separate (Journ. and Proc. Roy. Soc., N.S.W. Vol. xxxiii.). By F. B. Guthrie, F.C.S. *From the Author.*

Public Library of New South Wales, Sydney—Report of the Trustees for the Year 1899. *From the Trustees.*

The Surveyor, Sydney. Vol. xiii. Nos. 8-9 (Aug.-Sept., 1900). *From the Editor.*

Australasian Journal of Pharmacy, Melbourne. Vol. xv. No. 177 (Sept., 1900). *From the Editor.*

Field Naturalists' Club of Victoria—Victorian Naturalist. Vol. xvii. No. 5 (Sept., 1900). *From the Club.*

University of Melbourne — Calendar for the Year 1901: McCoy's Prodromus of the Palæontology of Victoria. Decade v. (1877). *From the University.*

Department of Mines, South Australia—Record of the Mines of S.A.: Report on the Gold Discovery at Tarcoola (1900). By H. Y. L. Brown, F.G.S. *From the Author.*

Department of Agriculture, Perth, W.A.—Journal. August, 1900. *From the Secretary.*

New Zealand Institute, Wellington—Transactions. Vol. xxxii. (1899). *From the Auckland Museum.*

British Museum (Nat. Hist.), London—Illustrations of the Botany of Captain Cook's First Voyage round the World in H.M.S. Endeavour in 1768-71. By the Right Hon. Sir Joseph Banks, Bart., K.B., P.R.S., and Dr. Daniel Solander, F.R.S. With Determinations by James Britten, F.L.S. Part i. Australian Plants (Folio, 1900). *From the Trustees.*

Cambridge Philosophical Society—Proceedings. Vol. x. Part vi. (1900). *From the Society.*

Geological Society, London — Quarterly Journal. Vol. lvi. Part 3 (No. 223: Aug., 1900). *From the Society.*

Linnean Society, London—Journal. *Zoology*. Vol. xxviii. No. 179 (July, 1900). *From the Society.*

Manchester Museum, Owens College, Manchester—Notes from the Manchester Museum. No. 6 (Publication No. 30; 1900): Report for the Year 1899-1900 (Publication No. 31). *From the Museum.*

Marine Biological Association of the United Kingdom, Plymouth—Journal. New Series. Vol. v. T.p., &c. (1897-99); Vol. vi. No. 1 (July, 1900). *From the Association.*

Royal Society, London—Proceedings. Vol. lxvi. No. 433 (Aug., 1900): Further Reports to the Malaria Committee, 1900. *From the Society.*

Royal Dublin Society, Dublin—Economic Proceedings. Vol. i. Part 1 (Nov., 1899): Scientific Proceedings. Vol. ix. (n.s.) Part 1 (Oct., 1899); Scientific Transactions. Series ii. Vol. vii. Parts 2-7 (June, 1899-April, 1900); Index to Scientific Proceedings and Transactions from 1877 to 1898. *From the Society.*

Madras Government Museum—Bulletin. Vol. iii. No. 2 (1900). *From the Superintendent.*

Perak Government Gazette. Vol. xiii. Nos. 24-25 (July-Aug.,

Washington Academy of Sciences — Proceedings. Vol. ii. pp. 41-246 (July-August, 1900). *From the Academy.*

Museo Nacional de Buenos Aires—Comunicaciones. Tomo i. No. 6 (May, 1900). *From the Museum.*

Archiv für Naturgeschichte, Berlin. lxiii. Jahrgang. ii. Band. 2 Heft. 2 Hälfte (1897): lxvi. Jahrgang. i. Band. 2 Heft (1900). *From the Editor.*

Medicinisch-Naturwissenschaftliche Gesellschaft zu Jena — Jenaische Zeitschrift. xxxiii. Band. Heft iii. u. iv. (1900). *From the Society.*

Zoologischer Anzeiger, Leipzig. Band xxiii. No. 621 (Aug., 1900). *From the Editor.*

Muséum d'Histoire Naturelle, Paris—Bulletin. Année 1900. Nos. 2-4. ; Nouvelles Archives. 4me Série. Tome i. 1er-2me Fasc. (1899). *From the Museum.*

R. Università di Torino—Bollettino. Vol. xv. Nos. 367-376 (Jan.-June, 1900). *From the University.*

College of Science, University of Tōkyō, Japan—Journal. Vol. xiii. Part 1 (1900). *From the University.*

PHOSPHORESCENT FUNGI IN AUSTRALIA.

By D. McAlpine.

(Communicated by J. H. Maiden, F.L.S.)

(Plates xxxi.-xxxii.)

The phenomenon of phosphorescence or luminosity in fungi has long been known, but the cause of it is still in dispute. Plenty of specimens displaying it were met with in the suburbs of Melbourne during May, and by calling attention to the fact someone with the time and opportunity may be induced to investigate the phenomenon. My principal reason, however, for dealing with the subject now is to bring forward some fresh material which, if it does not throw any new light upon the matter, may at least remove some sources of error.

The following account of the phenomenon is given by Dr. Cooke in his "Introduction to the Study of Fungi":—"Several Agarics have this property, of which the largest number, for any locality,

ings of a Naturalist," says:—"There is a species of the genus *Agaricus* which has been observed to be vividly luminous. It is very common in the Australian woods in the vicinity of Sydney about the localities of the South Head Road and among the scrubs and forests on the approach to the headlands of Botany Bay, and emits a light sufficiently powerful to enable the time on a watch to be seen by it.

"I have frequently gathered this fungus, and on placing it in a dark room found that it has retained the luminous power for two successive nights; the phosphorescence becoming fainter on the second, disappears entirely by the third night. The whole of the plant shines with a pale, livid and greenish phosphorescent glow."

The naturalist Drummond has likewise vividly described the phenomenon in "Hooker's Journal of Botany" for 1842 and 1843, in letters written from Swan River, West Australia. He says:— "Two species of *Agaricus* grow parasitic on the stumps of trees and possess nothing remarkable in appearance by day, but by night they emit a most curious light, such as I never saw described in any book. The first species in which I observed this property was about two inches across and growing in clusters on the stump of a *Banksia* tree near the jetty at Perth, W.A. When this fungus was laid on a newspaper, it emitted by night a phosphorescent light, enabling us to read the words round it, and it continued to do so for several nights with gradually decreasing intensity as the plant dried up." The other species was remarkably large, measuring 16 inches in diameter, and weighing about five pounds. The specimen was hung up to dry in the sitting-room, and in passing through the apartment in the dark it was found to glow. "No light," he says, "is so white as this, at least none that I have ever seen. The luminous property continued, though gradually diminishing, for four or five nights, when it ceased on the plant becoming dry. We called some of the natives and showed them this fungus when emitting light, and the poor creatures cried out 'chinga,' their name for a spirit, and seemed much afraid of it."

On another occasion he saw at a distance a tree in the forest aglow, and he imagined that it must have been set in a blaze by lightning. "On making my way to it, I found that the light was produced by a remarkable Agaric which grew, tier above tier, up the trunk of a dead *Eucalyptus occidentalis*." The species was different from those previously described.

The descriptions of the phenomenon here given by two good observers on the spot will serve as a general introduction to the subject.

Fungi possessing this property.

The number of phosphorescent species is not large, only about 21 being determined with certainty, and they are generally natives of warm climates and belong mostly to the family *Agaricaceæ*. Of these no less than five are confined to Australia, and fifteen of them altogether are known in our island continent. Only some of those detected by Drummond have been determined, and a number probably await investigation.

The following list gives the known species, chiefly according to Zopf,* and I have added the distribution of those found in Australia. This will enable collectors to seek for those at present unknown or undescribed.

1. *Armillaria mellea*, Vahl. Europe, America, and Australia;

7. *P. lampas*, Berk. Victoria, West Australia, Tasmania; on languid but not dead stems of *Grevillea*.
8. *P. nidiformis*, Berk. West Australia; on the ground.
9. *P. noctilucens*, Lev. Manilla; on tree stems.
10. *P. olearius*, Dec. S. and S.E. Europe; among roots of Olive trees.
11. *P. phosphoreus*, Berk. Tasmania; on roots of trees.
12. *P. prometheus*, B. & C. Hong Kong; on dead wood.
13. *Collybia cirrhata*, Pers. Germany, Britain.
14. *C. longipes*, Bull. Germany, Britain, Victoria, Queensland.
15. *C. tuberosa*, Bull. Germany, Britain, Queensland.
16. *Fomes annosus*, Fr. Europe, America, Queensland.
17. *Polyporus grammocephalus*, var. *emerici*. Queensland, on trunks; New Guinea.
18. *P. sulphureus*, Fr. Europe, Asia, Africa, America, Queensland, Tasmania.
19. *Corticium coeruleum*, Fr. New South Wales, Queensland, Britain; said to be phosphorescent.
20. *Xylaria hypoxylon*, Grev. Europe, Australia; common.
21. *X. polymorpha*, Grev. Europe, Australia; common.

In the Honey Agaric (*A. mellea*) and the species of *Xylaria* it is only the mycelial threads which are phosphorescent, and the brilliant luminous appearance often seen in mines is due to the so-called rhizomorphs of the same or similar fungi. It is curious to note that the fructification which arises from these mycelial threads and is the perfectly developed form should not exhibit luminosity.

Tulasne,* writing in 1848, remarks that four species only of luminous Agarics appear at present to be known, viz., *Pleurotus gardneri*, *P. igneus*, *P. olearius*, and *P. noctilucens*, Lev., whereas at least twenty-one are now known and probably several are unrecorded for Australia.

* "Sur la Phosphorescence des Champignons." Ann. Sci. Nat. ix. p. 338 (1848).

In the original descriptions of the Australian species notes are often given as to the nature of the luminosity, since this could only be observed in the fresh state. A few of these remarks are here reproduced.

Pleurotus candescens—"Its luminosity is of a silvery shine and very apparent; it is partially restored to it when moistened again" (Mueller).

Pleurotus gardneri—"The whole plant gives out at night a bright phosphorescent light, somewhat similar to that emitted by the larger fire-flies, having a pale greenish hue. From this circumstance and from growing on a palm, it is called by the inhabitants 'Flor de Coco'" (Gardner).

Pleurotus illuminans—"We have now before us a luminous mushroom, by which in a dark room last night we were able to read distinctly the headlines of several newspapers" (Collector to Mueller).

Pleurotus phosphoreus—This species was so phosphorescent that Mr. Gunn, who discovered it in Tasmania, was able to read by its light, and it remained luminous six days or more.

While the observations regarding the nature of the light and the general effect produced are valuable, still there are various points requiring minute inspection on which even good observers were in error. Thus De Candolle, who first made known the

They certainly occur on the trunks of living trees, although the bark of the particular portion to which they are attached is dead, but that is probably due to the destructive action of the mycelium of the fungus.

Observations on Pleurotus candescens, F.v.M. & Berk.

Specimens of this fungus are very common during April and May in the neighbourhood of Melbourne, and a few observations were made on the phenomenon of phosphorescence this year.

Specimens were detached from a Tea-tree trunk on the afternoon of 6th April and retained their phosphorescence for at least a week.

The luminosity was confined to the gills, with the exception of the downy material (mycelium) at the base of the stem, from which, however, it disappeared in about two days. Portions of the cuticle were removed from the pileus, also the white flesh, but there was not the slightest trace of phosphorescence. The white spores were shed in great abundance, but they showed no signs of luminosity.

The phenomenon was exhibited during the day as well as at night, for when specimens were taken into a dark cellar they shone equally well.

The effect of moisture was also tested by immersing a piece of the gills in water. After immersion for an hour and a half, no perceptible effect on the luminosity was observed except, perhaps, it was just a shade duller.

The light emitted was a whitish glimmer with a faint suggestion of blue, but the phosphorescent light is not the same in all species of fungi. In some species it is more bluish, in others more greenish or greenish-yellow, and in a third more of the white light. The gills gave a decided acid reaction when fresh and in the full glow of luminosity.

In order to test the effect of the luminous glow on the photographic plate, Mr. A. J. Campbell, F.L.S., kindly tried a specimen. He exposed an ordinary photographic plate for an hour above one of the luminous fungi, measuring about 4 or 5 inches in diameter. The result was that the plate was distinctly "fogged" from the action of the light and was not so affected when exposed without the fungus. The plate was masked with a leaf design, leaving the centre part exposed.

In some of the luminous bacteria the blue and violet rays of the spectrum predominate, and they have consequently been photographed by their own light. B. Fischer has also demonstrated that the light from streak cultures of these microbes is sufficiently strong to illuminate and photograph adjacent objects, such as a watch.*

Supposed cause of the phenomenon.

As a consequence of respiration or the combustion of carbon compounds, heat is liberated in all living plants, but the development of light only rarely takes place. As we have seen, only a few fungi become luminous, and it is found that as the respiration becomes feebler the light decreases in intensity and ceases entirely at death.

The production of light is also usually confined to certain portions of the organism, and may occur either in the vegetative portions or the fructification. In *Armillaria mellea*, for instance,

As far as has been determined, the conditions which influence the production of light are the following :—

1. It is only in the living organism that the phenomenon occurs, and the greater the vital activity the more marked is the phosphorescence. Brefeld* observed in the mycelial strands of *A. mellea* that only the youngest and softest portions were phosphorescent, while the older brown and hard strands were no longer capable of it.

2. Oxygen is necessary, for as soon as it is withdrawn the luminosity ceases, but it reappears when the air is restored. In pure oxygen the light does not become more intensive.

3. Phosphorescence is also dependent on the temperature. There is a minimum below which the light immediately ceases, and a maximum of luminosity beyond which temperature the light decreases until the heat is sufficient to kill the fungus, and then the luminosity is gone for ever. It appears that the minimum, the optimum and the maximum may vary in the same fungus according to its vital activity.

4. As regards moisture, the luminosity does not appear to be affected by wet or dry weather. I immersed a small portion of the gills of *P. candescens* in water for an hour and a half, and the light was practically the same at the end of that time.

Phosphorescence has been proved to be due to minute organisms—the photogenic bacteria—in the case of sea fish and animal flesh, but although bacteria are present on the gills of this fungus they have nothing to do with the phosphorescence. It will be noted that the luminous bacteria occur on flesh and the carcases of fish and are living organisms, just as the fungi are only phosphorescent while alive. It was concluded by Fabre† that phosphorescence is the result of the respiratory activity of the fungus, but that would hardly account for the phenomenon being restricted to certain parts nor for its being confined to so few forms of fungi. Even if due to a temporary increase of oxidation the exciting cause of such increase would require explanation.

* Schimmelpilze iii., p. 171.

† Ann. Sci. Nat. Series iv. Vol. 4 (1858).

It is a debated question whether the light proceeds from within the organism or from excreted luminous metabolic products. The researches of Radziscewski* seem to afford a reasonable and simple explanation of the luminosity, and they support the latter view. He found that certain organic substances such as the aldehydes and aldehyde-ammonia derivatives, as well as fatty oils, have the faculty of becoming luminous in alkaline solutions when oxygen is present. Such compounds (*e g.*, fatty oils) are known in the phosphorescent fungi, and if they are united in an alkaline solution with the oxygen obtained in respiration, the cause of the luminosity might thus be explained. Oxidation products or acids are formed from the luminous materials by the vital activity of the organism and the luminous organs in *P. candescens*, viz., the gills, were decidedly acid. The metabolic products are known as *phosphorescents*, and, uniting with oxygen, they evolve light outside the organism.

Luminosity is a better term for the phenomenon than phosphorescence, since it is not of the same nature as true phosphorescence. The luminous fungi glow without previous exposure to the sun, and the property cannot be excited by mere heat, as in the case of certain mineral substances such as phosphorite. Further, it is not due to the formation of some readily oxidizable compound of phosphorus such as phosphuretted hydrogen in the

to guide the nocturnal moths, so the luminous light of these fungi will guide the flies and beetles in the dark direct to the spore-bearing portion. In addition to the light, there is a strong odour, at least in this particular species, and so the night-flying insects will be attracted just as the day-loving insects are guided by the bright colours and the penetrating odours of other members of this large family.

Technical Description.

Pleurotus candescens, F.v.M.—Glowing Pleurote.

Caespitose, imbricated, with strong smell, phosphorescent. Pileus up to 6 inches across, fleshy, soft, sub-dimidiate, at first convex and horizontal, then becoming puckered, conchate, generally concave beneath or sometimes above, glabrous, moist, even, satiny, yellowish to brownish or becoming lavender, flesh white, cuticle may become separable. Stem short, stout and thickened upwards, from 2-3 inches long, firm, obliquely ascending, lateral or excentric, white to yellowish, downy at base.

Gills decurrent, moderately crowded, broad, white, with a yellowish tinge.

Spores white, elliptical, $7\frac{1}{2}$–$8\frac{1}{2}$ × $4\frac{1}{2}$–$5\frac{1}{2}$ μ.

Beaumaris, Victoria; on trunks of Tea-tree or on the ground arising from the roots. April–May, 1900.

There is a considerable amount of variation in this species. The colour may vary from yellowish to brownish or even be entirely lavender in specimens of the same size. In the self-same tuft, the stems may be lateral or excentric or practically central, while the pilei may be conchate or round and slightly convex. This variation is largely caused by the extensive overlapping and the necessity of each one accommodating itself to its surroundings. From 8 to 12 may spring from a common base, and spreading out horizontally and overlapping each other, they must vary in shape, especially when some of them have a diameter of 6 inches. They are found on the trunks of living trees, although usually the particular spot from which they spring is dead, but this is probably caused by the mycelium destroying the tissues.

REFERENCES TO AUSTRALIAN PHOSPHORESCENT FUNGI.

BENNETT, G.—"Gatherings of a Naturalist in Australasia," p. 59. London (1860).
Reference to a luminous Agaric.

BERKELEY, M. J.—Hook. Journ. Bot. ii., p. 426 (1840).
Pleurotus gardneri described.

—— Hook. Lond. Journ. Bot. iii., p. 185 (1844).
Pleurotus nidiformis described.

—— Hook. Lond. Journ. Bot. iv., pp. 44-45 (1845).
Pleurotus lampas = *P. noctilucus* described.

—— Hook. Journ. Bot. vii., p. 572 (1848).
Pleurotus phosphoreus described.

—— "Introduction to Cryptogamic Botany," p. 265. London (1857).

—— "Flora of Tasmania—Fungi." Hooker's Botany of the Antarctic Voyage. Pt. iii. Vol. ii., p. 244 (1860).
Pleurotus phosphoreus referred to.

—— "Australian Fungi." Journ. Linn. Soc. xiii., p. 155 (1873).
Pleurotus illuminans and *P. candescens* described.

COOKE, M. C.—"Fungi: their nature, influence and uses," p. 110. London (1875).

—— Grevillea x., p. 96 (1882).
Polyporus grammocephalus var. *emerici* described.

—— "Handbook of Australian Fungi." London (1892).

ON SOME NEW SPECIES OF COCCIDÆ FROM AUSTRALIA, COLLECTED BY W. W. FROGGATT, F.L.S.

By E. Ernest Green, F.E.S., Government Entomologist, Ceylon.

(Communicated by W. W. Froggatt, F.L.S.)

(Plate xxxiii.).

Rhizococcus viridis, sp.nov.

Living insect green (Froggatt). Dried examples and those preserved in alcohol dark purplish-brown. Strongly convex above, with deep transverse corrugations. Under side somewhat concave, the margins clasping the twig upon which the insect rests. Oblong-oval to broadly oval according to age. Divisions of abdominal segments strongly marked, the margins of the four terminal segments produced into rounded lobes: the anal lobes small, but strongly chitinous and darker coloured (fig. 1). Antennæ with six joints only, the 6th longer than the preceding two together (fig. 2). Antennal formula 6, (1, 3), 2, (4, 5). Legs well developed; normal. Tarsus nearly as long as tibia; the two together about equal to femur and trochanter combined. Digitules fine hairs. Spiracles rather small and inconspicuous. Anal ring with six hairs. Derm with minute circular trilocular pores (fig. 1*a*) and scattered longish sharply-pointed spines on both dorsal and ventral surfaces, larger and more crowded on margins of terminal four or five segments. Caudal setæ (on anal lobes) comparatively short, only three or four times the length of adjacent spines. Total length 2·50 mm.; breadth 2·0 mm.

Young larva (viewed as embryo within body of female), oblong-oval, tapering behind. Antennæ 6-jointed, 6th longest. Dorsum with series of longish blunt spines.

Hab.—Mittagong, N.S.W., (Froggatt Coll. No. 318); on young twigs and leaf-stalks of *Acacia decurrens*.

Subfamily ASTEROLECANIINÆ.

ANTECEROCOCCUS, gen.nov.

Adult female not forming a complete test separable from the insect; but with body closely covered with adherent waxy matter. Antennæ and legs atrophied, the latter consisting merely of a tubercle surmounted by a claw. Anal lobes prominent, each bearing a long seta; and between them a chitinous dorsal prolongation as in *Cerococcus* and *Olliffia*. Derm with figure-of-8-shaped pores.

Male puparium as in *Cerococcus*.

Adult male not known.

ANTECEROCOCCUS PUNCTIFERUS, sp.nov.

Dried examples of the adult female (fig. 3) are irregularly broadly oval, abruptly narrowed behind: dorso-ventrally depressed: with a median longitudinal rounded ridge on thorax, on each side

After maceration the insect is seen to be pyriform (fig. 5): the thoracic parts broadly oval: the abdomen abruptly narrowed and constricted near its base, terminating in a pair of prominent conical anal lobes (fig. 9). Mouth-parts large: mentum dimerous. Antennæ rudimentary; divisions confused and impossible to define: truncate, with a few stout hairs at apex (fig. 6). Legs 6, atrophied; each consisting of a stout claw on a rounded chitinous tubercle (fig. 7). Stigmata large and conspicuous. Anal ring with eight stout flattened hairs, finally tapering towards the extremity. There are also some hair-like spines arising from the walls of the anal tube. Anal lobes (fig. 9) prominent: a well defined strongly chitinous patch on the inner edges: each bearing at extremity a long stout seta besides several stout curved spines. Dorsad of the lobes is a triangular chitinous extension of the body, the free extremity rounded and projecting between the lobes, its base continued in a chitinous transverse band which curves inwards and partly encircles the anal ring. Derm closely set with 8-shaped pores of two sizes (figs. 8*a* and 8*c*). The larger pores are grouped at definite spots on the dorsum and give rise to the tufts of glassy filaments seen on the living insect. Viewed in profile, at the edge of the body, they are seen to be sunk in shallow cylindrical pits. The smaller 8-shaped pores are scattered over the whole surface of the body and presumably secrete the waxy matter that closely invests the insect. At each of the stigmatic areas, inside the group of large pores, is a band of small circular pores, with minute compound orifices (fig. 8*d*) extending over both dorsal and ventral surfaces. There are other small circular pores, with larger simple orifices (fig. 8*e*), scattered irregularly over the surface. And on a level with the point where the abdomen is constricted, are two pairs of circular bodies (fig. 8*f*), the function of which is obscure. They appear to be seated just below the derm of the dorsal surface. They are chitinous, funnel-shaped, with the base multiperforate like a sieve. They are possibly homologous with the multilocular glands of *Cerococcus*. Length 1·25 to 1·50 mm.; breadth (across thorax) 1 mm.

Male puparium whitish or very pale fulvous: thinly coated with granular wax, as in the adult female. Thoracic area with two deep longitudinal furrows. Posterior extremity with a large circular valve (see fig. 4) placed horizontally. Length 1·25 mm.

Hab.—Bathurst, N.S.W., (Froggatt Coll. No. 317); on *Pittosporum eugenioides.*

DESCRIPTION OF PLATE XXXIII.

Figs. 1-2—*Rhizococcus viridis.*

Fig. 1.—Abdominal extremity of adult female: dorsal view.
Fig. 2.—Antenna of adult female.

Figs. 3-9—*Antecerococcus punctiferus.*

Fig. 3.—Adult female, with waxy covering: dorsal view.
Fig. 4.—Puparium of adult male.
Fig. 5.—Adult female, after maceration: ventral view.
Fig. 6.—Antenna of adult female.
Fig. 7.—Foot of adult female.
Fig. 8.—Pores from derm of adult female. *a.* Larger 8-shaped pores. *b.* Larger 8-shaped pores, in profile. *c.* Smaller 8-shaped pores. *d.* Small circular pores, from stigmatic area. *e.* Small circular pores, with simple orifices. *f.* Sieve-like bodies, from base of abdomen.
Fig. 9.—Abdominal extremity of adult female: dorsal view.

AUSTRALIAN LAND PLANARIANS: DESCRIPTIONS OF NEW SPECIES AND NOTES ON COLLECTING AND PRESERVING.

No. 2.

By Thos. Steel, F.L.S., F.C.S.

(Plate xxxiv.)

Part i.—Descriptions of new species, &c.

Since the publication of my previous paper on Australian Land Planarians (11, p. 104)* I have obtained several interesting undescribed species, descriptions of which, with other notes, are contained in this paper.

For specimens of the following species from Western Australia I am indebted to Mr. Chas. G. Hamilton, who collected and sent them to me preserved in spirit. They are of particular interest as being the first recorded land planarians from that part of Australia, and from the fact of two of the species being found living in a dry sandy district which would be considered extremely unfavourable for such creatures.

In describing these land planarians I have thought it best to adhere to the old generic classification in the meantime, though Prof. von Graff, in his recently published great monograph of the group, (12) has proposed the new genus *Artioposthia* for a number of species which have hitherto been included in the genus *Geoplana*.

Some points in Graff's arrangement of these may require modification, through, amongst other reasons, his not having had an opportunity of examining specimens of some of the species. At

*The numbers in brackets refer to the Bibliography at the end of the paper.

a future date I propose to deal with all the described Australian forms in detail, so as to, as far as possible, bring into accord the work of different observers.

The type specimens of all the species described in this paper are deposited in the Australian Museum, Sydney.

GEOPLANA FUSCO-DORSALIS, n.sp.

(Pl. xxxiv., fig. 1.)

Ground colour of dorsal surface uniform dark grey, with a narrow median line of pale grey. Anterior tip brown. Ventral surface uniform rich cream colour. Eyes numerous, single row round anterior tip, then very thickly grouped in about seven horizontal rows extending for at least 8 to 9 mm. backwards. Length of spirit specimens 50 mm. by 4 mm. at widest part.

The peripharyngeal opening 27 mm. from anterior tip; genital opening not clearly distinguishable.

Hab.—Near Perth, Western Australia; under chips, &c. (Mr. Chas. G. Hamilton).

The above are the colours as noted from the living specimens by Mr. Hamilton. In spirit they remain apparently unaltered.

In general shape this planarian belongs to the group of which *G. quinquelineata*, Fletch. & Hamil., may be taken as represen-

thinly sprinkled. Then there is a similar [illegible] of partially dotted space, and then a band of [illegible] [illegible] together to form a continuous margin. Eye [illegible] round anterior tip, then [illegible] grouped [illegible] for several mm.

Length of spirit specimen [illegible] [illegible] specimen [illegible] Another specimen [illegible] opening [illegible] to the young of this species [illegible] composed of aggregated [illegible] margins, and [illegible] species [illegible] large specimen [illegible] The length of [illegible]

[illegible]

This species [illegible]

[illegible]

GEOPLANA MELANOCHROA, n.sp.

(Pl. xxxiv., fig. 3.)

Specimen in spirit. Dorsal surface uniform black. Ventral surface black at margins, with a pale central band of about one-third of the total width. The whole ventral surface has a purple tinge.

Eyes quite invisible until a specimen was cut in sections, when they were found to extend in a single row round anterior tip and right down the sides in a sparse row without side grouping, but with an occasional straggler out of the line.

Length 16 mm. by 1½ mm. in width. Peripharyngeal aperture 9 mm. from anterior tip, with genital opening 12½ mm., or just midway between peripharyngeal aperture and posterior end.

Hab.—Armadale, Darling Ranges, Western Australia; in ightly timbered country, in sheltered gullies, under pieces of ironstone (Mr. C. G. Hamilton).

This curious little planarian bears some resemblance to *G. atrata*, mihi, (11, p. 105) but is readily distinguished by the absence of the black ventro-median line, and by the greater width of the black marginal bands on same surface.

GEOPLANA GRAMINICOLA, n.sp.

what widely apart. No side grouping. Length when crawling 13 to 18 mm. by less than 1 mm in width.

In a spirit preserved specimen which is 16 mm. in length, the peripharyngeal aperture is 9 mm. from the anterior end and the genital opening 12½ mm.

Hab.—Petersham, near Sydney: common in my garden living amongst grass. I got this elegant little planarian feeding on dead slugs which I had casually left under a damp sack lying on grass in the shade of my house. By keeping up this mode of trapping them I secured specimens during the greater part of the year, but they are most abundant in the summer months. During October and November, 1889, on a very limited area of ground, some five or six square yards, I captured under sacks over 200 individuals.

I have not met with this species anywhere else, but if, as is certainly the case in my garden, it burrows amongst the roots of grass, it might readily be overlooked. It may possibly be an introduced form, though I have failed to find a description of it in Prof. von Graff's monograph before-mentioned. The species which appears to come nearest to it is *Geoplana gamuhin*, Graff, from North Celebes, but differs in being several times larger and in having a sharply defined dorso-medial band.

Geoplana scaphoidea, nom.nov.

Geoplana elegans, Steel (11, p. 111, Pl. vi., fig. 2), non *Planaria elegans*, Darwin (1, p. 244); non *Geoplana elegans*, Darwin, Muller and Schultze, (2, p. 29); non *G. elegans*, Darwin, Graff (12, p. 328).

When I gave the name *G. elegans* to a land planarian from Queensland, I was unaware that it was preoccupied by Darwin for a species from Rio de Janeiro, and having since ascertained the fact, I now desire to re-name the species as above.

Geoplana quinquelineata, *Fletch. & Hamil.*, var.nov.

The typical *G. quinquelineata* (Pl. xxxiv., fig. 7) has the lines varying greatly in different individuals in intensity of

Brittlebank (5, p. 48) records instances in which he observed land planarians in Victoria, feeding on, in one case, a wood-louse, and in the other the larva of a beetle. Dendy also found a *Geoplana* in the act of preying on a beetle, (6, p. 132; 7, p. 68; and 9, p. 116).

Spencer (15, p. 86) speaks of the voracity of land planarians in general, and of *G. Spenceri*, Dendy, in particular, he having seen this species catch and suck the inside out of a beetle.

Quite recently Scharff (13, p. 33), in describing a new European *Rhynchodemus* of extremely large size, mentions the fact of his finding it feeding upon a snail. A remarkable feature about this species as figured is the unusually far forward position of the apertures.

For some years past I have myself had numerous opportunities of studying land planarians in a state of nature and in captivity, and I have on very frequent occasions seen them feeding on insects of various kinds, earthworms, slugs, *Peripatus*, &c., and one species on another; but I have never seen one attempt to eat vegetable matter of any kind, either fresh or decayed. The suctorial oral tube, while eminently well adapted for imbibing the juices and soft tissues of insects, &c., is not at all suited for the ingestion of vegetable fibre or decayed wood, and I think that there can be no doubt that Moseley's explanation of Darwin's

In those land planarians possessing more than two eyes, these organs are usually very numerous, amounting in some cases to several hundred. They consist of minute black dots, only occasionally sufficiently large to be visible to the naked eye, extending in a single row round the edge of the anterior tip usually expanding into a more or less crowded group on each side just behind the tip, and then thinning out again to a single row on either side which may extend right down to the posterior end. In some species there is little or no side grouping, the eyes being arranged in a single row extending backwards to a variable distance. Those planarians possessing two eyes only (*Rhynchodemus*, &c.), have them situated one on each side, just behind the anterior tip. The eyes appear to be merely sense organs serving to distinguish light from darkness, thereby enabling the animal to avoid the light when it so desires. I do not believe that the eyes possess any degree of true vision in the sense of enabling the planarians to distinguish objects.

As pointed out by Dendy (10, p. 42) the eyes may be well displayed by crushing the anterior tip of a living specimen between a glass slip and coverglass, and examining under the microscope. Small or young individuals may also be prepared by soaking in oil of cloves, after previous preparation in strong spirit, when they become translucent, and when mounted in Canada balsam show the eyes to great advantage, besides giving an excellent view of the alimentary canal and some other interesting details of structure.

When moving about they hold up the anterior end, curving the sides downwards so that a cross-section would have the outline of a horseshoe, and delicately sweep the tip about, gently touching any objects that may be in the way. If a small article dipped in weak spirit, or the finger be held near the tip, they become aware of its presence without coming in contact with it, and quickly turn away so as to avoid it. Doubtless the ciliated pits situated in a row beneath the eyes at this part, serve, as suggested by different observers (3, p. 145; 8, p. 44), as olfactory organs. By their aid the worm probably becomes aware of the presence of objects to be avoided and also of its prey.

Though land planarians are properly nocturnal in their habits, some species are by no means adverse to moving about in daylight during dull, moist weather. I have specially noted this habit in the case of *Geoplana cærulea*, Moseley, which I have found at Wentworth Falls and elsewhere actively crawling along by day. My brother, Mr. John S. Steel, has noticed the same fact about this species at Mount Bauple in Queensland. Dendy has recorded the occurrence of the same habit in other species, notably *G. Sugdeni*, Dendy.

The tenacious slime which is so freely secreted all over the body surface is no doubt of service as a protection from possible enemies by rendering them unpalatable, as well as constituting an important means of capturing their prey.

Dendy has recorded his observations on the rejection of living planarians by domestic fowls (7, p. 69). I have noticed that fowls will readily eat planarians which have had the slime coagulated by preservation in spirit. This treatment would, of course, by coagulation, completely neutralize the objectionable properties of the slime.

Land planarians are, of course, hermaphrodite, the ♀ and ♂ organs opening into a common genital atrium the external

So far as I have been able to ascertain by reference to literature, Dendy is the only observer so far who has noticed the method of laying the egg-capsule. He describes the presence of a large aperture in the back of a specimen of *Geoplana triangulata*, Dendy, after the event (9, p. 116). Commenting on this, Graff (12, pp. 239-240, *footnote*) merely remarks that he supposes we have here to do with a rupture of the dorsal wall of the genital atrium.

The egg-capsule in some species attains large dimensions. I have an example of that of *G. Spenceri*, Dendy, which is 10 by 7 mm.

Although I have occasionally found egg-capsules laid in my vivaria by captive specimens, I have not until recently been able to confirm Dendy's observation on this curious habit. For the opportunity of doing so I am indebted to Mr. H. Stuart Dove, of Launceston, Tasmania, who sent me a fine specimen of *G. Mortoni*, Dendy. When the specimen reached me it had just deposited the capsule, which was still of a fresh brown colour. On the dorsal surface of the worm, just behind the position of the genital opening, was a rent or tear of the tissues through which projected the peculiar comb-like lobes of the genital organs. The protrusion of these shows that the dorsal wall of the genital atrium is torn to permit of the exit of the capsule.

Pl. xxxiv., fig. 10, shows the appearance of this interesting specimen. The opening caused by the exit of the capsule appears to heal very quickly, certainly within two or three days, as otherwise I must have noticed it on other occasions.

In writing to me regarding a specimen of *G. Dovei*, mihi, which was about to deposit a capsule, Mr. Dove specially mentioned that the integument on the dorsal surface just over the capsule was very tense and appeared ready to give way, and other observers besides myself have noticed the same appearance. From these facts and the obvious improbability of the large capsule being able to make its exit through such a small passage as the genital opening, I am inclined to think that Dendy's surmise as to the above being the normal mode of laying is correct.

The period of incubation for the egg-capsule after laying appears to be very variable. I have seen some hatch after being laid a few days only, while others have remained for seven to eight weeks in the vivarium before the young emerged.

(*b*). *Collection and preservation.*—When collecting land planarians they are best put into small tin boxes or wide-mouthed bottles, the covers or corks of which fit fairly close, but are readily removed. These should be loosely filled with fresh leaves such as those of docks, dandelions, fresh moss or such like, but anything wet, or aromatic leaves, should be avoided In such boxes of leaves they will live for a considerable time in cool weather, and I have so carried them alive from New Zealand to Sydney and thence to Melbourne, and have received by post in Sydney from friends in Tasmania, Queensland, Victoria, and South Australia.

Though planarians can be fairly well preserved by putting them direct into ordinary methylated spirit, they are displayed to much greater advantage when prepared in the manner described below. When simply placed in spirit the slime with which their bodies are invested is at once coagulated, forming a

Formaline solution made by diluting 3 volumes of the strong formaline of commerce to 100 volumes with water, has proved an admirable preservative, and has the great advantage of not coagulating the slime like spirit. In a general way, however, I prefer the latter medium, as on the whole the results attained when it is properly used seem to me rather the more pleasing. Though formaline in some cases induces rather more colour change than spirit, and is a somewhat disagreeable medium when the specimens have to be much handled during examination, it may be used with confidence if preferred. When the specimens have to be preserved in the field, or the collector is unable to do more than put them direct into the preservative, formaline should be used. It should not be made any stronger than indicated above, and should be used plentifully.

In preserving these worms—and I may here state that earthworms, leeches, and slugs may be treated in precisely the same manner—I first of all kill them with weak spirit. The strength which I prefer is 1 of ordinary methylated spirit to 15 of water. The worms are dropped into plenty of this contained in a shallow dish, and die in 10 to 15 minutes without the slime becoming coagulated, or the worm becoming distorted or broken. I now take them up one at a time by means of a pair of weak-springed flat-bladed forceps, and with the aid of a damp cloth and my wetted fingers gently draw the blades of the forceps along the worm's body, slightly compressing the body between them, and in this way scrape off the slime, wiping the forceps on a piece of rag. With a little care this can be done readily and safely even with the most delicate planarians, and leaves them beautifully clean and free from the objectionable slime. It is necessary to keep the fingers moist, otherwise the worm will stick and become damaged.

Having the formaline or strong spirit (80 to 85 per cent.) in another shallow dish, the worm is gently laid in it, and by means of fingers and forceps flattened and straightened out and prevented from shrinking too much as the spirit hardens it. A few moment suffices for this, and it is now left while the others are being similarly treated. Finally the specimens are transferred with formaline or spirit to the tubes or bottles in which they are to be preserved. Small specimens are more conveniently handled during hardening in a pool of spirit poured on a sheet of flat glass. Specimens for histological purposes may after killing and cleaning as above, be laid out in corrosive sublimate solution in alcohol, allowed to soak for an hour or so, and then transferred to spirit. They may thus be obtained nicely straightened and in good order for cutting into sections.

In the paper before-mentioned I described experiments on the use of chloroform, kerosene, &c., as preservatives, and it may be of interest to here record the results of these trials. Chloroform proved quite unsuitable, owing to its great volatility and want of hardening power. Kerosene, after prolonged contact, gradually displaced the water in the tissues, giving the specimens a translucent horny appearance, and completely spoiling them. The carbolised oil recommended by Haly* as a general preservative, is a very messy medium, and of no use at all for planarians. I understand that Haly has now abandoned its use in favour of formaline. I have tried acetone, but the results of its use are in no way better than with spirit.

I cannot recommend any better medium than spirit, used as described above, or formaline if more convenient.

of the spirit through the corks, and if the jar is placed in darkness, enables the specimens to be preserved indefinitely without any depreciation in colour. I mention this fact specially, as more than one of my friends who have preserved land-planarians have erroneously attributed loss of colour caused by exposure to light to bleaching by the spirit. Spirit, I have found by careful observation, will not cause any further alteration in colour after the fugitive body colour has been dissolved, if the specimens are kept in darkness. I have specimens preserved in this manner which have remained quite unchanged for seven or eight years.

In conclusion I desire to express my thanks to the friends already mentioned who have furnished me with specimens, to Professor W. A. Haswell, F.R.S., &c., for his kindness in giving me the use of his copy of Professor von Graff's Monograph, and to Mr. J. P. Hill, B.Sc., F.L.S., for histological and other assistance.

As it is extremely desirable that specimens of these interesting worms from as many parts of the colonies as possible should be secured for study and description, I shall be very pleased to hear from anyone willing to assist in the work of collection.

Postscript *(6th Oct., 1900).*

GEOPLANA MEDIOLINEATA, Dendy, var. SIMULARIS, n.var.

(Pl. xxxiv., fig. 6.)

A few days after this paper was read, I received from Mr. J. W. Mellor, of Fulham, near Adelaide, South Australia, a fine collection in a living state of the planarian of which I had only previously seen one perfect and several damaged specimens from near Adelaide, and two damaged examples from Western Australia, and which I had determined to be a variety of *G. quinquelineata*, Fletch. & Hamil.

Mr. Mellor's collection includes a large and graduated series which clearly proves that the form in question is really a variety of *G. mediolineata*, Dendy; and I am desirous, therefore, of correcting the determination given above at the earliest opportunity. From a careful examination of all the specimens now available, I am satisfied that they belong to one species, and that its affinities are with *G. mediolineata* rather than with *G. quinquelineata* as I at first supposed.

The collection sent me by Mr. Mellor may be divided into three series as follows :—

(*a*). Typical specimens of *G. mediolineata*, Dendy, (7, p. 76, pl. vii., figs. 1-2, and this paper, pl. xxxiv., fig. 4), with bold median line as usual and the lateral diffuse lines at anterior tip running back for a short distance.

(*b*). Examples of Dendy's three-lined variety of the same species (*loc. cit.*, p. 77, pl. vii., figs. 3-3*a*., and this paper, pl. xxxiv., fig. 5). The lateral lines extend in some individuals from end to end, while in others they fade out at varying distances back.

(*c*). A very distinct variety having five continuous dorsal lines (pl. xxxiv., fig. 6) and which constitutes the form which I propose to distinguish as *G. mediolineata*, var. *simularis*, to denote its resemblance to *G. quinquelineata*. From the latter species and its var. *accentuata* the present variety is distinguished by having the lines at unequal distances apart, the lateral lines being close to the median, while the marginal lines have about twice that space between them and the lateral lines, and are also about

think it may be safely identified as an example of series (*b*) as above.

Between these three forms there are numerous intermediate examples which carry us step by step from the typical *G. mediolineata* to the extreme var. *simularis* with five continuous lines. Several small specimens of var. *simularis* differ from the others in having all the lines diffuse, the spaces of body-colour closely speckled with brown, and the marginal lines distinctly split longitudinally so as to form seven lines in all, but these need not be specially considered, as they are obviously mere extreme variations.

All the specimens have the lines of a bright red-brown and anterior tip of same colour. The body-colour varies from bright sulphur-yellow to pale cream. Ventral surface white.

Hab.—Through the good offices of Mr. A. H. C. Zietz, F.L.S., &c., of the South Australian Museum, I have received specimens from near Adelaide, South Australia; from Miss C. A. Selway, Gilberton; from Miss Eimer, found in her shade-house at Norwood; and, as already mentioned, from Mr. J. W. Mellor. Mr. C. G. Hamilton sent me from Armadale, Darling Ranges, Western Australia, in spirit, two specimens which, though somewhat damaged, I have no doubt belong to the new variety.

While dealing with the relationship of the above species, it may be well here to mention that Graff (12, p. 374, pl. v., figs. 15-16) gives an account of specimens of *G. quinquelineata* from Victoria, sent to him by Dendy. His fig. 16 is the form which I named var. *accentuata* (*loc. cit.*), and fig. 15 is the same variety with broad diffuse paired stripes similar to those in the specimen from Milton, N.S.W., before mentioned.

REFERENCES.

1.—Darwin, C., Annals & Mag. Nat. Hist., Vol. xiv., 1844.

2.—Mueller & Schultze, "Beiträge zur Kenntniss der Landplanarien," Abh. d. Nat. Ges. zu Halle. 4r. Band, 1856.

3.—Moseley, H. N., Phil. Trans. Royal Soc. London, Vol. 164, 1873.

4.—Fletcher & Hamilton, Proc. Linn. Soc. New South Wales, 1887.

5.—Brittlebank, C. C., Victorian Naturalist, Melbourne, Vol. v., 1889.
6.—Dendy, A., Victorian Naturalist, Vol. vi., 1891.
7.———— Trans. Royal Soc. Victoria, Melbourne, 1890.
8.———— Proc. Royal Soc. Victoria, Melbourne, 1891.
9.———— Report Aust. Assoc. Adv. Science, Vol. vi., 1895.
10.—Spencer, W. Baldwin, Proc. Royal Soc. Victoria, 1890.
11.—Steel, T., Proc. Linn. Soc. New South Wales, 1897.
12.—Graff, L. von, Monog. der Turbellarien. ii. Tricladida Terricola (Landplanarien), Leipzig, 1899.
13.—Scharff, R. F., Journ. Linn. Soc. Zoology, London, Vol. xxviii., 1900.

EXPLANATION OF PLATE XXXIV.

Fig. 1.—*Geoplana fusco-dorsalis*. Dorsal aspect (nat. size).
Fig. 2.— „ *arenicola*. *a*. Dorsal, *b*. ventral aspect (× 2).
Fig. 3.— „ *melanochroa*. Ventral aspect (× 3).
Fig. 4.— „ *mediolineata*. Dorsal aspect of typical form. *a*. Anterior portion, *b*. middle (× 4).
Fig. 5.— „ *mediolineata*. Dorsal aspect of three-lined form, middle of body (× 4).
Fig. 6.— „ *mediolineata*, var. *simularis*. Dorsal aspect, middle of body (× 4).
Fig. 7.— „ *quinquelineata*. Dorsal aspect, typical form, middle of body (× 4).
Fig. 8.— „ *quinquelineata*. Dorsal aspect, diffuse-lined form, middle of body (× 4).
Fig. 9.— „ *graminicola*. Dorsal aspect (× 3).

OBSERVATIONS ON THE TERTIARY FLORA OF AUSTRALIA, WITH SPECIAL REFERENCE TO ETTINGSHAUSEN'S THEORY OF THE TERTIARY COSMOPOLITAN FLORA.

Part II.

On the Venation of Leaves and its Value in the Determination of Botanical Affinities.

By Henry Deane, M.A., F.L.S., &c.

(Plates xxxv.-xxxviii.-xxxviii. *bis.*)

If a botanist, visiting a part of the world hitherto unexplored and failing to find the flowers or fruits of the plants occurring there, were to content himself with collecting leaves and thereupon to proceed to classify his specimens in natural orders and genera and to give them specific names, what would be thought of him? He might have ardently devoted himself to the task of collecting, have exposed himself to hardships and dangers in so doing, and he might have spent months or years at the work of classifying and describing his collection, and publishing his results, but in the end what would be the value of all his labours? To say the least of it, he would only earn ridicule if he were spending his own money; if, on the other hand, his expenses were provided by others, he would be scouted as an impostor, and he would run great risks of being put either into a gaol or a lunatic asylum. Yet it is difficult to distinguish this method from that adopted for the most part by palæontologists in dealing with fossil leaves. It is very rarely that fruits are found, and when they are, the structure is to a large extent obliterated, so that leaves are almost entirely depended on. Thousands of species have been created and their affinities declared on the evidence afforded by leaves alone, and often with an imperfect knowledge of the existing flora of the region and a preconceived desire to prove a certain theory.

The character of leaf-venation would prove an unerring guide to the determination of fossil plants if those characters were arranged through the vegetable kingdom on any orderly plan. If the leaves of all plants of a particular natural order possessed a common general character, and if those characters became more more defined and specialised as the genera were dealt with—if, finally, the specific differences were less than the generic differences, and still less than those of the ordinal differences, leaves would be as important a guide in the work of botanical classification as the parts of the flower, and the value of any particular fossil leaves would only depend upon the perfection of their preservation, as the affinities of those in which the substance and venation remain intact could then be absolutely determined.

One does not need to go far through the vegetable kingdom to find that leaves are not constructed on this methodical system, and almost immediately it will be recognised that not only do the leaves of different species of the same genus often differ very greatly from one another, but that their venation seems to be laid

have at least something in common, even if they are not made on exactly the same plan; whereas those names refer only to the common species with which we from our childhood have been familiar. Even the genus *Acer*, most of the species of which have leaves constructed on the palmate or five-lobed pattern, shows very considerable variation; while the leaves of the different species of the genus *Quercus* show innumerable varieties of design, so great a divergence, indeed, that without following them through all gradations and intermediate links it might be wondered whether they could ever in the past have originated in one ancestral stock. The leaves of the same species seem occasionally to be made on a different pattern, as, for example, in *Q. aquatica*, a North American species, where in some cases the lateral or secondary veins run out to the margin of the lamina and terminate in teeth, while in other varieties they do not reach the margin at all, which is entire, but become lost in the general reticulation. *Q. cinerea* is another example.

The genus *Quercus* furnishes an excellent example of leaf-variation, and I have taken the opportunity of offering a number of examples for illustration. The genus is one which at the present day is widely distributed and which is, according to Ettingshausen, responsible for a great many of the fossil forms found in the Miocene beds of Australia. It, therefore, possesses a special interest. The number of species known to the authors of the "Genera Plantarum" and recognised by them is 300. In Plate xxxv. a sufficient number are figured to prove the great variety of types. As before mentioned, *Q. aquatica* and *Q. cinerea* show great varieties within the species. Reference may be made to Ettingshausen's* *Beiträge zur Tertiarflora von Java*, where remarkable variations of these two species are figured.

I have tabulated below the variations shown in Plate xxxv., grouping the leaves first, according to their shape; secondly, according to the character of the secondary veins: thirdly, according to the tertiary system. The texture of the leaf, whether mem-

* Sitzungsbericht der K. Akad. der Wissensch. 1 Abth. 1883, Vol. 87.

branous or coriaceous, flat, convex or with recurved margins, might also be used as a basis of classification. It may be seen that as all these different characters pass into one another by almost imperceptible gradations, there might be tabulated an enormously large number of combinations sufficient to include most dicotyledonous leaves, provided only that digitate, palmate and penniveined leaves are excluded.

SHAPE—

Entire—ovate, obtuse	*Q. elliptica.*
„ obovate	*Q. Phellos.*
„ ovate, lanceolate	*Q. lancifolia.*
„ „ „	*Q. amherstiana.*
„ „ „	*Q. fenestrata.*
„ broadly ovate, lanceolate	*Q. philippinensis.*
Serrate, regularly not deeply lanceolate	*Q. oxyodon.*
„ irregularly lanceolate	*Q. pseudococcifera.*
„ „ „	*Q. lancifolia.*
Irregularly toothed, broadly ovate	*Q. alnifolia.*
„ „ ovate	*Q. suber.*
„ „ „	*Q. coccifera.*
Coarsely toothed, lanceolate	*Q. castaneæfolia.*
„ „ ovate	*Q. Xalapæ.*
Regularly toothed, cuneate	*Q. Libani.*
Coarsely sinuate-toothed, rhomboidal	*Q. coccinea.*
Crenate, ovate	*Q. montana.*

Wavy and reaching margin .... *Q. stellata.*
" " " .... *Q. coccinea.*
" " " .... *Q. pedunculata.*
Curved and reaching margins .... *Q. oxyodon.*
" " " .... *Q. alnifolia.*
Curved and reaching the next above .... *Q. amherstiana.*
" " " " .... *Q. philippinensis.*
Curved and indistinct towards margin .... *Q. elliptica.*
" " " " .... *Q. Phellos.*
Irregular and indistinct... .... *Q. virens.*

TERTIARY VEINS—

At right angles to secondary, or nearly so......... *Q. castaneæfolia.*
" " " " ... . *Q. oxyodon.*
" .... *Q. amherstiana.*
" .... *Q. Libani.*
" .... *Q. Prinos.*
" .... *Q. Xalapæ.*
" .... *Q. pubescens.*
" .... *Q. montana.*
" " " " .... *Q. alnifolia.*
" " " " .... *Q. suber.*
At right angles to midrib .... *Q. philippinensis.*
" " " .... *Q. lancifolia.*
Reticulate, but more or less arranged in lines ... *Q. sessiliflora.*
Coarsely reticulate .... *Q. stellata.*
" " .... *Q. coccinea.*
More finely reticulate .... *Q. oloides.*
" " . ... *Q. pseudococcifera.*
" " .... *Q. Phellos.*

In Plate xxxvi. I give some illustrations of leaves of the genus *Eucalyptus*, showing the remarkable variation in structure existing in the leaves of different species. Probably there are few people in Australia who would not express themselves sure of recognising a "gum tree" leaf when they saw it: they might even feel themselves insulted if doubt were expressed as to their power of doing so, and yet it is quite clear they could not express off-hand their idea of the typical *Eucalyptus* leaf. It may be a surprise to many to find on what different plans the vein system of the leaves of different species is arranged, and how impossible it is to pick out any one variety and say that is the *Eucalyptus* type. The secondary veins afford a great many different varieties. Observe for instance:—

E. coriacea and *E. stellulata* with their longitudinal veins; *E. Siberiana* and others with secondary veins placed at an acute angle with the midrib.

Follow the series down till the secondary veins become almost transverse.

E. microcorys, unlike most Eucalypts, has quite thin leaves.

The shape of the leaves might be thought to be fairly uniform and that they would all fall under the description of linear, lanceolate, acuminate, more or less falcate, and oblique at the base. All these characters break down in some of the specimens. Then it has been supposed that the intramarginal vein would be a pretty sure guide. It is, however, found in *Myrtaceæ* generally, in some *Proteaceæ*, in many *Apocynaceæ*, in many species of *Ficus*, and in genera belonging to many other natural orders. It is further to be remarked that in *Eucalyptus* itself its position is very variable, so that while in some leaves it is a considerable distance from the margin, in others it is so close to the edge as to be barely distinguishable.

In Plate xxxvii. are shown various Australian representatives of the Natural Order *Laurineæ*. It will be seen that the

Compare *Cupania serrata*, *Harpullia alata* and *Akania Hillii* with *Xylomelum pyriforme*, *Lomatia ilicifolia*, *L. Fraseri* and *Orites excelsa*.

On the other hand, note the differences between different leaves of the same species, *Akania Hillii* (Pl. xxxviii. *bis*, figs. 5-8), and between the leaves of *Lomatia Fraseri* and *L. ilicifolia*, plants which have such a multitude of intermediate forms that it becomes difficult, if not impossible, at times to decide under which species specimens are to be grouped (Pl. xxxviii. *bis*, figs. 1-3).

The pinnæ of many of the Australian *Sapindaceæ* and the leaves of some Australian *Saxifrageæ*, and the pinnæ of others, correspond to a remarkable degree with those of certain species of *Quercus*, a fact which Ettingshausen has apparently overlooked.

The leaves of *Doryphora sassafras* and those of the toothed variety of *Atherosperma moschata* (*Monimiaceæ*) are built on the same plan as the toothed variety of leaves of *Myrsine variabilis* (*Myrsinaceæ*). There is no one point in which they can be said to belong to a different type. A comparison of different leaves off the same plant of *Myrsine variabilis* affords an example of remarkable specific variation (Plate xxxviii., figs. 3-4).

As examples of leaves almost identical occurring in different Natural Orders, those of *Pennantia Cunninghamii* (*Olacineæ*) may be compared with *Villaresia Moorei* (*ib.*), and what is still more remarkable the leaves of *Vitis sterculifolia* and *Endiandra discolor* (*Laurineæ*) show a resemblance even to the glands or domatia which exist in the angles formed by the secondary veins with the midrib.

Compare also *Helicia ferruginea* (*Proteaceæ*) with *Olearia argophylla* (*Compositæ*) in Plate xxxviii. *bis*, figs. 11-12.

There are innumerable other examples of leaves of different genera and orders being so much alike that it is impossible to distinguish them, but the above examples are perhaps sufficient to prove the fact of the existence of such similarities. While searching for the affinities of fossil leaves the difficulty is generally not in finding resemblances but in making a selection, as the resemblances are often so numerous.

Judging from a paper read by Ettingshausen before the Imperial Academy of Vienna in the year 1854,* no one could recognise the difficulties more than that author. In this paper he mentions how attention had been called to the subject of leaf venation two years before by Leopold von Buch at a meeting of the Berlin Academy of Sciences, and regret therein expressed that investigation with a view to classification of leaf characters had not been seriously attempted. Ettingshausen points out in his paper the impossibility of carrying out any systematic classification, and he mentions how the forms and venation of *Ficus* and *Vochysia*, *Cinnamomum* and *Strychnos*, *Mertensia* and *Ceanothus* or *Zizyphus*, *Fagus* and *Dipterocarpus*, *Salix* and certain *Lythrarieæ*, *Jacaranda* and *Mimoseæ*, *Nyssa*, *Diospyros* and *Pittosporum*, *Santalum* and *Sapotaceæ* entirely correspond. On the other hand, he shows how we come across the most heterogeneous types of leaves in one and the same natural order or even the same genus, as for example, in *Bignoniaceæ*, *Saxifrageæ*, *Büttneriaceæ*, *Euphorbiaceæ*, in *Ficus*, *Sterculia*, &c.

It would appear, therefore, that the matter scarcely calls for any argument seeing that this great authority in fossil leaf naming acknowledged at the outset how utterly unreliable leaf characters

sound in principle, when it consists in comparing fossil remains from successive horizons with existing plants growing in the same or in contiguous regions, but it is quite a different thing when the attempt is made to prove that certain fossil forms belong to groups non-existent in this particular part of the world and found on the other side of the equator, and perhaps in another hemisphere.

EXPLANATION OF PLATES.

Plate xxxv.

Fig. 1.—*Quercus elliptica.*
Fig. 2.— ,, *Phellos.*
Fig. 3.— ,, *aquatica.*
Fig. 4.— ,, ,,
Fig. 5.— ,, ,,
Fig. 6.— ,, *lancifolia.*
Fig. 7.— ,, ,,
Fig. 8.— ,, *amherstiana.*
Fig. 9.— ,, *fenestrata.*
Fig. 10.— ,, *philippinensis.*
Fig. 11.— ,, *oxyodon.*
Fig. 12.— ,, *suber.*
Fig. 13.— ,, *alnifolia.*
Fig. 14.— ,, *pseudococcifera.*
Fig. 15.— ,, *castaneæfolia.*
Fig. 16.— ,, *Libani.*
Fig. 17.— ,, *Xalapæ.*
Fig. 18.— ,, *coccinea.*
Fig. 19.— ,, *Prinos.*
Fig. 20.— ,, *montana.*
Fig. 21.— ,, *sessiliflora.*
Fig. 22.— ,, *pubescens.*

Plate xxxvi.

Fig. 1.—*Eucalyptus coriacea.*
Fig. 2.— ,, *stellulata.*
Fig. 3.— ,, *Sieberiana.*
Fig. 4.— ,, *amygdalina.*
Fig. 5.— ,, *obliqua.*
Fig. 6.— ,, *pilularis.*
Fig. 7.— ,, *microcorys.*

Fig. 8.—*Eucalyptus polyanthema.*
Fig. 9.— ,, ,,
Fig. 10.— ,, *paniculata.*
Fig. 11.— ,, *populifolia.*
Fig. 12.— ,, *largiflorens.*
Fig. 13.— ,, *longifolia.*
Fig. 14.— ,, *maculata.*

Plate xxxvii.

Fig. 1.—*Cinnamomum Burmanni.*
Fig. 2.— ,, *Tanala.*
Fig. 3.— ,, *Oliveri.*
Fig. 4.—*Cryptocarya triplinervis.*
Fig. 5.— ,, *cinnamomifolia.*
Fig. 6.— ,, *australis.*
Fig. 7.— ,, *microneura.*
Fig. 8.— ,, *obovata.*
Fig. 9.—*Litsæa hexanthus.*
Fig. 10.— ,, *dealbata.*
Fig. 11.—*Endiandra glauca.*
Fig. 12.— ,, *Muelleri.*
Fig. 13.— ,, *pubens.*
Fig. 14.—*Rhodamnia trinervia.*
Fig. 15.—*Mallotus philippinensis.*

Plate xxxviii.

Fig. 1.—*Doryphora sassafras.*
Fig. 2.—*Atherosperma moschata*

NOTES ON THE BOTANY OF THE INTERIOR OF NEW SOUTH WALES.

By R. H. Cambage, L.S.

(Plates xxxix-xl., fig. 3.)

Part I.—From the Darling River at Bourke to Cobar.

In following my duties as a Mining Surveyor in the Western District, opportunities have been afforded of observing the flora in certain parts of the interior of New South Wales, and though I do not claim to have made a complete botanical survey, but have only noted the principal trees in passing along, I have thought that the information thus acquired might be acceptable. My endeavour has been to prepare it in such a way as to be useful to a student in the field passing over the same country, and to serve as a record for future reference.

I am indebted to Mr. J. H. Maiden, F.L.S., Director of the Botanic Gardens, and Mr. R. T. Baker, F.L.S., Curator of the Technological Museum, for assistance in identifying some of the plants.

The country dealt with in Part I. of this paper extends from Bourke to Cobar, but I propose to subsequently speak of the principal trees met with thence to the Bogan, Lachlan, and Murrumbidgee Rivers.

Starting at Bourke on the Darling River we have *Eucalyptus rostrata*. Schl., (River Red Gum), following the banks of the stream, where is also *Nicotiana glauca*, a tobacco plant introduced from South America. On the level river country a conspicuous tree is *Eucalyptus microtheca*, F.v.M., (Coolabah), which is easily identified by its rough grey bark all over the trunk, and its perfectly smooth white limbs. Another tree very similar in general appearance, as both have pale leaves and a drooping habit, is *E. largiflorens*, F.v.M., though a little inspection soon enables one to separate them, as the latter has the grey bark covering the branches as well as the trunk. Moreover the fruits are quite different, those of *E. microtheca* being readily identified by the short calyx and exserted valves. Both are considered to be "Box" trees. There seems to be a variety of *E. largiflorens* with greener and broader leaves than the type, but the fruits are identical.

All these trees—*E. microtheca*, *E. largiflorens* and its variety, according to my observations—grow only on what is known as the river or black soil country, and never away on the hills. They are of crooked growth, and average about 30 to 40 feet high. Over the country which is now being described, *E. microtheca* was only found extending as far as 12 miles south of Bourke, ceasing with the black soil, though it goes northward through Queensland; while *E. largiflorens* was noted again on the Bogan 30 miles above Nyngan, and also on the Lachlan at Condobolin.

Leaving the Darling River the road taken from Bourke was towards Cobar, which is south about 100 miles. For the first 9 miles no other Eucalypts were noted except *E. microtheca* and *E. largiflorens*, the next to appear being *E. populifolia*, Hook.,

Acacia stenophylla, A. Cunn., following a damp course.

Acacia Oswaldi, F.v.M., (Dead Finish or Miljee).

Ventilago viminalis, Hook., (Supple Jack). This vernacular name arose from the fact that the branches and stems of these trees often entwine, thereby presenting some similarity to the vines known as Supple Jack on the coast.

At 9 miles, on a low quartzite ridge, are :—

Acacia aneura, F.v.M. (Mulga) which is about the principal fodder tree between Bourke and Cobar in times of drought.

Grevillea striata, R.Br., (Beefwood, from the wood being prettily marked with medullary rays).

Atalaya hemiglauca and *Owenia acidula*, F.v.M., (Colane or Gruie).

E. populifolia and *Geijera parviflora*, Lindl., (Wilga) a beautiful shade tree common all over the western district, and one of those few trees which has only one vernacular name. Often heated controversy arises in the bush owing to one species having about half-a-dozen vernacular names, or the one vernacular name is sometimes applied to several trees in different localities.

Again near the road is *Apophyllum anomalum*, F.v.M., (Currant or Warrior Bush).

Just before reaching the 12 mile post *Eucalyptus microtheca* and *E. largiflorens* cease, and the former is seen no more. Then we have :—

Eremophila Mitchelli, Benth., aboriginal name "Budtha," and sometimes called Sandalwood from the fragrance of the wood, but not to be confused with the Sandalwood of Western Australia, *Santalum cygnorum*. "Budtha" is one of the strongest scented woods of the western district, and the trees are commonly up to nine inches in diameter; but, unfortunately, when they attain that size they generally show a strip of decay up one side which seriously impairs their usefulness.

Capparis Mitchelli, Lindl, (Wild Orange, as it bears a fruit somewhat similar to an orange in shape and size, and moreover the trees are thorny).

Casuarina Cambagei, Baker, (Belah). This is the only Casuarina wood that does not show the medullary rays, although they are to be seen in the upper branches. In the Bourke to Cobar district the Belah has a very glaucous appearance, which is possibly one of the reasons why it was confused with *C. glauca* in the past. Towards the Lachlan the branchlets are greener.

Acacia excelsa, Bentham (Ironwood, from the hardness of the wood, which is brittle and inclined to splinter).* Mature trees have a clean trunk and drooping foliage (Pl. xl., fig. 3), but the young trees are covered with branches on the trunk. A curious feature of many interior trees is the protection afforded by spreading growth and numerous branches in young stages as compared with that of after years. Several species have this peculiarity, among others being *Acacia excelsa*, *Grevillea striata*, *Capparis Mitchelli*, and perhaps most of all *Flindersia maculosa*, F.v.M. In the Agricultural Gazette of New South Wales (Vol. x., Part 11), Mr. W. S. Campbell, F.L.S., has drawn attention to this matter in an interesting illustrated article.

Acacia aneura is next noticed.

Near the 12 mile post is a clump of trees of a species of Acacia commonly called Gidgea. This is a well known north-western tree, as the timber is much sought after for stockwhip handles, owing to its durability, hardness and dark colour. Fence posts of Gidgea on the Hungerford-road, north of Bourke, said to have [illegible]

passing this clump 12 miles south of that town, no more is seen on the road to Condobolin.

So far I have not been able to satisfactorily identify this species, and have reason to think that there has been some confusion with *Acacia homalophylla* (Yarran). Its nearest affinity certainly seems to be with this species. In the "Flora Australiensis" (Vol. ii., p. 383) the phyllodia are spoken of as being "very finely striate with parallel veins only to be seen under a lens." Now this could not refer to Gidgea, as the veins in the leaves are quite distinct; while in Yarran there are often none visible, a difference generally noticeable even in dried specimens. Again *A. homalophylla* is found in Victoria, but it is usually understood in the west that Gidgea does not occur much south of the Bourke district. In fresh specimens there are two very simple and decisive tests for these trees. One is to damp the leaves, and Gidgea will soon proclaim itself by the smell; the other is to double the leaf between the thumb and finger, and if a Yarran it will snap right off and seldom hang by even the slightest fibre, while the veins in the Gidgea leaf will prevent it from snapping at all. This test is no use in dried specimens, as both will snap readily. Gidgea wood is darker, stronger and heavier than Yarran, but in the bark and general appearance the trees are somewhat similar. Gidgea is the larger tree of the two, and when compared with Yarran of mature growth is the more umbrageous.

In botanical specimens there is no doubt it is difficult to separate these two species, though in the field they are seldom confused. Gidgea foliage has always appeared to me as fairly dark green, and Yarran more of a yellow-green; though curiously in the herbarium the leaves of the former dry nearly white, and those of Yarran often appear dark beside them. Still here they can generally be identified even in very old specimens, as a stray leaf of the Gidgea may have been doubled or twisted in the pressing and have retained its bent form. Should a Yarran leaf be doubled in pressing it will snap.

Assuming that both these trees have been placed under *A. homalophylla*, an assumption which may be disputed, the question then arises which is the type. The evidence on this point is in favour of Yarran. In the first place the species was described by Allan Cunningham, who must certainly have seen Yarran in his exploration trip across the Liverpool Plains and on to the Darling Downs in 1825. The country from Bourke northward in which Gidgea grows was not visited by a white man till Sturt reached it in 1828, so that it seems possible that Cunningham never saw Gidgea. But if he did, he would scarcely have described the western tree, and have ignored the species through which he had been passing for at least 150 miles.

According to Mr. F. M. Bailey, F.L.S., Government Botanist of Queensland, there is an Acacia with a strong smell towards the western boundary of that colony on the Georgina River, and evidently to distinguish it from the original Gidgea (or Gidgee as it is often spelt) of Bourke to Charleville this tree is called Georgina Gidgee (*Acacia georginæ*, Bail.). In describing the phyllodia of this tree ("The Queensland Flora," Part ii., p. 495) Mr. Bailey writes, "texture thick, hard and brittle." Now this differs from the Bourke tree, as in it the leaves are neither thick nor brittle. The pods so far as seen are also quite different.

I understand that Mr. Baker is now dealing with this matter,

Dodonæa viscosa, var. *attenuata*, A. Cunn., (Hopbush).

Heterodendron oleæfolium.

Atalaya hemiglauca.

Eucalyptus populifolia.

Pittosporum phillyræoides, DC., a graceful tree up to 20 ft. high, with drooping foliage and yellow fruit. "Berrigan" is said to be the aboriginal name for it on the Lachlan.

At 16 miles a new Eucalypt with pale leaves appears, *E. intertexta*, Baker, and continues practically the whole way to Condobolin, crossing to the south of the Lachlan. It is known variously as "Gum," "Coolabah," "Yellow Box," "Red Box," "Bastard Box," and is one of the largest trees in the west. It gets its name of Gum from its upper bark on the trunk and branches being white and smooth, while the lower is light brown and flaky, but the hardness of the wood, which is red and difficult to split, proclaims its affinity to the Box trees.

Between the 16 and 43 mile posts, besides *Eucalyptus populifolia* and *E. intertexta*, which continue all the way, and are the only Eucalypts, there are :—

Capparis Mitchelli.

Loranthus linearifolius, Hook., (a mistletoe with pink flowers in May, growing on *Flindersia maculosa*).

L. Exocarpi, Behr. (a mistletoe with yellow flowers in May, growing on *Acacia aneura*).

Heterodendron oleæfolium, rather a dwarf form known as Blue Bush from its glaucous leaves. This species is altogether of diminutive size about Bourke and Cobar, and becomes much larger towards the south-east.

Acacia Burkittii, F.v.M., (Needle Bush), growing in clumps of a few acres, about 6 or 8 feet high.

At 27 miles are :—

Acacia homalophylla, (A. Cunn.), rather a drooping form of Yarran, but not *Acacia pendula*, A. Cunn.

A. excelsa.

Grevillea striata.

Eremophila Mitchelli, a common tree throughout the western district.

Geijera parviflora.

Acacia Oswaldi.

Apophyllum anomalum.

Atalaya hemiglauca.

At 30 miles are :—

Hakea leucoptera.

Sterculia diversifolia, G. Don, (Currajong).

Cassia artemisioides.

At 40 miles are :—

Ventilago viminalis.

Santalum lanceolatum, R.Br. This is a tree with light brown bark and very pale wood, often called "The Blacks' Medicine Tree," from the fact that the bark soaked in water was formerly used by the aborigines for medicinal purposes.

About 8 miles west of this point is *Fusanus acuminatus*, R.Br., (Quandong), growing on Gundabooka Mountain, the formation of which appears to be Devonian Sandstone.

Leaving the main road at 43 miles, going south-westerly to Wilgaroon Station and returning to the main road at 55 miles from Bourke or 43 miles from Cobar, the trees and shrubs noted, in addition to those of the last 20 miles, are :—

From 43 miles north of Cobar to Mount Drysdale at 22 miles are :—

Sterculia diversifolia.

Acacia aneura.

Eucalyptus terminalis, F.v.M., (Bloodwood), found on Mount Dijou 5 miles east, where I collected it in June, 1892, but I can hear of it nowhere south of this.

At 41 miles are :—

Fusanus acuminatus.

Jasminum lineare, R.Br., a climber.

Lyonsia eucalyptifolia, F.v.M., one of the largest western climbers.

At 40 miles are :—

Celastrus Cunninghamii, F.v.M.

Beyeria viscosa, Miq.

Eucalyptus Morrisii, Baker, the most stunted form of Mallee in the Cobar district; and considered the easiest to eradicate.

E. viridis, Baker, (Narrow-leaf or Whipstick Mallee).

Acacia doratoxylon, A. Cunn., (Currawong).

Acacia decora, Reichb., (Silver Wattle).

At 33 miles from Cobar *Flindersia maculosa* ceases on this road, though it extends westerly.

At 29 miles the first White Pine, *Callitris robusta*, R.Br., is met. In coming from Bourke to Nyngan in the train the first of this species is seen between Coolabah and Girilambone at about 85 miles south-easterly from Bourke. This is about east of the point where it is met on the Bourke to Cobar road at about 70 miles south of Bourke.

Atalaya hemiglauca ceases near the 30 mile post from Cobar, and is seen no more on this trip.

Most of the trees mentioned between the 16 to 40 miles from Bourke continue to Mount Drysdale.

I was informed that five miles west of Drysdale there are a few acres of *Acacia harpophylla*, F.v.M., (Brigalow), but was unable to go and see it.

From Mount Drysdale to Cobar, 23 miles, *Alstonia constricta* was no more seen, and I think its habitat is northerly. The following were noted :—

Helichrysum Cunninghamii, Benth.	*Geijera parviflora*,
Eucalyptus Morrisii,	*Acacia Oswaldi*,
Acacia doratoxylon,	*Apophyllum anomalum*,
Acacia decora,	*Eremophila Mitchelli*,
A. excelsa,	*Acacia homalophylla*,
Eucalyptus intertexta,	*Heterodendron oleæfolium*,
E. populifolia,	*Dodonæa viscosa*, var. *attenuata*,
Grevillea striata,	*Acacia Burkittii*,
Acacia aneura,	*Canthium oleifolium*,
	Callitris robusta.

The above are given in the order in which they were met with.

At the 20 mile post :—

On the left are about a dozen acres of Mallee, *Eucalyptus oleosa*, F.v.M., often called Red Mallee from the colour of the wood.

Then there are :—

Fusanus acuminatus,

Sterculia diversifolia, an excellent fodder tree, but limited in quantity),

Hakea leucoptera,

it became evident that the rounded form is most common in the northern locality, and that the leaf assumes the flattened shape as the species comes south-easterly. Or comparing it with climatic conditions, the rounded leaf is general in the hotter part and the flat leaf in the cooler, though I can only speak of the country now being described. The pods and yellow flowers are the same throughout.

At 12 miles :—*Acacia homalophylla*,
Eucalyptus intertexta,
E. populifolia.

At 7 miles :—*E. oleosa*.

At 6 miles :—*Casuarina Cambagei*.

At 5 miles :—*Geijera parviflora*,
Acacia aneura,
Eremophila Mitchelli,
Callitris robusta,
Exocarpus aphylla,
Capparis Mitchelli,
Acacia colletioides, A. Cunn., (Pin Bush),
A. hakeoides, A. Cunn., (in some western localities called Black Wattle),
A. excelsa,
Dœmia quinquepartita, F.v.M., a climber.

At 3 miles :—*Hakea leucoptera*,
Eucalyptus dumosa, A. Cunn.,
E. viridis,
Fusanus acuminatus.

At 2 miles :—*Pittosporum phillyrœoides*,
Myoporum deserti,
Ventilago viminalis, (not seen again after passing Cobar).

The distance travelled from Bourke to Cobar is about 100 miles, and altogether nine Eucalypts were seen, viz. :—

E. rostrata (only along the river), *E. microtheca*, *E. largiflorens*, *E. populifolia*, *E. intertexta*, *E. Morrisii*, *E. viridis*, *E. oleosa* and *E. dumosa* in the order named. About midway and five miles east of the road is *E. terminalis*.

Four of these species are what are known as Mallees, a name applied to those dwarf Eucalypts which throw out a cluster of stems from one root, and also grow in clumps or scrubs from a few acres up to several hundred. They prefer slightly elevated land (not necessarily high ridges), never being found on river flats. In this they are the exact anthithesis of *E. microtheca* and *Acacia pendula* (Myall or Boree). The leaves are full of oil, and around Cobar where these trees are plentiful there is certainly a great future in store for the oil industry. As the trees are short the leaves are easily accessible, and pruning them stimulates the growth, so that it might almost be said the supply is inexhaustible.

The Mallees are all easily separated by their fruits, but a bushman would recognise *E. Morrisii*, Baker, as being the shortest, having fairly broad leaves, rather rough flaky brown to grey bark, getting whiter near the top, the softest wood, and being more easily torn out by the roots than any of the others. The fruits are large, often in threes, and might be confused with some forms of *E. tereticornis*, Sm., or *E. viminalis*, Labill. The trees average from 10 to 15 ft. high, with a diameter from 2 inches to 6 inches. The wood reminds one of *E. dealbata*, F.v.M., and in other ways seems to have affinities with it, but I have seen both growing on one hill, and they are quite distinct.

spreading habit. This species also grows into fairly large trees 50 feet high, and is then known as Big or Giant Mallee. Its fruits are rather large, being slightly barrel-shaped, with the greatest diameter in the middle. For utility, its wood, which is hard, stands before any of the other three.

E. dumosa, A. Cunn., is known as White Mallee from its having white smooth bark to the ground. It is generally found growing with *E. oleosa*, and these two form Mallee scrubs, sometimes associated with *E. viridis*, but the latter will often form a scrub by itself, as also will *E. Morrisii* to a less extent.

E. dumosa and *E. oleosa* might be confused in the field through growing together and the great similarity in their leaves. Their fruits at once separate them, those of *E. dumosa* being generally the larger and not constricted at the rim. A characteristic difference in general appearance is that the stems of *E. dumosa* are erect and white, while those of *E. oleosa* are more spreading, slightly crooked, and have the lower parts covered with brown flaky bark (Pl. xxxix., figs. 1-2). Both have hard wood. It may here be mentioned that no other Mallees except these four were met with north of the Lachlan on this road.

One feature noticed all through was that all these species prefer sedimentary formation, generally Silurian slate, and rarely grow on igneous rocks.

The only Casuarina noticed between Bourke and Cobar was *C. Cambagei*, Baker.

The Acacias were *A. stenophylla*, *A. Oswaldi*, *A. aneura*, *A. excelsa*, *A.* sp., (Gidgea), *A. Burkittii*, *A. homalophylla*, *A. doratoxylon*, *A. decora*, *A. colletioides* and *A. hakeoides*.

Generally speaking the country traversed is level, but along the southern half towards Cobar a few hills of sedimentary origin rise a few hundred feet above the surrounding plain.

The view from one of these tops is very different from anything to be seen near the coast, and less beautiful. Still it is well worth seeing, and there is a weird charm in the great expanse of wilderness which appears on every side.

Other hills can be seen rising here and there at distances up to 30 or 40 miles away. From the base of one to the other the land is so level that in a good season it looks like a green carpet extending for miles, and the heads of the waving Mallee remind one of a field of maize. But too often the carpet seems horribly worn into great patches of red and brown, and the drought-stricken land, denuded of all grass and herbage, glares through the straggling foliage.

A FISH DISEASE FROM GEORGE'S RIVER.

By R. Greig Smith, M.Sc., Macleay Bacteriologist to the Society.

In the middle of August there was received from the Fisheries Department the carcase of a bream which had been found floating, in a dying condition, on the surface of George's River, a few miles south-west of Sydney. The epidermis showed several slightly hæmorrhagic patches. The lateral blood vessels and adjacent portions of the muscles were congested; the liver was mottled and pulpy; the peritoneal cavity contained cheesy masses. The stomach was much congested and contained a reddish viscid fluid. The intestine was slightly congested near the anus, and contained a dark bile-coloured fluid. The other organs were apparently normal, and the muscles and alimentary canal were found to be free from parasites.

In the bacteriological investigation, plate cultures were made with media which had been inoculated with the several parts of the carcase. The muscles were proved to be sterile, while the heart blood, the spleen, and the liver contained three organisms. The first of these was a gelatine-liquefying fluorescent bacterium (*Bact. fluorescens*), which was not investigated further. The second was a rather large bacterium, which slowly liquefied gelatine and exhibited bipolar staining. A pure culture of this organism was distributed in normal saline, and a few drops of the suspension was injected into the muscles of two carp. Beyond showing a scar at the point of inoculation, these were unaffected and were apparently healthy after a month's observation. The organism was not investigated further.

The third organism was a smaller bacterium which rapidly peptonised gelatine and stained bipolarly. It occurred in practically pure culture in the intestine, and with the other two bacteria in the organs. When suspended in normal saline and inoculated into the muscles of a carp, it strongly affected the experimental fish. In 24 hours its movements were slow; in 40 hours (the morning of the second day) it was found floating upon its side near the surface of the water of the aquarium, respiring rapidly and swallowing and ejecting air. It was apparently in a dying condition. Four hours later it had sunk to the bottom of the water and was respiring very slowly, still lying upon its side. On the morning of the third day it was found dead, floating upon the surface of the water. Death had probably taken place in 52 hours after inoculation.

No lesions were observed on the epidermis with the exception of a hæmorrhagic streak between the rays of the caudal fin. The muscles were very hæmorrhagic on the side upon which the carp had been inoculated, and which had been downwards for practically 24 hours. The muscle at the site of inoculation (midway between the anal fin and the posterior end of the dorsal fin) was a shade less hæmorrhagic than the surrounding muscle. The organs, with the exception of the kidneys, were apparently normal; these were pulpy and swarmed with the inoculated bacteria. A plate culture from this organ showed the bacteria

not infected by way of the alimentary canal, but by an accidental wound or a bite from another fish.

The organism was cultivated upon the usual media with the view of determining its name or its allies, and the actual appearances, etc., are here recorded for future guidance.

Description of the Bacterium.

Shape, etc.—It is a short rod with rounded ends. The size is variable, ranging from 0·45 to 0·7 μ in breadth, and from 1·2 to 1·5 μ in length. Longer forms may occur. In the animal tissues it has a distinct capsule. The organism stains feebly with methylene-blue, better with thionin-blue, and strongly with carbol-violet or carbol-fuchsin. With the latter stains many of the forms appear more or less bent and vibrion-like. When stained with thionin-blue, with dilute fuchsin, or with carbol-fuchsin, followed by washing in alcohol, the younger organisms exhibit bipolar staining; the older and longer organisms stain in three places, at the poles and centrally. This latter appearance is probably caused by two organisms whose ends are close together while enclosed within a thin capsule. The cells are actively motile, the motion being produced by a single terminal flagellum, whose length varies from 1½ to 3 times the length of the cell. The organism does not appear to form spores.

Gelatine plate.—In 24 hours, at 22° C., the organism forms in the medium circular liquefied areas about 1·5 mm. in diameter. The colony shows a white central point and a white marginal ring. When magnified sixty-fold, there is seen an irregular brownish-black granular centre, then a clear portion containing large floating granular clumps, and finally similar large granules clustered around the margins. The appearance is like a colony of *Vibrio choleræ*. The edge of the colony is smooth In 48 hours the colony widens to 5 mm., and the contents appear uniformly turbid.

Gelatine stab.—The medium is liquefied along the line of inoculation in 24 hours in a tubular manner, and an air-bubble appears at the top. The liquefied medium is turbid, and there is a white deposit but no film. In 48 hours the medium is almost completely liquefied, the fluid is turbid, and there is a slight film. After a time (7 days) a turbidity extends from the surface to about 0·5 cm. downwards, the lower portion of the liquefied medium is clear, and there is a white granular deposit; no film is apparent.

Glucose-gelatine or lactose-gelatine.—No gas is produced in either of these media.

Agar plate.—The surface colonies are circular, translucent white, raised, and moist glistening. When magnified they are seen to be circular, finely granular, with isolated coarse granules in the centre; the margin is homogeneous and the edge smooth. The deep colonies when magnified are seen to be yellowish brown, coarsely granular, and irregularly shaped, an arrow-head pattern being most usually seen.

Agar slope.—At 22° the growth is translucent white and raised, with a slightly lobed or straight margin and smooth edge. It is spread out and irregular at the base when in contact with the condensed water. The latter is turbid and contains a sediment. At 37° the organism refuses to grow unless a considerable quantity of material is used for sowing. The growth is, however, never so luxuriant as at 22°.

The white colour of the colonies, the liquefaction of the gelatine, and the motility of the organisms, show that the bacterium has its closest allies in a group which consists chiefly of harmless water bacteria. The tendency to produce slightly curved or vibrion forms is characteristic of the phosphorescent bacteria which form a subdivision of this group. The organism, however, does not produce phosphorescence when grown in sea water, in sea water with 1% peptone, in sea water gelatine or upon sterile fish muscle. Excluding the phosphorescence, the other characters show an affinity with the subdivision and, as far as the rapid liquefaction of the gelatine is concerned, with one of the members *Bacillus luminosus*, Beijerinck, which is identical with an organism described by Katz as *Bacillus argenteo-phosphorescens liquefaciens* and renamed *Vibrio luminosus* by Lehmann and Neumann. As the organism does not appear to have been previously described, I propose the name, following the nomenclature of Lehmann and Neumann, *Vibrio bresimœ* (low Latin—Bresmia, the bream).

Mr. D. G. Stead exhibited specimens of a "land-crab" (*Cardisoma* sp.) from Tanna, New Hebrides, known to the Tannese as "To-ba." Also the remains of four fresh-water crayfishes (*Astacopsis bicarinatus*) taken from the stomach of a Murray Cod (*Oligorus macquariensis*). He also called attention to the large numbers of the "southern craytish" (*Palinurus Edwardsii*) which have been sold in Sydney lately, a few of which had been caught in Port Jackson.

Mr. North exhibited the type of *Eremiornis carteri*, a new genus and species of bird from North-west Australia, recently described by him ("Victorian Naturalist," Vol. xvii., p. 78); also two specimens, including the type, of *Platycercus macgillivrayi*, from the Burke District, North Queensland, described by him on page 93 of the present month's number of the same publi cation. Likewise the eggs of the following species:—Great Palm Cockatoo, *Microglossus aterrimus*, Gmel.; Banks's Black Cockatoo, *Calyptorhynchus banksii*, Lath., Yellow-tailed Black Cockatoo, *C. funereus*, Shaw; Western Black Cockatoo, *C stellatus*, Wagl.;

in his paper. Also a photograph of a group of White Box, *Eucalyptus albens*, near Cowra, showing the effect of the heavy snow storm in July; these particular trees suffered more than others growing at higher levels.

Mr. Fred. Turner exhibited, and offered some observations on a collection of thirty-five Australian or introduced plants which many dairy farmers and managers of butter factories suspect of causing the bad flavour (fishiness) in butter that has been so noticeable in Australian dairy produce in London and the colonies of late, and has resulted in so much controversy in the newspapers by dairy bacteriologists and produce merchants. The plants were collected in nearly all the Australian Colonies, and were forwarded to Mr. Turner for identification and report. The list of the plants is as follows :--

RANUNCULACEÆ: *Ranunculus lappaceus*, Sm., *R. plebeius*, R.Br., *R. parviflorus*, Linn.—CRUCIFERÆ: *Barbarea vulgaris*, R.Br., *Cardamine dictyosperma*, Hook., *C. tenuifolia*, Hook., *Blennodia filifolia*, Benth., *B. trisecta*, Benth., *B. nasturtioides*, Benth., *B. lasiocarpa*, F.v.M., **Capsella bursa-pastoris*, Mœnch., **Lepidium sativum*, Linn., *Thlaspi cochlearinum*, F.v.M., **Sinapis arvensis*, Linn., **Sisymbrium officinale*, Scop., **Senebiera didyma*. Pers.—LEGUMINOSÆ: *Trigonella suavissima*, Lindl., **Melilotus parviflorus*, Desf., **M. alba*, Willd.—UMBELLIFERÆ: *Apium australe*, Thou., *A. leptophyllum*, F.v.M., *Daucus brachiatus*, Sieb., **Anethum fœniculum*, Willd.—COMPOSITÆ: *Bidens pilosa*, Linn., **Tagetes glandulifera*, Sch., **Hypochæris glabra*, Linn., **H. radiata*, Linn., **Cryptostemma calendulaceum*, R.Br., **Inula graveolens*, Desf.—LABIATÆ: *Mentha satureioides*, R.Br., **Stachys arvensis*, Linn.—CHENOPODIACEÆ: **Chenopodium ambrosioides*, Linn., *C. carinatum*, R.Br., **C. murale*, Linn., *Salicornia arbuscula*, R.Br.

Introduced plants are indicated by the asterisks.

The Acting Director of the Botanic Gardens, on behalf of Mr. J. H. Maiden, exhibited two well-grown pot-plants in flower—*Dendrobium gracilicaule*, F.v.M., var. *Howeanum*, Maiden, from Lord Howe Island, distinguished from the common New South Wales form by the stouter pseudo-bulbs, the larger flowers, and the lighter colour of the flowers without the red markings on the outside of the sepals; and *Anthurium Scherzerianum*, Schott, var. *album*, from Guatemala.

Mr. Baker showed a specimen in spirit of the phosphorescent fungus, *Pleurotus candescens*, F.v.M., from New South Wales.

Mr. R. Greig Smith showed cultures and micro-photographs of the Bacterium described in his paper.

Mr. Deane, on behalf of Mr. P. E. Williams, exhibited a remarkable quartzite stone axe, with a hafting groove, obtained from an old aboriginal burying-ground on the River Darling near Wilcannia. The strong winds of the past four years had gradually removed the sand to a depth of 6 feet, whereby bones, and the tomahawk exhibited, were brought to light.

Mr. Palmer exhibited a fine aboriginal grinding-stone turned up by tho plough on the site of an old camping-ground at Lawson, Blue Mts.; in the absence of nardoo this stone may perhaps have been used for crushing the capsules, or grinding the seeds of one of the Eucalypts which are much sought after by the Gang Gang Parrots. Also a smaller grinding-stone and two small stone knives

WEDNESDAY, OCTOBER 31st, 1900.

The Ordinary Monthly Meeting of the Society was held at the Linnean Hall, Ithaca Road, Elizabeth Bay, on Wednesday evening, October 31st, 1900

Mr. Henry Deane, M.A., F.L.S., &c., Vice-President, in the Chair.

DONATIONS.

Department of Agriculture, Brisbane.—Queensland Agricultural Journal. Vol. vii. Part 4 (Oct., 1900). *From the Under-Secretary for Agriculture.*

Geological Survey, Brisbane—Annual Progress Report for the Year 1899. *From the Director.*

Royal Geographical Society of Australasia (Queensland Branch) —Queensland Geographical Journal. 15th Session. Vol. xv. (1899-1900) *From the Society.*

Australian Grasses (with Illustrations). By Fred. Turner, F.L.S., F.R.H.S., &c. Vol. i. (8vo. Sydney, 1895). *From the Government Printer, Sydney.*

Australian Museum, Sydney — Miscellaneous Publications. No. 6 (1900). *From the Trustees.*

Department of Mines and Agriculture, Sydney—Geological Survey: Records. Vol. vii. Part 1 (1900): Agricultural Gazette of New South Wales. Vol. xi. Part 10 (October, 1900). *From the Hon. the Minister for Mines and Agriculture.*

One Separate (from Agricultural Gazette of New South Wales; Miscellaneous Publication, No. 408). By D. McAlpine. *From the Author.*

Royal Anthropological Society of Australasia, Sydney—Science of Man. Vol. iii. n.s. Nos. 8-9 (Sept.-Oct., 1900). *From the Society.*

Royal Society of New South Wales, Sydney—Abstract of Proceedings, September 5th and October 3rd, 1900. *From the Society.*

Australasian Journal of Pharmacy, Melbourne. Vol. xv. No. 178 (Oct., 1900). *From the Editor.*

Department of Agriculture, Victoria—Guides to Growers. Nos. 45-47 : Five Reports by Messrs. A. A. Brown, J F. Howell, R. Crowe, D. McAlpine, A. N. Pearson, and H. W. Potts (1899-1900): First Steps in Ampelography. By Marcel Mazade (1900). *From the Secretary for Agriculture.*

Field Naturalists' Club of Victoria—Victorian Naturalist. Vol. xvii. No. 6 (Oct., 1900). *From the Club.*

Royal Society of Victoria, Melbourne—Proceedings. Vol. xiii. Part 1 (Aug., 1900). *From the Society.*

Statistics of the Seven Colonies of Australasia, 1861 to 1899 (1900). By T. A. Coghlan, Government Statistician, Sydney. *From the Author.*

Department of Agriculture, Perth, W.A.- Journal. Vol. ii.

Manchester Literary and Philosophical Society—Memoirs and Proceedings. Vol. xliv. Part iv. (1899-1900). *From the Society.*

Medical Press, London. Vol. cxx. No. 3201, Educutional No. (Sept., 1900). *From the Editor.*

One Separate: "Bryozoa from Franz-Josef Land," &c. (from Journ. Linn. Soc., Zoology, Vol. xxviii.). By A. W. Waters, F.L.S. *From the Author.*

Royal Gardens, Kew—Hooker's Icones Plantarum. Vol. vii. Part iii. (Aug., 1900). *From the Bentham Trustees.*

Royal Microscopical Society, London—Journal, 1900. Part 4 (Aug., 1900). *From the Society.*

Royal Society, London—Proceedings Vol. lxvi. No. 434 (Aug., 1900). *From the Society.*

Zoological Society, London—Proceedings, 1900 Part ii. (Aug.): List, &c., 1900. *From the Society.*

Royal Society of Edinburgh—Proceedings. Vol. xxii. (Sessions 1897-98, 1898-99): Transactions. Vol. xxxix. Parts ii.-iv. (1897-1899). *From the Society.*

Canadian Institute, Toronto—Transactions. Vol. vi. Parts 1-2 (Nos. 11-12; December, 1899). *From the Institute.*

Academy of Natural Sciences, Philadelphia—Proceedings, 1900. Part 1 (Jan.-Feb.). *From the Academy.*

American Academy of Arts and Sciences, Boston—Proceedings. Vol. xxxv. Nos. 20-22 (April, 1900). *From the Academy.*

American Naturalist, Cambridge. Vol. xxxiv. No. 404 (Aug., 1900). *From the Editor.*

American Philosophical Society, Philadelphia—Proceedings. Vol. xxxix. No. 161 (1900). *From the Society.*

Tufts College, Mass.—Tufts College Studies, No. 6 (Feb., 1900). *From the College.*

U.S. Department of Agriculture, Washington: *Division of Biological Survey*—Bulletin. No. 13 (1900); North American Fauna. No. 18 (1900): *Division of Entomology*—Bulletin. Nos. 24-25, n.s. (1900): *Division of Vegetable Physiology and Pathology*—Bulletin. No. 22 (1900). *From the Secretary of Agriculture.*

Perak Government Gazette, Taiping. Vol. xiii. Nos. 26-32 (Aug.-Sept., 1900). *From the Government Secretary.*

Archiv für Naturgeschichte, Berlin. lxvi. Jahrgang. i. Band. 3 Heft (1900). *From the Editor.*

Entomologischer Verein zu Berlin—Berliner Entomologische Zeitschrift. xlv. Band. 1 u. 2 Heft (1900). *From the Society.*

Gesellschaft für Erdkunde zu Berlin—Verhandlungen. Band xxvii. (1900). Nos. 2-4: Zeitschrift. Band xxxiv. (1899). No. 6. *From the Society.*

Medicinisch-naturwissenschaftliche Gesellschaft zu Jena—Jenaische Zeitschrift. xxxiv. Band. 1 Heft (1900). *From the Society.*

One Separate:—"Der Monotremen- und Reptilien-Schädel (1900). Von Dr. V. Sixta. *From the Author.*

Zoologischer Anzeiger, Leipzig. xxiii. Band. Nos. 622-624 (Aug.-Sept., 1900). *From the Editor.*

Société Royale de Géogra hie d' Anvers—Bulletin. Tome xxiv.

Société Royale des Sciences, Upsal—Nova Acta. Seriei iii. Vol. xviii. Fasc. ii. (1900). *From the Society.*

University of Upsala: Geological Institution — Bulletin. Vol. iv. Part 2 (No. 8, 1899). *From the University.*

Asiatic Society of Bengal, Calcutta—Journal. Vol. lxix. Part i. No. 1; Part ii. No. 1 (1900): Proceedings, 1900. Nos. v.-viii. (May-Aug.). *From the Society.*

College of Science, Imperial University of Tokyo—Journal. Vol. xiii. Part 2 (1900): Calendar for the Year 1899-1900. *From the University.*

Museu Paulista, S. Paulo—Revista. Vol. iv. (1900). *From the Museum.*

Department of Agriculture, Cape of Good Hope—Report of the Marine Biologist for the Year 1899. *From the Under-Secretary for Agriculture.*

South African Museum, Cape Town—Annals. Vol. ii. Part ii. (June, 1900). *From the Trustees.*

TASMANIAN LAND PLANARIANS.

Descriptions of New Species, &c.

By Thos. Steel, F.L.S., F.C.S.

(Plate xli.)

The first naturalist to collect a land planarian in Tasmania was Darwin, who found one species on the island on the occasion of the visit of the Beagle in 1832, and subsequently described it as *Planaria Tasmaniana* (1, p. 244).* Since then the only additions to our knowledge of the land planarians of Tasmania have been made by Dendy in 1892-3 in his papers read before the Australian Assoc. Adv. Science and the Royal Society of Victoria, and by Graff (10) who worked on material supplied by Dendy. Disallowing *Geoplana balfouri*, Graff, as being really synonymous with *G. Tasmaniana* (Darwin), and *G. Wellingtoni*, Dendy, which is very doubtfully identified by Graff (10, p. 369) from Dendy's description of a single probably immature specimen,

The following is a list of the names. Those marked A occur also in Australia:—

Geoplana Tasmaniana (Darwin).
" " var. *flavicincta*, Steel.
" *typhlops*, Dendy,
" * *Adæ*, Dendy .... A
" † " var. *fusca*, Dendy.
" † *walhallæ*, Dendy .... A
" *diemenensis*, Dendy.
" *Mortoni*, Dendy.
" * *munda*, Fletcher & Hamilton............ A
" * *variegata*, Fletcher & Hamilton .... A
" *Dovei*, Steel.
" *lyra*, Steel.
" *sanguinea* (Moseley) .... A
" *Sugdeni*, Dendy .... A

Through the zeal of Mr. H. Stuart Dove, of Launceston, I have been supplied with numerous consignments of planarians collected in the neighbourhood of that town, and also at Table Cape in N.W. Tasmania. The specimens were sent to me by Mr. Dove both alive and preserved, from time to time during the last few years, and his kindness in this matter has enabled me to offer this contribution to our knowledge of a somewhat neglected branch of the Tasmanian fauna.

All the species found so far can be included in the old genus *Geoplana*, no example of a *Rhynchodemus* having yet been met with. There is nothing specially distinctive about the land planarians endemic to Tasmania. They bear a general resemblance to the forms occurring in Australia. Considering the relationship of Tasmania to the mainland, this is to be expected, and, as a matter of fact, the species peculiar to Tasmania do not differ in character from those common to Tasmania and Australia,

---

* Signifies that I have not seen Tasmanian examples, and † that I have not seen specimens from any locality.

more than local species occurring in individual regions of Australia, differ from one another.

The type specimens of the species described in this paper are deposited in the Australian Museum, Sydney.

GEOPLANA DOVEI, n.sp.

(Plate xli., fig. 1.)

Geoplana Lucasi, Dendy, (6, p. 180, and 8, p. 421, but *not* 3, p. 43; 4, p. 74; nor 5, p. 40, pl. iv. fig. 4), Graff (10, p. 350, in so far as reference is made to the Tasmanian form, but not to the Victorian form); *non* Spencer (9, p. 91).

Geoplana Lucasi was originally described and figured by Dendy from specimens obtained in Victoria. Subsequently he doubtfully ("provisionally at any rate") identified examples of a land planarian procured in Tasmania as belonging to the same species Graff, in commenting on specimens of the Tasmanian worm sent to him by Dendy, contrasts the dorsal markings with those figured and described by Dendy for the Victorian species, and throws doubt on the specific identity of the two. Graff does not appear to have had any of the Victorian species before him, and hesitated about constituting the Tasmanian form a distinct species. He says:—

dorsal line and being of similar shape is by no means sufficient reason for considering them as identical; indeed many of the described species of land planarians from all parts of the world differ less from one another than do the forms under consideration.

As I am satisfied that Dendy's description of the Victorian *G. Lucasi* cannot be made to cover the Tasmanian species, I have deemed it advisable to rename the latter, of which the following is a description :—

Dorsal surface to the eye appears of a uniform dark purplish-brown colour, with a narrow median very dark brown line just distinguishable, extending from end to end in some individuals but disappearing about midway between the ends in others. Under the lens the dark brown of the surface is resolved into very numerous inosculating stipplings, mainly longitudinal, on a groundcolour of somewhat paler brown, and extending uniformly over the whole surface. Ventral surface uniform pale brown with a pinkish tinge.

The body is flattened and leaf-shaped, about equal width for the greater part, curving abruptly to a blunt point posteriorly, and produced at anterior end to a pointed tip.

Eyes large and conspicuous, in a single row round the anterior tip and continued without any grouping, in a row for a few mm. down the sides.

Length in spirit 43 mm. by 7 mm. broad. Peripharyngeal aperture 25 and genital opening 32 mm. from anterior tip. An example when crawling was about 57 mm. in length by 8 mm. wide, and when contracted at rest became shortened considerably and from 10 to 12 mm. wide.

Hab.—Table Cape, Tasmania (Mr. H. Stuart Dove).

In shape and general character this species seems to me to be related to *G. Mortoni*, Dendy, but the markings on both surfaces of the latter are very distinctive.

I have pleasure in associating the name of Mr. Dove with this handsome species in recognition of his excellent work in the collection of land planarians in Tasmania.

Early in September, 1899, an individual of this species which Mr. Dove collected at Table Cape, deposited an egg-capsule, which, together with the planarian, he sent on to me. The box containing the specimens was some 8 days in transit, during which time the capsule hatched. I found two young, together with the adult, in a living and healthy state. The following is a description of the young :--Dorsal surface dark purplish-brown, with stipplings resembling those of the adult. One specimen was 11 and the other 10 mm. in length. The latter has the median dorsal space almost free from pigment, forming a pale brown band with only a few scattered brown specklings, and no indication of the black median line. In the longer specimen this pale space is very faintly indicated, the pigmentation being nearly uniform all over, and the black line is plainly visible exactly as in the adult, extending for about half-way from anterior tip.

Ventral surface in both thickly speckled with brown spots which have a tendency to crowd into the median space and towards the margins, and thus to form three longitudinal bands with less speckled space between.

When alive the cross section of these young was exceedingly angular, the back being raised in the form of an acute longitudinal ridge, giving the little creatures a very slug-like aspect, but otherwise the shape resembled that of the adult. Eyes very large

young ones hatched out in 8 days after, the period of hatching is probably just within that time. The second capsule was found by Mr. Dove already deposited, lying beside the adult, and occupied a similar time in transit. The darker pigmentation of the young individual from this capsule and its somewhat greater length, together with the fact of the capsule having been already laid when found, lead me to infer that it was some days older than the others.

The capsules of this species are about 4 mm. in diameter.

GEOPLANA LYRA, n.sp.

(Pl. xli., fig. 2.)

Colour of dorsal surface light yellow, with three sharply defined lines of dark brown. Starting from anterior end the lines are of equal width until just over the peripharyngeal opening, when they become somewhat broader, the median one more so than the others, and continue thus as far as the genital aperture, when they again resume their original proportions.

The group of lines is well in the centre of the body, having a clear space between the lateral ones and the sides of about $\frac{1}{3}$ of the total width of the body, and somewhat resembling the arrangement of the strings of a stringed musical instrument. At both ends the lines merge into one another. The anterior tip is coloured brown.

Ventral surface of a somewhat paler yellow than the dorsal, without markings.

Eyes as usual, in a single row round anterior tip and in a straggling line with moderate grouping for some distance down the sides.

In a spirit specimen, which is 25 mm. long by 3½ mm. at widest part, the peripharyngeal opening is 15 and the genital aperture 20 mm. from anterior tip. The same example when crawling had a length of about 40 mm. Another individual was 50 mm. in length when crawling. In repose the body is drawn up to a broad strap-shape.

Hab.—Table Cape, Tasmania (Mr. H. Stuart Dove and Mr. Easton).

In spirit the yellow colour dissolves, leaving the body white, the brown lines remaining unchanged.

From the description alone this species might be confused with the three-lined form of *G. mediolineata*, Dendy (4, p. 77), but it is readily distinguished by its totally different shape and by the very central position of the group of lines.

GEOPLANA TASMANIANA (Darwin).

(Pl. xli., figs. 3 and 4.)

Planaria Tasmaniana, Darwin (1, p. 244); *Geoplana Tasmaniana*, Fletch. & Hamil., (2, p. 361), Dendy (6, p. 178; 7, p. 369; 8, p. 421), Graff (10, p. 370); *Geoplana balfouri*, Graff (10, p. 375, Pl. v., figs. 31-33). Other references are given by the authors cited above.

In 1844 Darwin (1, p. 244) somewhat meagrely described this species, the only one with which he met in Tasmania. For many years no further land planarians appear to have been collected in Tasmania until in 1893 Dendy (6, p. 178) gave a full description, amongst other species, of a form which he considered to be identical with that originally made known by Darwin. Graff (10, pp. 370 and 375), having material sent by Dendy before him, concludes that the latter was mistaken in his identification, and

am of the opinion that Dendy was correct in his identification, and that Graff has fallen into an error in re-naming the species.

Darwin is clearly speaking of the dorsal surface as a whole when he says in his description :—"Colour dirty 'honey-yellow,' with a central dark brown line bordered on each side with a broader line of pale 'umber-brown.'" So that we have here a description of a worm having a dirty honey-yellow dorsal surface margined by broad umber-brown stripes, and with a dark brown median line (Pl. xli., fig. 3). This description tallies exactly with the larger proportion of the specimens of this species which I have seen, as do also the arrangement of the eyes, the dimensions of the body, and the position of the orifices, as stated by Darwin. Graff has evidently misread this description, for he refers to *G. Tasmaniana* as having one band or stripe, whereas Darwin very plainly mentions three.

Dendy (*loc. cit.*) speaks of there being five dorsal lines or bands, comprised of the median line, lateral bands and marginal stripes, and states that on preservation in spirit the margins of the dorsal surface with their stripes become turned in to form lateral surfaces. In my experience the above marginal stripes should more properly be spoken of as submarginal, for they are distinctly situated in the living animal just beneath the dorsal edge, and indeed I should without hesitation have termed them ventro-marginal. I shall, however, speak of them in the sequel as submarginal. Certainly the species is somewhat variable in its markings, the narrow submarginal stripes not being constantly present, while the broad lateral bands are most erratic in their extent. I possess a series of specimens, in some of which not only are the submarginal stripes quite absent, but the lateral bands are also suppressed save for about one-third of the length of the body at the anterior end, the remainder of the dorsal surface, with the exception of the dark median line, which is a constant character, being of a speckled colour, which is exactly expressed by Darwin's term "dirty honey-yellow." This form bears a good deal of resemblance to the Australian species *G. ornata*, Fletch. & Hamil. In others the lateral bands are broadened out so as to cover the whole dorsal surface except a narrow strip of dirty yellow on either side of the median line, the submarginal stripes being also strongly defined. Between these two extremes there are many gradations.

A medium specimen in spirit is 30 mm. in length, with the peripharyngeal opening 19 and the genital aperture 24 mm. from anterior tip. A small example 20 mm. in length has the relative positions of the apertures 11 and 14½ mm. respectively.

On two occasions examples of this species sent me by Mr. Dove from Trevallyn Hills deposited egg-capsules. The first was laid about June 2, 1899, and failed to hatch. The other deposited in July of same year hatched out in my vivarium about 4th September, some seven weeks after being laid. In both cases the capsules were between 2 and 3 mm. in diameter. From the one which hatched there emerged four young, three of which were about 13 mm. in length when fully extended, the fourth being only about half that size. The ground colour of dorsal surface was milk-white, sprinkled all over with brown speckles, tending to arrange themselves in an irregular median band.

Ventral surface also milk-white, with no markings. Eyes in a single row round the entire margin, with no side grouping.

Two of the little ones fed on flies which I disabled and gave to them, and remained alive for between 2 or 3 weeks, but eventually

A spirit specimen 27 mm. long by 4 mm. wide at broadest part, has the dorsal pigmented area 3 mm. in width and bordered by ½ mm. of yellow margin.

The eyes, which are in a single row with a few scattered individuals, are rendered conspicuous by being situated on the light-coloured margin which extends all round at the tips as well as the sides, completely framing the coloured area of dorsal surface. The submarginal stripes are entirely absent, the ventral surface being uniform white or pale yellow.

In same specimen the peripharyngeal opening is 15 and the genital 20½ mm. from anterior tip.

Hab.—Trevallyn Hills near Launceston, Tasmania (Mr. H. Stuart Dove).

In a small collection of land planarians from Rotorua, New Zealand, given to me by my friend Mr. C. Cooper, of Auckland, there is one small specimen, probably immature, of an apparently undescribed form, which bears a remarkable resemblance to the above variety. This specimen in formaline has a very dark brown narrow dorso-median line with dark brown on either side, paler next median line, and darkest next marginal space. Marginal space cream-white. Marginal spaces each ¼ of total width, coloured area ½ of same. Length 12 mm. by 2 mm. in width. Peripharyngeal aperture 9 mm from anterior tip, genital opening not visible. Anterior tip brown, not inclosed by light margin, differing in this respect from the above-described Tasmanian form. Eyes in a single somewhat straggling row, with bare indications of grouping.

The only external point of difference between this little planarian and a small example of *G. Tasmaniana* var. *flavicincta*, is that the pale margin does not extend round the anterior tip.

Geoplana diemenensis, Dendy.

(Plate xli., fig. 8.)

G. diemenensis, Dendy (6. p. 179: 8. p. 421): *Artioposthia diemenensis*, Graff. (10, p. 404. Pl. v., figs. 20-24).

This species was not figured by Dendy, and by his description alone it is not easy to distinguish from *G. Tasmaniana*. Graff, however, gives excellent figures of specimens sent to him by Dendy, and from these identification can be readily made.

I have received from Mr. Dove a small series of examples from Trevallyn Hills and Table Cape, in which are included individuals having the range of markings illustrated by Graff.

Graff's figures 20 and 21 (*op. cit.*) have reference to specimens of the typical form from Mount Wellington, while 22-24 represent examples of Dendy's "slight variety" found by Professor Spencer at Parattah. I have not seen specimens of the "slight variety," but all my examples with markings corresponding to Graff's figures thereof correspond in size and stoutness with the typical form (Pl. xli., fig. 8). I have some doubt if more than one species is not here involved, but Dendy and Graff appear to be satisfied on the point.

GEOPLANA MORTONI, Dendy.

(Plate xli., figs. 6 and 7.)

Geoplana Mortoni, Dendy, (6, p. 181; 8, p. 421), Graff (10, p. 349).

Mr. Dove has sent me specimens of this fine species from Trevallyn Hills near Launceston, and Table Cape. Those from the former locality are all of the finely speckled forms (Pl. xli., fig. 6), and in the specimens from the latter place the markings are coarser. Some of the Table Cape specimens, in addition to the

An egg-capsule of this species from a specimen collected at Trevallyn Hills, and laid during transit from Launceston, was about 4 mm. in diameter. It hatched out a few days after being placed in my vivarium, when 11 young emerged on 1st August, 1900. These were 5 mm. long, by about 1 mm. broad at widest part. The colour of both surfaces is milk-white. The anterior tip both above and below is coloured brown through aggregation of brown speckles similar to those on adult. These specklings are on both surfaces continued sparingly down the margins, forming ill-defined marginal bands for a variable distance back. Remainder of surface very sparingly dotted with specklings. Orifices not visible.

Eyes conspicuous, in a single row round the entire margin of the body, somewhat closer together at anterior than posterior end.

The appearance of the parent worm after laying the above capsule is described and figured in Part ii. of my foregoing paper on Australian Land Planarians. Pl. xli., fig. 6, in present paper shows the position of the opening. It should be mentioned that the peculiar comb-like glandular genital organs there mentioned would, according to Graff's definition, cause this species to be included in his genus *Artioposthia*. In his monograph he leaves the species in the genus *Geoplana*, and does not mention the structure of the genital organs.

In all external characters my specimens agree precisely with Dendy's description, and with Graff's figures in so far as they go.

Geoplana sanguinea (Moseley).

This common and widely distributed Australian species has not hitherto been recorded from Tasmania. Dendy (7, p. 370) at first identified a form devoid of eyes as *G. alba*, Dendy (*syn. G. sanguinea*), but subsequently (6, p. 184, and 8, p. 420) described it as a distinct species under the name of *G. typhlops*. The distinguishing external feature between the two species is the possession or otherwise of eyes. All the examples from Tasmania examined by Dendy had no eyes, and hence were assigned by him to *G. typhlops*.

From Table Cape Mr. Dove has sent me alive a number of specimens in which the eyes are plainly visible, and which in the arrangement and size of these, and all external respects, are identical with Australian examples of *G. sanguinea.* From the same place, and also from his garden at Launceston, he has forwarded specimens which careful anatomical examination shows to be devoid of eyes and which are otherwise externally indistinguishable from *G sanguinea,* thus answering in all respects to Dendy's description of *G. typhlops.*

The largest specimen of *G. sanguinea* from Table Cape, after preservation in spirit, is 103 mm. in length, and has the peripharyngeal opening 63 mm., and the genital aperture 72 mm. from anterior tip. Of the specimens of *G. typhlops* from same locality the three largest specimens, in spirit, have the following dimensions stated as above, the first of these was 115 mm. in length when crawling :—

Total length.	Peripharyngeal aperture.	Genital opening.
90	64	76 mm.
83	54	66 mm.
80	54	67 mm.

It is interesting to find these two forms thus associated with one another.

5. Dendy, A., Trans. Royal Soc. Victoria, Melbourne, 1891.
6. ——— Proc. Royal Soc. Victoria, Melbourne, 1893.
7. ——— Report Australian Assoc. Adv. Sc , Vol. iv., 1892.
8. ——— Report Australian Assoc. Adv. Sc., Vol. v., 1893.
9. Spencer, W. Baldwin, Proc. Royal Soc. Victoria, 1890.
10. Graff, L. von, Monog. der Turbellarien. ii. Tricladida Terricola (Landplanarien), Leipzig, 1899.

EXPLANATION OF PLATE XLI.

Fig. 1.—*Geoplana Dovei*, n.sp. Dorsal aspect (nat. size).
Fig. 2.— ,, *lyra*, n.sp. Dorsal aspect (× 2).
Fig. 3.— ,, *Tasmaniana* (Darwin). Dorsal aspect (× 2).
Fig. 4.— ,, ,, ,, Ventral aspect of another individual showing double pharyngeal tube (× 2).
Fig. 5.— ,, *Tasmaniana* (Darwin), var. *flavicincta*, n.var. Dorsal aspect (× 2).
Fig. 6.— ,, *Mortoni*, Dendy. Dorsal aspect showing opening caused by exit of egg-capsule (× 1½).
Fig. 7.— ,, *Mortoni*, Dendy. Dorsal aspect, banded form (× 1½).
Fig. 8.— ,, *diemenensis*, Dendy. Dorsal aspect (× 1½).

The outline beneath each figure represents the section of the body at middle.

STUDIES IN AUSTRALIAN ENTOMOLOGY.

No. X.

Description of a new Tiger Beetle from Western Australia.

By Thomas G. Sloane.

Tetracha greyanus, n.sp.

Stout, cylindrical, parallel-oval, apterous. Prothorax with a lateral carina on anterior part of sides; elytra subtruncate, strongly and roughly punctate towards base, smooth (with very minute punctures in derm) towards apex, the smooth apical part with a discoidal row (4 or 5) of fine punctures, a submarginal row (3) on sides and a marginal row along apex.

Upper surface metallic; elytra green on punctate part, becoming bluish on smooth apical part; prothorax green, with middle of disc, middle of anterior margin, and a small spot at middle of basal margin bluish-black; head green, middle of front and occiput bluish-black; underparts of prothorax green, bluish in middle;

posterior transverse impression very strongly impressed; a narrow lightly reflexed carina on sides extending back from anterior angles almost to middle. Elytra oval (9 x 5·8 mm.), parallel on sides, arcuate-truncate at apex with external angles widely but shortly rounded, strongly punctate on basal half, roughly granulate amongst the punctures near the base, the puncturation finer and the derm not granulate posteriorly; apical declivity smooth (minute punctures and transverse rugæ noticeable under a lens). Ventral segments rugulose; rugæ of three basal segments longitudinal on sides; last segment narrowly bordered, median apical notch of ♂ very small. Length 15-18, breadth 5·3-6·8 mm.

Hab.—Carnarvon District (Shark's Bay), W.A.; several specimens sent to me by Mr. C. French.

This species by its cylindrical and parallel form, connate elytra, absence of wings and the small median notch of the apical ventral segment in the ♂ is allied to *T. (Megacephala) cylindrica*, Macl., but differs from that species, and also from *T. (Megacephala) frenchi*, Sl., and *T. (Megacephala) spenceri*, Sl., by its shorter elytra more truncate at apex, and the well developed lateral carina on the anterior part of the prothorax. The features given above as characteristic of *T. cylindrica* and its allies seem to be those which chiefly serve to differentiate that group of species from the other unicolorous Australian species, of which *T. pulchra*, Brown, is the type, which are species of quite different facies, the elytra being shorter, more depressed, much more ampliate at the base as compared with the prothorax, and less parallel on the sides. *T. (Megacephala) howitti*, Casteln., is a species I have not before me; in many ways it seems intermediate between *T. cylindrica* and its allies, and *T. pulchra* and its allies. The Rev Thos. Blackburn, who knows *T. howitti*, believes it to be an apterous form. See pp. 634-635 for some interesting notes from his pen on this species.

Note.—M. Fleutieux has formed the genus *Pseudotetracha*, for species of which *T. cylindrica* is the type, but the genus is so characterised that I doubt whether any species except *T. cylindrica* and *T. frenchi* would find a place in it, and in any case if these species are to be removed from *Tetracha* it seems to me impossible to consider *T. pulchra* as more closely allied to species such as *T. australasiæ*, Hope, than to *T. cylindrica*, as is implied by leaving the two former together in a different genus from the last. For these reasons I have not adopted the genus *Pseudotetracha*.

TETRACHA FRENCHI, Sloane.

Megacephala frenchi, Sl., P.L.S.N.S.W. (2), viii. 1893, p. 25.

M. Fleutieux has published the opinion that *T. frenchi*, Sl. = *T. (Megacephala) howitti*, Casteln.,* but I am sure he is mistaken in this. There was formerly in the Howitt Collection at the Melbourne University a specimen labelled *Megacephala howitti*, Casteln., which was evidently one of the two original specimens brought from Cooper's Creek by Mr. A. W. Howitt (the other having been given to M. de Castelnau by Dr. Howitt).† This specimen now appears to be lost, for Mr. C. French, who recently looked for it at my request, could not find it; but I saw it in November, 1893‡, and published a note dated 24th January, 1894, recording the great difference in facies between *T. (Megacephala) howitti*, Casteln., (as represented in the Howitt Collection), and *T. frenchi*. This note appeared previously to M. Fleutieux's, and was written only to help to re-establish the validity of de

tion were taken at Cooper's Creek, and I am perfectly satisfied from memory that the specimen in the Howitt Collection named as *T. howitti* was nothing like *T. frenchi*. I had not a specimen of the *Tetracha* taken by the Callabonna Expedition with me to compare with it [*i.e.*, *T. howitti*], but it struck me as being certainly the same thing." In conclusion, I would draw attention to the evident and important differences between *T. frenchi* and the description of *T. howitti* which, in themselves, seem to effectually dispose of any possibility of their being synonymous :—(*a*) The great difference in facies as shown by de Castlenau's measurements of *T. howitti* (7 x 3 lines), *T. frenchi* (about 9 x 3 lines); the elytra in *T. howitti* "short of an oval form," in *T. frenchi* long, parallel and cylindrical. (*b*) The differences in colour—"the buccal parts, the base of antennæ, legs and last two segments of the abdomen of a light yellowish-brown" in *T. howitti*; while in *T. frenchi* the two apical segments of the abdomen are black (as is the whole of the abdomen excepting the lateral parts of the three basal segments, which are metallic-green or purple), the legs being entirely black, the antennæ piceous-brown.

As I have not a specimen of *T. howitti*, and am unable to see one, I asked the Rev. Thos. Blackburn to send me a note on it, and the following is the substance of his communication, founded on the *Tetracha* brought from Cooper's Creek by Mr. Zietz :—"This insect agrees extremely well with Castelnau's description of *T. howitti* (*e.g.*, elytra short of an oval form, . . . basal five joints of antennæ light yellowish-brown . . . legs yellowish-brown . . . last two segments of abdomen brown, becoming yellow on the last). Your *T. frenchi* is utterly different The species I have as *T. howitti* is about as unlike *T. frenchi* as one *Tetracha* can be unlike another." He also has added a comparison with *T. murchisoni*, Fleut., as follows :—"It is a shorter and considerably more convex form—the convexity of the elytra much greater from scutellum to apex (looked at from the side) and also from one lateral margin to the other (looked at from behind). The pronotum of *T. howitti* does not differ much in outline, but its sides are evidently more sinuate behind the middle. The lateral carina of the pronotum is non-existent. The elytra are distinctly more nitid, and their basal puncturation is less close and limited to the basal one-third of the area. The antennæ are much darker (after about the fourth joint), and the legs of a distinctly less clear testaceous colour. In the ♂ the apical ventral segment is scarcely sinuate at its apex."

DESCRIPTION D'UN ECHANTILLON DE KEROSENE SHALE DE MEGALONG VALLEY, N.S.W.

PAR PROF. BERTRAND, LILLE.

(*Communicated by W. S. Dun.*)

Origine de l'Echantillon.—L'échantillon dont je vais donner la description m'a été adressée par M. Dun du Geological Survey de la New South Wales le 31 Mai 1899. Il avait été recueilli dans la Megalong Valley près Katoomba, Comté de Cook, New South Wales.

Caractères macroscopiques.—Charbon schisteux, noir mais avec reflet légèrement satiné, compact et très tenace. La cassure verticale est irrégulière, écailleuse comme celle d'un schiste ardoisier difficilement fissile. Cette cassure verticale est noire, légèrement satinée par de petits traits horizontaux vitreux, très courts et extrêmement fins, qui se détachent sur le fond terne. De plus la cassure montre de grands filets horizontaux de charbon brillant craquelé ou chagriné. Ces filets se relient entre eux par des lignes faiblement inclusées du même charbon brillant. Filets et lignes brillantes donnent immédiatement l'impression d'une pénétration de la masse par la matière bitumineuse.

Les cassures horizontales sont irrégulières. Elles ont une tendance marquée à devenir largement conchoïdes. Elles sont très finement chagrinées, noires, légèrement satinées, avec points noirs vitreux très petits et très souvent aussi avec de petites plaques de charbon brillant craquelé

La section verticale, faite à l'émeri, est brun noir, très foncée, avec reflets roux. Elle est terne. A la loupe la stratification est reconnaissable. Elle est très fine. La section horizontale faite a l'émeri est brun noir, foncée à reflets roux. Elle parait homogène sauf le long des traits noirs qu'y dessine les affleurements des lames de charbon brillant.

Pour gagner un peu de brièveté dans les nombreuses comparaisons que j'aurai à faire par la suite, je procéderai comme le font les vendeurs de charbons et les fabricants de gaz. Je désignerai les échantillons de kerosene shale de chaque localité comme s'ils représentaient effectivement des variétés distinctes de ce mineral et de même qu'on dit du Lens, du Carvin, pour désigner les charbons de ces concessions minières, je dirai du Megalong Valley, du Blackheath, pour désigner le kerosene shale des extractions de Megalong Valley, de Blackheath, etc.

Le Megalong Valley resemble beaucoup au Blackheath. Dans ce dernier les petits échantillons* donnent l'impression de plaquettes plus minces résultant de la présence de fentes horizontales plus ouvertes. La teinte noire du Blackheath est plus terne. Sa tranche verticale est plus rousse. La stratification est un peu plus soulignée. Les infiltrations bitumineuses qui coupent plus obliquement la masse sont plus visibles.

Le Mount Victoria est plus terreux et plus schisteux. Il est noir mat, complètement terne même sur les cassures verticales fraiches. Les points vitreux y sont très petits et très espacés dans le fond mat. Les cassures horizontales suivent la séparation des lits. Leur surface est granulée. Les tranches verticales et horizontales sont noir mat. La teinte rousse y est très peu visible. La fissilité de la masse est plus accusée. Le Mount Victoria est

J'ai crû devoir insister quelque peu sur ces différences des aspects macroscopiques, les seuls accessibles pour le mineur, parce qu'elles sont la traduction extérieure de faibles variantes de structure, dues elles-mêmes à de petites modifications toutes locales dans le mode de formation des divers amas du kerosene shale. Ce sont des nuances qu'un mineur exercé perçoit inconsciemment et qui lui apprennent peu à peu, par la pratique journalière à reconnaître d'un coup d'oeil les variétés d'un charbon.

Description des plaques minces.

(a) Ensemble des coupes vues à un grossissement de 30 diamètres.

A un grossissement faible, l'aspect d'ensemble des coupes verticales du Megalong Valley est le même que celui des coupes correspondantes du Blackheath et du Mount Victoria. Ces coupes montrent un empilement de *Reinschia australis* dans une matière générale brun foncé. Les thalles généralement plats sont bien isolés, largement écartés les uns des autres, couchés horizontalement et très affaissés, mais sans trace de compression. D'après ce premier aspect on prévoit que les thalles ne forment qu'une fraction relativement peu élevée de la masse.

Dans le Megalong Valley la préparation est fréquemment coupée par des lignes rouge brun plus transparentes. Elles sont horizontales ou peu inclinées, ondulées. Elles s'effilent et se ramifient à leurs extrémités. C'est leur transparence relative qui les signale à l'observateur. Elles correspondent aux filets brillants des grandes fractures verticales. Il est impossible à ce grossissement faible de préciser la nature de ces lames rouges et de dire si ce sont des lames végétales imprégnées de bitume ou bien en bitume libre pur ou modifié. Ce n'est pas là le bitume agissant massivement et altérant les thalles voisins comme aux points de pénétration de la matière bitumineuse dans le Joadja Creek. La matière interposée entre les thalles est généralement rouge brun, plus transparente que celle du Blackheath. Dans certains îlots ou les thalles adultes se raréfient, cette matière devient noire et fuligineuse comme celle du Mount Victoria. En ces points, elle prend une charge particulièrement forte en menus débris de parois végétales noir brun, très foncés, mais qui ne sont pas encore arrivés à l'état de fusain.

Les petits cristaux blancs tardifs qui picotent la matière brune du Megalong Valley sont moins abondants que dans le Blackheath.

L'aspect d'ensemble des préparations horizontales du Megalong Valley est aussi celui des préparations correspondantes du Blackheath et du Mount Victoria. Quelques thalles plus gros, coupés horizontalement à des niveaux variés, sont séparés par de larges plages rouge-brun dans lesquelles on voit les coupes isolées de thalles plus jeunes et plus petits; ou bien encore dans lesquels transparaît la surface polaire de thalles coupés tangentiellement et visibles à travers une épaisseur variable de tissu fondamental. Les thalles sont moins bien conservés que ceux du Blackheath. On dirait qu' ils ont subi une légère contraction. Ce caractère est bien plus accentué dans le Mount Victoria. Je n'ai pas vu de fragments presque fusinifiés se détachant nettement sur la coupe comme dans le Blackheath. Dans le Mount Victoria les plages interposées entre les thalles de *Reinschia* sont beaucoup plus noires, rendues presque fuligineuses par leur charge en menus débris de parois végétales très altérées.

Sur les coupes horizontales, les gros thalles m'ont montré une légère tendance à un allignement général. La direction de cet al-

Sur les sections horizontales, les gros thalles très pâles sont entourés de petits thalles. Tous paraissent se toucher directement et se presser directement tant est grande leur densité relative par rapport à la dilution locale qu'ils ont fait subir à la gelée fondamentale. Cette gelée n'est pas visible à 30 diamètres sur les coupes horizontales. Des lambeaux de parois végétales noir brun, presque fusinifiés se détachent nettement parmi les thalles.

(b) Etude des sections à un grossissement de 300 à 400 diamètres.

Examinée à un fort grossissement, sur des coupes verticales transverses,* la matière interposée entre les thalles du Megalong Valley paraît nettement formée de deux parties : Une substance ou gelée primitive brun foncé peu transparente, dans celleci est enfermée un autre matière rouge brun plus transparente qui s'y présente en îlots et en filets ondulés. L'abondance de cette matière rouge brun est une caractéristique du Megalong Valley.

La substance brune initiale est la gelée brune humique que j'ai rencontrée dans tous les charbons humiques et dans les charbons d'algues, c'est à dire dans les schistes bitumineux et dans les bogheads. Elle est un peu moins colorée que celle du Blackheath. Elle est beaucoup plus foncée que celle du Mount Victoria. Elle a manifestement subi une imprégnation bitumineuse. Elle est chargée de nombreux bactérioïdes. Elle contient un grand nombre de menus fragments de parois végétales noir brun. Ces derniers y sont inégalement répartis. Les parcelles completement fusinifiées y sont rares.

La plupart des bactérioïdes sont bullaires. Ils mesurent de 0·3 μ à 0·8 μ. Quelques-uns sont pleins fortement colorés en brun ou en noir. Dans ce dernier cas, il est à peu près impossible de les distinguer d'un très petit cristal de pyrite.

---

* Les coupes verticales faites dans la direction que j'avais déterminé comme transverse ont été obtenues beaucoup plus minces que les coupes radiales prélevées sur les mêmes cubes de taille. La cohésion de la matière est donc sensiblement différente dans les deux sens.

La matière rouge brun s'y présente en îlots et en fils ondulés. Ces derniers forment un réseau plus transparent à mailles allongées horizontalement qui se détache dans la gelée fondamentale. Les îlots se relient directement aux fils réticulés, mais pour relever ce caractère il faut le relever spécialement. Les îlots ont la configuration de masses molles affaissées. On peut donc hésiter sur la nature de ces îlots et se demander d'abord si ce ne sont point là des thalles altérés plus profondément modifiés que les thalles résinoïdes.* La gélose y serait plus fortement colorée par le bitume imprégnant, les canalicules seraient effacés. Je n'ai pas trouvé de transition entre ces îlots et les thalles gommeux. Cette absence de transition entre deux états d'altération d'un même corps gélosique et la liaison directe des ilôts aux fils du reticulum m'ont fait rejeter l'interprétation de ces îlots comme thalles altérés dont la structure aurait disparu. D'autre part ce ne sont point non plus des gouttes de bitume individualisées comme celles du Schiste du Bois d'Asson.† La matière rouge brun contenant quelques bactérioïdes, au moins dans sa surface, j'arrive à conclure qu'il s'agit d'une pénétration de matière bitumineuse dans un réseau de déchirures de la gelée fondamentale peut-être partiellement obturées par un exsudat. La matière bitumineuse n'a pas altéré les thalles. Elle ne pénètre pas directement dans ceux-ci. Ceux-ci ne sont pas brisés et raccornis comme cela se voit le long des

Les lames gélosiques sont passées a l'état de corps jaune d'or. La gelée humique à bactérioïdes s'est teintée en brun.

En quelques points la matière rouge brun se présente sous forme de plaques horizontales plus épaisses, isolées ou reliées au réseau général. Dans ces lames j'ai souvent observé les traces d'une organisation figurée. Elles indiquaient qu'une lame de bois ou de liége, convenablement humifiée, ou encore un lambeau de feuille, avait retenu le bitume et s'en était imbibé; mais je n'ai pu fournir cette preuve pour tous les amas. L'existence de lames bitumineuses sans substratum figuré initial reste donc possible.

La répartition de l'infiltration bitumineuse est inégale. En quelques points, grace surtout à la présence des grandes lamelles ligneuses ou subéreuses imbibées elle peut atteindre un chiffre très élevé * J'ai trouvé jusqu' à 0·111 comme coefficient vertical de l'intervention du bitume dans des carrés mesurant 150 μ de coté.

L'abondance relative de la matière rouge brun plus transparente fait que le fond sur lequel se détachent les thalles de *Reinschia* parait rouge dans le Megalong Valley. On aperçoit bien une infiltration bitumineuse rouge dans le Blackheath. Elle est beaucoup plus restreinte. Celle du Mount Victoria est encore plus réduite.

La gelée humique et son réseau bitumineux forme la partie dominante du Megalong Valley.

Les organites contenus dans la gelée humique du Megalong Valley sont :

Des thalles de *Reinschia australis.*

Quelques spores.

Des grains de pollen.

Quelques lames cuticulaires.

Plus les menus fragments de parois végétales diversement altérées qui ont été déjà signalés.

---

* Il est impossible pratiquement delimiter la lame ligneuse ou subéreuse qui sert de substratum à un amas bitumineux du reste de cet amas, et quand on fait les relevés de l'intervention du bitume on compte comme bitume tout ce qui est à l'état d'amas rouge brun. En procédant ainsi on exagère beaucoup l'intervention bitumineuse

Les quatre premières catégories de ces organites sont à l'état de corps jaunes.

Le degré d'intervention des thalles de *Reinschia* dans le Megalong Valley est caractérisé par les nombres ci-après :

Nombre des rangées de thalles comprises dans 1^{mm} de hauteur... 45

Nombre des thalles rencontrés sur 1^{mm} de longueur horizontale transverse 22

Nombre des thalles rencontrés sur 1^{mm} de longueur horizontale radiale 17

Nombre des thalles contenus dans 1 mm^3 16830

Le coefficient vertical d'intervention des thalles est 0·449*; il peut s'abaisser jusqu'à 0·248.

Le coefficient horizontal d'intervention des thalles est 0·682†; il peut descendre jusqu'à 0·328.

Le coefficient d'intervention en volume est 0·370. En quelques points ou les gros thalles se raréfient j'ai vu ce coefficient s'abaisser jusqu'à 0·141.

La fréquence de chaque catégorie de thalles est indiquée par les nombres ci-après :

Jeunes thalles 82 savoir :

56 thalles moyens dont 6 jeunes, 14 trés jeunes, 36 extrêmement

D'après ces relevés on voit que les thalles plats interviennent largement dans la masse. Il y a un certain nombre de petits thalles cérébriformes. L'élément dominant est formé par les jeunes thalles moyens. Nous avons trouvé ces caractères dans le Mount Victoria et dans le Blackheath. De même que le Mount Victoria le Megalong Valley présente beaucoup de thalles orangés.

Les thalles agés sont souvent déchirés et représentés par des lambeaux séparés.

Tous les gros thalles et la plupart des jeunes thalles montrent des gravures très nettes dans leur matière gélosique. Ces gravures sont ici particulièrement nombreuses, très accusées et souvent parallèles aux couches d'épaississement de la gelée. Elles paraissent être d'origine bactérienne; elles ont été faites du vivant de la plante. Dans un grand nombre de ces gravures j'ai observé des micrococcoïdes mesurant de 0·15 μ à 0·5 μ et des filets bruns continus. Micrococcoïdes et filets bruns sont [illegible] plus fortement teintés que les protoplastes cellulaires du *Reinschia* et que la matière rouge brun très claire qui remplit la cavité des thalles. Micrococcoïdes et filets sont pleins. La forme sphérique des micrococcoïdes jointe à leur localisation dans les gravures des thalles permet de voir dans les micrococcoïdes des restes d'organismes bactériens teintés par localisation du bitume. La présence des filets bruns continus oblige, en l'absence de preuves directes, à apporter de très grandes réserves à cette interprétation bactérienne des micrococcoïdes. Les gravures des thalles débouchent souvent à leur surface. On les observe encore sur les thalles orangés.

Les thalles gommeux à structure effacée sont peu nombreux.

Les thalles cérébroïdes rouges, à gros canaux, sont extrêmement rares. Je n'en ai observé que quelques exemplaires dans un ensemble de coupes terminales dont la surface totale couvrait [illegible] mm². C'est un caractère que j'avais relevé en [illegible] dans le Blackheath et le Mount Victoria.

J'ai trouvé [illegible] spores par millimètre carré* et [illegible] grains de pollen.** Pollen et spores ne sont visibles que sur les coupes

* [illegible]

** [illegible]

verticales. J'ai rencontré quelques sacs polliniques dont tous les grains étaient demeurés contigus. Les grains de pollen isolés très affaissés ont contribué à stratifier la masse. Les spores sont à l'état de corps jaunes, les grains de pollen sont colorés en brun clair.

Les cuticules ont été rencontrées de loin en loin à l'état de minces lames jaune d'or isolées.

En résumé le kerosene shale reçu de Megalong Valley appartient au type du Blackheath. C'est un kerosene shale à thalles très isolés, posés à plat et trés affaissés dans une gelée brune abondante.

L'infiltration bitumineuse plus abondante que celle du Blackheath a accru la transparence du fond ou les thalles sont engagés et lui a donné une coloration rouge-brun. A cette bituminisation plus forte correspond une augmentation sensible de l'intervention du charbon brillant dans la constitution de la masse. Il s'agit d'une infiltration de bitume dilué.

La gélose joue encore un rôle important dans le Megalong Valley mais elle n'est plus l'élément dominant de ce charbon. La preponderance appartient au melange de gelée brune et de bitume. Le Megalong Valley s'éloigne des charbons gélosiques ou Bogheads pour se rapprocher des cannels; c'est une forme de

Les différences observées entre le Megalong Valley et le Blackheath sont très faibles. Elles sont à peu près celles que présentent deux morceaux d'un même lit près en deux points peu éloignés. Il diffère un peu plus du Mount Victoria parceque l'infiltration bitumineuse de ce dernier est moins forte et que la charge de sa gelée brune en fragments brun noir ou fusinifiés est plus grande.

Le Megalong Valley diffère beaucoup plus du Mort's Upper Tunnel. Dans ce dernier la gélose des thalles dilue si fortement la gelée humique que tous les thalles semblent se toucher. Les thalles adultes y sont aussi plus nombreux. Les menus fragments de parois végétales presque fusinifiées y sont fréquents. Le Mort's Upper Tunnel est un boghead car son élément dominant est la gélose. Cette différence qui arrive à modifier les caractères macroscopiques de la roche correspond simplement au point de vue géogénique, à une abondance plus grande des algues par rapport à la gelée humique qui se précipitait en même temps. Cette différence peut se produire localement en certains points d'un même banc.

LIST OF PAPERS

DEALING WITH THE MICROSCOPIC STRUCTURE OF THE NEW SOUTH WALES KEROSENE [illegible].

1.—BERTRAND C.E.—Conferences sur les Charbons de Terre: Les Bogheads à Algues. *Bull. Soc. Belge de Géol., Pal. et d'Hydrologie.* 1893 [1894], vii., pp. 45-[illegible]. pls. 4, 5.

2.——— Nouvelles Remarques sur le Kerosene Shale de la Nouvelle-Galles du Sud. *Comptes rendus Acad. Sci.*, 1896, cxx[illegible], pp. 615-617.

3.———Nouvelles Remarques sur le Kerosene Shale de la Nouvelle-Galles du Sud. *Bull. Soc. d'Hist. Nat. Autun*, 1896, [illegible], pp. 114.

4.———Nouvelles Remarques sur le Kerosene Shale de la Nouvelle-Galles du Sud. *Ann. Soc. Géol. Nord.* 1897, xx[illegible]., pp. 141-144.

5.—BERTRAND (C.E.).—Caractéristiques du Kerosene Shale dans les Gisements du Northern Coal Field et de Newcastle, Nouvelle-Galles du Sud. *Assoc. Franç. Av. Science*, 1897, pp. 341-346.

6. ———Premières Notions sur les Charbons de Terre. *Bull. Soc. Industrie Min.*, 1897, xi., Liv. 3, pp. 551-597.

7.— ———Caractéristiques d'un Charbon à Gaz, trouvé dans le Northern Coalfield de la Nouvelle-Galles du Sud. *Comptes rendus Acad. Sci.*, 1897, cxxv., pp. 984-985.

8.— ———Les Charbons humiques et les Charbons de Purins. *Trav. et Mem. Univ. de Lille*, 1898, vi., Mém. No. 21.

9.————Conférences sur les Charbons de Terre. *Bull. Soc. Belge. Géol.*, 1897, xi., Mém., pp. 284-310.

10.————Caractéristiques du Kerosene Shale dans les Gisements du Northern Coalfield et de Newcastle, Nouvelle-Galles du Sud. *Comptes rendus Assoc. Franç. Av. Sci.*, 1897 [1898], pt. 2, pp. 341-346.

11.————Charbons humiques et Charbons de Purins. *Comptes rendus Assoc. Franç. Av. Sci.*, 1897 [1898], pt. 2, pp. 328-340.

11 *bis.*———Note sur l'origine de la houille. *Comptes rendus* Soc. Ind. Min., Jan., 1898, pp. 17-19.

12.————Nouvelles recherches sur les Charbons. Les Charbons humiques et les Charbons de Purins. *Ann. Soc. Géol. du Nord.*, 1899, xxviii., pp. 26-47.

8 —BERTRAND (C.E. AND RENAULT (B.).—Note sur la Formation schisteuse

18—NEWTON (E. F.).—On "Tasmanite" and Australian "White Coal." *Geol. Mag.*, 1875, ii. (2), pp. 337-342, pl. 10.

19.—RENAULT (B.).—Sur quelques Micro-organismes des combustibles fossiles. *Bull. Soc. Ind. Min.*, 1899, xiii. (3), Liv. 4, pp. 865-1169, Atlas. [Australian Bogheads, pp. 1010-1017.]

20.—SEWARD (A. C.).—Coal: its structure and formation. *Science Progress*, 1895, ii., p. 355.

21.————Fossil Plants, i. (8vo. London, 1898), pp. 178-183.

22.—ZEILLER (R.).—Eléments de Paléobotanique. (8vo., Paris, 1900), pp. 34-35.

vigorous cultures of these fresh plates of peptone-glucose-gelatine* were prepared. The examination of the colonies that developed upon these plates made it evident that there were two varieties of yeast. As passage through a solid medium tends to alter the characters of the yeast, the two kinds were isolated by the dilution method from a vigorous culture in Hansen's peptone-glucose fluid.

Pure cultures of the two yeasts obtained in this way showed that they both belonged to the group which has for its type *Saccharomyces membranifaciens*. The first of these yeasts consisted of round, oval or sausage-shaped cells, containing one or two refractile granules. In wine must it formed a strong crumpled film and a slight flocculent precipitate. Many cells of the film contained two, and occasionally three, round or flattened spores. The second of the yeasts formed in wine must a slight transparent film and a bulky white sediment. The cells were chiefly oval, indistinctly vacuolated, and contained one or two refractile granules. Only a few of the film cells contained two spores. Neither of the yeasts induced a visible fermentation.

I was informed that the turbidity which arises in the wine would probably not be found to have a bacterial origin since the cloudiness is never so pronounced as when the wine is attacked by bacterial diseases. From this it was to be

same time two small bottles of unpasteurised wine were placed under observation. All the bottles were incubated at 22° C. for three weeks, when it was found that the pure yeasts, although still alive, had not grown, and it was evident that they were not responsible for the trouble. The unpasteurised wine was turbid, and had a thin surface film and a slight precipitate. During the experiment it was noted that a growth occurred first upon the surface of the unpasteurised wine and then spread downwards. An examination of the film revealed a zoogloea mass of bacteria. The same kind of bacteria were obtained in the body of the wine. On re-examining the original bottle of wine from which the yeast had been obtained, it was seen that a film had by this time formed, and it consisted entirely of bacteria similar to those found in the experimental bottles. The film had probably been present at an earlier stage, but it had been so slight as to be obscured by the dark colour of the glass of the bottle.

The investigation had so far advanced as to indicate the infecting organism, and in order to obtain it in pure culture, plates of nutrient agar and peptone-glucose-gelatine were prepared, but no growth appeared in these. The fact that the organism does not grow upon these media explains how I did not obtain it in my earlier experiments. Neither did it grow on nutrient glucose-gelatine. Colonies were, however, obtained upon plates of nutrient agar to which about one-third of the volume of pasteurised wine had been added while the agar was fluid. The colonies appeared at the end of four days when grown at 22° C. as small white points. When magnified sixty-fold they appeared round, brownish-black, and finely granular; the edges were smooth and the colonies had each a dark central point (the starting point of the colony). Within a central zone of half the radius of the colony the granulation was darker than at the margin. The deep and subsurface colonies appeared irregular, rough and slightly moruloid. The colonies enlarged, becoming round, white and glistening like a drop of wax. The middle of the colony was raised. The bacterial colonies derived from the original bottle of wine and from the wine from the storage cask, were identical in appearance and structure, while films showed the same organism.

When taken from the surface of the wine and examined in the fresh and moist unstained condition, the organisms are seen to be of two kinds. One is thin and refractile, the other broader and non-refractile. The thinner cells are generally divided into two and rarely three parts, while the broader cells appear homogeneous. The thinner cells are about 0·7 μ broad and vary from 2 to 3 μ in length; the broader cells are from 0·8 to 1·0 μ broad and 0·9 to 1·5 μ long. The parts of the thinner cells showed up well with aqueous eosin, and when so stained appeared as spheres within a common capsule. Little protuberances that suggested buds appeared attached to the end of a few of the broader cells. This would indicate that the organism is a yeast, although at first sight the minute size makes it probable that it is a bacterium. If it is a yeast, then the endocellular eosin-staining spheres are probably spores. The appearances of the stained and balsam mounted films, however, indicate that the organism is a bacterium. Although little prominences could be occasionally seen on a few cells in stained films, yet they were so few and of so doubtful a nature that they could not be taken to prove the organism to be a yeast. In one case a distinct bud appeared on a cell at the margin of a colony grown upon a wine-gelatine film. When,

size, and may therefore be called short stout rods. In fluids other than wine, as for example glucose-yeast-water, the organisms are thinner and show stained parts of unequal size. A rounder form is obtained when cultivations are made in nutrient gelatine, to which a little wine, say one-fourth of the volume of gelatine, is added. In this medium they appear as coccobacteria, varying in size from 0·5 : 0·6 μ to 0·6 : 0·9 μ. When stained with blue they appear like diplococci, the unstained central part being a narrow unstained line, on either side of which are hemispherical stained portions.

The organism grows best in media which contain the products of yeast activity, such as wine or yeast-water. When floated upon the surface of wine it grows freely, though slowly: on the other hand, when submerged in the mass of the wine it grows very slowly indeed. In the latter case a slight surface film appears before growth occurs in the fluid. This is reversed when, instead of wine, yeast-water with or without glucose is employed. In this medium a delicate film appears after the medium has become turbid and a white sediment has collected at the bottom of the liquid. During the course of the experiments it became evident that the greater the air surface of the wine in the experimental bottles the quicker did the turbidity appear. The inference from this is that the bacterium in wine behaves as an aerobe, and it is probable that the slight aeration which the wine receives in the process of bottling stimulates the growth of the organism which has been restricted by the anaerobic condition of the wine in the large casks. The aerobic character was proved by placing the organism under the practically anaerobic conditions that obtain in Buchner's tubes (from which the oxygen is removed by alkaline pyrogallate). Yeast-water inoculated with the bacterium and placed under these conditions failed to produce a growth.

The temperature best suited to its development would appear to be about 25° C., as growth was feeble at 17° and comparatively quick at 22°. At 37° the organism refused to grow.

In endeavouring to discover the constituents of wine which favour the development of the bacterium I employed nutrient agar, to which various substances had been added. In the first place, alcohol obtained by distilling wine produced a slight growth. A similar growth was also obtained by the addition of levulose and of lactose. Maltose, dextrose and sucrose were without influence, while the addition of a small quantity of tartaric acid (to make 0·3%) to the carbohydrates hindered the development of the organism. Dextrin, glycerol and succinic acid were likewise inoperative. No growth took place in Hansen's peptose-glucose with or without the addition of gelatine. From these experiments it would appear probable that the one particular constituent of wine that is active in stimulating the growth of the organism is alcohol.

The obligate aerobic character of the bacterium, and the tendency that it has for forming films upon the surface of fluids, are suggestive of the acetic bacteria. With the object of determining the production of acid a culture was smeared over a plate of wine-litmus-lactose-agar and incubated at 22°. The litmus in the vicinity of the culture was reddened. As this might be due to the formation of lactic acid from the lactose or acetic acid from the alcohol, a plate of nutrient agar, to which had been added alcohol (wine distillate) and chalk, was smeared with a culture of

however, negatived by obtaining the infecting organism from the later samples which had been carried in bottles sealed with sterilised corks.

With regard to the remedy for the trouble, it is apparent that since the organism is in the bulk of the wine and does not obtain access during the process of bottling, the employment of sterile bottles, corks, etc., would be of no avail. The wine itself must be treated so as to kill off the organism, and the only legitimate method of doing this consists in the pasteurisation of the wine. According to an experiment conducted with infected yeast-water, the lethal temperature was found to be between 66° and 72° C. A further experiment showed that the organism was killed and the yeast-water remained bright after an exposure for 5 minutes to 66° C. But the organism may be killed at a lower temperature when it is present in wine, and in an experiment where the surface of wine was infected previous to pasteurisation a temperature of 43° C. when maintained for 5 minutes was found to be sufficient.

In the experimental bottles of wine which I exhibit, both bottles were filled to the shoulder with wine, and then pasteurised for 15 minutes at 70° C. One was inoculated by floating a small loop taken from a wine film upon the surface. The other bottle was kept as a contrast to show the differences brought about by the organism. In five days, at 22° C., a strong film had spread over the surface. On shaking, the film broke up and settled to the bottom of the wine. On the seventh day the wine was decidedly turbid; on the fourteenth day it was very turbid, and had a strong sediment and surface film. The contrast bottle remained clear.

NOTE.—The conditions under which the above pasteurisation experiments were performed are not precisely such as would obtain in ordinary practice where the bacteria are suspended throughout the body of the wine. A fresh sample was obtained for the purpose of ascertaining the temperature requisite to kill the organisms in naturally infected wine. The original wine (Chablis) had, however, been blended and sent to the London market, but in its place another variety (Reisling, 1897) which undergoes a similar clouding, was forwarded to the laboratory. This wine was made at the same time and from grapes grown upon the same kind of soil as the Chablis, the only difference being in the variety of grape employed. On its arrival at the laboratory, the wine had begun to show a turbidity, but this did not affect the experiment, as a check unpasteurised flask was always kept for purposes of comparison. A preliminary experiment showed that the bacteria were killed when exposed to a temperature rising in five minutes from 40° to 45° C. In the second experiment, the temperature rose one degree in five minutes, and those flasks which were heated from 42° to 43° C. were subsequently found to have been sterilised. The flasks which had been heated at 40°-41° C. showed a growth-turbidity four days later than those heated at a lower temperature; at 41°-42° C. the turbidity appeared seven days later. In wine which had been heated below the lethal temperature, a surface film appeared as the original sediment deposited. The film increased, and the wine became turbid throughout its mass. Those flasks which were heated at and above the pasteurisation-temperature (43° C.) remained clear after depositing their sediment. It is, therefore, immaterial whether the bacteria are suspended in wine or floated

CONTRIBUTIONS TO A KNOWLEDGE OF THE FLORA OF AUSTRALIA.

PART III.

BY R. T. BAKER, F.L.S., CURATOR, TECHNOLOGICAL MUSEUM, SYDNEY.

(Plate xlii.)

STERCULIACEÆ.

*KERAUDRENIA HOOKERIANA, *Walpers*.—Girilambone, N.S.W. (W. Bäuerlen). The specimens agree fairly well with the published descriptions of the species, except some minor differences such as calyx-lobes coloured pink and not acute; capsule more angular, sometimes even winged; stamens and staminodia quite persistent, even until after the seed has dropped out of the capsules.

LEGUMINOSÆ.

HARDENBERGIA ALBA, sp.nov.

Leaflets solitary, ovate-acuminate or obtuse, or terminating in an acute point, rounded at the base, about 3½ inches long and 1½ inches broad, thin, almost membranous, never coriaceous, shining above, pale green in colour, of a still lighter shade underneath, lateral veins and reticulations only faintly marked, articulate on a petiole about 9 lines long. Stipules small, recurved, striate. Flowers numerous, on long axillary panicles, measuring sometimes up to 6 inches, on slender pedicels about 5 lines long, mostly in pairs; standard emarginate, 3 lines broad, 4 to 5 lines long, *white*

---

* Species marked with an asterisk have not previously been recorded from New South Wales.

in colour, as well as the keel and wings Calyx 1½ lines long, glabrous. Pod sessile, about 1½ inches long, 4 lines broad; seeds oblique.

Hab.—Country beyond the River Darling (Mrs. Helena Forde).

This plant differs more especially from *H. monophylla*, Benth., in the character of its leaves, which are light in colour, of a uniform shape, membranous and showing scarcely any reticulation. The flowers also are white wherever the plant grows, whether in cultivation or its native habitat. It is now cultivated in many private gardens, and is a very attractive plant with its long panicles of white flowers.

The flowers of *H. monophylla* are consistently blue, whilst the leaves are variable in shape and size, coriaceous and strongly reticulated—characters as above stated quite absent from *H. alba*.

In regard to its habitat Mrs. H. Forde writes me :—"I never found the plant growing on the frontage to the River Darling, but procured it from the *back country* going towards Lake Victoria, where many plants were found that did not grow on the river frontages"

Another point of difference from *H. monophylla* that might be mentioned is that it often occurs as a shrub showing scarcely any tendency to a climber (Rev. T. A. Alkin, M.A.)

ACACIA LINEATA, *A. Cunn.*—Mulwala, N.S.W., the most

Australia. The local name would imply that it occurs in Queensland, but it has never yet been recorded for that colony.

The pods, seeds and funicles are identical with Mueller's figure of the species in his *Iconography of Acacias*, so that if his material is correctly matched the identity of Bauerlen's material cannot be doubted.

A. LUNATA, *Sieb.*—Bentham (*B.Fl.* ii. 373), when describing this species, states that the seeds are close to the upper suture of the pod. Specimens having this feature have been obtained on Forest Reserve, No. 1, Mulwala, N.S.W., by Mr. Wyburd. This is the first time I have ever found pods corresponding to Bentham's description. The Acacia generally passing under the name of *A. lunata* is *A. neglecta*, Maiden & Baker.

A. CAMBAGEI, sp.nov. ("Gidgee" or "Gidgea").

(Plate xlii.)

A medium-sized tree, with pendulous branchlets, the foliage of a pale or glaucous hue; branchlets angular. Phyllodia *falcate, lanceolate*, obtuse or slightly acuminate, up to 5 inches long and from 5 to 9 lines broad, with numerous fine parallel veins, two or three more prominent than the rest; thin or membranous. Peduncles about three lines long, slender, in axillary clusters of about 6, each bearing a *globular* head of about 12 flowers. Sepals broad, spathulate, ciliate on the upper edge, free and less than half as long as the petals.

Petals glabrous. Pod flat, straight, about 3 lines long and 4 lines broad, veined, valves thin, not contracted between the seeds. Seeds ovate, longitudinal or slightly oblique; funicle short, filiform, not folded nor dilated.

Hab.—Bourke, N.S.W., and northward to Queensland (R. H. Cambage).

It differs in herbarium material from "Yarran" (*A. homalophylla*), in its larger and glaucous phyllodes, and the distinct venation, and also in the shape of its pods, and in the shape of the funicle. The phyllodes are somewhat similar to those of *A. excelsa*, Benth., in venation and colour, as also in the timber and pod. In botanical sequence it is placed between that species and *A. harpophylla*, F.v.M.

It has been usual in the past to regard in herbaria this tree ("Gidgee") and "Yarran" as one and the same species, and in the botanical literature of the Acacias they are designated as *A. homalophylla*, A. Cunn. It would appear, however, in the field that the "Gidgee" and "Yarran" are never confounded by settlers, the two trees, as they remark, "being quite different."

Mr. R. H. Cambage, L.S., who has given recent attention to these particular trees, and who has repeatedly disputed their being specifically the same, has procured sufficient material and evidence to convince me that the two should be separated; and I now propose the name of *A. Cambagei* for the Wattle known over a large tract of the interior of New South Wales as "Gidgee."

Bentham's description of *A. homalophylla*, A. Cunn. (*B.Fl.* ii. 38?), and Mueller's figure (*Iconography of Acacias*) of that species undoubtedly refer to "Yarran," which I have myself collected in several parts of this colony. The two species can be easily separated in dried specimens, the phyllodes being quite dissimilar, as well as the pods, funicle and timber.

In the field "Gidgee" is separated from "Yarran" by the offensive odour of its phyllodia and timber, which in wet weather

The largest and most conspicuous Acacia in the Western plains of this colony, attaining a height of 60-80 feet and a diameter of 3 feet. The timber is hard, close-grained, interlocked, of a deep red or darkish colour, and possesses a beautiful figure, so that it is one of the most ornamental of our timbers.

It can hardly be doubted but that it must have been collected by early botanists in this colony, but Bentham only records it from Queensland (*B.Fl.* ii. 390).

A. CARDIOPHYLLA, *A. Cunn.*—Eight and a half miles east of Wyalong, N.S.W., the farthest south-west locality (R. H. Cambage).

MYRTACEÆ.

RYLSTONEA CERNUA, *Baker.*—In the Society's Proceedings for 1899 (p. 643), Messrs. Maiden and Betche endeavoured to show that my genus *Rylstonea* is identical with their species *Verticordia Darwinioides*, and in a tabulated form give the affinities and differences of individual characters of the two shrubs, stating at the same time that it was still from incomplete material that their deductions were made. Even in their own tabulation of specific characters it will be seen that the differences are well marked in the two plants, and certainly warrant their being separated.

That I "did not refer to the obvious affinity" of my plant to theirs was because there appeared so much uncertainty and doubt surrounding their plant that no one could say what it was eventually going to be, for according to their own admission "it was an abnormal and stunted form" they had to work upon. Complete material of their plant is still wanting to discuss its merits to systematic rank.

The statement that the calyx of *Verticordia* is hemispherical is challenged, and *V. Wilhelmii*, F.v.M., is quoted as proof to the contrary. Bentham's words are (*B.Fl.* iii. 14) — "Calyx hemispherical, turbinate rarely cylindrical," but in no instance does he state definitely concerning a species that the calyx is cylindrical.

Previous to describing the genus, specimens of a very large number of species of *Verticordia* were received by me from Western Australia, and they all possess a common facies and an inflorescence as described by Bentham, *i.e.*, flowers usually pedicellate in the upper axils, forming often broad terminal leafy corymbs, or simple leafy spikes or racemes—a statement quoted by Messrs. Maiden and Betche (*loc.cit.*) in order to show that "the inflorescence is extremely variable, and not a character in which the separation of a new genus could be based."

It is on the shape of the calyx and the calyx-lobes that my genus is based,—characters upon which Cunningham founded the cognate genus of *Homoranthus*, and which determination was supported by Bentham in his "*Genera Plantarum*," and from these features alone I think I am justified just as much as Cunningham in establishing a new genus on my material.

Messrs. Maiden and Betche state that the calyx of *Verticordia Wilhelmii* is cylindrical, and on this statement it is proposed to remove this species from its previous classification and place it in my genus under the name of *Rylstonea Wilhelmii*, F.v.M.

Bentham's note (*B.Fl.*, iii. p. 19) in connection with this species is significant; he states:—"This single species differs from all others of the genus in inflorescence and the shape of the calyx, and its lobes form an approach to those of *Homoranthus.*"

and all the species of *Verticordia* in the west and north-west of the Continent.

The two genera will thus include species that have well marked and distinct generic characters and a distinct geographical distribution.

It is now proposed that the genus *Rylstonea* should include the following species :—

Genus Rylstonea, *Baker*.

Proc. Linn. Soc. N.S.W. 1898, p. 768.

R. cernua, *Baker*, Proc. Linn. Soc. N.S.W., 1898, p. 768.

Hab.—Widdin, Mt. Corricuddy, near Rylstone, N.S.W.

R. Darwinioides, *Maiden & Betche* (Syn. *Verticordia Darwinioides*, Maiden & Betche), Proc. Linn. Soc. N.S.W., 1899, p. 643.

Hab.—Dubbo, N.S.W.

Tentatively placed under *Rylstonea*, as in my opinion, when complete material is obtained, the species will stand under this genus.

R. Wilhelmii, *F.v.M.* (Syn. *V. Wilhelmii*, F.v.M.), Trans. Vic. Inst. 122, and B.Fl. iii. 19.

Hab.—South Australia.

Eucalyptus amygdalina, *Labill.*—As far west as Isabella, Burroga, N.S.W. (R. H. Cambage).

E. Smithii, *Baker* ("White Top.")—Wingello, N.S.W., in the gullies, on the Nandi Road, Sutton Forest (some very fine trees); also plentiful on the Nepean Water Reserve.

E. dextropinea, *Baker*.—Wingello, N.S.W., plentiful in the gullies; its most northern locality. Its timber is locally used extensively, and it is considered to be of excellent quality, being known as "White Mahogany." It, however, has no connection with *E. acmenioides*, the "White Mahogany" of the Coast.

E. oreades, *Baker*.—This smooth-barked "Mountain Ash" of the Blue Mountains occurs at the Caves between Wentworth and Katoomba. Round the water holes near these caves are some tall Eucalypts, every one of which is this species (Henry G. Smith).

E. DELEGATENSIS, *Baker*.—On the lower ranges of the Snowy Mountains, at an elevation from 4,000-5,000 feet (W. Bäuerlen).

E. BICOLOR, *A. Cunn.* ("Bastard Box").—It has been customary amongst botanists in recent times to place this species of Cunningham's under Mueller's *E. largiflorens*. Now Cunningham was the first to discover also this latter species, which he named *E. pendula*. Mueller, however, considering Cunningham's description to be indefinite, gave it the name of *E. largiflorens*.

Thinking it would be very strange if a great collector and botanist, such as Cunningham undoubtedly was, should have given two names to one and the same tree, I had occasion recently to investigate the matter, and the material now in my possession shows conclusively that he (Cunningham) had two distinct trees in his mind's eye when he recorded them.

I am much indebted in this instance to the writings of the late Dr. Woolls for finding the particular trees of *E. bicolor*. In his "Contributions to the Flora of Australia" (p. 232) he gives the locality Cabramatta, where will be found trees that exactly coincide with Cunningham's description of *E. bicolor*, and in no way agree with *E. largiflorens*, F.v.M. (*E. pendula*, A. Cunn.), of the interior. I and others have now seen both trees in the field

the fruits are only half the size of those of the eastern species. The oils also are quite different.

Bentham (*B. Fl.* iii. p. 214) acknowledges Cunningham's species *E. bicolor*, and places as its synonyms *E. pendula*, A. Cunn., and *E. largiflorens*, F.v.M., but he, of course, only had dried material to work upon. However, there can be no doubt now that Cunningham was correct in his determinations, and in the light of our present knowledge it is proposed in future to acknowledge both his species, *E. bicolor* and *E. pendula*, and in the latter case his name takes priority over Mueller's name of *E. largiflorens*.

The localities for the two species are :—

E. BICOLOR, *A. Cunn.*—Cabramatta (Rev. W. Woolls); banks of South Creek (R. T. Baker and W. Bäuerlen).

E. PENDULA, *A. Cunn.*, (Syn. *E. largiflorens*, F.v.M.)—Dubbo to Bourke (W. Bäuerlen); Lachlan River, Condobolin (R. H. Cambage).

EUCALYPTUS BRIDGESIANA, *Baker*.—The most westerly locality I have now to record for *E. Bridgesiana*, Baker, is Mulwala, where it grows in the Forest Reserve No. 1591 (G. Wyburd).

Since publishing my description of this species in the Society's Proceedings for 1898, Messrs. Deane and Maiden continue to refer it in their series of "Notes on Eucalypts" to *E. Stuartiana*, F.v.M. These authors state that "the figure of *E. Stuartiana* in the 'Eucalyptographia' is one of the happiest of the delineations of that work, and is simply unmistakaole" (Proc. Linn. Soc. N.S.W., 1899, p. 628). With all deference to the late learned Baron, it must be said that his description refers to a Victorian tree with a red timber and with a red stringy bark. Specimens of these have been obtained from Victoria, and can be seen in the Technological Museum, Sydney, where they are placed in juxtaposition to those of *E. Bridgesiana* of this colony, which has a white, woolly, persistent bark, and which bark yields an oil (Proc. Linn. Soc. N.S.W. 1898, p. 166), while the timber is light-coloured. These characters alone are sufficient in my opinion to warrant a separation of the trees, and one could never put such Museum specimens possessing diametrically opposite characters (given above) under the same specific name with any degree of correctness.

That Mueller had the Victorian tree in his mind's eye is shown by his many references to his *E. Stuartiana* as a tree closely allied to, if not the same species as *E. pulverulenta*, Sims, for in his "Eucalyptographia" he states:—"*E. pulverulenta* is distinguished from *E. Stuartiana* only in its foliage . . . the bark of *E. Stuartiana* and *E. pulverulenta* are very much alike." Such statements can only apply to the Victorian Eucalypt, as there is no resemblance in *E. Bridgesiana* to *E. pulverulenta*. Mr. A. W. Howitt, a co-worker with Mueller in the Eucalypts and one who collected the original material, holds that it was the Victorian Apple on which *E. Stuartina* was founded. Messrs. Deane and Maiden state (*loc. cit.*) that they recently proceeded to the National Herbarium to study the specimens there and to confer with the Curator, Mr. J. G. Luehmann, long Baron von Mueller's principal assistant. and one who knows best the late botanist s view on this and many other points. After carefully investigating the matter they saw no reason to refrain from accepting the plate in the "Eucalyptographia" as faithfully

recently to include under *E. amygdalina*, Labill., such well defined species as *E. radiata*, Sieb., *E. dives*, Schauer, *E. fastigata*, Deane and Maiden, *E. regnans*, F.v.M., and *E. Smithii*, Baker.

The research on the genus Eucalyptus now in progress at this Museum has proved conclusively that the above are all true and distinct species if diagnosed on what appears to be a natural classification; and it is on such grounds that I hold that *E. Stuartiana* and *E. Bridgesiana* are also distinct species.

E. HEMILAMPRA, *F.v.M.*, "Mahogany."

(Syn. *E resinifera*, Sm., var. *grandiflora*, B.Fl. iii. p. 246)).

An examination of the full material of this Eucalypt proves that Mueller was correct in his systematic placing of the tree, and now it is proposed to suppress its varietal position and restore it to the specific rank as originally proposed by its author. It has a wide range and its products and specific characters are constant throughout its distribution. From Manly (Woolls) to the Clyde River (W. Bäuerlen).

It is easily distinguished in the field from *E. resinifera*, Sm., by its large fruits and pale pinkish timber.

E. EXIMIA, Schauer ("White Bloodwood").—Generally considered a mountain species, but it is to be found on all the ridges at Gosford and in odd places on the Hawkesbury side (R.T.B.) and almost to Berowra (R. H. Cambage).

LORANTHACEÆ

LORANTHUS GRANDIBRACTEUS, *F.v.M.*—The first recorded specimens of this species from this colony were obtained by Mr. W. Bäuerlen at Tibooburra, and he has now collected it at Girilambone growing on *Eucalyptus populifolia*, F.v.M. This is so far the limit of its eastern range.

COMPOSITÆ.

BRACHYCOME CILIARIS, *Benth.*—New Angledool (A. [illegible]). Previously recorded only from the southern interior of the colony, but it probably extends over the whole western area

GOODENIACEÆ.

GOODENIA STEPHENSONII, *F.v.M.*—Woodford, Blue Mountains (W. Bäuerlen); Bylong (R. T. Baker).

EPACRIDEÆ.

BRACHYLOMA DAPHNOIDES, *Benth.*—Barmedman, N.S.W., its most westerly recorded locality (R. H. Cambage).

ASCLEPIADEÆ.

*GYMNANTHERA NITIDA, *R. Br.*—Tumbulgum, N.S.W. (W. Bäuerlen).

PENTATROPIS QUINQUEPARTITA, *Benth.*—New Angledool (A. Paddison). Moore, in his "Handbook of the N.S.W. Flora," gives the range of this species as "Western Plains except the North." It is now shown to extend to the Queensland border.

CHENOPODIACEÆ.

CHENOPODIUM ATRIPLICINUM, *F.v.M.*—New Angledool (A. Paddison). Previously recorded only from the southern interior (Darling River).

MONIMIACEÆ.

*STEPHANIA ACULEATA, *Bail.*—Tumbulgum, Tweed River; Moonambah and Wardell, Richmond River (W. Bäuerlen).

LILIACEÆ.

DIANELLA COERULEA, *Sims.*—New Angledool (A. Paddison).

JUNCACEÆ.

†LUZULA CAMPESTRIS, *DC.*, var. BULBOSUM ACCEDENS, *F. Bert.*—Bodalla, N.S.W. (W. Bäuerlen).

†L. CAMPESTRIS, *DC.*, var. AUSTRALASICA, *F. Bert*—Snowy Mountains (W. Bäuerlen).

†JUNCUS PLANIFOLIUS, *R.Br.*, var. GEMINUS, *F. Bert.*—Rylstone (R. T. Baker).

*J. POLYANTHEMA, *F. Bert.*, var. PARVA.—Homebush (J. H. Maiden); Carlton, near Sydney (Mrs. Clarke).

†J. PRISMATOCARPUS, *R.Br.*, var. GEMINUS, *F. Bert.*—Rylstone (R. T. Baker).

CYPERACEÆ.

* ELEOCHARIS DIETRICHIANA, *Boekel.* — Wonnaminta (W. Bäuerlen); Lismore (W. Bäuerlen).

*SCIRPUS MULTICAULIS, *F.v.M.*—Cobham Lake (W. Bäuerlen).

*S. CERNUUS, *Vahl* (Enunc. ii. 245—Cosmop. *forma pymæa*).—Snowy Mountains (W. Bäuerlen).

‡CLADIUM JAMAICENSE, *Crantz* (Inst. i. 362 = *germanicense*).—Tumbulgum (W. Bäuerlen).

§CAREX CANESCENS, *L.*, var. ROBUSTA, *Blyth.*—Snowy Mountains (W. Bäuerlen).

---

* Species marked with an asterisk have not previously been recorded from New South Wales.

† Variety not previously recorded for N.S.W.

‡ New for Australia.

§ Variety not recorded for N.S. Wales, new for Australia.

FUNGI.

MORCHELLA CONICA, *Pers.*—Pinnacle Mountain, 20 miles south of Forbes, growing on sandy soil of Devonian sandstone (R. H. Cambage); previously recorded from near Tamworth (Proc. Linn. Soc. N.S.W. 1896, p. 503).

*PHYSARUM RUFIBASIS, *B. & Br.*—Ballina, N.S.W.; on moss (W. W. Watts).

*CLAVARIA FUSIFORMIS, *Sow.*—Tumbulgum (W. Bäuerlen); on the ground.

§FOMES HOLOSCLERUS, *Berk.*—Tumbulgum, Richmond River, on a living root (W. Bäuerlen).

*F. INFLEXIBILIS, *Berk.*—Bellinger River, on decaying logs (Mr. Williams); Tumbulgum (W. Bäuerlen).

*F. APPLANATUS, *Fr.*—Never Never, Rylstone, on dead wood (R. T. Baker).

*F. HEMITEPHRUS, *Berk.*—Blue Mountains, on dead wood.

*POLYPORUS SULPHUREUS, *Fr.*—Tumbulgum, Tweed River, on decaying branches (W. Bäuerlen).

*P. MELANOPUS, *Fr.*—Tumbulgum, Tweed River, on decaying logs (W. Bäuerlen).

§TRAMETES GIBBOSA, *Fries.*—Berowra, on dead timber (R. T. Baker).

*POLYSTICTUS LILACINO-GILVUS, *Berk.*—Tamworth, on dead timber (D. A. Porter).

*POLYSTICTUS BRUNNEOLEUCUS, *Berk.*—Strathfield, on decaying timber (W. Lewis). This is the first record of this species from the mainland of Australia, being previously only recorded from Tasmania.

EXPLANATION OF PLATE XLII.

Acacia Cambagei, sp.nov.

Fig. 1.—Flowering twig.
Fig. 2.—Pod with seeds.
Fig. 3.—Bud.

---

* Species marked with an asterisk have not previously been recorded from New South Wales.

§ Species new for Australia.

ON SOME NEW SPECIES OF EUCALYPTUS.

By R. T. Baker, F.L.S., Curator, Technological Museum, Sydney.

(Plates xliii.-xlvi.)

E. intermedia, sp.nov.

"Bloodwood" or "Bastard Bloodwood."

(Plate xlvi., fig. 1.)

A medium-sized tree with a *light brown fibrous* bark.

Leaves lanceolate, acuminate, about 6 inches long, and 1-1½ inches wide or more, pale on the underside; lateral veins oblique, fine, numerous, parallel; intramarginal vein quite close to the edge.

Flowers mostly in large terminal corymbs. Calyx turbinate, 4 lines in diameter, 3 lines long, on a pedicel of about 4 lines. Ovary flat-topped. Stamens all fertile; anthers parallel, opening

nature of the timber, bark, oil and fruits which have not the marked recurved rim of that species.

From *E. eximia* it differs in having pedicellate fruits, a stringy flaky bark, a pinkish timber, and in its chemical constituents.

Dr. Woolls was cognisant of the differences existing between these species, for in his "Flora of Australia" (p. 238) he states:—"At the Clarence and Richmond Rivers the 'Bloodwood' prevails to a great extent, and the workmen reckon two kinds – the one with smooth and the other with rough bark. . . It seems probable that the Mountain 'Bloodwood' (*E. eximia*), which overhangs the valley of the Grose, is different from the Bloodwood of the north." As stated above, other botanists have always regarded the northern "Bloodwood" as identical with the Sydney and southern "Bloodwood"; but Dr. Woolls is the only one who connected it (the northern one) with *E. eximia*, Schau., and recent observations also show it to have affinities with that species.

Its physical characters, however, are so evenly balanced between the two that it is decided to give it specific rank.

It differs from *E. terminalis*, F.v.M., the "Bloodwood" of the interior, in its bark, timber and oil; and from *E. trachyphloia* in its larger fruits, bark and chemical constituents.

Its fruits are exactly identical in size and shape with those of *E. maculata*, but it resembles this spotted gum in no other characters.

This tree is constant throughout an extensive range, as it was found many years ago at Barney's Wharf, Cambewarra, by W. Bauerlen, who forwarded specimens to the late Baron von Mueller, who considered it a hybrid between *E. corymbosa* and *E. maculata*, but of course he only had dried material upon which to base his opinion.

The timber of both the southern and northern trees is similar in colour, hardness and other characters, and the chemical constituents of the oil show no variation.

Timber.—A pale-coloured timber, hard, straight-grained, and easy to work. It is much closer in texture than the Sydney

Bloodwood, *E. corymbosa*, Sm. The figure is occasionally not unlike that of *E. maculata*, Hook. Gum veins are not infrequent. It is considered a good durable timber, and superior to that of *E. corymbosa*, Sm. It has quite a metallic ring when the fractured edges of a piece are rubbed together.

Oil.—The yield from this oil is ·125 per cent. It consists very largely of pinene, 58 per cent. of the oil distilling below 172° C.; only a trace of eucalyptol could be detected. The specific gravity of the crude oil at 15° C. = ·8829. The specific rotation of the crude oil $[a]_D = + 11{\cdot}2°$. This oil differs from the oil of the Bloodwood of the Sydney district, inasmuch as the latter is lævo-rotatory to about the same extent. The rotation of the oils from *E. corymbosa*, Sm., and *E. eximia*, Schau., and this species varies in about equal proportions, that of the oil of this species being about half-way between those of *E. corymbosa* and *E. eximia*, although the constituents of the oils of the three species differ but slightly, being largely pinene.

E. ANGOPHOROIDES, sp.nov.

"Apple-Top Box."

(Syn. *E. Bridgesiana*, Baker, *partim.*)

(Plate xlvi., figs. 4*a*, 4*b*, 4*c*.)

acuminate. Ovary domed. Stamens all fertile; anthers parallel, opening by longitudinal slits.

Fruits hemispherical to slightly pear-shaped, 2 lines in diameter and under 4 lines long; rim thick, sloping outwards—a ring just below the edge; valves generally 4, exserted under 1 line.

Hab.—Colombo, N.S W. (W. Bäuerlen); Towrang, N.S.W. (R. T. Baker).

The herbarium material of this species is so similar to that of *E. Bridgesiana* that on my first examination it was included under that species (Proc. Linn. Soc. N.S.W., 1896.)

My field observations since that date, and the acquisition of further material such as timber and oil, have convinced me that the two trees are quite different, and should not be included under the same name. Mr. W. Bäuerlen, indeed, who has known the trees for very many years, has always held that the two were different in specific characters.

E. Bridgesiana is known vernacularly as "Apple" and "Woolly-butt," but this tree as "Apple-top Box." As stated above, the foliage, fruits and flowers certainly resemble those of the former species, but there the similarity ends. The bark is a true box-bark, but the timber is quite unlike that of a box.

It differs from *E. Cambagei*, Deane & Maiden, in the superiority of its timber and the inferiority of its oil, and the shape of its fruits; and from *E. nova-anglica*, Deane & Maiden, in the bark, colour of timber, and oil.

It has little affinity with such Boxes as *E. hemiphloia*, F.v.M., *E. Woollsiana*, Baker, *E. conica*, Deane & Maiden, *E. pendula*, A. Cunn., (*E. largiflorens*, F.v.M.), although it appears to be a connecting link with these and what are known as *Bastard Boxes* such as *E. Cambagei*, Deane & Maiden, and *E. bicolor*, A. Cunn.

It is quite limited in its distribution, and presents no difficulty of determination in the field.

The bark has not an essential oil as pertains to *E. nova-anglica* and *E. Bridgesiana*.

Although it has a regular light-coloured grey box bark, yet the appearance of the tree is more like that of an "Apple tree" (*Angophora*), hence the local name of "Apple-Top Box."

Timber.—A pale-coloured, soft, specifically light timber, open in the grain, and perhaps to be regarded as porous. It has not the broad sapwood of *E. Bridgesiana*, Baker. It seasons well, and is suited for cabinet work, as it closely resembles in colour, weight and texture the timber of *Angophora intermedia*, DC. It is much superior to that of *E. Bridgesiana.*

Oil.—The yield of oil from this species is ·185 per cent. A large quantity of phellandrene is present, also some pinene, and 26 per cent. of eucalyptol was found in the rectified oil (fraction representing 70 per cent. of the crude oil). The specific gravity of the crude oil at 15° C. was ·9049; the specific rotation of the crude oil = $[a]_D$ —12·7, the lævo-rotation being due to the phellandrene. The constituents of this tree differ greatly from those obtained for *E. Bridgesiana*, Baker, a species which in appearance it somewhat resembles. The oil of *E. Bridgesiana* is of excellent quality, while that of this species is of little commercial value, irrespective of the small yield (H. G. Smith).

E. WILKINSONIANA, sp. nov.

(Syn. *E. hæmastoma*, var., F.v.M., Eucalyptographia, Dec. ii.; *E.*

Fruits hemispherical, 5 lines in diameter, rim thick, *red*; valves slightly exserted, acute.

Hab.—Dromedary Mountain (C. S. Wilkinson, F.G.S.); Colombo (W. Bauerlen); Barber's Creek (H. Rumsay); Sutton Forest (R. T. Baker).

This is the "Stringybark" variety of *E. hæmastoma*, Sm., mentioned by Baron von Mueller in his Eucalyptographia under that species.

It was first observed in this colony by the late Government Geologist, Mr. C. S. Wilkinson, F.G.S., at Dromedary Mountain at an elevation of 1,500ft. above sea level, and named for him by Mueller as stated above.

It differs, however, from *E. hæmastoma*, Sm., in the nature of the timber, texture and venation of leaves, bark and chemical constituents of the oil and kino; and it is on these differences that it is now raised to specific rank.

The bark and timber ally it to *E. eugenioides*, "White Stringybark," and in botanical sequence it is placed next to that species.

The oil resembles that of *E. lævopinea*, Baker, but no other characters connect it with that species.

The fruits, and particularly the oil, differentiate it from the other "Stringybarks," such as *E. capitellata*, Sm., *E. macrorhyncha*, F.v.M., *E. eugenioides*, Sieb., *E. dextropinea*, Baker.

The red rim of the fruits has evidently been the cause of the misplacing of this species, but it is well known now that this is a character common to a number of Eucalypts.

It is a feature quite absent from *E. lævopinea*, Baker. In fact the fruits of the two species are so very different that the trees could not be synonymised with any degree of correctness in specific naming. The bark, leaves, venation and timber of these trees also differ.

E. lævopinea, Baker, has a hard, compact bark right out to the branchlets, whilst this tree has a light-coloured, loose stringy bark, not extending out to the limbs.

It is quite distinct in specific characters from the two stringy barks described in this paper, viz., *E. nigra* and *E. umbra.*

Timber.—Pale-coloured, very hard, close-grained, heavy. In transverse and compression tests, it stands higher than that of any of the other Stringybarks above enumerated. It is evidently an excellent timber, and is strongly recommended for forest conservation.

Oil.—The yield of oil from this species averages about ·9 per cent. It consists very largely of lævopinene. A small quantity of eucalyptol is present in the oil at some time of the year, and a small quantity of phellandrene at others. This terpene alters much, occurring in small quantities in many Eucalyptus oils at certain seasons of the year, whilst at other times it is absent. The specific gravity of the crude oil was ·894 at 15° C.

The specific rotation of the crude oil $= [a]_D - 23{\cdot}9$; no phellandrene was detected in the January oils.

No less than 86 per cent. of the oil distilled below 170° C. The lævo-rotation of this oil is due to the lævopinene present, The odour of the oil after saponification of the small quantity of ester present is exactly that of oil of turpentine (H. G. Smith).

E. OVALIFOLIA, sp. nov.

" Red Box."

minate. Ovary flat-topped. Outer stamens sterile. Anthers parallel, opening by pores at the truncate end.

Fruits small, 2 lines long and 1½ lines in diameter, rim thin, contracted slightly at the orifice, valves not exserted.

Hab.—Bathurst, Rylstone and Camboon (R. T. Baker); Hargraves (A. A Suttor); Gerogery (L. Mann).

A medium-sized or rather stunted tree growing in poor, sandy, rocky soil. The bark can hardly be said to be smooth, and neither is it altogether a box-bark such as that of *E. albens*, Miq., or *E. hemiphloia*, but rather between a box and a smooth bark. The upper parts of the trunk and limbs are quite smooth.

It is allied to *E. melliodora* in the shape and venation of the leaves, and perhaps in the exterior character of the bark, but it has not the yellow stain on the inner surface of the bark such as obtains in *E. melliodora.* It differs, however, from that speeies in the shape of the fruits, colour of timber, and chemical constituents of its oil.

It differs from the typical *E. polyanthema*, Sch., of Victoria, which has a persistent box-bark right out to the branchlets, larger and orbicular-shaped leaves, and larger fruits. The oils of the two species are not at all identical, but there is a resemblance in their timbers.

It differs from *E. conica*, Deane & Maiden, in having a smoothish bark, and in the shape of the leaves and fruits, and chemical constituents of the oil; nor can it be confounded with *E. pendula*, A. Cunn., (*E. largiflorens*, F.v.M.) which has a box-bark, and fruits and leaves quite different from this species.

The timber, leaves and bark differentiate it from the Lignum-vitæ, *E. Fletcheri*, Baker, of St. Mary's and Thirlmere.

In botanical sequence it is placed next to *E. Dawsoni*, Baker, as it approaches this tree in the colour of its timber, and occasionally in the shape of the leaves, but differs in every other respect.

There appears to be no reference to this tree in the writings of Dr. Woolls and Mr. A. G. Hamilton, both of whom wrote on the

Mudgee Flora, so that it must have escaped their observations, as it occurs at Hargrave, mid-way between Mudgee and Wellington.

Timber.—When growing on poor ironstone ridges the tree becomes rather stunted and the stem has a tendency to barrel, so that it yields only small specimens of timber. It is red-coloured, hard, close and straight-grained, and very durable in the ground. It is suitable for all kinds of heavy work.

Oil.—The yield of oil from this species is ·27 per cent. It contains much phellandrene and but a minute quantity of eucalyptol at time of distillation. It is not, however, a commercial oil. It is distinctly different from *E. polyanthema*, Schau., of the south, which gives a commercial oil rich in eucalyptol.

The specific gravity of the crude oil at 15° C. is ·9058. The specific rotation of crude oil $[a]_D = -9{\cdot}93°$.

There is very little difference in the constituents of this oil and that of *E. Fletcheri*, Baker, the "Lignum-vitæ" or Black Box at St. Mary's, as they both contain the same constituents in practically the same amount (H. G. Smith).

E. FLETCHERI, sp.nov.

"Lignum-vitæ," "Box."

Ovary flat-topped. Outer stamens sterile. Anthers parallel, opening by terminal pores.

Fruits conical, about 4 lines long, 3 lines broad; rim thin, and mostly in mature fruits with a notch; capsule sunk.

Hab.—South Creek, St. Mary's (R. T. Baker and N. V. Fletcher); banks of the Nepean River (Rev. Dr. Woolls); Thirlmere (W. Cambage).

It is named after the late Norman Fletcher, B.A., a promising young botanist much interested in Eucalypts, who, in company with the author, some years ago discovered trees of this species at South Creek, St. Mary's, near the railway bridge.

A tree apparently restricted in its geographical distribution to the watershed of the Nepean River of this colony.

The late Dr. Woolls was very probably the first to collect material of this tree for botanical determination, and he forwarded it to Mueller under the local name of "Lignum-vitæ" (Eucalyptographia, Dec. iii). This latter author, working on morphological grounds, confounded it with the Victorian "Red Box," *E. polyanthema*, Schau. The dried specimens of the two species are very much alike in the shape of the leaves and fruits, but the trees differ considerably in other characters. For instance, the Victorian "Red Box" has a persistent *box-bark* right out to the branchlets, and a *dark red timber*, while its leaves are larger than those of this species. The New South Wales tree has *thick, rough, flaky bark*, and *the wood, which is of a brown colour* towards the centre, is very hard and tough, as recorded by Dr. Woolls (Fl. Aus. p. 236). The two timbers alone are sufficient to differentiate the trees, whilst their essential oils possess quite distinct chemical constituents.

This is another example showing how essential it is that field observations are required in order to determine correctly the specific rank of Eucalypts.

It generally occurs on the banks of rivers and creeks, growing along with *E. bicolor*, A. Cunn., but this latter species, although having a somewhat similar bark and timber, is quite different in the fruits, leaves, venation and oil.

In colour of timber and bark it appears to stand apart from the western "Boxes," such as *E. Woollsiana*, *E. conica*, Deane & Maiden, *E. albens*, Miq., but in fruit and shape of leaves it resembles *E. populifolia* and *E. polyanthema*, Schau., whilst it only approaches *E. conica* in the shape of the fruit.

The leaves are thinner than those of *E. populifolia*, and have not the lustre so distinctive of that species.

It differs also from this latter species in the shape of its fruits, and in its timber and oil.

E. quadrangulata, Deane & Maiden, has a lighter-coloured timber, sessile fruits with extended valves, and lanceolate leaves.

Timber.—It is well described by Dr. Woolls (*loc. cit.*). No doubt owing to its good qualities it has been extensively cut by timber-getters, as it is quite rare now in its original habitat (Nepean), but is more plentiful at Thirlmere. It is worthy of propagation.

Oil.—The yield of oil from this species is ·294 per cent. It contains much phellandrene and but a minute quantity of eucalyptol. In constituents and characters this oil differs but little from that obtained from "Red Box," *E. ovalifolia*, Baker, of Rylstone. The specific gravity of the crude oil at 15° C. is ·8805. Specific rotation of crude oil $[a]_D = -14{\cdot}2°$. The lævo-rotation is due to phellandrene. It is not a commercial oil. The amount

Sucker-leaves lanceolate, alternate, 2-3 inches long, ½ to ¾ inch broad. Mature leaves under 6 inches long, on a petiole less than ½ inch; narrow-lanceolate, tapering to a fine recurved point, mostly of a thin texture, of a light yellowish-green, sometimes slightly shining; venation obscured, impressed on the upper surface; lateral veins few, intramarginal vein removed from the edge.

Peduncles axillary, from 2-12 lines long. Flowers few. Calyx about 1 line in diameter, tapering into a short stalk. Operculum hemispherical, acuminate, and often shorter and *more obtuse than shown in the plate.* Ovary flat-topped. Stamens all fertile; anthers parallel; connective large and long, attached at base to the filaments.

Fruits small, 1 line in diameter, hemispherical to slightly pear-shaped; rim thin, slightly contracted, valves not exserted.

Hab.—Girilambone, Cobar and Trangie (W. Bäuerlen); Nyngan and Murga (R. H. Cambage).

This tree is a half-barked "Box," and allied in bark and timber to *E. populifolia*, *E. albens* and other cognate box-trees.

Of all the box-trees described this species has probably the narrowest leaves. The fruits are small, and somewhat approach in shape those of the Green Mallee, *E. viridis*, Baker; but the bark, timber, and chemical constituents of the kino and oil differentiate it from that species.

The leaves have a shining surface occasionally, as pertains to *E. populifolia*, F.v.M., or *E. Behriana*, F.v M. It differs from *E. microtheca* in the valves of the fruit not being exserted, in the colour of the wood, and in the bark and chemical constituents.

From *E. hemiphloia* it differs in the nature of its timber, oil, buds and leaves; from *E. pendula*, A. Cunn., in the venation and shape of the leaves, the shape of the fruits and constituents of the oil, and particularly in its timber, and it has a more erect habit than this species. *E. populifolia* has much wider leaves, but the bark of the species is very similar, but is not associated in any other respect with this species.

Mr. W. Bäuerlen states "that it is usually associated with *E. populifolia*, the Green Mallee (*E. viridis*, Baker), and the Grey Mallee (*E. Morrisii*, Baker), on which account it is called 'Mallee Box.' I have never seen it in mallee form, and as a result of my inquiries it appears that it does not grow in that form."

Of described species it is most closely allied to *E. hemiphloia* and other "Boxes" in oil, kino, and botanical characters.

It differs from *E. conica*, Deane & Maiden, in height, bark, timber, oil and fruits.

Although the two species are not easily separated on herbarium material, they are never confounded in the field.

Timber.—Hard, close-grained, interlocked, heavy, durable timber, of a brownish colour. Useful for bridge-decking, posts, railway sleepers, and general building purposes. It is in great request at the Cobar mines for shoring the roofs.

Kino.—Turbid in cold aqueous solution, but the turbidity is removed on boiling. The constituent present besides tannin is "eudesmin" (H. G. Smith). The kino is plentiful even on trees not in any way injured (W. Bäuerlen).

Oil.—The yield of oil from a large number of distillations was

E. UMBRA, sp.nov.

"Stringybark," "Bastard White Mahogany."

(Plate xliv.)

A tall tree, attaining sometimes a height of 100 feet, with a dark-coloured stringybark.

Sucker-leaves opposite, sessile, cordate, ovate, acuminate, thin, pale-coloured on underside; venation more pronounced on the underside; upper surface shining; over 3 inches broad and under 6 inches long. Mature leaves lanceolate, falcate, large, up to 9 inches long and 1½ inches broad, *pale-coloured on both sides, coriaceous;* venation very distinct; lateral veins distant, spreading, oblique; marginal vein removed from the edge.

Flowers in short axillary peduncles, 6-9 in the umbel. Calyx 1 line long, on a pedicel about 2 lines long. Operculum hemispherical, shortly acuminate. Ovary flat-topped. Anthers kidney-shaped.

Fruits in the early stage pilular and under 3 lines in diameter, and the rim thin and valves sunken, but in the mature stage inclined to be pear-shaped, with a diameter of 5 lines, and a very thick red rim.

Hab.—Wardell, Dundoon, and Tumbulgum (W. Bäuerlen); Peat's Ferry, Military Road (R. T. Baker); Tinonee (J. H. Maiden); Gosford (J. Martin); Cowan Creek and Milton (R. H. Cambage); Eastwood (R. T. Baker).

The early fruits of this species have a remarkable resemblance to those of *E. acmenoides*, Schau.; in fact, so much so, that in herbarium material the two have very probably on this character been confounded in the past. The two species differ, however, considerably in the shape, texture, colour and venation of the leaves, as well as in the mature fruits, which have a broad rim.

E. acmenoides, Schau., has thin leaves with a pale undersurface, the leaves undoubtedly resembling those of an *Acmena (Eugenia*, as now understood). But those of *E. umbra* are of a uniform colour on both sides, longer and broader, and with a very marked venation much like that of *E. patentinervis*, Baker.

The sucker-leaves are quite distinct from those of *E. acmenoides*, which also has a lighter-coloured bark, but a superior timber.

E. acmenoides is well figured by Müeller in his "Eucalyptographia," and this species can be from the above description easily distinguished from it, so that it is not considered necessary to give a drawing here.

In botanical sequence it should be placed it next to *E. acmenoides*, from which species, however, it also differs in the chemical constituents of its oil, as well as in the nature of its timber and bark.

The broad sucker-leaves differentiate it from any described species of Stringybark, to which division of the Eucalypts it undoubtedly belongs; and, as stated above, this is one of the characters which separate it from *E. acmenoides*, Schau.

Mr. W. Bäuerlen gives the following description of this tree as observed by him in the northern scrubs :—

"Height 40-80 feet; diam. 2-4 feet. Bark stringy, used for bark. Timber usually pale-coloured, much like that of *E. acmenoides*, which tree it resembles also in the bark and general appearance, but is easily distinguished from it by its broader and thicker leaves, with a more bluish colour; especially by the very broad young leaves, somewhat yellowish in colour and conspicuously veined, while those of *E. acmenoides* are much smaller, narrower, thinner, and of a deeper green colour; in fact, much

and is altogether a much inferior timber to White Mahogany, *E. acmenoides*—a fact well-known to the timber-getters.

Oil.—The yield of oil from this species is ·155 per cent. No phellandrene was found, but much dextropinene was present. It contains but a minute quantity of eucalyptol.

The characteristic constituent of the oil from this species is an acetic acid ester. The specific gravity of the crude oil at 15° C. was ·8963·

The specific rotation of the crude oil was $[a]_D + 41{\cdot}5$.

The saponification figure for the ester was 35·8.

Another consignment of the leaves of this species was received a month later, and gave practically identical results, showing again the constancy of the constituents in the oils of the same species.

The yield of oil was ·169 per cent.; specific gravity crude oil = ·8901·

Specific rotation crude oil was $[a]_D + 43{\cdot}8$, and the saponification figure for the ester was 35·3 (H. G. Smith, F.C.S.).

E. NIGRA, sp.nov.

"Black Stringybark."

(Plate xlvi., fig. 3.)

A tall tree with a black stringy bark.

Leaves lanceolate, scarcely falcate, occasionally oblique, mostly under 4 inches long and under 1 inch wide, of a dull green colour; venation only faintly marked on the upper surface, but very distinct on the lower; lateral veins oblique, distant; intramarginal vein removed from the edge.

Peduncles axillary, short, under 4 lines, bearing a cluster of from 8-12 small flowers. Calyx hemispherical, under 2 lines in diameter, on a short pedicel. Operculum hemispherical, acuminate, about 1½ lines long when mature. Ovary flat-topped. Anthers very small, parallel, filaments very slender.

Fruits about 4 lines in diameter, hemispherical to pilular, rim variable, thin, or truncate and even domed occasionally, valves slightly exserted.

Hab.—Richmond River District (W. Bäuerlen); Cook's River Sydney (H. G. Smith).

From *E. Wilkinsoniana*, Baker, and *E. macrorhyncha*, F.v.M., it differs in fruits, timber and chemical constituents of the oil. From the Stringybark, *E. umbra*, Baker, of this paper it differs in the shape of the sucker leaves and chemical constituents of the oil, although the immature fruits of these species are somewhat similar.

E. eugenioides, Sieb., and *E. capitellata*, Sm., approach each other very closely in morphological characters, and there often seems to be a gradation between the two, but, nevertheless, the two species are quite distinct; and so in this, although there also appears some similarity in the fruits of this species and *E. eugenioides*, yet the two differ in too many characters to be the same species.

The sucker-leaves are not unlike those of *E. capitellata*, whilst the buds are similar to those of *E. eugenioides*. The fruits approach somewhat in shape those of the latter species, with which it has probably been confounded in the past when determined on dried specimens.

If it were not for the distinctive character of the timber and oil I should certainly have made it a variety of *E. eugenioides*, but the former product is of too poor a character to be associated

Oil.—Yield very small, only 3½ oz. from 534 lbs. of leaves, in fact too small to make a fractional distillation. It has thus the smallest yield of the stringy-barks next to *E. capitellata* (H. G. Smith).

EUCALYPTUS LACTEA, sp.nov.

"Spotted Gum."

(Plate xlvi., fig. 5.)

A fair-sized tree with a dirty, flaky bark, which occasionally is smooth.

Sucker leaves ovate; leaves of mature trees lanceolate, up to 6 inches long and varying in breadth up to 9 lines, straight or falcate, not shining, of the same shade of green on both sides; petiole under 1 inch. Venation fairly well marked, veins oblique, spreading, the distinct intramarginal vein removed from the edge. Oil dots numerous.

Peduncles axillary, with few flowers (5 to 7) in the head, occasionally only 3. Calyx hemispherical. Operculum hemispherical, shortly acuminate. Ovary flat-topped. Stamens all fertile; anthers parallel, opening by longitudinal slits.

Fruits hemispherical to oblong; rim with valves domed and almost touching, thus leaving only a slight aperture to the ovary; or the rim thin and the valves exserted and widely distended.

Hab.—Mount Vincent, Ilford (R. T. Baker); Oberon Road, O'Connell (R. T. Baker, R. H. Cambage); Southern Road, Wingello (R. T. Baker, H. G. Smith); along the main Western Road, Blackheath and Mt. Victoria (R. T. Baker).

In the field this tree might be confounded with *E. viminalis* or *E. hæmastoma*, as both these Eucalypts have a similar although variable bark.

The bark of this species, however, never has the horizontal "scribble" insect markings almost invariably occurring on *E. viminalis*, Labill., and *E. hæmastoma*. It has similarly shaped leaves in all its stages of growth, whilst the sucker-leaves of *E. viminalis* are narrow, cordate-lanceolate, sessile.

The fruits differ from those of *E. viminalis* in shape, rim, and direction of valves. The trees too are not found near water, as pertains almost invariably with *E. viminalis*, but on dry, stony ridges. It differs also from that species in the constituents of its oil.

It resembles *E. maculosa*, Baker, in the shape of the fruits, but differs from it in the timber, bark and oil constituents. It differs from *E. hæmastoma*, Sm., in timber, fruits, leaves and chemical constituents of the oil; and from *E. aggregata*, Deane & Maiden, in bark, fruit, oil and habitat.

Its specific characters differentiate it from any of the other smooth-barked species. Of the rough-barked Eucalypts its fruits are often not unlike those of *E. fastigata*, Deane & Maiden, and *E. Smithii*, Baker.

The specific name refers to the copious exudation of a milky substance from the stem when the tree is cut at certain seasons of the year.

Timber.—A very pale-coloured, whitish timber, fissile, only used for fuel, much softer than that of *E. hæmastoma.*

Oil.—The yield of oil was ·541. It is not a commercial oil at present, as it contains but a very small quantity of eucalyptol. No phellandrene is present.

The specific gravity of the crude oil was ·8826 at 15° C.; of the

lines broad, mostly 3 inches long, acuminate, often with a recurved point; midrib raised on the underside, giving the leaf a strong resemblance to that of an *Olea*, not shining; intramarginal vein removed from the edge, lateral veins oblique, spreading, finely marked, only occasionally distinctly pronounced; petiole about 3 lines long. Oil glands very numerous.

Peduncles axillary, short, 2-3 lines long, angled, with from 8-12 flowers. Buds in the early stage of development angular, surrounded *by numerous acuminate*, glabrous, ribbed, whitish *bracts*, short, 1 to 1½ lines long, glaucous. Calyx conical, tapering into an exceedingly short pedicel. Operculum obtuse, or only very slightly acuminate, hemispherical. Ovary flat-topped. Stamens all fertile; anthers parallel, opening by longitudinal slits.

Fruits hemispherical to pear-shaped, 2 lines in diameter, glaucous; rim thin, slightly contracted, valves deeply set, not exserted.

Hab.—West Wyalong (R. H. Cambage, L.S.).

This Eucalypt is one of the Mallees occurring between the Lachlan and Murrumbidgee Rivers, where it is known as "Blue Mallee," to distinguish it from its congeners.

The dried herbarium material is not easily separated from that of *E. viridis*, Baker, *E. Woollsiana*, Baker, and *E. conica*, Deane & Maiden, as the fruits of all these species are almost, if not identical; but this Eucalypt differs principally from them in never attaining tree-form, and in respect of its floral bracts.

Other points of difference are the angular buds, its glaucous character, shape of the leaves and quadrangular branchlets, whilst the most marked distinctive character of all is its oil, the yield and chemistry of which place it amongst the most valuable of our trees famous for the medicinal qualities of their oils.

Amongst other Mallees, it differs (1) from *E. gracilis*, F.v.M., and *E. dumosa*, A. Cunn., in the shape of the fruits and leaves, and the constituents of the oil; (2) from *E. oleosa* in the absence of the long, well-exserted valves, leaves and chemical constituents.

In botanical sequence it is placed next to *E. viridis*, as the fruits and leaves mostly resemble that species.

E. Woollsiana and *E. conica* have much broader leaves, and are classified amongst the "Box" group of Eucalypts.

Oil.—The oil obtained from this species is one of the best for Eucalyptus oil distillation growing in this colony. The yield of oil, obtained from material sent from Wyalong in the beginning of December, was 1·35 per cent. The crude oil was only slightly coloured, being of a lemon tint. The odour reminded one of eucalyptol, and volatile aldehydes are but present in minute quantities. Free acid and ester are also comparatively small. 91 per cent. of the crude oil was obtained boiling below 183° C., and this contained 57 per cent. eucalyptol; no phellandrene was present, but pinene was detected. The crude oil was slightly lævo-rotatory, due to the presence of aromadendral (the previously supposed cuminaldehyde). The specific gravity of the crude oil at 15° C. was ·9143, and of the rectified oil ·9109, this comparatively low specific gravity being due to the presence of such a small amount of constituents having a high boiling point. The specific rotation of the crude oil was $[a]_D - 2·13°$ (H. G. Smith).

EXPLANATION OF PLATES XLIII.-XLVI.

Plate XLIII.

E Woollsia nov

PLATE XLV.

E. Fletcheri, sp.nov.

Fig. 1.—Sucker leaf.
Fig. 2.—Twig with buds.
Fig. 3.—Flowering twig.
Fig. 4.—Stamens (enlarged).
Fig. 5.—Anther (enlarged).
Fig. 6.—Cluster of fruits.

PLATE XLVI.

Fig. 1.—Fruit of *E. intermedia*, sp.nov.
Fig. 2.—Fruit of *E. Wilkinsoniana*, sp.nov.
Fig. 3.—Fruits of *E. nigra*, sp.nov.
Fig. 4*a*, 4*b*, 4*c*.—Fruit and leaves (mature and sucker) of *E. angophorides*, sp.nov.
Fig. 5.—Fruit of *E. lactea*, sp.nov.
Fig. 6*a*, 6*b*.—Fruit and leaf of *E. ovalifolia*, sp.nov.
Fig. 7.—Fruits of *E. polybractea*, sp.nov.
Fig. 8—Buds (showing bracts) of *E. polybractea*, sp.nov.

NOTES AND EXHIBITS.

Mr. R. Etheridge, Junr., on behalf of Mr. Percy E. William exhibited, and contributed the following Note on, a disc of sla This stone, 12 inches in diameter and 1½ inches in thickness, w

lateral secti

as it was ground fell upon a flat rock in front of the cleft. It resembled, viewed from above, a primitive grindstone. The nardoo was poured down the cleft, and was crushed between the wheel (turned by hand) and the two rollers. This is the description of the *modus operandi* as given to Mr. Norman Grant, Dist. Engineer for Roads and Bridges, by the licensee of East Toorale Hotel, who found it *in situ*. The flat rock and the mechanism of the two cylinders had been injured by a bush fire; and it is very much to be regretted that the discoverer did not bring away the rollers. I hope some day to be able to visit the place myself and be able to make a sketch of the ground. As far as I have been able to learn, this is the only mill partaking of the nature of machinery, made by natives, yet discovered. Mr. Norman Grant kindly presented me with the stone in 1897, knowing the interest I take in the ethnology of this country.

Mr. Froggatt exhibited specimens of wheat stalks damaged when young by the attacks of an Aphis. Also specimens of the two Coccids described in Mr. Green's paper read at the last Meeting; and a collection of individuals of *Icerya purchasi*, from which a number of the useful ladybird beetles *Novius cardinalis* had bred out.

Mr. R. T. Baker exhibited an extensive series of herbarium and timber specimens, and essential oils in illustration of his papers. Also a sample jar of the bottled fruits of the quandong (*Fusanus acuminatus*, R.Br.) as preserved by settlers in the interior.

Dr. John Hay sent for exhibition a beautiful spike of the flowers of a new hybrid orchid, *Cymbidium Loweanum-eburneum*.

Mr. Dun showed a series of specimens and microscopic preparations of kerosene shale from various localities, to illustrate Professor Bertrand's paper.

Mr. E. Betche, of the Botanical Gardens, exhibited a series of interesting plants, including *Soliva sessilis*, Ruiz & Pav., a weed introduced from South America, which was recorded last year for the first time in Australia and is now found to be spreading in the Port Jackson district; and the following five rare species of *Epacris* from the Blue Mountains and the Moss Vale district—*E. apiculata*, A. Cunn., *E. robusta*, Benth., *E. Hamiltoni*, Maiden & Betche, an unrecorded variety of *E. Calvertiana*, F.v.M., and an unrecorded variety of *E. purpurascens*, R.Br.

Mr. Fred. Turner exhibited and offered some observations upon—

(1) *Homeria aurantiaca*, Sw., a South African iridaceous plant which has become acclimatised in certain South Australian pastures, and is suspected of poisoning stock. The South Australian Government Dairy Bacteriologist forwarded him the plant for identification and report. Although this is the first time he has received the plant from S. Australia as a suspected poison plant, both *H. aurantiaca* and *H. collina* have established themselves in certain pastures in Victoria, and he had received authentic accounts of stock dying from eating these plants which were sent to him for identification and report. Both these species of *Homeria* have long been cultivated in Australian gardens.

(2) *Euphorbia eremophila*, A. Cunn.—Mr. G. G. Bunning, Darr River Downs, North Western Queensland, forwarded him specimens for identification, and stated that this plant had lately

WEDNESDAY, NOVEMBER 28TH, 1900.

The last Ordinary Monthly Meeting of the Society for the Session was held at the Linnean Hall, Ithaca Road, Elizabeth Bay, on Wednesday evening, November 28th, 1900.

Mr. Henry Deane, M.A., F.L.S., &c., Vice-President, in the Chair.

DONATIONS.

Department of Agriculture, Brisbane—Queensland Agricultural Journal. Vol. vii. Part 5 (Nov., 1900). *From the Secretary for Agriculture.*

Department of Mines and Agriculture, Sydney—Agricultural Gazette of New South Wales. Vol. xi. Part 11 (Nov., 1900). *From the Hon. the Minister for Mines and Agriculture.*

Botanic Gardens and Domains, Sydney—Annual Report for the Year 1899. *From the Director.*

Royal Society of New South Wales, Sydney—Abstract of Proceedings, November 7th, 1900. *From the Society.*

The Surveyor, Sydney. Vol. xiii. No. 10 (Oct., 1900). *From the Editor.*

Royal Anthropological Society of Australasia, Sydney—Science of Man. Vol. iii. n.s. No. 10 (November, 1900). *From the Society.*

Australasian Institute of Mining Engineers, Melbourne—Proceedings, First Ordinary Meeting, 1900. *From the Institute.*

Australasian Journal of Pharmacy, Melbourne. Vol. xv. No. 179 (Nov., 1900). *From the Editor.*

Field Naturalists' Club of Victoria—Victorian Naturalist. Vol. xvii. No. 7 (Nov., 1900). *From the Club.*

Department of Agriculture of Western Australia—Journal. Vol. ii. Part 4 (October, 1900). *From the Secretary.*

Western Australia: Aborigines Department—Reports for the Financial Years ending 30th June, 1899, and 30th June, 1900: Report by Dr. Milne Robertson upon certain Peculiar Habits and Customs of the Aborigines of Western Australia (8vo. Perth, 1879). *From the Victoria Public Library, Perth.*

New Zealand Institute, Wellington—Transactions. Vol. xxxii. (1899). *From the Institute.*

Conchological Society of Great Britain and Ireland—Journal of Conchology. Vol ix. No. 12 (Oct., 1900). *From the Society.*

Manchester Literary and Philosophical Society, Manchester—Memoirs and Proceedings. Vol. xliv. Part v. (1899-1900). *From the Society.*

Royal Microscopical Society, London—Journal, 1900. Part 5. *From the Society.*

Royal Society, London—Proceedings. Vol. lxvii. No. 435 (Oct., 1900). *From the Society.*

Proceedings of the Fourth International Congress of Zoology.

Washington Academy of Sciences — Proceedings. Vol. ii. pp. 247-340 (October, 1900). *From the Academy.*

Museo de La Plata — Anales: Seccion Geológica y Mineralógica ii. Première Partie (1900). *From the Museum.*

Museo Nacional de Montevideo—Anales. Tomo iii. Fasc. xiv. (1900). *From the Museum.*

Geological Survey of India, Calcutta—Memoirs. Vol. xxix. (1899); Palæontologia Indica. Series xv. Himálayan Fossils. Vol. iii. Part 1 (1899). *From the Director.*

Medicinisch-Naturwissenschaftliche Gesellschaft zu Jena—Jenaische Zeitschrift. xxxiv. Band. Heft ii. u. iii. (1900). *From the Society.*

Naturforschende Gesellschaft in Basel — Verhandlungen. Band x. Heft 1 (1892); Bd. xi. Heft 2-3 (1896-1897); Bd. xii. Heft 1-3 and Supplement (1898-1900). *From the Society.*

Zoologischer Anzeiger, Leipzig. Band xxiii. Nos. 625-627 (Sept.-Oct., 1900). *From the Editor.*

Société Royale Linnéenne de Bruxelles—Bulletin. 26me Année. No. 1 (Nov., 1900). *From the Society.*

Journal de Conchyliogie, Paris. Vol. xlviii. No. 2 (1900). *From the Editor.*

Societas Entomologica Rossica, St. Petersburg — Horæ. T. xxxiv. Nos. 3-4 (1900). *From the Society.*

Académie Royale des Sciences à Stockholm — Briefe von Johannes Müller an Anders Retzius von dem Jahre 1830 bis 1857 (1900). *From the Academy.*

La Nuova Notarisia, Padova. Serie xi. Ottobre 1900. *From Dr. G. B. De Toni.*

South African Museum, Cape Town—Annals. Vol. ii. Part iii. (Sept., 1900) *From the Trustees.*

ON THE AUSTRALIAN FAIRY-RING PUFF-BALL.

(*Lycoperdon furfuraceum*, Schaeff.)

By D. McAlpine.

(*Communicated by J. H. Maiden, F.L.S., &c.*)

(Plate xlvii.)

In February, 1898, it was officially brought under my notice that several bowling-greens in the neighbourhood of Melbourne were being much injured by having bare patches in the form of rings. The caretakers of the greens at first considered these to be due to grubs in the soil, but when the various remedies they tried failed (such as a dressing of soot, which really aggravated the mischief), then my services were called into requisition. On visiting the bowling-greens, the otherwise bare patches were found to be studded with knobby or flattened puff-balls, the rolling and working of the green preventing their further development above ground, and the numerous distinct and fairly large circles showed at a glance that here we had to do with the well known "Fairy-rings," although not hitherto observed or at least

denser growth, so that the circles or segments of circles were readily seen at a distance. Since, however, the bowling-greens exhibited the most marked effects of the fungus within a limited area, I will confine myself to a description of them.

In a plan accompanying a preliminary report* on the subject, drawn to scale by an officer of the Public Works Department, one of the bowling-greens is shown with nine rings more or less distinct, within an area of about one-quarter of an acre. They varied in size from 8-24 feet in diameter; sometimes they were solitary, at other times they formed a chain of three, or one might be within another. They were visible to the ordinary observer at a glance, because the circle was comparatively bare of grass, and in some instances so bare as to be more like a footpath than anything else.

Rings are sometimes divided into two classes—one with a ring only, and the other with dead grass in the centre; but here they were all distinct rings with the exception of one, in which there seemed to be the blending of two segments of different circles, forming a pear-shaped outline, with a smaller complete circle in the centre of the larger segment.

In this particular green, the caretaker had observed the rings some nine years ago when they were very small, and he was surprised to find them growing bigger every year. It was suggestive that wherever these rings appeared the self-same fungus was found, even at such distances as the lawn at Flemington and the Caulfield Racecourse. Where growing normally there were no bare patches, but in the bowling-green this was a characteristic feature, which may be explained from the different conditions prevailing there. The regular watering of the green encouraged the growth of the fungi, but the constant rolling flattened them out whenever they appeared at the surface, so that only small and distorted specimens could be obtained. The

* Department of Agriculture, Victoria—Report by Mr. D. McAlpine, Government Vegetable Pathologist, on Fairy-Rings, and the Fairy-Ring Puff-Ball (Melbourne, 14th May, 1898).

consequence was that the underground or vegetative portion of the fungus spread all the more and enveloped the roots of the grass, so that the soil was literally permeated by the destructive mycelium of the fungus. In this way the destruction of the roots of the grasses would be all the more complete, and no doubt the constant cutting and mowing would hasten the decay of the already languishing grass. In ordinary fairy-rings, too, the circle is almost always imperfect, because some accidental obstacle usually intervenes to prevent the mycelium spreading equally outward. But in the bowling-green the circles were complete, because the conditions there were so uniform. Every portion of the green was regularly watered, rolled and cut, so that the mycelium had an equal chance to spread all round.

In seeking to account for the presence of the puff-ball in the bowling-greens around Melbourne, the fact was elicited that only in those cases where fresh sheep-manure had been used as a top-dressing did the rings appear. I also observed that in the early morning the puff-balls were uprooted and scattered by birds (Minahs and Starlings particularly), and no doubt they would carry the spores and even the spawn with them to other places. This happened in the reserved spot where the puff-balls grew to maturity.

In suggesting a remedy, I discouraged the use of fresh natural

which sometimes injure the lawns in suburban districts. The Fairy-ring Fungus (*Marasmius oreades*, Fr.) has not been found in Australia.

The ring-forming habit belongs also to lichens and mosses, as well as some of the higher plants, and in Sweden a grass popularly known as *elf dansar* has become celebrated as the ring-forming plant.

L. furfuraceum, Schaeff., is here newly recorded for Australia, although from the perplexing synonymy given by Saccardo* it might seem as if already recorded.

L. pusillum, Batsch, was determined by Kalchbrenner† in 1875, and was sent to him from Rockhampton, in Queensland. Then Dr. Cooke, in his "Fungi Australiani," gives the same species for New South Wales and West Australia in addition to Queensland, and Saccardo,‡ in his "Sylloge Fungorum," makes *L. furfuraceum*, Schaeff., the equivalent of *L. pusillum*, Batsch.

These two species differ considerably in size, but the main distinction lies in the former having a well developed sterile base, while in the latter it is obsolete. So important is this difference from the systematic point of view, that in the modern classification of the genus *Lycoperdon* there are two leading divisions recognised—(1) those with sterile base rudimentary or absent, and (2) those with sterile basal stratum well developed. *L. pusillum*, already recorded for Australia, belongs to the first division, while *L. furfuraceum* belongs to the second division; and there is no doubt upon this point, since Schaeffer in his original drawings of the species shows the sterile base as in our own specimens. The present species has also been verified by Massee of the Royal Gardens, Kew, England, so that its identity may be regarded as established. In the suburbs of Melbourne this species was very common indeed, and could be gathered by hundreds in the season, and it is somewhat surprising that it should have been so long overlooked, unless it is one of those

* Syll. Fung. Vol. vii. p. 110 (1888).

† Grevillea, Vol. iv. p. 74 (1875).

‡ *Loc. cit.*

cases where fungi suddenly appear in a new locality. It should be looked for in the other colonies as well, and it would be interesting to note if, under different conditions of soil and climate, heat and moisture, it grows in the form of "fairy rings."

LYCOPERDON FURFURACEUM, Schaeff.—Fairy-ring Puff-ball.

Gregarious, often in clumps of three to five, often distorted from mutual pressure, with the odour of mushrooms. Subglobose to oblong, either depressed or somewhat elongated, variable in size, up to 2¼ in., broad and high, usually sessile on a broad base of attachment with numerous root-like fibres, or tapering slightly towards base, and with distinct brown root in young state.

Outer peridium at first quite distinct and easily peeled off, creamy-white, studded all over with closely crowded minute conical spikes, which ultimately disappear, especially on top. Inner peridium white at first, smooth, elastic, membranaceous, finally of a greyish-green colour, dehiscing irregularly at top, or forming at first an elongated oval slit.

Cellular sterile base well-developed, about half the height of peridium, compact, generally flat; pale seal-brown in mass; filaments yellowish-green, flexuous, non-septate, very sparingly branched, 2-2½ μ broad, stained yellowish by potassium-iodide-

peridium in old specimens may spread out in lobes after the spores are dispersed, completely exposing the compact sterile base. The spores did not exceed 4 μ in diameter. When young and about the size of hazel-nuts, the puff-balls were creamy-white and spiky, and quite firm to the touch, being fixed in the soil by a single distinct root. Then towards maturity they become cinnamon-brown on top, soft and springy to the touch, and the original root has branched considerably, the puff-ball being fixed by a mass of dark-brown root-like fibres. At this stage the spikes can be easily rubbed off, if they do not fall away naturally.

EXPLANATION OF PLATE XLVII.

Lycoperdon furfuraceum, Schaeff.

Fig. 1.—Small puff-ball, with single root springing from base.
Fig. 2.—Nearly mature puff-balls.
Fig. 3.—Top of puff-ball showing dehiscence at first.
Fig. 4.—Outline section showing sterile base.
Fig. 5.—Threads of capillitium (× 1000).
Fig. 6.—Filaments of sterile base (× 1000).
Fig. 7.—Group of spores (× 1000).

NOTES ON THE BOTANY OF THE INTERIOR OF NEW SOUTH WALES.

By R. H. Cambage, L.S.

(Plate xl., fig. 4.)

Part II.—From Cobar to the Bogan River above Nyngan.

Leaving Cobar, and taking the road south-easterly towards Nymagee, it is found that for about 4 miles most of the timber has been removed owing to settlement in connection with the gold and copper mining carried on in the neighbourhood. Then the following trees are noticed :—

Eucalyptus intertexta (Gum or Coolabah), *E. populifolia* (Bimble Box), *E. viridis* (Whipstick Mallee), *Apophyllum anomalum* (Currant or Warrior Bush), *Capparis Mitchelli* (Wild Orange), *Canthium oleifolium* (Lemon Bush), *Geijera parviflora* (Wilga), *Eremophila Mitchelli* (Budtha), *Acacia excelsa* (Ironwood), *A.*

still on the road from Cobar to Nymagee, there are *Acacia Oswaldi*, *A. homalophylla*, *Canthium oleifolium*, *Eucalyptus populifolia*, *Apophyllum anomalum* (Currant Bush, of whose fruits bronzewing pigeons and emus are fond).

Here and elsewhere was noticed a shrub up to 6 and 8 feet high, in general appearance much like *Eremophila Mitchelli* (Budtha), but always growing in shrub-form, with sticky leaves. It was collected in various places between Bourke and Nymagee, but owing to the season of the year (May and June) no specimen was obtained that would enable the plant to be identified.

Other trees passed were—*Pittosporum phillyræoides* (Berrigan or Little Whitewood), *Geijera parviflora* (Wilga), *Eucalyptus oleosa*, and *E. intertexta* (Gum). Wherever this last-named tree grows it can be readily identified at a distance by its pale foliage and white limbs. In going west in the train this species is first sighted a few miles on the Dubbo side of Nyngan, and thereafter is generally in view till the river soil of the Darling is reached at about 20 miles from Bourke. It seems to avoid what is known as river country and grows only in the dryer parts. Near Girilambone I have known it mistaken for *E. melliodora*, A. Cunn., (Yellow Box), chiefly owing to the colour and smoothness of the bark on the upper part of the trunks, but with a little practice the two species can be separated by the foliage alone. *E. melliodora* does not grow at Girilambone so far as I know. In the colour of the sap and of the wood there is a distinct difference.

Next in order were—*Eremophila Mitchelli*, *Hakea leucoptera*, *Heterodendron oleæfolium*, *Acacia colletioides*, *Cassia eremophila*, *Acacia aneura*, *Exocarpus aphylla*, *Fusanus acuminatus*, *Casuarina Cambagei* (still with pale branchlets), *Acacia hakeoides*, *Eucalyptus dumosa*, *Capparis Mitchelli*, *Eucalyptus Morrisii* (a Mallee, on a pebbly hill growing with *Acacia doratoxylon* and *A. decora*), *Callitris robusta*, *Helichrysum Cunninghamii* (a shrub), *Eucalyptus viridis*, *Dodonæa viscosa*, var. *attenuata*, *Myoporum deserti*, and a mallee scrub of *Eucalyptus oleosa*, *E. dumosa*, and *E. viridis*.

Leaving the Nymagee road, and turning north-easterly from the Restdown Mines to Mount Boppy, a distance of about 20 miles, we first have about 2 miles of mallee scrub made up of *Eucalyptus oleosa*, *E. dumosa*, and *E. viridis*. It is in such places as this that the Mallee Hen (*Leipoa ocellata*, Gould) scratches up a mound of loose earth, 3 or 4 feet high and 12 to 15 feet in diameter, in which she lays her eggs in the spring, leaving them to be hatched by the heat of the sun. I am informed that when the young emerge from the shell, they at once start to scratch, feet uppermost; the effect being that the dust accumulates under their backs, and they are gradually raised to the surface of the mound, when they are at once ready to run off.

After passing the mallee there are—*Callitris robusta*, *Apophyllum anomalum*, *Acacia doratoxylon*, *A. colletioides*, *A. aneura*, *Myoporum deserti*, *Fusanus acuminatus*, *Bertya Cunninghamii*, Planch., (Broom Bush), *Bossiaea* sp. (locally known as Stick Bush; no flowers were procurable), *Capparis Mitchelli*, *Acacia homalophylla*, *Eucalyptus populifolia*, *E. intertexta*, *Hakea leucoptera*, *Acacia hakeoides*, *Heterodendron oleaefolium* (locally called Apple Bush, though in most places it is known as Rosewood), *Acacia Burkittii*, *Cassia eremophila*, *Eremophila longifolia*, *Canthium oleifolium*, *Geijera parviflora*, *Acacia decora*, and *A. excelsa*.

At about 9 or 10 miles is the "Mulga Tank," so named because

commonly assumes the mallee-form. *E. Morrisii* is found taking the highest ridges.

Turing south-easterly from the gold mines near Mount Boppy, through Geweroo and Trowell Creek stations to the Nyngan and Nymagee road, we pass *Acacia excelsa*, *Callitris robusta*, *Eremophila Mitchelli*, and *Tecoma australis*, R.Br., a climber often called Bignonia, with a beautiful white bell-shaped flower. In the month of May this creeper is in full flower, spreading over the trees on the ridges, and has a most attractive appearance.

Near here are hills of igneous formation, the first so far noticed, and it is interesting to mark any change in the vegetation.

On a ridge of quartz felspar porphyry we expect to find *Eucalyptus Morrisii*, as it has been found in all similar situations on hills of sedimentary origin However, we look here in vain, but in its place is *E. dealbata*, A. Cunn. This is the most north-westerly locality in which I have ever found this species. There has been some difference of opinion among botanists as to whether this is a distinct species or only a western form of *E. tereticornis*. My observations are that *E. tereticornis* grows all along the coast north and south of Sydney, and that typical trees extend westerly, missing the higher Triassic sandstone of the Blue Mountains, but occurring again on the Devonian, Silurian, and igneous formations, in various situations on high and low land, but chiefly good soil, right out to Orange. After passing Molong, it is usually confined to the valleys. Here *E. dealbata* makes its appearance, on hills of Devonian sandstone, and this tree continues right out to Mount Hope, and possibly beyond, capping hills of Devonian sandstone and quartzite, Silurian slate, and igneous formation with considerable impartiality, but invariably taking the high land. It has an extensive range north and south Where the formation is granite one of the species is sure to be represented.

E. tereticornis follows for some distance keeping the low land, but ceases long before *E. dealbata*. The former is known as Forest Red Gum, and the latter as Mountain and Cabbage Gum, though this last name is more often applied to *E. hæmastoma*, Sm., a white brittle gum which grows on the Silurian slate ridges about Bathurst and Orange, &c., where it is also called Brittle Gum and Brittle Jack. In general appearance *E. tereticornis* and *E. dealbata* are wonderfully alike, both in bark and foliage, even in detail. The woods are also very similar, though that of *E. dealbata* is softer, and not considered so lasting; still this is often attributed to the fact that one grows on a good flat, and the other on a ridge with shallow soil.

E. dealbata is the smaller tree, and out west will often grow in mallee-form—that is, with 8 or 10 stems from one root, and from 12 to 15 feet high, though it never forms extensive scrubs like typical mallee, and the wood is not so tough.

The opercula of *E. tereticornis* are generally long and pointed, but in *E. dealbata* they present more variation than those of most Eucalypts, being sometimes quite pointed, and at others almost hemispherical, varying in length from about half an inch to two lines. Taking the fruits of *E. tereticornis*, it is found that they are pedicellate and domed, while those of *E. dealbata* are almost sessile and truncate, and therein, together with a difference in the opercula, seem to lie the chief distinguishing marks; yet in the districts where both trees grow it is undeniable that these conditions to some extent seem to graduate from one to the other, as it is possible to find trees with fruits both sessile and domed. Ignoring intermediate stages, and taking *E. tereticornis* of Parra-

me that there is no gradation between these species, although they have strong affinities. Nor does there appear to be any between *E. rostrata* and *E. dealbata*, for in one instance, where a hill of Devonian sandstone comes down near the Lachlan River, I have collected *E. rostrata* on the river side and *E. dealbata* on the hill side of the road, about a chain apart, both being typical trees.

Messrs. Deane and Maiden (Proc. Linn. Soc. N.S.W. Vol. xxv. p. 466) have withdrawn specific rank from this species, and described it as a variety of *E. tereticornis*. The result of several years of close observation of these trees in the field has been to incline me to support their action.

Altogether the question of the relationship between *E. tereticornis* and *E. dealbata* seems a difficult one, and my object in offering the foregoing remarks is in order that they may be considered in any future investigation regarding the western Eucalypts.

Another tree now appearing for the first time is *Casuarina quadrivalvis* (She Oak and Mountain Oak). In the west this species appears on most of the igneous hills, and seems to prefer them, though it is not restricted to the one formation. Being much in demand as fodder for stock, the continued dry seasons have had the effect of causing this tree to be almost exterminated from some of the hills. It differs from the other Casuarinas of the west in its pendulous foliage and very large cones.

Amongst others noticed in passing along were :—

Acacia doratoxylon, which, with very few exceptions, grows on hills, though it does not discriminate between geological formations; it is one of those Acacias that has its flowers arranged in spikes, and not the usual capitula or flower-heads.

A. homalophylla, which likes a stiff soil on flats.

A. decora, often called Silver Wattle, although it has not the feather leaf usually associated with the term of Wattle. This species prefers slightly elevated land, and in the month of September has a profusion of beautiful "Wattle" blossoms.

In addition to the foregoing were—*Acacia Oswaldi*, *A. hakeoides*, *Cassia eremophila*, *Canthium oleifolium*, *Dodonæa viscosa*, var. *attenuata*, *Heterodendron oleæfolium*, and also an Acacia which, from herbarium specimens, seems to be *A. dealbata*, Link, though I observed it over a distance of at least 200 miles, and never associated it with that species. It is a small tree, ranging from about 6 to 8 feet high, with a diameter from 4 to 6 inches. Its foliage is the usual feather leaves of the "Wattle," but these are green, and scarcely ever have that glaucous appearance so well known on the Silver Wattle (*A. dealbata*) of the cold highlands. Still I noticed this feature in a slight degree towards the Murrumbidgee, but never north of the Lachlan. From its colour it looks more like a dwarf form of *A. decurrens*. It has an extensive range, as will be seen from my notes which follow, where it will be referred to as *A. dealbata* (green variety).

Next in order were—*Eucalyptus viridis*, *E. oleosa*, *E. dumosa*, *Casuarina Cambagei*, *Bertya Cunninghamii*, and *Acacia colletioides*.

At about a dozen miles from Mount Boppy another Eucalypt is added to the list. This is *E. Woollsiana*, Baker, a box tree with clean greyish-white limbs, and generally an erect useful trunk. This species is perhaps more plentifully distributed in the western district than any other Eucalypt, and in places grows

tree which is best known in the western district as White Box is *E. albens*, with pale bark and glaucous leaves, but its habitat is under the western fringe of the high mountain spurs running from the Great Dividing Range, avoiding the cold country, and extending westward along slight undulations to the low plain country proper. Here it ceases, but is met and overlapped by *E. Woollsiana.* All along, and near these points of contact, the latter is called Black Box, to distinguish it from *E. albens.* It is also a darker tree, having dark green and slightly glossy leaves. In times of drought sheep will eat the leaves of *E. albens*, especially after they have been cut a day or two, but they object to the leaves of *E. Woollsiana.* This species grows on all formations, whether igneous or sedimentary, but is seldom or perhaps never found on the tops of high hills. It is, however, very partial to low ridges, and also grows plentifully on low land. Though it is to be found on the banks of the Lachlan, it is not, on a typical black soil river flat, so much in its regular habitat there as *E. microtheca* or *E. largiflorens.*

After passing *Acacia homolophylla*—some with broad and some with narrow leaves—*A. doratoxylon*, *Eucalyptus tereticornis* var. *dealbata*, *Pittosporum phillyrœoides*, *Fusanus acuminatus*, and *Acacia decora*, a low slate ridge is reached where there is much Mallee—*E. oleosa*, *E. dumosa*, and *E. viridis.*

Next to be noticed is another Eucalypt now seen on this road for the first time, *E. sideroxylon*, A. Cunn., (Ironbark). This is the common Ironbark of the interior, and chiefly belongs there, although it occurs near the coast as at Liverpool, St. Mary's, &c. It is rarely found growing at an altitude exceeding 2000 feet above sea level. By some it is called Black and by others Red Ironbark owing to the colour of the bark and wood respectively. The aboriginal name is Mugga. In going west it is first met with on the western line beyond Kerr's Creek, and on the Orange to Forbes line beyond Molong, so that it covers much the same country as *E. dealbata*, and also prefers ridges. In the west this species bears a profusion of blossoms in the months of April and May. The flowers are generally creamy-white, but on some trees they are red. It is fairly plentiful between the Macquarie and Murrumbidgee Rivers, occurring in patches, and shows a decided preference for a sedimentary formation. The timber is used freely for railway sleepers, and although it is considered one of the finest woods of the west, it is not to be compared for general use with *E. paniculata*, Sm., the Grey or White Ironbark of the coast. The wood of the latter tree is tough, while that of the western one is comparatively dry and brittle.

In view of the prominence given to the question of hybridization of Eucalypts by Messrs. Deane and Maiden in the Proc. Linn. Society, Vol. xxv. p. 111, where they deal with *E. affinis* which grows among *E. sideroxylon* and *E. albens*, also with another tree growing among *E. siderophloia* and *E. hemiphloia* at Homebush and Liverpool, it occurred to me that if cross-fertilization exists between the above trees, the same sort of thing may take place in other species, notably between *E. sideroxylon* and the western box tree recently described by Mr. R. T. Baker as *E. Woollsiana*. Knowing that I should meet with these two species growing together in several places, I decided to make diligent search for trees which would answer the required conditions of hybrids. After coming into the Ironbarks and Box a few hundred yards, three trees were found which seem intermediate in every respect between *E. sideroxylon* and *E. Woollsiana* both in the colour

The chief flowering time of *E. sideroxylon* in the west is about April and May, but flowers can generally be found before and after those months. I have seen it flowering at Cabramatta in January and July. *E. Woollsiana* flowers about February and March, but flowers of this species have been collected in May in the same locality as that in which *E. sideroxylon* was then flowering. I am not able to state the flowering time of the tree which looks half Box and half Ironbark, for though a very few flowers were collected in June, buds were found in September which seemed to indicate that the trees would be flowering in October. The scarcity of the tree makes it difficult to arrive at a definite conclusion in the matter.

It is no part of my purpose to try and prove that the tree in question is a hybrid, but simply to offer observations which may assist in settling the question. I have handed specimens to Mr. Deane, who will probably investigate this species.

About 7 miles west of Trowell Creek House is a quantity of Acacia trees up to 25 and 30 feet high, chiefly long stems, with a diameter of about 4 inches. Leaves are slightly silvery and somewhat resemble those of *Heterodendron oleæfolium*. They are growing on scrubby sedimentary formation in company with *Phebalium glandulosum* and a Leptospermum. Only flowers in a very young state were procurable in May. These were arranged in racemes. The tree was not seen again, and I have not been able to identify the species.

Going on past Trowell Creek Homestead to the Nyngan Road there are, *Eucalyptus intertexta*, *E. populifolia*, *Myoporum deserti*, *Phebalium glandulosum* (a shrub), *Geijera parviflora*, *Eremophila Mitchelli*, *E. longifolia*, *Canthium oleifolium*, *Acacia hakeoides*, *A. colletioides*, *A. homalophylla*, *Eucalyptus oleosa*, *E. viridis*, *Heterodendron oleæfolium*, and *Hakea leucoptera*.

Crossing the Nyngan to Nymagee Road, at a point about 36 miles from Nyngan, and going to Honeybugle Station, a distance of about 20 miles easterly, the following are noticed :—*Callitris robusta*, *Capparis Mitchelli*, *Sterculia diversifolia*, *Eucalyptus oleosa*, *E. dumosa*, *E. viridis*, *E. populifolia*, *E. intertexta* (but not so plentiful as formerly, as the eastern margin of its habitat between the Bogan and Lachlan is now nearly reached), *E. Woollsiana*, *Casuarina Cambagei*, *Fusanus acuminatus*, *Acacia homalophylla*, *A. decora*, *A. colletioides*, *A. doratoxylon*, *A. Oswaldi*, *A. dealbata* (green variety), *Geijera parviflora*, *Heterodendron oleæfolium*, *Eremophila Mitchelli*, *E. longifolia*, and *Hakea leucoptera*.

For the last 50 miles no trees were seen of *Acacia aneura* or *A. excelsa*, but both were found, in limited quantity, growing a few miles north of Honeybugle homestead, near an old copper mine in which were found some very good specimens of hornblendic rock.

Going easterly from Honeybugle homestead to Mudall on the Bogan River, a distance of about 16 miles, amongst other trees passed in driving along, were *Callitris robusta*, *Eremophila Mitchelli*, *E. longifolia*, *Heterodendron oleæfolium*, *Eucalyptus populifolia* (Bimble Box, sometimes spelt Bimbil; I understand that Bimble is the aboriginal name for this tree), *Sterculia diversifolia*, *Eucalyptus intertexta*, *Geijera parviflora*, *Canthium oleifolium*, *Apophyllum anomalum*, *Acacia excelsa* (some very good trees for two or three miles), *A. Oswaldi*, *A. doratoxylon*, *A. decora*, *A. homalophylla*, *Capparis Mitchelli*, *Hakea leucoptera*, *Myoporum deserti*, *Fusanus acuminatus*.

dread of the coach-driver in wet weather), though it may be many miles back from the river. For years I have been trying to work out the question whether Myall and Boree referred to one or two trees. Both names are used variously throughout the western division of New South Wales, though generally certain districts keep to the one name. In some cases *Acacia homalophylla* is locally called Yarran or Myall, while *A. pendula* is known as Boree; or *A. homalophylla* is called Yarran, *A. pendula* Boree, and *A. salicina*, Lindl., Myall, which is quite wrong. These mistakes often arise through people being indifferent as to proper local names. Generally *A. homalophylla* is known as Yarran, and *A. pendula* as either Myall or Boree, or sometimes as both. My conclusion is that Myall is the name usually given in localities north of the Bogan and extending into Queensland, and is possibly the aboriginal name given by some northern tribe; while Boree is the tribal name of the aborigines (approximately) south of the Lachlan River; and that both refer to the same tree. Many and varied have been the replies to my questions when trying to get at the truth of this matter, and quite recently I was told in all seriousness that there is a difference between Myall and Boree; but so slight that no white man can tell what it is. The wood of *A. pendula* is much sought after owing to its fragrance, but in this feature it has a strong rival in *A. homalophylla*, though for strength and durability Myall is superior. It is a splendid fodder tree, but is fast being exterminated, as all the young plants are browsed down by sheep and cattle. Horses, however, seem to spare them, as small trees are quite common in paddocks not accessible to sheep and cattle. In fact, almost the only places now in which young Myalls can be seen are horse paddocks and within the railway fences.

In speaking of Gidgea in a former paper (Part i.), it was mentioned that the leaves of that tree would not snap if doubled up while green, but that the leaves of *A. homalophylla* (Yarran) would. It may here be mentioned that the leaves of *A. pendula* (Myall) snap just as readily as those of Yarran under similar conditions, and are very like them in texture. Still the constituent parts must be different, as stock much prefer Myall to Yarran, and it is not at all rare to find a young plant of the latter.

Passing along towards the Bogan the following were noticed :—*Geijera parviflora*, *Casuarina Cambagei*, *Hakea leucoptera*, *Heterodendron oleæfolium*, *Pittosporum phillyræoides*, *Acacia colletioides*, *Eucalyptus populifolia*, *E. largiflorens*, *Eremophila Mitchelli*, *E. maculata*, F.v.M., (Native Fuchsia).

On the banks of the Bogan River at this point are *Eucalyptus rostrata* (River Red Gum), also *Casuarina Cambagei* (Belah), which from its situation might very easily be mistaken for *C. Cunninghamiana*, Miq., (River Oak), but from observations made at different points, there seems to be no River Oak growing on any part of the Bogan, though it is to be found on the upper portions of probably every other western river.

From Cobar to Mudall by the road travelled is about 120 miles, and 12 different species of Eucalypts were noticed :—*E. intertexta*, *E. populifolia*, *E. viridis*, *E. oleosa*, *E. dumosa*, *E. Morrisii*, *E. tereticornis*, var. *dealbata*, (*E. dealbata*, A. Cunn.), *E. Woollsiana*, *E. sideroxylon*, *E. largiflorens*, *E. rostrata*, and the Ironbark Box.

The species of Casuarina were—*C. Cambagei*, and *C. quadrivalvis*.

STUDIES ON AUSTRALIAN MOLLUSCA.

Part III.

(Continued from p. 513.)

By C. Hedley, F.L.S.

(Plate xlviii.)

Mathilda rosae, n.sp.

(Plate xlviii., figs. 13, 14.)

Shell tall, turreted, upper whorls subulate, final whorls spread; perforate only for the last two whorls. No perfect shell has come under my notice. A survey of a series of imperfect shells of different ages suggests that fourteen or fifteen whorls are the proper complement. The youngest of the series has a heterostrophe apex; as growth proceeds the upper whorls are lost. Colour white (bleached?). Suture channelled. Sculpture: longitudinal sculpture is missing; on the last whorl are six sharp, elevated, equidistant keels, their own breadth apart, the anterior keel the largest, the posterior least, on preceding whorls are fewer keels. Base smooth and flattened. Aperture very oblique, nearly triangular, beaked on the exterior angle by the extension of the peripheral rib, patulous at the junction of base and pillar. The specimen drawn, of which only seven whorls remain, is 5 mm. in length and 2·8 in breadth.

Hab.—Balmoral Beach, Middle Harbour, near Sydney; several dead shells picked out of shell sand by the late Mrs. C. T. Starkey.

The perforation, want of longitudinal sculpture and enlargement of the peripheral rib, amply distinguish this species from its congeners. It probably lives in deep water.

The species is named in memory of my late friend Mrs. C. T. Starkey, who made a large collection of the minute marine shells of this neighbourhood. By her desire the collection was handed to me for study, after her decease.

DRILLIA HOWITTI, Pritchard & Gatliff.

Pritchard & Gatliff, Proc. R. Soc. Vic. xii., 1899, p. 101, pl. 8, f. 2.

A perfect adult specimen from Sydney Harbour is contained in Mrs. Starkey's collection. I have determined it by comparison with authentic specimens received from Mr. Gatliff. Dr. Cox has also shown me specimens dredged in 25 fathoms off Sydney Heads.

BITTIUM MINIMUM, T. Woods.

Tenison Woods, Proc. R. Soc. Tasmania, 1877, p. 123.

(Fig. 20.)

A specimen of this species has been handed to me by Mr. J. Brazier, who took it in an old bottle from 8 fathoms off the Bottle and Glass Rocks, Sydney Harbour, May 22nd, 1886.

For the better recognition of this species I add a drawing of a Tasmanian specimen, kindly lent to me for the purpose by Miss Lodder.

Fig. 20.

diameter apart, cross the shell obliquely. A few faint and distant spiral grooves are perceptible. Traces are present of a thin, membranous yellow epidermis. Aperture auriculate, slightly oblique. Outer lip sinuous, neither thickened nor reflected, columella broadened at the base and a little reflected. A small but sharp and deeply seated plication occurs on the body whorl, anterior to the centre; immediately in front of this, and deeper still, is a second, ill-developed fold. A heavy callus is spread on the body whorl. Length 3 mm.; breadth 1·16 mm.

Hab.—The material at my disposal was collected in the neighbourhood of Sydney by Mr. J. Brazier as follows:—(1) East side of Double Bay; one specimen; type: (2) Under a small stone on a "samphire flat" at the head of Tambourine Bay, in company with *Ophicardelus*; three living specimens: (3) In shell sand, Cook's Landing Place, Botany Bay; two dead shells.

Type to be preserved in the Australian Museum.

The species before us is the second of the genus *Leuconopsis*, founded by Hutton (Trans. N. Z. Inst., xvi., 1883 [1884], p. 213) for the reception of *Leuconia obsoleta*, Hutton, (Journ. de Conch. xxvi, 1878, p. 43, and Man. N.Z. Moll., p. 34). The name implies affinity to *Leuconia*, but to my mind the genus nearest approaches *Auriculus*.

Capt. Hutton, whose kindness to his fellow naturalists is unfailing, has sent me, for comparison with *L. inermis*, a set of co-types of *L. obsoleta* from Auckland, New Zealand. One of these, whose length is 2·24 and breadth 1·44 mm., is figured (fig. 16) for comparison with the novelty. The Australian species differs by its feeble columellar plait, more whorls, absolutely larger size, but relatively more slender proportions.

BLAUNERIA LEONARDI, Crosse.

Crosse, Journ. de Conch. xx., 1872, pp. 71 and 357, pl. xvi., f. 4.

(Plate xlviii., fig. 9.)

The auriculoid genus *Blauneria** has not hitherto been recognised as a constituent of the Australian fauna. Specimens are

*Shuttleworth, Berner. Mitth. Diagn. Neue Moll., 1854, p. 148.

now before me, gathered by Mr. J. Brazier from Blackwood Bay, Mount Adolphus Island, and from Warrior Island, both in Torres Straits. Though twice the size I find no specific difference between these and examples from Noumea of Crosse's species. The individual figured is from Warrior Island, and measures 4·5 mm. in length, and 1·5 mm. in breadth.

Stenothyra australis, n.sp.

(Plate xlviii., fig. 10.)

Shell small, subperforate, ovate, thin, translucent, very glossy, rather flattened. Colour brown. Whorls five, rapidly increasing, separated by an impressed suture. Last whorl disproportionately large, comprising three-quarters of the whole shell, bulging at the periphery, rapidly and deeply descending for the last half whorl. The earlier whorls are distantly, spirally grooved, the later are quite smooth. The translucency of the shell presents by optical illusion a subsutural band. The strangled aperture is nearly perpendicular, regularly oval, bevelled and thickened within. Length, 3·68 mm.; breadth, 2·32 mm.

Hab.—The half dozen examples seen were handed to me for description by Mr. J. Brazier. He procured them from drifted mangrove leaves on the beach at Bowen, Queensland.

Operculum—? Long. 1½, diam. 1 mill. *Hab.*—In Australia. Mus. Cuming.

S. frustillum has, I believe, not been again collected, noted in literature or illustrated. Since Cuming's localities are so dubious, the species may not really be Australian. *S. australis* differs from *S. frustillum* in size, the number of whorls, the proportion of the last whorl, the sculpture and the colour.

DAPHNELLA TASMANICA, T. Woods.

T. Woods, Proc. R. Soc. Tasmania, 1875, p. 145.

(Fig. 21.)

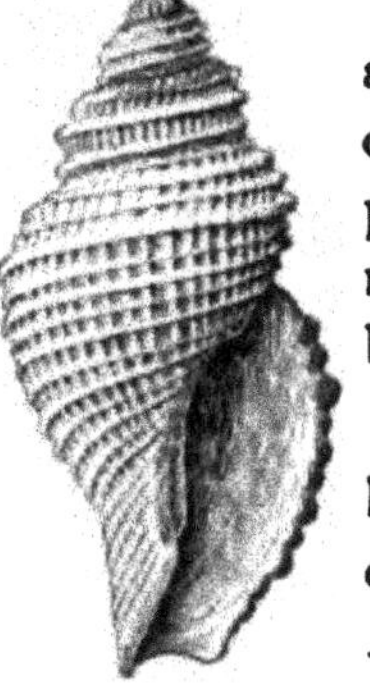

Fig. 21.

The Rev. H. D. Atkinson has kindly lent me a specimen, which, I understand, is one of the original lot of this species. From it the accompanying figure was made. It seems probable to me that *Clathurella lamellosa*, Sowerby,* was based on the young of this.

To this species Messrs. Pritchard and Gatliff have united *Mangilia jacksonensis*, Angas † A comparison of the descriptions of Woods and Angas shows that such synonymy could only rest on mistaken identification.

HYDATINA TASMANICA, Beddome.

Akera tasmanica, Beddome, Proc. R. Soc. Tasmania, 1882, p. 169.

(Fig. 22.)

Fig. 22.

To the kindness of Miss M. Lodder I am again indebted for the opportunity of illustrating an unfigured species. Prof. Tate, upon inspecting this drawing, pointed out to me that the species would be better referred to *Hydatina*.

*Sowerby, Proc. Mal. Soc. II., 1876, p. 28, pl. iii., f. 11.

†Pritchard & Gatliff, Proc. R. Soc. Vic., xii., 1900, p. 175.

TROPHON BRAZIERI, T. Woods.

T. Woods, Proc. R. Soc. Tasmania, 1875, p. 136.

(Fig. 23.)

An authentic specimen furnished by Mr. J. Brazier afforded the accompanying illustration.

EUTHRIA CLARKEI, T. Woods.

T. Woods, Proc. R. Soc. Tasmania, 1875, p. 138.

(Fig. 24.)

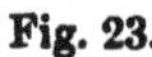

Fig. 23.

The present figure is derived from an example in Miss Lodder's collection.

SCHISMOPE PULCHRA, Petterd.

Petterd, Journ. of Conch. iv., 1884, p. 139.

(Fig. 25.)

Miss Lodder kindly lent me the original of my sketch.

IRAVADIA AUSTRALIS, n. sp.

(Plate xlviii., fig. 12.)

Shell subcylindrical, rimate, decollate; remaining whorls four and a half, rapidly increasing; covered by a dense and rather glossy brown epidermis. Sculpture: the last whorl is encircled by eight sharp, spiral, nearly equidistant keels, about four times their own width apart. On the penultimate whorl there are five of such. These are perpendicularly crossed by fine raised hair lines, which bead the keels and lattice the interstices. Aperture white within, a little oblique, ovate, subchanelled posteriorly, peristome entire, reflected, fortified without by a thick, outstanding varix. Length, 5·3 mm ; breadth, 2·25 mm.

Hab.—Bowen, Queensland. Half a dozen examples from fresh water (J. Brazier).

Type to be preserved in the Australian Museum.

The above is the first of the genus *Iravadia** to be recorded from Australia. I have not the advantage of seeing Indian shells, but, according to published figures, the Australian species seems to differ by more cylindrical form.

Both *Stenothyra* and *Iravadia* may be added to the list of Torresian types; in other words, are stock which Australia derived from Malaysia.

PETTERDIANA THAANUMI, Pilsbry.

Pilsbry, The Nautilus, xiii., Ap., 1900, p. 144.

(Plate xlviii., fig. 11.)

I am indebted to the author of this species for co-types obtained near Cairns, Queensland. With these I have compared a series from the Bellenden Ker Range in the same neighbourhood. The type is 3·3 mm. in length and has four whorls; the shells from

* Blanford, Journ. As. Soc. Bengal, Pt. ii., 1867, p. 56, Pl. 2, ff. 13, 14. Nearly allied if not identical seems *Hydrorissoia*, Bavay, Journ. de Conch. xliii., 1895, p. 90.

the Bellenden Ker are larger, the example figured herewith being 4·75 mm. long, have an extra half whorl and the aperture is more expanded. In other respects the shells are alike, so that I conclude that the specimens from the Bellender Ker represent the adult state of the species.

CALLOMPHALA GLOBOSA, n.sp.

(Fig. 26.)

This species differs from *C. lucida* by being smaller, higher in proportion, and closely engraved with numerous fine, spiral striæ. Maj. diam., 2·4; minor diam., 2·1; height, 2mm.

Hab.—Near Darnley Island, Torres Straits, in 30 fathoms, three specimens collected by J. Brazier.

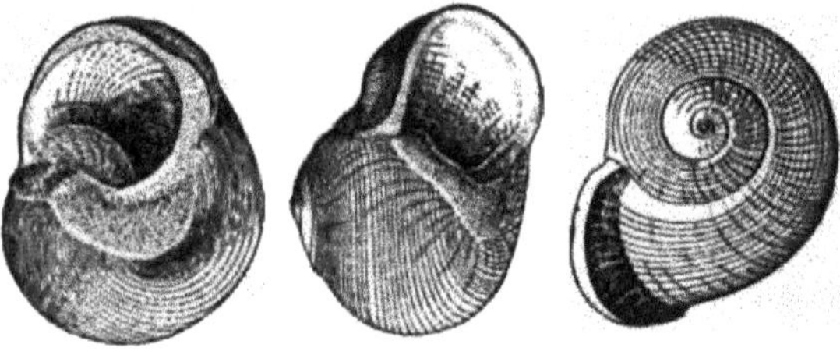

Fig. 26.

STYLIFER PETTERDI, Tate and May.

Tate & May, Trans. R. Soc. S. Aust. xxiv., 1900, p. 96; *S. robusta*, Petterd, Journ. of Conchology, iv., 1884, p. 140; *non* Pease, 1860.

(Fig. 27.)

Fig. 27.

Under the name of *Stilifer tumida*, Petterd, Mr. Brazier exhibited to this Society, August 29th, 1894, a shell from the neighbourhood of Wollongong, N. S. Wales. This name has, I believe, not been published by Petterd. The Wollongong shell which I have lately examined is certainly *S. robusta*, Petterd. As the species has never been illustrated, I now present a drawing of a Tasmanian example kindly lent me for the purpose by Miss Lodder.

ENDODONTA CONCINNA, n.sp.

(Plate xlviii., figs. 1-3.)

Shell minute, thin, subdiscoidal, spire scarcely elevated, umbilicus broad and deep. Colour pale yellow, faintly and obscurely mottled with brown. Whorls four, rounded, excavated at the suture and a little flattened at the base. The first half whorl is tilted from the plane of adult growth. Umbilical crater a third of the diameter of the shell, exposing previous whorls, umbilical suture deeply enfolded. Sculpture: no spirals exist; the first half whorl is bare of radial sculpture, which commences abruptly as fine, sharp, raised hair lines, about twice their diameter apart, sinuate at the suture as usual in the group; of these I count 98 on the last whorl; their interstices have a few very fine lines, parallel to the main sculpture. Height, ·56; major diameter, 1·48; minor, 1·24 mm.

Hab.—Bundaberg, Queensland; collected by Dr. T. H. May.

Type to be preserved in the Australian Museum.

Chlamys fenestrata, n.sp.

(Plate xlviii., figs. 17-19.)

Shell small, very solid, convex, almost equilateral and equivalve, inferior margin well rounded. Both auricles well developed, angular, with a few faint radii. Colour rose with irregular white splashes. Sculpture: the central third of the valve is occupied by a deep and broad furrow flanked by two broad and high ridges. On either side of these are two comparatively insignificant ridges with their complementary furrows. Each main radius and sulcus has three or four secondary radii, those in the hollows smaller and wider apart. The whole surface of the valve is microscopically ornamented by fine, close, raised threads which cross the radials at right angles. Fig. 19 shows a portion of the central furrow highly magnified.

The interior is stained deep rose. The adductor scar is deep. The impress of the main radii extend as furrows to the umbo, while the secondary radii merely denticulate the margin. The left valve develops a few pectinidial teeth. The chondrophore lies within the margin of the hinge plate. On either side of it and beneath the ligament is a narrow, vertically striated tract. Beneath this again is a slight cardinal rib. Height, 20 mm.;

ASAPHIS CONTRARIA, Deshayes.

Psammobia contraria, Deshayes, Catalogue des Mollusques de l'île de la Réunion, 1863, p. 11, Pl. i., ff. 20-21.

(Plate xlviii., figs. 4-8.)

Nearly forty years ago Deshayes received a single right valve of this remarkable shell from Bourbon. For its reception Dr. von Martens afterwards instituted the section *Heteroglypta*.* Mr. J. Brazier has handed to me a specimen with both valves in natural apposition, and slightly larger than the type, which he collected in shell sand at the mouth of the Nambuccra River, N.S.W. The extreme rarity of the shell, its unexpected occurrence in Australia, and the brevity of published information are sufficient reasons for again figuring and describing it. Its inclusion in *Asaphis* is due to Dall.†

Shell small and thin, oblong, rather compressed, equivalve, a little inequilateral, slightly gaping behind. Colour pale yellow. Umbones sharp, almost touching, transversely directed. Ligament short, elevated above the shell. Lunule minute, sharply excavated, lanceolate. The regular growth lines, which can be traced everywhere, are masked by an elaborate superficial ornament like the pattern of *Cytherea pectinata*. The tip of the beak has a few regular, concentric ridges, past which the sculpture is differentiated into four areas: a small anterior, a large medio-anterior, a still larger medio-posterior and a small posterior. The anterior panel, limited by an imaginary line drawn from the beak to the middle of the anterior margin, has a few, widely separated, well developed, longitudinal ridges. To form the next panel each ridge turns at right angles, narrows to a fine thread and runs a direct course to the ventral margin; in this medio-anterior panel there are about twenty of such narrow, close, oblique, raised threads, which become smaller and closer as they approach the

*Martens, Moll. Maur. 1880, p. 331.

† Dall, Trans. Wagner Inst. iii., pt. v. 1900, p. 981.

umbo. The medio-posterior panel extends from a line vertically below the beak to a line from the beak to the centre of the posterior margin, and is occupied by a dozen rather strong, widely spaced, straight, radiating ridges. The posterior panel contains low, broad and strong folds, meeting the radiating ridges at an acute angle.

The radial sculpture slightly crenulates the ventral margin and furrows the interior. Pallial sinus rounded, deep, extending to the centre of the valve. Lateral hinge teeth absent Cardinals: in the right, one, in the left, two: the right and posterior left notched and furrowed; the anterior left large and fitting into a deep socket in the opposite valve. Length, 16; height, 11; breadth of conjoined valves, 7·5 mm.

EXPLANATION OF PLATE XLVIII.

Figs, 1-3.—*Endodonta concinna*, Hedley; in various aspects.
Figs. 4-8.—*Asaphis contraria*, Deshayes; in different aspects, with details of hinge and muscle scars.
Fig. 9.—*Blauneria leonardi*, Crosse.
Fig. 10.—*Stenothyra australis*, Hedley.
Fig. 11.—*Petterdiana thaanumi*, Pilsbry.
Fig. 12.—*Iravadia australis*, Hedley.
Figs. 13-14.—*Mathilda rosæ*, Hedley; in different aspects.
Fig. 15.—*Leuconopsis inermis*, Hedley.

NOTE ON AN ECHIDNA WITH EIGHT CERVICAL VERTEBRÆ.

By R. Broom, M.D., B.Sc.

In a paper communicated to this Society on the muscles of the shoulder girdle in the Monotremes, Dr. W. J. S. McKay* has shown how very variable *Echidna* is in regard to the number of its dorsal, lumbar, sacral and caudal vertebræ. In a series of sixteen specimens there were found no fewer than eleven different arrangements of the vertebræ; while in a seventeenth specimen there was a rib more on the right side than on the left, and yet the formula of neither side agreed with that of any of the other sixteen specimens. The dorsals vary from 14 to 17; the lumbars from 2 to 4; the sacrals from 3 to 4; and the caudals from 10 to 14. All the specimens agreed, however, in having 7 cervical vertebræ.

On recently looking over some of my Echidna specimens to see if I could find any distinct reptilian characters in the cervical vertebræ, I was somewhat surprised to come across a specimen in which the eighth vertebra, which ought to have been the first dorsal, is provided with a pair of quite rudimentary ribs and is to be thus really a cervical vertebra.

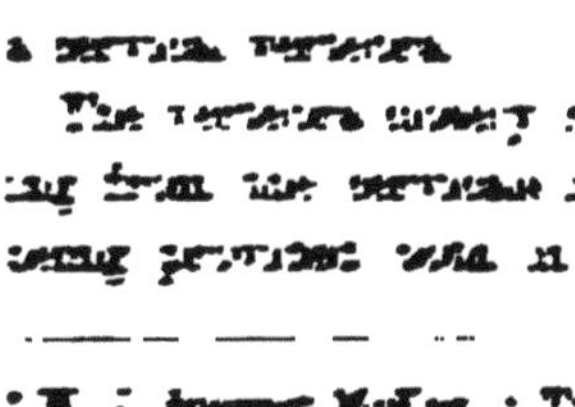

The vertebra closely resembles the normal first dorsal—differing from the cervicals in the greater length of the spine and in being provided both in front and behind with well developed

* W. J. Stewart McKay, "The Morphology of the Muscles of the Shoulder Girdle in Monotremes," P.L.S.N.S.W. [illegible]

zygapophyses. The dorsal vertebræ in *Echidna* usually have the arches perforated for the passage of the spinal nerves, but this character is generally absent in the first dorsal, and in one specimen in my possession is absent in the first five dorsals. In this abnormal eighth vertebra there is no indication of a perforation in the arch or even of a distinct notch for the nerve.

In *Echidna*, as in *Ornithorhynchus*, the ribs have almost completely lost the double-headed character so well seen in their Theriodont ancestors and retained in the large majority of mammals. In the first rib there is usually an indication of the double-headed condition. In this abnormal eighth vertebra there is no difficulty in recognising the double articulation of the rib with the centrum and with the transverse process. Usually the first rib articulates not only with the first dorsal, but also with the seventh cervical vertebra; the short ribs of this specimen articulate with the eighth vertebra alone.

The rib of the right side tapers to a point, but the left rib is slightly dilated at the end and forms an articulation with the side of the first fully developed rib.

The rib of the ninth vertebra, which is the first to meet the sternum, does not articulate with the anterior broad part of the manubrium sterni as normally, but with its hinder part, almost as the second rib does usually.

ON THE OSSIFICATION OF THE VERTEBRÆ IN THE WOMBAT AND OTHER MARSUPIALS.

By R. Broom, M.D., B.Sc.

(Plate XLIX.)

During the course of development there may be recognised many ancestral characters which are either quite lost or greatly obscured in the adult forms. In the adult of the human species only the dorsal vertebræ have distinct ribs, yet during the early stages of development there are clear indications of elements homologous with ribs in the first three sacral vertebræ and in at least the seventh cervical. As in the ancestral reptilian condition well developed pleurapophysial elements were attached to practically all the vertebræ, it is but natural that further indications should be met with in the development of the vertebræ of the lower mammals.

On looking into the mode of ossification of the vertebræ in the marsupials, I have come across one or two interesting points in which the condition differs from that ordinarily found in the higher mammals.

Cervical Vertebræ.—The atlas vertebra in the marsupials, as is well known, differs from that in the higher mammals in the lower piece being in some forms very small and in many others quite absent. In the monotremes the inferior element is well developed as in the higher forms, and in the more primitive marsupials such as the Dasyures, a well developed *osseous* element unites the two arches below. Though this is the condition in the majority of the Polyprotodonta, the intermediate piece is very small in certain species of *Perameles*, and even in *Thylacinus* is quite narrow. In the smaller Diprotodonta (e.g. *Petaurus breviceps*) a small intermediate piece is present which almost exactly resembles that in certain species of *Perameles*. In

Trichosurus the two arches meet inferiorly without a third element, but occasionally a rudimentary inferior element is present. In the smaller wallabies the arches meet as in *Trichosurus*. In the larger wallabies and kangaroos, in the wombat, in *Phascolarctus*, and in the large extinct Diprotodonts, there is a more or less wide gap between the lower ends of the arches bridged by fibrous tissue. It would seem as if in the smaller and ancestral Diprotodonts the intermediate piece had become gradually reduced in size until it became lost, and that as the Diprotodonts increased in size the arches became again separated, the place of the lost intermediate piece being taken by fibrous tissue.

The axis vertebra is very similar in structure to that in the higher forms. There is, however, one interesting point of difference in the development, in that whereas in man, and probably most of the higher mammals, the odontoid process is ossified from a pair of centres, in the marsupials there is but a single median centre as in the centra of the more normal vertebræ.

The cervical vertebræ from the 3rd to the 7th are ossified from three centres—one for the body and one for each arch. I have been unable to find any ossification which could be regarded as a costal element.

Dorso-lumbar Vertebræ.—The dorsal vertebræ are developed

process developed from the arch to which is articulated an independent additional element—an autogenous transverse process. According to Flower* the transverse processes of the anterior lumbar vertebræ of the pig are originally autogenous elements though coalescing very early with the rest of the vertebræ, and in certain cetaceans the transverse processes of the lumbar vertebræ are autogenous elements. In many reptiles, and especially in those reptiles from which the mammals appear to have sprung—the Theriodonts—all the trunk vertebræ have costal elements, and in the lower trunk or lumbar region these elements are articulated to the vertebræ exactly as are the autogenous processes in the wombat. Fig. 4 represents the upper side of a lower trunk vertebra of *Cynognathus*, and if this be compared with the fourth lumbar of the wombat (fig. 3) it will be seen that the two agree closely except in the different degree of development of the parts, and that there is no reasonable doubt but that the rib-like appendages of the vertebra of *Cynognathus* are homologous with the autogenous transverse processes of the vertebra of the wombat.

Sacro-caudal Vertebræ.—In the human subject the term "sacrum" is applied to the anchylosed five vertebræ which support the pelvic bones. Here there is no difficulty in defining the limits of the sacral series of vertebræ, and in many other mammals the difficulty is no greater. There are many forms, however, in which not only are a different number of vertebræ anchylosed in different individuals, but where even in the one individual the number increases as age advances. According to Flower, "a more certain criterion is derived from the fact that some of the anterior vertebræ of the sacral region have distinct additional (pleurapophysial) centres of ossification, between the body and the ilium. To these, perhaps, the term *sacral* ought properly to be restricted, the remaining anchylosed vertebræ being called *pseudo-sacral*, as suggested by Gegenbaur." If this criterion, however, be applied to the sacrum of the marsupials it

* W. H. Flower, "Osteology of the Mammalia."

will be found at once to break down, since in most marsupials the upper caudal vertebræ have well developed autogenous transverse processes which are undoubtedly serially homologous with the pleurapophysial centres of those vertebræ which support the pelvic bones.

If the fifth vertebra of the sacro-caudal series of a half-grown wombat be examined (fig. 9) it will be seen to be made up of a well developed flattened centrum and a feebly developed arch, with on each side a rather large, flattened autogenous transverse process. The transverse processes articulate mainly with the centrum, but also with the arches. A similar description would apply to the 6th, 7th, and 8th vertebræ, but on reaching the 9th the transverse process is found to articulate only with the centrum. On passing forwards the 4th vertebra is found to be very similar to the 5th, while the 3rd differs only in the slightly increased development of the arch and of the transverse processes. The second vertebra, which is usually regarded as a true sacral vertebra, has the transverse processes strongly developed for articulation with the ilia. In the first vertebra of the sacro-caudal series the elements are exactly the same as in the fifth vertebra and only differ in being larger, and in having the arch proportionately more largely developed, and in having the lateral elements specialised for the support of the

... es. In fig. 5 is s[illegible] ...iew of the first sac o-

In *Trichosurus* the first two vertebræ have large autogenous lateral elements, and a considerable number of the succeeding vertebræ have small autogenous transverse processes.

As the lateral pleurapophysial elements of the first sacral vertebra are thus seen to be homologous with the autogenous lateral elements of the succeeding vertebræ, it becomes quite impossible to draw a distinction between sacral and caudal vertebræ by the criterion above referred to.

In the manati and the beaver among Eutherians the transverse processes of the caudal vertebræ are developed autogenously; but Flower doubts if "this circumstance alone is sufficient to entitle them to be considered as costal elements." From the marsupial condition it is manifest that the caudal transverse elements are homologous with the lateral elements of the first sacral vertebra, and as it is pretty well established by comparative anatomy that the lateral elements of the sacrum are modified ribs, we are forced to the conclusion that the autogenous transverse processes of the upper caudal vertebræ in the marsupials are really costal elements.

EXPLANATION OF PLATE XLIX.

Fig. 1.—Front view of 3rd lumbar vertebra of half-grown wombat.

Fig. 2.—Front view of 4th lumbar vertebra of half-grown wombat.

Fig. 3.—Upper view of 4th lumbar vertebra of half-grown wombat.

Fig. 4.—Upper view of a lower trunk or lumbar vertebra of *Cynognathus* (reduced; modified from Seeley).

Fig. 5.—Front view of 1st sacro-caudal vertebra of half-grown wombat.

Fig. 6.—Front view of 1st sacro-caudal vertebra of *Deuterosaurus* (reduced; modified from Seeley).

Fig. 7.—Under view of 1st sacro-caudal vertebra of half-grown wombat.

Fig. 8.—Under view of 5th and 6th sacro-caudal vertebræ of half-grown wombat.

Fig. 9.—Under view of 2nd, 3rd, and 4th sacro-caudal vertebræ of half-grown wombat.

Fig. 10.—Under view of first four sacro-caudal vertebræ of young *Didelphys* (× 4).

CONTRIBUTION TO THE BACTERIAL FLORA OF THE SYDNEY WATER SUPPLY, II.

By R. Greig Smith, M.Sc., Macleay Bacteriologist to the Society.

Many pathogenic bacteria have been separated from water or from mud lying beneath well and river waters by inoculating the water, either before or after concentration, or the mud, directly into the tissues of animals in which the non-pathogenic organisms disappear whilst the pathogenic bacteria multiply and produce their characteristic lesions. Several pathogenic bacteria have been discovered in this way, as for example, *Bac. cuniculicida* and *Bac. hydrophilus fuscus*. This method, however, is only applicable in the separation of organisms pathogenic to animals, and is of no use for organisms such as *Bact. typhi* and *Vibrio choleræ*, which are parasitic in man only. The former organism is perhaps the most to be feared in the water of this country. The latter is fortunately still a stranger. Before going further, it may be said that the separation of *Bact. typhi* is extremely diffi-

typhi itself. On this account the most that can be done is to eliminate as far as possible the harmless bacteria and investigate the others. Many processes have been recommended for this purpose, and of these some have been shown to be worthless, whilst others are apparently of some value. The methods depend upon the inability of the water bacteria to grow upon special media or ordinary media with the addition of (1) phenol, (2) acids, (3) alkalies, or (4) salts.

The addition of phenol or carbolic acid to ordinary media or to special media has been most frequently advised, and in enumerating the various methods a beginning may be made with it. Dunbar has shown that the growth of the typhoid bacterium is interfered with by the presence of 0·116% of phenol in water or nutrient media, while 0·144 checks the growth entirely. In publishing these facts he showed the uselessness of some methods in which percentages of phenol greater than 0·116 were employed for the separation of *Bact typhi*. A method devised by Vincent is not open to Dunbar's objection.

Vincent's method.—Five drops of a 5% solution of phenol are added to tubes containing 10 c.c. of nutrient bouillon, and into these small increasing quantities of the water under examination are introduced. The tubes are incubated at 42° C., and any that show a growth are used for the preparation of plates of solid media. The colonies that grow in the plates are examined and the bacteria determined by the appearances and reactions of subcultures. Bandi* also advises the employment of phenol, but at a comparatively high temperature.

Bandi's method.—200 c.c. of bouillon are added to 2 litres of water and the mixture is incubated for 5 hours at 45° C. Twelve drops of the mixture are then added to 10 c.c. of peptone-free bouillon, together with 5 drops of 5% phenol. From the culture twelve successive transfers are made into the carbolised medium, the cultures being maintained at 45° C. The twelfth subculture

* Bandi, Centralblatt für Bakt. 1 Abt. xxv., 596, Ref.

is inoculated into peptone-free bouillon without phenol, and after growing at 45° is used to inoculate ordinary gelatine in which the *typhi* colonies grow slowly. A method very similar, but at a lower temperature, is employed by Jordan* for the separation of *Bact. coli* and the elimination of other glucose fermenting bacteria, *e.g.*, *Bact. cloacæ.*

Jordan's method.—A quantity of the suspected water is incubated with acid meat extract and with phenol in the proportion of 1 to 1000 of fluid. The incubation is continued for 12 to 18 hours at 38° to 40° C., when the culture is used to infect plates of litmus-lactose-agar. The colonies that redden the medium are picked out and the bacteria proved by their faculty for coagulating milk, producing indol in bouillon, fermenting glucose and not peptonising gelatine. The employment of litmus-lactose-agar considerably lightens the work of separation. The use of phenol by itself has not found so much favour among bacteriologists as a mixture of phenol and hydrochloric acid which was introduced by Parietti.

Parietti's method.—A solution is prepared containing 5 grm. phenol, 4 grm. hydrochloric acid, and 100 grm. distilled water. Tubes containing 10 c.c. nutrient bouillon are treated with 3, 6, and 9 drops of the prepared solution and are then incubated for 24 hours at 37° C. to make sure that no organisms have gained

he has modified the method to ensure the growth of *Bact. typhi*. Instead of taking for further examination the turbid bouillon containing the greatest number of drops of phenol solution, he takes the tube next lower in order; on the other hand, to obtain *Bact. coli commune* he uses the tube containing most disinfectant. The details of the method are here given briefly.

Hankin's modification.—To five tubes each containing 10 c.c. of ordinary nutrient bouillon are added 0, 1, 2, 3 and 4 drops of Parietti's solution and a few drops of the suspicious water. A water which contains comparatively few bacteria is filtered through a procelain candle and the microbes upon the porcelain surface brushed into a few c.c. of sterile water, which is used for infecting the bouillon. The tubes are incubated for 24 hours at 37°; at the end of this time an observation is made and one of the series is taken for further cultivation. The turbid tube containing the greatest number of drops of disinfectant is discarded, excepting when *Bact. coli commune* is sought for. Generally the turbid tube next in order is taken unless it has a thick scum upon the surface, or it has a growth visible only in the deeper layers of the liquid, or when bubbles of gas are apparent in the medium. In the latter cases the choice should fall upon a tube which is uniformly turbid. The tube that is taken is used to start a fresh series of bouillon cultures, the lowest member of which has the same number of drops of Parietti's solution as the tube taken. The process of selection is repeated upon these tubes after a 24 hours' incubation. A third series may be prepared, but usually one of the second series is used for making a smear culture on a fairly dry agar slope. Any suspicious colonies that grow on the sloped surface are inoculated into litmus-agar tubes and further examined.

Hilbert* criticised Hankin's method unfavourably, so far as it relates to the separation of *Bact. typhi*. He found that drops of a *typhi* suspension prepared by diluting a 24 hours' bouillon

---

* Hilbert, Centralblatt für Bakt. 1 Abt. xxvii., 526.

culture 100,000,000 times, produced cultures of the organism in tubes containing 3 drops of Parietti's solution. In the presence of *Bact. coli commune*, however, *Bact. typhi* could not be found even in suspensions diluted only 100 times. It is apparent, that in diluted Parietti's fluid, *coli* prevents the growth of *typhi*.

Wittlin* in testing the effect of 3 to 7 drops of Parietti's solution in 10 c.c. of fluid found that the ordinary water bacteria, whose presence in water is unimportant, either do not develop or grow very feebly. The latter fact was made evident by the motile bacteria becoming non-motile. On the other hand, the pathogenic bacteria, as well as those which normally inhabit the animal intestine, are quite unaffected by the acid phenol solution. The *coli* group are indifferent to 7 drops of solution per 10 c.c. Similarly the bacteria of the excrement, *e.g.*, *Staphylococcus aureus*, *Streptococcus erysipelatis*, *Bact. pyocyaneum* and *Bact. ochraceum* are unrestricted in their development. After standing in the acid medium, motile bacteria gradually lose their motility, the saprophytes becoming motionless much sooner than the pathogenic organisms with the exception of *Bact. pyocyaneum*. Wittlin grew many bacteria in pure culture in carbolised bouillon and noted the growth on the first or second day. Development always occurred with those bacteria which are not influenced by the reagent, however small the number of cells originally taken.

Potassium iodide appears to be most serviceable in aiding the growth of *Bact. typhi*, or what is practically the same thing, checking concomitant bacteria to such an extent as to render the isolation of colonies of *typhi* an easier matter than would obtain in media devoid of this salt. It is used in conjunction with a potato medium which was introduced into bacteriology by Holz. The preparation of the potato medium, however, as recommended by Elsner, to whom we are indebted for the employment of potassium iodide, differs slightly from that recommended by Holz.

*Elsner's method.**—Potatoes are washed, pared, and finally grated; a litre of water is added for every 500 grams of the pared potatoes and the suspension is pressed through a cloth. The strained fluid is allowed to stand overnight and then it is filtered. The filtrate is boiled and again filtered. Ordinary gelatine is dissolved in the extract, when it will be found that every 10 c.c. of the potato-gelatine requires from 2·5 to 3 c.c. of tenth normal alkali to neutralise (using phenolphthalein as the indicator); 10 c.c. portions of the gelatine are put into tubes and sterilised. When required, 1 c.c. of sterile, 10% potassium iodide is added to the molten gelatine and plates are made after infection. On the plates, *Bact. coli communis* grows to the exclusion of other bacteria, as large colonies in 24 hours, while *Bact. typhi* produces small, clear colonies like drops of water in 48 hours. The latter are finely granular, while the former are brownish and coarsely granular.

Elsner used potassium iodide as a result of experiments conducted with various chemicals, salts, resins, alkaloids and amido-bodies. Of those which he tried—in number they came to many hundreds—potassium iodide together with an acid medium was found to be the best for differentiating between the typhoid and colon varieties of bacteria and for excluding others. This method is used chiefly for detecting *typhi* in faeces, but Remlinger and Schneider† have used it in testing waters and soils. Kübler and

* Elsner, Centralblatt für Bakt. 1 Abt. xvii., 591, Ref.

† Remlinger und Schneider, Centralblatt für Bakt. 1 Abt. xix., 244, Ref.

Neufeld* were able by this method to separate *Bact. typhi* from a polluted water. *Bact. coli* was not present, from which they concluded that the pollution had been caused by the access of the urine of typhoid patients.

Other writers have experimented with the selective action of salts, acids, and alkalies in separating *typhi* and *coli* from water. Uffelmann† found that the former could withstand considerable quantities of citric acid, acetic acid or alum. He advised the addition of 10% citric acid and 0·25% methylviolet to ordinary nutrient gelatine. Dunbar, however, objected to his method, as it did not distinguish between *coli* and *typhi*. Riedal‡ proposed the use of iodine trichloride. Köhler§ showed the action of various organic and inorganic acids, alum, phenol, the alkaline hydrates and stains upon the growth of *typhi*. Fermi‖ has recorded the resistance of various bacteria to increasing doses of certain chemicals. The bacteria are chiefly pathogenic, and although the table can be used for separating the bacteria of certain groups, yet there is no indication of the usefulness in isolating noxious from innocuous water bacteria that might have been expected from a paper which entailed a vast amount of work. There are some points, however, that may be noted, The micrococci (including the pyogenic varieties) may be separated from other bacteria by growth upon agar containing up to 10

added. Billings and Peckham* found that *Bact. typhi* and *coli commune* were more resistant to the antiseptic action of light than water bacteria.

Marpmann† has described a method which he employs for separating three noxious classes of bacteria from water. To the water is added an equal quantity of neutral bouillon, and the mixture is incubated at 30° C. for 24 hours in order to increase the numbers of the bacteria. At the end of this incubation period, a portion of the mixed culture is inoculated into nutrient gelatine containing 0·2% citric acid and maintained at 20°-22° C. The colonies that grow are examined for *Bact. typhi*. Another portion is inoculated into gelatine containing 2% sodium carbonate and kept at 10°-18° C. The colonies are examined for cholera vibrions as well as the cloaca bacteria, which include the putrefactive microbes commonly found in the animal intestine, the representative organism being *Bact. coli commune*. Another portion is inoculated into nutrient agar containing 2% sodium carbonate, kept at 30°-37° and examined for cadaver bacteria. The last two groups are really subdivisions of one which includes all intestinal (sewage, excrement, etc.) bacteria. They induce a foul or putrefactive fermentation of albuminoids and carbohydrates.

Theobald Smith‡ separates *Bact. coli commune* by adding quantities of the water varying from 0·1 to 1 c.c. to 10 fermentation tubes containing 1% glucose bouillon. In the presence of *coli* 40 to 60% of the volume of the tube contains gas in from 3 to 4 days. The reaction of the medium becomes strongly acid (5 c.c. tenth normal alkali are required to neutralise 10 c.c.). Plates are prepared from the most probable tubes before the end of a week. It is to be noted that a production of gas and acid, very similar to *Bact. coli commune*, is produced by varieties of *Proteus*, *Bact. cloacæ* and the *Bact. lactis aerogenes* group of bacteria.

---

* Billings and Peckham, Centralblatt für Bakt. 1 Abt. xix., 244, Ref.

† Marpmann, Centralblatt für Bakt. 1 Abt. xvii., 362.

‡ Theobald Smith, Centralblatt für Bakt. 1 Abt. xviii., 494, Ref.

of alum and sodium carbonate to the water. The suspension is centrifuged and the precipitate dissolved in a small quantity of 1·5% potassium hydrate; this does not affect the vitality of the bacteria which are then cultivated. From 96 to 100% of the micro-organisms present in the water can be obtained by this method.

The Sydney city water was tested by the various methods, and the organisms that were brought into prominence by the individual schemes were studied in detail. Taken as a whole, the bacterial flora that survives treatment with disinfectants is limited. In the tests the water, as a rule, was passed through a porcelain filter, and the bacteria brushed from the surface of the candle with a sterile stiff brush into 10 c.c. of sterile water. Various quantities of the suspension were used.

The three methods of Vincent, Bandi, and Jordan are very similar, and as the two latter are the more recent and improvements upon the former, they alone were tried. By Bandi's method the first tube of carbolised meat extract only became turbid, succeeding tubes failed to show a growth. There was only one kind of organism in the turbid tube. The temperature of 45° C. appears to be too high for the growth of species of *Bact. coli*, especially under such unfavourable circumstances as obtain in

at a later date with the equivalent of 2000 c.c. of water was likewise found to be sterile. This fact is rather curious, because a culture of *Bact. coli commune* isolated from faeces grew to a considerable extent in the medium at 42°. It probably indicates that the varieties of *Bact. coli* that are present in the water (and a variety was found in 800 c.c. on the same date as the repetition) are considerably weakened, so much so as to be unable to grow in Jordan's medium at 42° C.

With a water temperature of 13° C. (end of August) no growth was obtained in tubes containing 7 drops of Parietti's solution to every 10 c.c. of a mixture of bouillon and bacterial suspension, even when the equivalent of a litre of water was employed. Growth was obtained later in the year (middle of October) when the temperature of the water had risen to 17° C. To tubes containing 10 c.c. of bouillon were added 1·5 c.c. portions of bacterial suspension (representing 300 c.c. of water) and from 1 to 6 drops of Parietti's solution. The tubes were incubated at 37° C. A turbidity and surface film appeared in each, and as time went on the film became black. Plates which were infected with the 4 and the 6 drop growth produced colonies which appeared like drops of gum. This organism was proved to be *Bac. mesentericus niger*, Lunt, a recently discovered potato bacillus. Using the same quantities of bouillon and bacterial suspension growth occurred in the presence of 9 drops of Parietti's solution, but 9 drops appeared to be the limit for the quantity of suspension employed, as the tube with 10 drops remained clear. With regard to the ratio between Parietti's solution and the infected medium, the dropping pipette delivered 36 drops to the c.c. (grm.) and the bouillon mixture measured 11·5 c.c. The plates that were prepared with the cultures in the 7 and 9 drop tubes produced what appeared in each case to be a pure culture of Bacterium I. This organism appears to be *Bac. aquatilis communis*. In a test made at a later date, when the temperature of the water had risen to 22° C., colonies of *Bact. coli commune* were found in plates which had been prepared from a 24 and a 48 hours' culture in 8 drops of Parietti's solution per 10 c.c. of bouillon and suspension. The organism was found in 800 c.c. and upwards of water. Owing to there being so few organisms in the water that can withstand the action of Parietti's solution, transfers from the primary tubes, as advised by Hankin, were found to be unnecessary. *Bact. coli commune* grew quite freely.

In Elsner's potato gelatine, colonies were obtained on the plates infected with the equivalents of 20 c.c. of water and upwards, and on the potato medium the colonies appeared to consist of two kinds. They were found to contain the same organism, the difference in appearance having been induced by surface and sub-surface growth. The separated organism, Bacterium II., which appears to be an ally of *Bac. pinnatus*, grows sparing at 37° C., and, probably, would not have been separated by any method that necessitated the employment of this temperature.

Marpmann's citric acid gelatine entirely excluded the bacteria and in their place there was obtained a luxuriant culture of moulds, among which were *Mucor racemosus* and *Penicillium glaucum*. There were also many pink colonies of the common pink yeast. The action of 2% sodium carbonate in an agar medium at 37° C. was not tried because agar with 0·75% sodium carbonate proved to be too alkaline a medium for the organisms of the water. The latter amount of carbonate was, I believe,

from lactose in an agar medium. *Mic. pyogenes γ albus* and Bacterium III., both of which reddened litmus, were also separated.

In preparing the medium for Pake's method sodium formate is employed. This salt could not be obtained in Sydney, so it was prepared by distilling glycerine with oxalic acid, neutralising with sodium hydrate and recrystallising. Fermentation took place in all the tubes which contained the equivalent of amounts of water varying from 80 to 1600 c.c. Plates prepared from all the tubes showed the constant presence of a gelatine-liquefying bacterium, *Bact. cloacæ*, while a second non-liquefying organism, which proved to be *Bact coli commune*, occurred only in the largest quantity.

In reviewing the various methods for the separation of *Bact. coli commune*, the most serviceable appear to be Pake's and Parietti's. The easier method is undoubtedly the latter, and a ratio of 8 drops (equivalent to 0·22 c.c.) to 10 c.c. of bouillon and suspension appears to be the best for the purpose. *Bact. coli commune* was found in the Sydney water when quantities of 800 c.c. and upwards were employed, and when the water had a temperature of 22° C. It is to be noted, however, that the organisms have a low vitality, since they cannot withstand over 9 drops of Parietti's solution per 10 c.c., and also they succumb to carbolised meat-extract at 42°. I have separated *Bact. coli commune* from faeces with 12 drops per 10 c.c., and the separated organisms grew in carbolised meat-extract at 42°. The low vitality of the water *coli* is undoubtedly due to a prolonged existence in the water.

The cultural characters of the organisms which were isolated and which have not already been described in the first paper are described in the following paragraphs.

Micrococcus pyogenes γ albus, Rosenbach.

A non-motile coccus occurs in groups; the individuals measure 0·6-0·7 μ and are not decolorised when treated by the Gram method. On the gelatine plate the colonies grow as white points, which, when magnified, appear circular and opaque. In seven days the colonies, which remain punctiform, are surrounded by a zone of softened gelatine, and microscopically they are seen to have a rough outline and a margin like a yeast colony. The gelatine-stab becomes filiform and faintly tuberculate. On agar the stroke is porcelain-white and remains narrow; the margin is lobular, and the growth spreads irregularly at the base. There is a good growth upon potato, but it is indistinguishable from the medium. Bouillon becomes turbid and forms a coherent sediment, but no film. Milk is coagulated and the reaction becomes acid. The culture on litmus-lactose-agar reddens the litmus, but does not produce gas in the medium. Nitrate is reduced to nitrite, indol is formed in bouillon, and gas is not produced from glucose.

The organism was separated by Jordan's method on account of its reddening litmus. It appears to be *Mic. pyogenes γ albus*, Rosenbach.

Bacillus mesentericus niger, Lunt.

This member of the potato bacilli is a motile rod, with rounded ends, and measures 0·6 : 1·7-3 μ. It forms oval central spores, and is stained by Gram's method. The growth upon potato is very much folded and dry; it soon covers the entire surface, and becomes black in colour. The potato is also blackened. The

while the upper layers become turbid and a white wrinkled film forms on the surface. The film darkens, and a black pigment is diffused into the upper layers. A strong indol reaction was obtained. No nitrite is formed from nitrate. Milk is first thickened, then the casein is peptonised. The colouring matter is not quinone, since no blue colour is obtained with potassium iodide and starch in acid solution. The black pigment is not formed in gelatine or milk.

The organism was separated by means of Parietti's solution

Bacterium I.

The organism is an actively motile cocco-bacterium measuring 0·4 : 0·6 μ: it does not stain by the Gram method. When free to grow upon agar, the colonies become amœboid and translucent or iridescent white. The processes when magnified are seen to have a semilunar shaded tip like a finger-nail. The stroke on agar is porcelain-white, raised and glistening, the base is spreading, the margin irregular, and the edge smooth. In old cultures a brownish colour diffuses into the medium. Litmus-lactose-agar is unaffected. In gelatine the liquefying colonies are circular and crateriform with a white centre. Microscopically the centre is coarsely granular and the margin indefinite. The stab in gelatine becomes saccate, and the liquefied medium turbid, without film or precipitate. Gas is not produced from glucose. Bouillon becomes turbid and forms a sediment, but no film: the indol reaction was obtained. Nitrates are reduced to nitrites. Milk is coagulated at 37° with a neutral reaction. On potato a raised, irregular, dry, yellowish-white layer is formed, the colour becomes ivory-white and moist glistening.

This bacterium apparently differs in size only from *Bac. aquatilis communis*, Zimm. It was found in the tube containing 7 drops of Parietti's solution per 11·5 c.c. of suspension and bouillon 300 c.c. water. If the organism is the same as Zimmermann's, it seems peculiar that the disinfectant should have permitted the growth of what is described as the commonest water bacterium.

BACTERIUM II.

The organism is a short motile rod with rounded ends and measures 0·4 : 0·8-1·2 μ; it is decolorised by Gram's method of staining. On gelatine plate the colonies are round, white, and moist glistening. When magnified they appear rounded, light brown, granular, and smooth-edged.

The gelatine-stab is filiform with a restricted, lobular, flat nail-head. No gas is produced from glucose. The stroke on agar is white, moist glistening and slightly iridescent; it slowly widens laterally and spreads out at the base; the condensed water has no film. Litmus-lactose-agar is not affected. Bouillon becomes turbid and forms a sediment and a slight film. A slight indol reaction was obtained; nitrates were not reduced. Milk is not coagulated, and the reaction is unaltered. On potato the growth is irregular, slightly raised, pale buff in colour and with a fatty appearance; the growth becomes buff and moist glistening. The organism grows slowly at 37° C.

This bacterium was separated by Elsner's method. Its indifference to milk and glucose places it among the typhoid group of bacteria. Its nearest ally appears to be *Bac. pinnatus*, Ravenal. It has many points of difference from *Bact. typhi*.

BACTERIUM III.

transparent white with an irregular or amœboid margin; they may also be rounded, thicker and translucent white. Microscopically they are finely granular. The deep colonies are uncharacteristic. The stroke on agar is thin, moist glistening, white and iridescent: the growth rapidly spreads over the agar surface and gas bubbles form in the medium. Litmus-lactose-agar becomes red excepting at the surface, which remains a persistent blue; the red colour becomes bleached and gas bubbles are formed in the medium. Bouillon becomes turbid and forms a white sediment; a slight indol reaction was obtained. Nitrates are reduced to nitrites. Milk is coagulated with a faint acid reaction. On potato a brownish raised layer is formed and the medium slowly becomes of the same colour as the growth.

As the gelatine in stab culture was found to be fluid after a month's growth, the organism may be *Bact. Kralii*, Dyar. The gelatine-liquefying property of the organism is indicated also by the bleaching action upon litmus.

Bacterium cloacæ, Jordan.

The bacterium is an actively motile rod with rounded ends; it varies in length and breadth, measuring 0·6-0·7 : 0·8-1·4 μ; longer forms also occur. It is not stained by Gram's method. On gelatine the colonies are punctiform and microscopically *coli*-like, but they soon become shallow liquefied depressions. The stab in gelatine is napiform and becomes saccate; the upper layers of the liquefied medium are turbid, the lower layers clear with floating floccules. No film is formed. Gas is produced in glucose-gelatine. The colonies on agar are indefinite, white and watery, and surrounded by a cluster of small transparent islet colonies. Microscopically the structure appears homogeneous. The deep colonies are uncharacteristic. The stroke on agar is narrow, raised and translucent white; the agar surface is covered with a watery growth like a film of condensed water. The litmus in litmus-lactose-agar culture is bleached in the deep parts of the medium and deep blue at the surface. Bouillon becomes turbid and forms a slight precipitate and film; a slight indol reaction was obtained. Nitrates were strongly reduced to nitrites. On potato the growth was light brownish-yellow scarcely distinguishable from the potato, but afterwards it became darker, raised and spread over the surface. Milk was coagulated at 37° with a neutral reaction.

This organism was separated by Pake's method as well as by those of Bandi and Parietti. The first colonies on gelatine by Pake's method were like those of *Proteus mirabilis*, but the formation of cochleate strands from the margin of the colonies had ceased by the third transfer in ordinary media and the type became stable. It appears to be *Bact. cloacæ*, Jordan, a gelatine-liquefying form of *Bact. coli*.

Bact. coli commune, Escherich.

The bacterium is a short, stout motile rod with rounded ends; it measures 0·55-0·6 : 0·9-1·8 μ, and is not stained by Gram's method. The flagella are long and generally 5 to 9 in number, arising from places around the organism. The colonies on gelatine are punctiform, but became flat, spreading, white, and moist glistening with a raised centre. Microscopically the deep colonies are yellowish-brown, circular and finely granular with a smooth edge; when crowded upon the plate they appear zonate. The

white sediment and a film. The indol reaction was obtained, and nitrates were reduced to nitrites. Milk is coagulated with an acid reaction. On potato the growth is spreading, raised, moist glistening and yellowish-brown.

This organism, which is undoubtedly *B. coli commune*, was separated from 1600 c.c. of water by Pake's method, and also from 800 c.c. of water by 8 drops of Parietti's solution per 10 c c. of bouillon and bacterial suspension.

NOTES AND EXHIBITS.

Mr. T. Steel exhibited the distal half of a humerus of the extinct Marsupial, *Diprotodon australis*, Owen, from Darling Downs, Q.

Mr. Froggatt exhibited specimens of cherries from the Armidale district, showing the effect of the depredations of the Rutherglen Bug (*Nysius venator*, Bergr.); the pest, however, is amenable to treatment by the cyanide-fumigation process. Also commercial samples of carrot seed infested with the destructive small beetle, *Sitodrepa (Anobium) panicea*, Linn., the eggs of which were included with the seed when made up into packets.

Mr. J. J. Walker exhibited a specimen of, and contributed the following Note on, *Nacerdes melanura*, Schm., a European beetle not previously recorded from Australia. "This beetle, which belongs to the family *Œdemeridæ* of the Heteromerous section of the Coleoptera, is widely distributed throughout the maritime regions of Europe (including the south coast of England), but appears to be rare inland. I have found it plentifully on the coasts of Kent and Essex in early summer, also at Gibraltar.

Mr. W. S. Dun exhibited specimens of green carbonate of copper from the Narrabeen Shales of Long Reef. Attention was first drawn to these by Mr. Bensusan. Copper also occurs in the native form as separate particles or associated with carbonised plant remains. The carbonate originally occurred as native copper. Native copper has been found at lower horizons of the Narrabeen Shales in bores at Newington, Holt-Sutherland and Heathcote, and also at Bulli (vide T. W. E. David, Austr. Assoc. Adv. Sci., i., pp. 275-290).

Mr. J. P. Hill exhibited a Teleostean fish, an undetermined species of the viviparous genus *Cristiceps*, the ovary of which was seen to be packed with developing young. He also exhibited a series of drawings and photographs in illustration of his work during the past year on the development of Marsupials, including photographs of the early stages in the development of *Dasyurus viverrinus*; and photographs and drawings illustrating the evolution of the external form and the condition of the foetal membranes in *Trichosurus vulpecula*, *[illegible]*, and *Phascolarctos cinereus*. Also, on behalf of Prof. J. T. Wilson and himself, he exhibited the egg-shells of laid eggs of Echidna and Platypus, together with photographs of Platypus embryos and foetal specimens from the burrow.

Mr. G. A. Waterhouse exhibited three specimens of the butterfly, *Xenica hobartia*, Westw., caught at the National Park, on November 12th. Also specimens of *X. tasmanica*, Lyell, lately described from Tasmania.

Mr. Cambage exhibited herbarium and timber specimens, and photographs of typical trees and shrubs, illustrative of the vegetation of the interior of the Colony as described in his paper.

The Acting Director of the Botanic Gardens, on behalf of Mr. J. H. Maiden, exhibited flowering specimens of *[illegible]*, Lindl.; [illegible] *[illegible]*, [illegible] from Moulmein; *Cattleya* [illegible] [illegible] *Cattleya* [illegible].

Linden & Hudré) from Tropical America; and *Cyrtanthus parviflorus*, Hort., probably a form of *Cyrtanthus collinus*, Burchell, from S. Africa.

Mr. Palmer showed the stem of a sapling Eucalypt completely eaten through by borers. Also, from Port Mackay, examples of aboriginal work in the shape of a knitted scarf and various head dresses.

Mr. Musson sent for exhibition flowering stalks of *Acacia Baileyana*, F.v.M., from Richmond, which this year produced no fruit. The failure to produce fruit may have been due to the severe frosts which prevailed in the month of August, or perhaps also the pollinating agents were at fault.

Mr. Trebeck exhibited a branch of *Opuntia coccinellifera*, a prickly-pear without prickles, which he thought would make an excellent forage plant in the very dry districts of Western Queensland during severe droughts. Also a specimen of a nearly mature borer, *Phoracantha* sp., one of the longicorn beetles, in situ in a piece of Eucalypt timber.

WEDNESDAY, MARCH 27TH, 1901.

THE Twenty-sixth Annual General Meeting of the Society was held in the Linnean Hall, 23 Ithaca Road, Elizabeth Bay, on Wednesday evening, March 27th, 1901.

The Hon. James Norton, LL.D., M.L.C., President, in the Chair.

The Minutes of the previous Annual General Meeting were read and confirmed.

The President delivered the Annual Address.

PRESIDENTIAL ADDRESS.

During the year just expired, ten new members have been elected and two have resigned, but we have had no losses by death. There are now on the roll 118 ordinary members, 5 associate members, and 19 honorary and corresponding members. Although these figures do not show as large an increase as we had hoped for, it is a matter for congratulation that the Society has not retrograded.

I regret to have to mention that, on account of the pressure of other arduous duties, Professor Wilson has felt compelled to resign his membership of the Council. Mr. W. S Dun, F.G.S., was elected in his place, but, under section 6 of the Act of Incorporation of the Society, he now retires, and it rests with you to elect a President and six members of committee, in the place of myself, Dr. J. C. Cox, Dr. Thomas Dixson, Mr W. S. Dun, Professor W. A. Haswell, and Mr. Perceval R. Pedley, who retire by rotation, but are all eligible for re-election.

In the Bacteriological section, Mr. Greig Smith has been steadily at work, the results of which are shewn in the nine papers read by him before the Society during the year; and the Bacteriological plant is all in excellent condition.

The 100th Part of the Proceedings completing Vol. xxv., is in the press and will shortly be ready for issue. The publication of the "Catalogue of the Described Mosses of New South Wales," contributed by Rev. W. W. Watts and Mr. Whitelegge, has had to be deferred for the present, but it is hoped that the Society may be able to accomplish it during this year.

The caretaker's lodge has been repaired and much improved at an expense of £80 8s. 6d.

At the meetings of the Society there have been read 48 papers on various interesting subjects, including 12 dealing with ordinary botanical matters and 9 on bacteria—now admitted to belong to the vegetable kingdom.

It is not surprising that botanical research should form one of the leading features of a Society named after the immortal founder of a system magnificent in its simplicity.

Previously to his time, botany can hardly be said to have existed, for there was no intelligent scheme of nomenclature or classification; not infrequently flowers were called "roses, lilies," &c., and fruits "apples, pears," &c., without being in any way related to the orders in which those names occur; and trivial

element in which is the generative system of flowers, in many respects so strikingly like that of animals.

Botanists, especially beginners, who have been accustomed to use "Loudon's Encyclopædia of Plants," with its astounding mass of condensed information, can testify to the great assistance which it affords them at a very small sacrifice of time.

The despised aborigines of Australia, in addition to being astronomers, were undoubtedly the first Australian botanists, for, although they had no distinctive names for the various parts of flowers, and could not count the stamens or pistils on which Linnæus founded his system, yet they had a distinctive and probably descriptive name for every conspicuous plant which grew in their respective territories, and, as I understand, for many classes of plants also, such, for instance, as ferns, gum trees, and acacias. Banks, in his Journal, states that they had some knowledge of plants, as he could plainly perceive by their having names for them.

They were necessarily driven, by an inhospitable and uncertain climate, to turn to account, as a means of subsistence, every thing which could possibly help them to sustain life: to enable them to do this, they were compelled to take precautions against injury from the use of some plants which, through their poisonous properties, would otherwise have been seriously injurious if not fatal to them.

They knew also the medicinal and curative properties of plants, and by their aid often effected marvellous cures of diseases and wounds.

In illustration of what I have said of the classification of plants by the aborigines, I may state that a very intelligent half-caste native of Shoalhaven, some time since informed me that his people were accustomed to use a decoction made from a certain plant as a cure for diarrhœa, and that there were three different kinds of this plant. On his bringing me specimens, they turned out to be *Rubus mollucanus*, *R. parvifolius*, and *R. rosæfolius*.

This man also described the mode in which they prepared *Macrozamia* nuts for food, the principal thing necessary being the maceration of the nuts in water for several days, and he said that the arrowroot, which his master was then eating, was just like the preparation which his mother used to make for him out of these nuts when he was a boy.

Banks tells us that Captain Cook's people were assured by the natives that they ate the seeds of a plant, which appears to have been *Cycas media* (a near relation of the *Macrozamia*), and that, encouraged by this information, the officers ate some of them, but, being ignorant of the necessity for maceration, they were deterred from making a second experiment by a hearty fit of vomiting. The hogs, which were still shorter of provisions than the crew, ate these seeds heartily, and, about a week after, were all taken extremely ill of indigestion and two died, the rest being saved with difficulty.

For the first glimmerings of light upon the vegetation of Australia (Sir Joseph Hooker says) we are indebted to Dampier, who in 1688 visited Cygnet Bay. The genus *Dampiera* was named in his honour. His herbarium was, till lately, if it is not still, preserved at Oxford and contained 40 specimens, 18 of which (9 being Australian) were figured in the account of his

east coast of Australia; but these enthusiastic botanists made good use of the short time allowed them, and availed themselves with equal avidity of every other subsequent opportunity to increase their collections and their knowledge of Australian vegetation.

It is easy to imagine the delight with which they must have hailed the discovery of the wealth of a flora, then for the first time displayed to scientific and appreciative eyes, the greater part of which was absolutely endemic; unfortunately they were not here during the blooming season of our most conspicuous flowers, or while the whole neighbourhood was aglow with kaleidoscopic beauty, as it is in early spring.

One of the most conspicuous genera of the order *Proteaceæ* was named *Banksia* in honour of the discoverer, but the name of Dr. Solander, having been previously utilized for a West Indian solanaceous plant, does not appear among the genera of our flora.

After the death of Banks, the collections made by him and his assistant, together with considerable further collections, made during Cook's subsequent voyages by the Forsters, Mr. David Nelson, and Mr. William Anderson, were handed over to the British Museum, where, without having ever been published, they were hoarded for long years, as if they had been brought there solely for the purpose of being stored, and, as Bentham complains, became, with Cunningham's subsequent collections, practically unavailable for use.

From a note of Anderson's, we find that the leaves of *Leptospermum scoparium* were used as a substitute for Chinese tea, and were found to be of a pleasant taste and smell; hence the popular name of "Tea-tree," not "Ti-tree" as it is often written. *Forstera*, a Tasmanian genus of *Stylideæ*, was named in honour of the Forsters; *Nelsonia*, a genus of *Acanthaceæ*, in honour of Nelson; and *Andersonia*, a genus of *Epacrideæ*, in honour of Anderson.

In 1788 Mr. John White arrived in Sydney as surgeon-general to the settlement, and during seven years collected a considerable number of plants, and made drawings of others, which were published by Sir James Smith in his "Specimen of the Botany of New Holland," in "White's Journal of a Voyage to New South Wales," and in other works. In the "Specimen" some of these plants, including the Waratah, there called "*Embothrium*," were very creditably figured and colored.

In 1791, Captain Vancouver's expedition was accompanied by Mr. Archibald Menzies as botanist, who formed a good collection of the singular and extremely beautiful flora of King George's Sound.

In the same year the French expedition, under Captain D'Entrecasteaux, in the "Recherche" and "Espérance" in search of La Perouse, visited South Western Australia and Tasmania, where considerable collections were made by Monsieur J. J. Labillardière, who published figures and descriptions of 265 of the most interesting in his "Novæ Hollandiæ Plantarum Specimen," and described and figured others in his narrative of the voyage, which was accompanied by folio plates of several of the plants: the genus *Billardiera* was named after him.

About 1794, Colonel Paterson, who was in command of the New South Wales Corps, zealously devoted himself to investigating the botany of the colony and the northern part of Tas-

entire coast. He was accompanied by Dr. Robert Brown, the most keen and observant of all the botanists of Australia, who possessed the same extraordinary faculty of observation as was afterwards displayed by Gould in connection with the birds of Australia. His "Prodromus Floræ Novæ Hollandiæ" is regarded by all botanists almost with reverence, and he has been well called "The father of Australian botany." He was ably assisted in his work by Ferdinand Bauer, a draftsman whose name has been utilised to form the name "*Bauera*," and by Peter Good, a gardener, from whom the genus "*Goodia*" takes its name.

These gentlemen made collections not only at King George's Sound, but on the Blue Mountains, and at Bass Straits and Tasmania; and Sir Joseph Hooker, on their account, speaks of Flinders' voyage, as far as botany was concerned, as the most important in its results ever undertaken, and the results incomparably greater, not merely than those of any previous voyage, but than those of all similar voyages put together, and says that the Prodromus, though only a fragment, had for half a century maintained its reputation unimpugned of being the greatest botanical work that had ever appeared.

His extraordinary collection of specimens is admitted to be the foundation in England of the knowledge of Australian vegetation, and to show conclusively his power of observation, sagacity, zeal and industry, which, during his short visits, often exhausted the flora of the parts he touched at.

When Brown commenced his labours, the number of ascertained Australian plants amounted to 1,300, of which 1,000 had been collected for the most part by Sir Joseph Banks. To this collection Brown added nearly 3,000 species.

He was fortunate in not accompanying Flinders in his subsequent voyage in the "Porpoise," which, with the "Cato," was wrecked on the Cato reef, and in the "Cumberland," which afterwards called at Mauritius, where her commander, to the eternal disgrace of the French Government, was imprisoned for six and a-half years, in order, it is believed, to enable Peron's account of the French expedition next mentioned to be published before, and so take precedence in nomenclature, &c., of Flinders' account, which, with its valuable appendix containing Brown's general remarks on the botany of "Terra Australis," was thus delayed till 1814. Brown's Prodromus, however, with its valuable illustrations, had, fortunately, been published in 1810: *Flindersia*, a genus of fine trees of the order *Meliaceæ*, of particular interest to Queenslanders, was named in honour of Flinders, and *Brunonia*, a genus of *Goodenovieæ*, in honour of Brown.

Captain Baudin's expedition, in the "Geographe," "Naturaliste" and "Casuarina," before alluded to, fell in with Flinders' expedition on the south coast of New Holland, on 8th April, 1802, when, although their nations were at deadly war with each other, they, being protected by passports, met on the most friendly terms. The French expedition was accompanied by M. Leschenault de la Tour, as botanist, who made good collections of specimens, and Brown named the beautiful genus *Leschenaultia*, of Western Australia, after him.

In 1802, David Burton was also sent out by Sir Joseph Banks to collect plant specimens in New South Wales, and *Burtonia*, of the *Leguminosæ*, was named after him.

In 1816, Allan Cunningham, His Majesty's botanical collector

That meddlesome fire-brand, Dr. Lang, in his eagerness to bespatter an official, endeavoured to deprive Oxley of the honour of his discovery, but his unfounded statements have been most ably and conclusively refuted by the Right Honorable Sir Hugh M. Nelson, in his last year's address to the Queensland Geographical Society. A statue has been erected to the memory of the reverend libeller; but the noble fellow, whom he tried to belittle, rests in an unknown grave, and his remains will probably be carted away in the common ruck, to make room for the proposed new Sydney railway terminus.

Between the years 1818 and 1821, Cunningham accompanied Admiral King in some or all of his four voyages to King George's Sound, Dampier's Archipelago, Hobart Town, Macquarie Harbour, Torres Straits, and many intermediate places. He subsequently visited many parts of the interior, and, in February, 1831, returned to England; but, after the unfortunate loss of his brother Richard, in Mitchell's expedition of 1835, he accepted the office of Colonial Botanist, and, in 1839, died at the age of 48, in the Sydney Botanic Gardens, where a monument was erected to his memory. There can be little doubt that his life was shortened by the hardships which he had so long and so continuously undergone.

As Sir Joseph Hooker remarks, his botanical travels were the most continuous and extensive that have ever been performed in Australia, or. perhaps, in any other country: his vast collections were, for the most part, transferred to the British Museum.

In 1818, Oxley, accompanied by Fraser, conducted a second expedition into the interior, which was as fruitful in botanical results as the first.

In the same and following year, Captain Freycinet's expedition, in the French corvettes "Uranie" and "Physicienne," accompanied by M. Gaudichaud as botanist, also did good service, and in honour of the leader and of the botanist the name *Freycinetia Gaudichaudi* was bestowed on a Queensland plant.

Afterwards M. Lesson, in "La Coquille" and subsequently in "L'Astrolabe," did some botanical service, and, in 1823, Dr.

F. W. Sieber, of Prague, visited New South Wales and made considerable collections, which were sold : it is from him that we get the name *Siebera*, a by no means conspicuous genus, and this shews how thoroughly the botanical field of the country had been exploited.

In 1823, 1825, and 1829, the vicinity of King George's Sound, Wilson's Promontory, Cape Aird, and Lucky Bay, were explored botanically by Mr. William Baxter, a gardener sent out by private enterprise to collect seeds and plants: the genus *Baxteria* of the *Juncaceæ* was named after him.

Mr. Charles Fraser, already mentioned, a soldier of the 73rd Regiment, after his arrival in New South Wales, became an indefatigable collector and explorer, visited Swan River in 1826-7, and Moreton Bay in 1828, and wrote excellent accounts of those districts. He afterwards visited Tasmania and took charge of the Sydney Botanical Gardens.

In 1827, Robert Sweet published his "Flora Australasica," which is the first really artistic completed work wholly devoted to Australian botany, and contains 56 plates of Australian flowers beautifully drawn and coloured.

Sir Thomas Mitchell's four expeditions into the interior, owing to his great fondness for natural history, and excellent system of observation, resulted in valuable contributions to

Australia, solely for the purpose of discharging a religious duty, but, owing to his knowledge of botany, his connection with a leading horticultural establishment in England, and his love of observing and collecting, the results of his journey have proved extremely valuable.

He kept journals and formed a considerable herbarium, including many plants collected by Sir William Macarthur: his name is commemorated in that of the genus *Backhousia*.

Mr. Ronald Campbell Gunn, with his friend and companion Mr. Robert William Lawrence, of Tasmania, who died in 1832, commenced exploring the northern parts of that island, and afterwards, between 1832 and 1850, collected so indefatigably over a great part of Tasmania that there are few plants there which he did not see alive and collect: his collections were all transmitted to England in perfect preservation, accompanied by notes which displayed remarkable powers of observation and a facility for seizing important characters in the physiognomy of plants, such as few experienced botanists possess. The name *Gunnia* was bestowed in his honour on the only epiphytal orchid in Tasmania, now relegated to *Sarcochilus*.

In 1833, Baron Charles von Hügel, an Austrian, made considerable collections in the Swan River colony, and in 1837 commenced the publication of his "Enumeratio Plantarum," which, however, was never finished.

In 1838, the establishment of Port Essington was attempted for the fourth time by Sir Gordon Bremer, and Mr. Armstrong went to reside there as collector for the Kew Herbarium. John MacGillivray was stationed there for some time, and in 1842 he accompanied an expedition in H.M.S. "Fly" and "Bramble" sent out to make a further survey of the tropical coasts of Australia. In 1847 he accompanied an expedition in H.M.S. "Rattlesnake" to Port Curtis, Rockingham Bay, [illegible] Cape York and [illegible] Island, and Moreton Island, and owing to the death of Captain Stanley the narrative devolved upon him. He made numerous interesting observations on the vegetation of Australia.

In 1838, Dr. Ludwig Preiss, at Swan River, made collections of upwards of 2,000 species of plants, including *Cryptogamia*, which were sold after a complete account of them had been published by various authors in "Plantæ Preissianæ," edited by Dr. Lehmann in 1844-5.

In 1839, Mr. James Drummond, of Swan River, one of the most zealous of Australian botanic collectors, commenced operations, and continued his work for upwards of 15 years. His labours took the practical form of collecting and forwarding for sale in Europe the plants of his district, including a vast number of novelties, rivalling in interest and importance those of any other part of the world. Dr. Lindley's able sketch of the vegetation of the Swan River is founded chiefly on Drummond's collections.

Capt. Mangles, at this time, collected many species of Western Australian Plants, and John Bailey, the Colonial Botanist of South Australia, arrived and made collections of living plants and seeds; but, as the native flora of Adelaide is probably the least interesting of those of all the Australian colonies, he soon directed his energy more towards the introduction of useful foreign economic plants. Whether he introduced those which, in many parts of the neighbourhood have almost entirely superseded the native vegetation, I do not know, but the varieties and masses of foreign plants which clothe many hundreds of acres of land

expeditions hereafter mentioned; for, with well provisioned ships to fall back upon, and to store the treasures collected, the explorers enjoyed most of the comforts to which they had always been accustomed. It was a very different thing when, instead of merely skirting the coast, they pushed boldly into the mysterious unknown heart of the country, which hurled defiance at the puny invaders of its vast dreary solitudes: but the spirit of enterprise, insuppressible in the Anglo-Saxon race, impelled its members from time to time to press forward into the vast unknown, with their lives in their hands, and, notwithstanding failure succeeded by failure, fresh victims were always ready to take the places of those who had failed; ultimately perseverance and pluck were crowned with success, in spite of the merciless attacks of unfeeling savages, in spite of cruel thirst often prolonged for days together under a burning sun, and in spite of hunger, often allayed only by killing their cattle, horses and camels, reduced to skeletons, as these poor animals were, when the time for their sacrifice had arrived. In illustration of the distress to which explorers were sometimes reduced for want of food, I may mention that on one occasion Ernest Giles, being alone in the desert, and on the very verge of starvation, heard a faint squeak, and immediately saw and pounced like an eagle upon a dying wallaby, and ate it "living, raw, dying, fur, skin, bones, skull, and all," and thought he should never forget the delicious taste of that creature weighing about two ounces, and he could not help wishing that he had its mother and father to serve in the same way.

In 1840, Captain Eyre's perilous journey from Adelaide to Swan River produced little more than negative botanical results, as there appeared to be, between Streaky Bay and Lucky Bay, scarcely any vegetation at all. He was accompanied by an overseer and three black boys, two of whom, during his temporary absence, plundered the camp, shot the overseer, and decamped; but with the remaining boy, he managed, after terrible hardships, to reach his destination.

In 1842, Charles Stuart began collecting in Tasmania and New South Wales, and continued so doing for many years, paying special attention to New England.

In 1844, the necessary funds were raised. by private subscription, to enable Dr. Ludwig Leichhardt to conduct an overland expedition from Moreton Bay to Port Essington. After the lapse of two years and two months, when everyone had given up all hope of seeing the Doctor again, he and his party managed to reach Port Essington in an almost starving condition, after their terrible journey of 3,000 miles. Having a considerable knowledge of botany, Leichhardt's narrative is by far the fullest public detailed account of the tropical vegetation of the interior of Australia which we possess.

In 1846, he started on a bolder expedition, from Moreton Bay to Swan River, with a party, of which Mr. John F. Mann is the sole survivor, but the weather and circumstances seem to have been against him, and the party returned, a failure.

However, nothing daunted, he made a second attempt in 1846, and perished miserably with all his party, whose loss is one of the mysteries of the interior, which probably will never be solved. He was evidently thoroughly deficient in the faculty of organization, and too reliant on the luck which had carried him successfully through former troubles, and was eminently unsuited

bear his name. His great success in his expeditions procured him the title of "The father of Australian discovery."

The Gregory family were inveterate explorers. In 1846, A. C. Gregory, F. T. Gregory, and H. C. Gregory, without any assistants, commenced operations by making an expedition along the Western coast of Australia. In 1848, A. C. Gregory with C. F. Gregory and others, conducted a second expedition over 1500 miles of country, and in the same year, A. C. Gregory, accompanied by Governor Charles Fitzgerald and others, conducted an expedition from Perth to the Geraldine lead mine, when the Governor received a bad spear wound from the aborigines. In 1855, A. C. Gregory, accompanied by H. C. Gregory, F. Mueller (as botanist), and others, started from Brisbane on an expedition which lasted 16 months, and is more particularly referred to hereafter. In 1857, F. T. Gregory led an expedition to trace the Murchison River to its source, and in 1858, accompanied by James S. Roe (as botanist) and others on a second expedition. In 1857-8, A. C. Gregory with C. F. Gregory and others, conducted an expedition in search of Leichhardt, and in 1861, F. C. Gregory, accompanied by Mr. P. Wallcott (as collector in natural history and botany), conducted another expedition, promoted principally by English capitalists interested in cotton manufacture, who proposed to establish a new colony on the north-west coast, having for its special object the cultivation of cotton.

In 1848, Mr. E. B. Kennedy, accompanied by twelve men, started on an expedition intended to penetrate from Rockingham Bay to Cape York, which he accomplished, but at the expense of the lives of himself and nine of his party. William Carron and two others were by great good fortune rescued, almost at their last gasp: Carron published an account of the expedition, and the story of poor Kennedy's death, so touchingly described by the faithful Jacky Jacky. Carron was an excellent botanist, and, notwithstanding the disastrous termination of the expedition, was able to bring many fresh plants to our knowledge: he was afterwards employed in the Sydney Botanical Gardens until his death.

In most of these expeditions valuable botanical collections were made, and were afterwards dealt with by Mueller.

About 1847, Mr. William Archer, of Tasmania, who had collected an excellent herbarium, placed it at the disposal of Sir Joseph Hooker, and contributed liberally to the cost of its publication.

In 1848, Mr. James S. Roe, who had accompanied King in his expeditions of 1818 and 1821, and was afterwards appointed Surveyor-General of Western Australia, and took part in nearly every exploring expedition sent forth in that colony, conducted an expedition into the interior, extending over 1800 miles: he was evidently an enthusiastic botanist, and we find his name occurring frequently in the specific nomenclature of the plants of Western Australia.

In 1851, Mr. John Carne Bidwill, a surveyor, who in 1841 published an interesting book, "Rambles in New Zealand," and was for a short time Curator of the Sydney Botanical Gardens, afterwards went to Queensland, where he successfully used his opportunities for doing good botanical service, and was the discoverer of the Bunya Bunya, *Araucaria Bidwilli*. He formed an excellent herbarium, which was transmitted after his death

by Port Albert and Wilson's Promontory, having travelled 1500 miles and collected nearly 1000 species of plants.

In the next year he visited more of the mountains of the colony, many difficult regions of Southern Australia, Lake Albert, the Murray Lagoons, the Cobboras Mountains, the Snowy and Buchan Rivers, travelled 2500 miles, and collected upwards of 500 different plants

In 1844-5 he again visited the Australian Alps, the Avon Ranges, Mount Wellington, the Snowy Plains, the Bogong Range, Mounts Hotham and Latrobe, and the Murrigang Mountains, raising the Victorian flora to 2500 species, including Cryptogams.

In 1855 he accompanied A. C. Gregory in his celebrated expedition before mentioned, from Sydney across Northern Australia, visiting the islands on the east and north coasts, ascending the Victoria River, exploring the limits of the great desert, traversing Arnheim's Land, and reaching the mouth of the Albert in the Gulf of Carpentaria, and the Gilbert River, and then travelling south-east, crossing the Lynd, the Burdekin, the Suttor, the Belyando, the Mackenzie and the Dawson, and returning to Sydney, viâ Brisbane, without the loss of a member of the overland expedition.

This extraordinary journey was considered only second in point of interest and extent of unknown country traversed to Leichhardt's first expedition, especially as continuous and systematic collections and observations abounding in novelty and interest were made.

Mueller is said to have travelled, in his various expeditions, over 20,000 miles. Sir William Hooker refers to him as the "Prince of Australian botanists," and Bentham appears often to have deferred to his opinion on account of his having had the opportunity of seeing living specimens, though Bentham himself, on account of his knowledge, not only of Australian botany, but of that of the whole world, must have been a more competent classifier than one who dealt with the local flora only.

Mueller formulated the cortical system of grouping our gigantic myrtles after having found that woodcutters, by paying attention to the bark and timber, could distinguish the ordinary species more readily than scientists by the inflorescence.

Like most systematic botanists, Mueller seems to have had little or no love of horticulture, and consequently complaints were made that the Botanical gardens under his charge were neglected and in a slovenly condition, and ultimately they were taken out of his hands. Mr. W. R. Guilfoyle having been appointed Curator, and being a practical landscape gardener as well as a botanist, has since worked wonders in improving them, and has satisfied every one except the Baron, who was never able to forgive him up to the time of his death in 1896.

The numerous botanical publications which redound to the credit of the Baron's memory, are too well and too favourably known to make it necessary for me to mention them categorically.

It is not possible to overestimate the value of these works, particularly as, through extended settlements, ringbarking, overstocking and other causes, the original flora is being constantly destroyed, though not unfrequently replaced by importations of worthless or even injurious plants.

In 1854, Mr W. H. Harvey came to Australia for the express purpose of investigating the algology of its shores, and visited King George's Sound, Swan River, and Cape Riche, Melbourne, Tasmania and Sydney, forming a magnificent collection: many of

themselves seem to have exceeded in importance and interest those of all previous explorers.

In 1863 was commenced the publication of the "Flora Australiensis" by George Bentham, assisted by Mueller, the chief foundation of which was the vast herbarium of Sir William Hooker, but every other Australian herbarium of any importance was examined, and this involved so great an amount of time and trouble that the last of the seven volumes did not appear till 1878. The book was published under the authority of the several Governments of Australia and is unquestionably the most important Australian botanical work ever published. It contains a minute description of almost every Australian plant, whether conspicuous or otherwise, and it is not an easy matter to find any plant within a hundred miles of Sydney which is not described in it. Fortunately Bentham was able to complete it before his death, and it now stands an imperishable monument to his memory. The seven volumes contain complete descriptions of upwards of 8000 species of plants, with references to at least as many synonyms, the approximate number of recognised species of living plants in the whole world being, up to the end of the nineteenth century, according to Professor Vines, [illegible].

In order to emphasise the importance of this great work, Her Majesty Queen Victoria was advised to confer on Bentham the Companionship of the Order of St. Michael and St. George, and to raise Mueller, who already possessed that distinction, to the Knight Commandership of the Order.

The last volume of the Flora having been issued 20 years ago, it was considered extremely desirable that a supplement, containing all subsequent discoveries, should be issued; and as a large proportion of these discoveries belonged to Queensland, it was hoped that the matter would be entrusted to Mr. F. M. Bailey, Botanist to the Queensland Government, whose numerous publications, such as the "Synopsis of the Queensland Flora," "The Companion for the Queensland Student of Plant Life," "The Fern World of Australia," and its companion, "Lithograms of the Ferns of Queensland," comprising 191 plates of ferns copied by direct impression of the fronds off the stone, prove beyond doubt his eminent fitness for the work, for which he at once made preparations; but, as Mueller, with some show of reason, claimed the right to continue the work in which he had assisted Bentham, Bailey at once gave way on condition that Bentham's system of classification and nomenclature should be continued. As this did not satisfy Mueller, and the usual intercolonial jealousies unfortunately came into play, the matter was dropped; but Bailey has commenced the publication of a new work, incorporating all the Queensland species of plants comprised in Bentham's book, with all the recent Queensland discoveries, and leaving all the other States to take care of themselves: a most unfortunate check to future as well as present botanical studies!

The terrible hardships which had from time to time been undergone by explorers in endeavouring to penetrate the interior led to the belief that it was impossible to reach the centre, much more to cross from sea to sea; but in 1861 the intrepid Burke and Wills, who were by no means so fit to make the attempt as many others, started from Melbourne with a considerable party and a number of camels just imported by the Government from India. In their eagerness to push ahead, they, with King and Gray, left their party at Cooper's Creek and reached the Gulf of

The anxiety, caused by the non-return of Burke and Wills and of Leichhardt at the expected times, gave a great impetus to further explorations, and resulted in an expedition under Landsborough from Carpentaria, another under Walker from Rockhampton, and another under M'Kinlay from Adelaide, which reached within a few miles of the Gulf of Carpentaria.

Mueller, in an appendix to the account of Landsborough's expedition, gave a list of the plants known to exist at the Gulf, and remarked upon the general similarity of the inter-tropical productions to those of the extra-tropical parts of Australia, and thought it likely that no other country retained its similarity of features throughout so great an area and through so many degrees of latitude.

The feat of crossing Australia had now been accomplished six times, and the road across was beginning to be almost considered a beaten track.

In 1869, John Forrest conducted an expedition in search of Leichhardt, and, in 1870, accomplished a journey from Adelaide to Perth round the great bight, over the track which had been so nearly fatal to Eyre; and in 1874 he travelled from Perth to the central telegraph line, which had then been stretched right across the continent. The account of his last journey contains an appendix by Mueller of the plants collected.

Mr. Thomas Elder most patriotically fitted out three expeditions, and after having introduced 121 camels, started Colonel P. E. Warburton from Adelaide in 1872. The spirit of exploration, which had started the remarkable expeditions of Roe, Lefroy and Hunt, having now been greatly encouraged by Governor Weld, through whom the Messrs. J. and A. Forrest had successfully pushed their explorations further and further into the waste of salt swamps which filled the centre of the continent; after suffering great hardships Warburton was only just able to reach Roebourne on his way to Perth, after picking the bones of his last camel. His health was so greatly broken that he could do little to assist the publication of the account of his expedition. Dr.

Trimen, of the botanical department of the British Museum, dealt with the plants collected by him.

Between 1872 and 1876, Ernest Giles conducted five expeditions, in the last of which he crossed the continent from Perth to Melbourne, and in all collected many thousands of plants which, however, with the exception of those of the first two expeditions, were lost. Those saved were classified and named by Mueller, and included in an appendix to Giles' book, "Australia Twice Traversed."

The botanical map of Australia, so to speak, having now been fairly well drawn and filled up, with the exception of spaces here and there left to be dealt with by future workers, it is not necessary to go laboriously into the accounts of all botanical discoveries which have followed those before-mentioned, with the exception of one which will be mentioned later on; but it seems fitting to mention a few of the late botanical workers of Australia, among whom the first name which suggests itself to me is that of the Rev. Dr. Woolls, who was a most energetic and enthusiastic worker, was persistent in his endeavours to popularize botanical studies, and always willing to advise and assist others.

Among many minor publications, he was the author of three or four small works, which have been found exceedingly useful. One of the persons who worked with him, and was devoted to the

and these were accompanied by very complete botanical descriptions. The drawings were mostly by himself, but in some he was ably assisted by Mr. A. J. Stopps. Fitzgerald generously allowed the Government to publish his work under the name of "Australian Orchids," without any remuneration to himself, and piloted it through the press to the extent of the first volume and four more parts, when, to the great regret of his friends and all lovers of botany, he died in 1892.

One further part was afterwards published, with the able assistance of Mr. Henry Deane, M.A., F.L.S., Engineer-in-Chief for Railways, and Stopps; but the difficulty of carrying on the work, without the help of the author, was found to be insuperable, and the residue of the money, voted by Parliament, was devoted to the preparation and publication, through Mr. J. H. Maiden, F.L.S., Curator of the Botanical Gardens, Sydney, assisted by Mr. W. S. Campbell, F.L.S., of seven parts of an elegant little book called "The Flowering Plants and Ferns of Australia."

In 1892, the Governments of Queensland, New South Wales, Victoria, and South Australia, extended their patronage of botany to a new branch, and published at their joint cost Dr. M. C. Cooke's "Handbook of Australian Fungi," which was illustrated with 36 coloured plates.

The nature of the vegetation of Queensland seems to have greatly encouraged botanical research, for there have always been a considerable number of enthusiasts there. Besides Bailey, I may mention Mr. John F. Shirley, B. Sc., who, during 15 years preceding 1893, had gathered and mounted 2,500 species of plants, being about half of the known flora of the country; also the late Dr. Joseph Bancroft, M.D., whose medical proclivities encouraged him to investigate the chemical properties of plants, and who was then sanguine of success in the preparation of a valuable ophthalmic remedy from *Duboisia myoporoides*, from which a drug, at one time worth over one hundred pounds an ounce, was producible.

Dr. Thomas L. Bancroft, the son of Dr. Joseph Bancroft, appears to have followed in his father's footsteps.

In Victoria, too, the study of botany has been greatly encouraged by Mueller and the numerous publications from time to time issued under his authorship, but New South Wales, South and Western Australia and Tasmania have not been far, if at all, behind.

I cannot here refrain from mentioning, in terms of the highest commendation, two books published by the Department of Agriculture of New South Wales, under the authorship of Mr. Fred. Turner, F.L.S., F.R H.S., who succeeded Mueller as consulting botanist to Western Australia: "The Forage Plants of Australia" and "Australian Grasses." These books have attracted most favourable notice in many parts of the world, have led to the cultivation of Australian forage plants and grasses in many foreign places, and have induced persons, interested in pastoral pursuits, to devote their attention to our native pasturage, instead of, as heretofore, endeavouring to supersede it by imported and unacclimatised plants, generally worthless and often absolutely injurious.

The only other Australian book, and that a small and unpretentious one, which I shall mention here, is "Australian Botany,

plants (27 of them being from Botany Bay), and contains Solander's original descriptions and the habitats of the plants figured.

As it had, for some years past, been thought desirable by scientific men, that a thorough examination of Central Australia should be made by experts, Mr. W. A. Horn, in a most noble spirit of patriotism, agreed to pay all the necessary expenses of fitting out an expedition to the M'Donnell Ranges, now known as the Larapintine region, to be conducted by leading men from the various colonies.

The expedition, consisting of 16 men, 26 camels lent by the South Australian Government, and two horses, started in May, 1894, accompanied part of the way by the noble-minded patron, who, after satisfying himself that everything was working smoothly, returned on his lonely ride of 1,000 miles to Adelaide.

No expedition had ever been fitted out so thoroughly, or managed so carefully and with so little discomfort to its members, each of whom worked so enthusiastically at the branch under his particular charge, that the results as a whole were far more important than those obtained by any former expedition; but no vestiges of the archaic flora or fauna, which it had been thought possible might still exist there, could be found. The botanical report of the work of the expedition was prepared by Professor Ralph Tate, F.L.S., F.G.S., who, however, discovered little that was absolutely new: the volumes, containing the whole of the reports, were edited by Professor Baldwin Spencer, M.A. The number of species of plants, known previously to the expedition, to inhabit the Larapintine region, was 502, and that has now been increased to 614.

Space and time will not permit me to mention numbers of zealous botanists, in all the Australian colonies, who have devoted themselves to the furtherance of the interests of botany, but I propose to conclude by giving shortly such particulars as I have been able to obtain with respect to the various botanical establishments maintained at each of the Australian capitals.

SYDNEY.

Naturally, Sydney was the first place in Australia where public Botanical Gardens were established, but there seems to be a little haziness about the time at which this was done.

In 1788, Governor Phillip reported that he had 16 acres of land, situated in the neighbourhood of Sydney Cove (*i.e.*, at Farm Cove), under cultivation on the public account, and in "Tench's Complete Account of the Settlement at Port Jackson," published in 1793, it is stated that, contiguous to Sydney (no doubt at Farm Cove), Phillip had established a Government farm, at the head of which a competent person of his own household was placed, with convicts to work under him; hence the name of "Farm Cove" superseding the aboriginal name of "Wockanna-gully." It was probably here that the first attempt at cultivation in the Colony was made, but most of the farms (among which was that at Farm Cove) were successively abandoned.

In 1790, Phillip is said to have given to Nicholas Devine permission to occupy part of what is known as "the lower garden," which must have been the land before mentioned, for the purpose of establishing a farm: the lease was afterwards cancelled, and, in 1794, some land at Newtown granted to Devine by way of compensation, and the land leased to him then became, and continued for a long time to be known as, "The Government

visited Swan River in 1826-7, Moreton Bay in 1828, and afterwards Tasmania, and wrote excellent accounts of the vegetation of those districts. Although the name of Fraser, as Colonial Botanist, was included in the list of members of Oxley's first expedition, published in the appendix to his book, as a document enclosed in Governor Macquarie's instructions to him, it is stated in a biographical notice of Allan Cunningham, contained in Hooker's "Journal of Botany," that he met Fraser (for the first time?) at the depot established on the Lachlan River, and that Fraser was attached to the expedition for the purpose of collecting for Lord Bathurst.

In 1829, the boundaries of the land reserved for the Gardens, &c., were notified; they included, not only the site of the present Gardens, but also the outer Domain, the Garden Palace grounds, the present Government House grounds, Bennelong Point where Fort Macquarie now stands, the site of the old Government House, the Circular Quay, Obelisk Square, and what are now called Hyde, Phillip, and Cook Parks. The nucleus of the Gardens was then in existence, but a not inconsiderable portion of the reserved land was little more than a preserve for "Five-Corners" (*Styphelia viridis*) and "Geebungs" (*Persoonia lanceolata*), and the public were only admitted on sufferance.

Fraser died in 1831, when John McLean became acting-superintendent until the arrival in 1833, and also after the death, of Richard Cunningham, who had been appointed in Fraser's place, and who was murdered by the blacks in 1835.

Mr. Robinson, on the recommendation of the Macarthurs of Camden, was then appointed temporarily, until Allan Cunningham took his brother's place in 1837; but Allan Cunningham resigned in the same year and died in 1839.

In 1840, Mr. John Anderson, botanist to King's expedition to South America, &c., was appointed superintendent, and made considerable collections in the neighbourhood of Sydney.

On the death of Anderson in 1842, Mr. James Kidd was placed temporarily in charge; at this time the Gardens were under the management of the trustees of the Australian Museum, the Macarthurs, the McLeays, and the Kings being represented on the Board. These gentlemen recommended Mr. John Carne Bidwill, who was appointed Curator and duly installed. Although the Home Government claimed the right of appointment, they would probably have approved of Bidwill, but, before the news of his appointment reached home, the nomination by Dr. Lindley of Mr. Charles Moore had been approved, to the great annoyance of Sir William Hooker, Director of the Royal Botanical Gardens at Kew, who claimed the right to be consulted, although he had no objection to Moore personally, and became afterwards very friendly with him. Moore, having been allowed time to pay a farewell visit to his friends, did not arrive in Sydney till 1847.

Bidwill's friends, including Sir Stuart Alexander Donaldson, were very angry at his being ousted, and the trustees of the Museum retired from the management of the Gardens. Donaldson, with the view of starving Moore out, moved the reduction of the annual parliamentary vote for the maintenance of the Gardens to £150. Having failed in this, he tried to get them cut up and sold in allotments, and fortunately failed in this also.

Moore, during his tenure of office for nearly half a century, remodelled the greater part of the Gardens, and made so many alterations, additions, and improvements, that he may well be

Besides other rooms, it contains four, which each measure 40 feet by 30 or upwards, and is fitted up with all necessary cases for the storage and display of dried plants and their products, botanical books, &c.; the number of specimens, now amounting to upwards of 8,000, is still increasing.

Since the preparation of this address the collections have been thrown open for the use of the public, and students of Australian botany will find it well worth their while to visit the building, not only casually, but systematically.

The area of the Gardens, without taking the outer Domain into account, is now 93 acres.

BRISBANE.

In 1828, Fraser was sent from Sydney to establish a Botanical Garden for Brisbane, and selected a site containing 42 acres on the north bank of the River Brisbane. Fraser managed the Garden until 1855, when Sir R. R. Mackenzie, Mr. William Augustine Duncan, afterwards Collector of Customs in Sydney, and Mr. T. Jones, were appointed a committee of management, and Mr. Walter Hill, Colonial Botanist and Superintendent. Hill took a great deal of interest in the indigenous and foreign tropical and sub-tropical vegetation, and, from 1874, had the able assistance of Mr. Fred. Turner, until he became manager of the Acclimatization Society's Gardens, Bowen Park, in 1879.

The climate being suitable, these gentlemen made the Garden quite unique in its beauty, enchanced by no less than 40 species of magnificent palms thoroughly acclimatized, by gorgeous flowering trees and shrubs, by a grand avenue of bunyas running along the bank of the river, and by handsome sub-tropical indigenous trees, collected by them from the rich scrubs abounding in the Colony; but, notwithstanding the daily growing beauty of the place, an attempt was made in the early seventies to destroy it by extending the business part of the city over it; this vandalism was fortunately averted.

Hill retired in 1881, when Mr. James Pink was put in charge as head gardener, and was succeeded by Mr. Cowan, who had charge until 1889, when Mr. Philip McMahon was appointed Curator.

The great flood of 1893, in its haste to reach the sea, without winding round the Gardens, swept across the peninsula, on which they are situated, greatly damaged the noble bunya avenue and all the lower portion of the grounds, landed the Government gunboat in their midst, and left the place in a most deplorable state: a second flood carried the gunboat back into the river, but, of course, otherwise increased the general desolation, from which, however, the place has since entirely recovered.

One of the features, which greatly popularises the gardens, is a beautiful kiosk, where morning bathers can obtain excellent breakfasts, and ladies give delightful *al fresco* tea parties in the afternoon.

McMahon has continued enthusiastically to carry out the idea of making the Gardens, which now cover about 37 acres, as nearly ideally tropical as the climate will allow.

MELBOURNE.

In 1842, Mr. Hoddle, Surveyor-General of Port Phillip (now Victoria), was instructed to survey 50 acres of land in Melbourne, including the site of the present Spencer-street railway station, as a site for a Botanical Ga[illegible] but not [illegible] ther seems to

In [illegible] W. R. Guilfoyle was appointed Director in place of Mueller, who then returned to the original position of Government Botanist. At the instance of Guilfoyle thirty acres were added to the existing Gardens, and he commenced to remodel the whole, now measuring 81 acres, constructed lakes, islands, bridges, kiosks, etc., and immensely improved the whole; but the chief of his improvements were the careful labelling of all the plants under his care and the establishment of a museum of economic botany.

ADELAIDE.

In 1838 Mr. John Bailey, having been appointed Botanist to South Australia, arrived there and commenced the collection of indigenous plants, but there was as yet no Botanical Garden.

In 1855, at the first meeting of the Botanical Garden Committee, a site containing about 40 acres was determined upon. Mr. George Francis, [illegible], was appointed superintendent, and held office for ten years, when he resigned through ill-health, and on his death his position was offered to [illegible].

In 1857 the Gardens were thrown open to public inspection, and the first catalogue of plants was published next year.

In 1866, at the suggestion of [illegible] Mr. [illegible], afterwards [illegible] of the [illegible] Museum, [illegible] the Gardens [illegible] and a large variety [illegible] collection.

In [illegible] Dr. [illegible] [illegible] and [illegible] afterwards [illegible] a house was built for the reception of the [illegible] which was finished [illegible]. The work of [illegible] and [illegible] for [illegible] construction and other purposes was [illegible] of a large [illegible] and it was thought that [illegible] years the [illegible] would be [illegible] of [illegible].

The [illegible] of [illegible] and [illegible] a series of [illegible] the erection of the [illegible] House [illegible] a museum of objects of [illegible] a museum of economic botany, the introduction of [illegible] and [illegible] the [illegible].

acclimatisation, and distribution of exotic plants, and the cultivation of timber trees, have made the Adelaide Gardens not only popular, but of considerable value to the Colony generally. All the Curators seem to have been endued with such a spirit of zeal and energy as enabled them to overcome the difficult nature of the climate, under which they were obliged to carry on their operations.

Schomburgk died in 1891, and Mr. Maurice Holtze, F.L.S., then Superintendent of the Botanical Garden at Port Darwin, was appointed in his place.

The area of the Adelaide Gardens, including the adjoining Botanical Park, which is managed under the same trust, is about 130 acres.

Perth.

The Botanical Gardens of Perth adjoin the beautiful Government House Domain, and are managed by Mr. Daniel Feakes as Curator. Mueller was, up to the time of his death, Consulting Botanist, and Mr. Fred. Turner was appointed to succeed him. Dr. A. Morrison is Botanist to the Bureau of Agriculture.

Hobart.

The Botanical Gardens of Hobart are situated on the bank of

[illegible]

[illegible]

[illegible]

[illegible]

[illegible]

[illegible]

OWMENT (CAPITAL).

	£	s.	d.		£	s.	d.
n				Loan A	3,000	0	0
..	14,000	0	0	Loan B	5,000	0	0
),				Loan (portion of Loan C, secured with other money by mortgage for £24,000,	11,700	0	0
.	5,700	0	0				
	£19,700	0	0		£19,700	0	0

ERIOLOGY (CAPITAL).

	£	s.	d.		£	s.	d.
Sir				Loan (portion of Loan C)	11,400	0	0
of				Also £900, from interest, included in Loan C	900	0	0
aid				Loan D	1,050	0	0
nd							
y	11,400	0	0				
ed							
oal	900	0	0				
ed)							
to							
..	700	0	0				
ed)							
ed	350	0	0				
	£13,350	0	0		£13,350	0	0

rect. We have also seen the securities.

E. G. W. Palmer, } Auditors.
Duncan Carson, }

P. N. Trebeck, Hon. Treasurer.

BALANCE SHEET.

Dr. Linnean Society of New South Wales. Cr.

	£	s.	d.		£	s.	d.
				Dec. 31, 1900.			
To Balance in Bank	21	14	3	By Ground Rent, Rates and Taxes ...	79	11	10
,, Deposit Receipt and Interest	721	0	0	,, Bank Fee, 10s.; Exchange, 6s. 2d. ...	0	16	2
,, Interest	1,341	0	0	,, Printing	260	9	6
,, Subscriptions				,, Plates and Illustrations	143	17	1
99 Members £103 18 10				,, Salaries and Wages	478	0	0
Arrears 11 11 0				,, Advertisements	6	16	0
Exchange 0 2 0				,, Postage	17	17	9½
	115	11	10	,, Refreshments and Attendance ...	5	7	6
,, Sales (including 100 copies of Proceedings purchased by the Government)	114	1	8	,, Petty Cash	9	18	8½
				,, Telephone	9	0	0
,, Fees (Bacteriology) .	11	10	0	,, Bacteriology Account ... £721 0 0			
,, Repaid by Bacteriological Account ...	25	3	0	,, Interest 492 0 0			
					1,213	0	0
	2,350	0	9	,, Insurance	7	13	10
,, Cheque not presented	17	12	6	,, Renovations, &c. (Caretaker's House)	80	8	6
				,, City Treasurer's Building Fee .	1	0	0
				,, Freight and Charges	2	9	4
				,, Cash in hand	1	0	0
				,, Balance in Bank (including one cheque (£17 12s. 6d.) not paid in	45	7	0
	£2,367	13	3		£2,367	13	3

March 14th, 1901.

Audited and found correct,

E. G. W. Palmer, } Auditors.
Duncan Carson, }

P. N. Trebeck, Hon. Treasurer.

ERIOLOGY (INCOME).

	£	s.	d.		£	s.	d.
..	37	12	4	By Salaries, &c.	426	0	0
...	721	0	0	,, Repaid to Society	25	3	0
...	360	10	0	,, Advanced on Loan	1,050	0	0
...	123	0	0	,, Cabinet for Slides (Fanning's Account)	5	2	6
...	123	0	0	,, Gas... ... .., ...	7	6	5
...	123	0	0	,, Fee to Bacteriologist	7	13	4
..	123	0	0	,, Petty Cash	6	0	0
...	1	8	10	,, Bank Fees	0	10	0
				,, Interest	0	17	3
				,, Balance in Bank (including one cheque, £1 8s. 10d., not paid in)	83	18	8
	£1,612	11	2		£1,612	11	2

ect.

W. Palmer, } Auditors.
n Carson, }

P. N. Trebeck, Hon. Treasurer.

INDEX.

(1900.)

Names in Italics are Synonyms.

[illegible]

[Printed off May 14th, 1901.]

Lightning Source UK Ltd.
Milton Keynes UK
UKHW010618110219
337000UK00006B/211/P